REACTIONS OF MOLECULES
AT
ELECTRODES

REACTIONS OF MOLECULES AT ELECTRODES

Edited by

N. S. Hush

School of Chemistry
University of Bristol

WILEY–INTERSCIENCE
a division of John Wiley & Sons Ltd.,
London · New York · Sydney · Toronto

Library of Congress catalog card number 70–149570

ISBN 0 471 42490 0

Printed in Great Britain at the Pitman Press, Bath

Contributing authors

B. Case — *Central Electricity Research Laboratory, Kelvin Avenue, Leatherhead, Surrey, England.*

B. B. Damaskin — *Institute of Electrochemistry, Academy of Sciences of the U.S.S.R., Moscow, U.S.S.R.*

R. R. Dogonadze — *Institute of Electrochemistry, Academy of Sciences of the U.S.S.R., Moscow, U.S.S.R.*

M. Fleischmann — *Department of Chemistry, University of Southampton, Southampton, England.*

A. N. Frumkin — *Institute of Electrochemistry, Academy of Sciences of the U.S.S.R., Moscow, U.S.S.R.*

J. M. Hale — *Orbisphere Limited, 91 Route de la Capite, Cologny, Geneva, Switzerland.*

J. A. McCleverty — *Department of Chemistry, University of Sheffield, England.*

W. Mehl — *Orbisphere Limited, 91 Route de la Capite, Cologny, Geneva, Switzerland.*

M. E. Peover — *Division of Molecular Science, National Physics Laboratory, Teddington, Middlesex, England.*

D. Pletcher — *Department of Chemistry, University of Southampton, Southampton, England.*

B. J. Tabner — *Department of Chemistry, University of Lancaster, Lancaster, England.*

J. R. Yandle — *The Computer Centre, University of Birmingham, Birmingham, England.*

Contents

Introduction

The study of processes occurring at metal, semiconductor or insulator electrodes forms an important part of electrochemistry. At the same time, it overlaps at many points with work in other, often apparently unrelated, fields. To some extent, this is reflected in current trends in experimental electrochemistry towards combining a number of different techniques (especially spectroscopic) with electrical measurements in analysing reaction mechanisms. However, specialists in other areas of chemistry are often not aware of the possible relevance of developments in the theory of electrode processes, or of kinetic and thermodynamic data for such processes, to their own fields. The usefulness of electrochemical methods in preparative chemistry has also not yet been sufficiently realized. It is hoped that all the topics discussed by the contributors to this volume will be of interest to many chemists who are non-specialists in electrochemistry, as well as to electrochemists.

In many electrode processes, the reacting species adsorb on the electrode surface, and much effort has gone into the measurement and interpretation of adsorption of molecules and ions from solution in ionizing solvents on to a variety of surfaces. A large part of the basic theory and many of the experimental results are due to the work of A. N. Frumkin and his co-workers over a period of nearly fifty years; in Chapter 1, Damaskin and Frumkin summarize the main theoretical conclusions that have emerged from these studies. For a given surface, solvent and adsorbate, the important variable governing the extent of adsorption is the electric field strength at the adsorption site. The electrochemist is in the fortunate position of being able to describe the electrical properties of his adsorbing surface in some detail, and also of knowing whether or not the species he is studying has gained or lost electrons in the process of adsorption. He can also continuously vary the electrical properties of the surface over a reasonably wide range of charge, potential and field strength. This makes possible detailed tests of theoretical

ix

expressions for the dependence of free energy of adsorption on these and related parameters, as Damaskin and Frumkin discuss. These findings are clearly of interest also for the problems of adsorption from the gas phase. It may be remarked that interpretation of adsorption from the gas phase has, in recent years, relied increasingly on measurements of infra-red frequencies and intensities of adsorbed molecules. There is growing interest in measurements of this kind at the electrode-solution interface, which should lead to useful comparisons and more detailed information about the nature of the bonding involved.

For a large number of electrode processes, adsorption of reactant or product species either does not occur or can be rendered inappreciable by a suitable choice of solvent, electrolyte or temperature. The simplest of these are electron transfer processes, in which one or more electrons are exchanged between the electrode and species in solution. Provided that certain conditions are fulfilled (e.g., suppression of diffuse double-layer potential), the currently-accepted theory of electron exchange at electrodes predicts that the exchange rate constants will be essentially independent of the nature of the electrode. An important quantity governing both the equilibria and kinetics of the processes will therefore be the interaction energy of acceptor and donor species with the solvent. Solvation energies of radicals, molecules, monatomic and molecular ions have been discussed over a long period, and the current state of experimental information and theoretical interpretation is discussed by Brian Case in Chapter 2. Ion-solvent interaction involves both short-range forces (leading to "chemical bonding" in the first solvation sheath) and longer-range forces which can be regarded as resulting from polarization of a dielectric continuum. The short-range forces are difficult to calculate in any but the simplest of cases, and long-range interactions are equally difficult to estimate for non-spherical ions of uncertain internal charge distribution. Nevertheless, much progress has been made. What is required to develop this further is a more extensive body of experimental data on the solvation energies (real potentials) of ions of widely differing electronic structure, charge, size and geometry. There are a number of recent developments which should make this possible. The range of simple electron-transfer systems for which equilibrium potential measurements can be made has been greatly extended by work on aromatic hydrocarbon/hydrocarbon ion systems, and on their aza-aromatic analogues, and, more recently, on redox couples formed by organo-metallics. At the same time, first and higher ionization potentials are now easily obtainable by photoelectron spectroscopy, while for negative ions an increasing number of measurements of electron affinities (both positive and negative) is appearing. Within the next few years, we should therefore have a much more varied experimental basis for testing theories of ion-solvent interaction, and major developments in this field are to be anticipated. These will, of course, be equally of interest for the interpretation of thermodynamic and kinetic data for homogeneous processes in ionizing solvents.

The free energy of activation of simple electron-exchange processes, either homogeneous or heterogeneous, is discussed in terms of the solvent reorganization energy involved in the formation of the transition state. This is closely related to ion solvation energies, and its calculation poses similar problems about the choice of an adequate model. This is discussed by J. M. Hale in Chapter 4. For electron-transfer systems involving molecules rather than monatomic ions, internal degrees of freedom must also be taken into account. In fact, the distinction is often somewhat artificial, as an exchange such as Fe^{2+}/Fe^{3+} in aqueous solution can equally well be considered as an exchange between the molecular complex ions $Fe(H_2O)^2_6{}^+$ and $Fe(H_2O)^3_6{}^+$, provided that it is of the outer-sphere type. Hale provides useful calculations of contributions from both internal and external energies of rearrangement in forming the transition state for a variety of systems. The fundamental kinetic theory is discussed in more detail by R. R. Dogonadze in Chapter 3; the extension of this general approach to reactions of more complex type, namely those involving formation and breaking of chemical bonds, is also outlined. It is evident that the formalism can also be extended to discuss the kinetics of many different types of reactions in ionizing solvents, so that information gained from the electrochemical processes will ultimately be of fundamental interest for the interpretation of the solution kinetics of ionic and ion/molecule reactions. A feature of Dogonadze's treatment which is of general interest to electrochemists is his detailed discussion of the transfer coefficient. The interpretation of the macroscopic transfer coefficient α and its complement $(1 - \alpha)$ as measuring the effective occupation probabilities of ion and metal orbitals respectively by the transferring electron in the transition-state complex of an activated electron-exchange reaction provides a useful insight into mechanism.

There is a close connection between the theories of electron transfer (homogeneous or heterogeneous) in which the reacting ions are in solution, and of transfer of nearly-localized electrons by a hopping mechanism in ionic or molecular solids. The recent revival of interest in conduction in low-mobility semiconductors has resulted in much new experimental material becoming available, particularly for ionic solids. The solid-state analogue of the solvent reorganization energy is $2E_p$, where E_p is the small-polaron binding energy; indeed, on the simplest possible model, in which the medium is regarded as a dielectric continuum, the theoretical expressions for these quantities are identical. There is evidently a need for closer comparison of electrode and solid-state processes. For example, an important prediction of electron-transfer theory is that the transfer coefficient α for simple exchanges should be related to the overpotential η by the expression:

$$\alpha = \tfrac{1}{2}\left(1 + \frac{e\eta}{\chi}\right)$$

where χ is the solvent reorganization energy. This should hold provided that there is no appreciable adsorption on the electrode surface, and the diffuse double layer potential has been suppressed. The net current density j resulting from exchange of an electron with the electrode is

$$j = \vec{j} - \overleftarrow{j}$$

where $\vec{j}$ and $\overleftarrow{j}$ are the current densities for the electron uptake and reverse donation steps respectively. This is given by

$$j = 2j_0 \sinh\left(\frac{e\beta\eta}{2}\right)\left\{\exp - \frac{\beta(e\eta)^2}{4\chi}\right\},$$

and the corresponding conductivity is:

$$\sigma = \sigma_0\left(\cosh\left(\frac{e\beta\eta}{2}\right) - 4e\beta\eta \sinh\left(\frac{e\beta\eta}{2}\right)\right)\left\{\exp - \frac{\beta(e\eta)^2}{4\chi}\right\}.$$

For a low-mobility semiconductor, where the conduction mechanism is free small-polaron hopping, the net current density and conductivity in the presence of a static field E for a polaron jump distance a is given by expressions of precisely the same form as those for the electrode process, when the overpotential η is replaced by aE.* The expression in curly brackets represents the high-field term, which is responsible in the case of electrode/solution transfers for the predicted variation of transfer coefficient with potential. At present, there is no firm evidence for the existence of a high-field effect either for small-polaron processes in crystals or for electrode-processes. Dogonadze stresses the importance of this concept in the case of electrode reactions. The outcome of experiments designed to test the reliability of predictions of high-field effects will show how well-based are the fundamental assumptions common to theories of electron transfer in solid-state, heterogeneous and homogeneous solution media. Similarly, observations of high-frequency conductivity, corresponding to absorption of photons in the near infra-red or visible regions of the spectrum for electrode-solution transfers,† analogous to those observed in transition-metal oxides,‡ would provide independent values for solvent reorganization energies.

* H. G. Reik, *Solid State Commun.*, **8**, 1737, 1970.
† N. S. Hush, *Electrochimica Acta*, **13**, 1005, 1968.
‡ I. G. Austin, B. D. Clay and C. E. Turner, *J. Phys. Chem.*, **1**, 1418, 1969.

Studies of electron or hole injection into an insulating crystal from ions in solution, in which the crystal thus functions as an insulator electrode, provide an interesting link between solution and solid-state electron-transfer processes. W. Mehl (Chapter 7) outlines experimental results for systems in which the electrode is an aromatic hydrocarbon crystal, and inorganic redox systems are used as donors or acceptors. These are successfully interpreted in terms of quite simple concepts, which assume the validity of adiabatic electron-exchange theory for the interfacial reaction. This has also been extended, as Mehl discusses, to photochemical reactions in which charge transfer with excitons in the crystal or with electronically excited states of donor and acceptor ions in the solution is observed.

In aromatic hydrocarbon crystals, electronic bandwidths are usually so small that for calculations of energetics, holes and electrons can be effectively regarded as molecular cations or anions embedded in a molecular dielectric, which is polarized by the ion charge distribution. The crystal polarization energy is thus analogous to the real potential of the ion in an ionizing solvent. Ions of aromatic hydrocarbons can also be studied in solution, and the energetics of formation of the solvated ions, calculated from the standard potentials can give information either about ionization energies or ion real potentials. These, in turn, can be used in discussing the kinetics of electron exchange between a metal electrode and a hydrocarbon in solution. The kinetic data are discussed in Chapter 5 by M. E. Peover, with particular reference to temperature and salt effects, and the advantage of being able to compare a series of closely-related systems is particularly apparent here. The interpretation of energetics of similar aza-hydrocarbon/aza-hydrocarbon anion systems is discussed in Chapter 6 by B. Tabner and J. Yandle; while a general parallelism with the hydrocarbon systems is established, it appears from this work that we still have a good deal to learn about the real potentials of relatively simple heteroanions.

Organic electrode processes of more general kinds are discussed in some detail by M. Fleischmann and D. Pletcher in Chapter 8. The mechanisms of these reactions are in many cases highly complex, but it is evident that a good deal is now known about the nature of the primary steps, and in a number of cases of the succeeding steps. There are interesting correlations with homogeneous reactions of these molecules. This is a great ocean, on the shores of which the theoretician presumably stands shivering, trying to nerve himself for the plunge. We can only hope that the sun will ultimately shine; meanwhile, organic chemists would do well to note the great variety of selective electrochemical reactions of organic molecules now known, which one would expect to see more fully utilized in preparative work.

The recent extensions of equilibrium or near-equilibrium measurements on redox couples to organometallic systems are outlined by J. McCleverty in Chapter 9. This is an almost inexhaustible field, and much information is now available on new and in some cases unusual redox systems. An interesting and welcome feature of the inorganic chemist's approach is the combination of electrochemistry with infra-red, U.V., N.M.R. and E.S.R. spectroscopy in determining the course and nature of the reactions. One hopes, however, that in future experimental work in this field, an attempt will be made to standardize the reference electrodes used for particular solvents, that there will be more uniformity in the choice of type and concentration of electrolyte, and that the measured potentials will be more precisely defined. From the point of view of the electrochemist, the main interest here is the broadening of the field of simple electron-exchange processes and the coupling of electron-transfer and dimerization noted in many systems studied. In addition, the electronic structures of the complexes can be varied in many ways, so that there is scope for interpretation of the dependence of rates and equilibria on electronic properties, which up to the present has been confined largely to π-electron systems. It is fortunate that in the last years, theoretical techniques for calculation of electronic properties and structure of molecules and radicals have rapidly developed, so that fairly detailed interpretation of the electrode behaviour of organometallic molecules will be possible.

N. S. HUSH

ADSORPTION OF MOLECULES ON ELECTRODES

B. B. Damaskin and A. N. Frumkin

Adsorption of organic substances at the electrode/solution interface is one of the important factors determining the kinetics and mechanism of electrode processes. First of all we should mention electroorganic synthesis reactions, the role of organic additions in the processes of electrodeposition of metals and alloys and the inhibition of metallic corrosion by organic inhibitors. In some cases it is necessary to have data on adsorption of organic substances on the surface of the dropping mercury electrode for successful use of polarography. Moreover, the investigation of adsorption phenomena at the electrode/solution interface is of interest by itself in connection with further development of the theory of the electric double layer structure. These circumstances are responsible for the interest shown lately in the investigation of adsorption of organic substances on electrodes.

Most of the theoretical and experimental investigations of adsorption have been carried out on the mercury electrode, using mainly the methods of surface tension and differential capacity measurements. The investigation of adsorption of organic substances on solid electrodes involves a number of

1

experimental difficulties. Thus there are no methods available at present which would permit us to determine the absolute surface tensions of solid electrodes, whereas the interfacial impedance measurements, giving quantitative data on the state of the surface of a liquid electrode, for a number of reasons are of limited applicability in the case of many solid electrodes (e.g. platinum metals). Quantitative investigations of adsorption of organic compounds on solid metals have become possible only in the past 10–15 years, following the development of the radioactive tracer technique, the pulse-potentiostatic, galvanostatic and some other methods. Even now, adsorption measurements on solid metals are performed with less accuracy than on the mercury electrode. This is due both to limitations of the methods and to specific properties of solid surfaces.

Since adsorption of organic compounds at the electrode/solution interface changes the wetting of the metal by the solution, affects its mechanical properties (e.g. hardness, creep, etc.), and in most cases leads to an appreciable change in the rate of electrochemical reactions, all these phenomena can be used for a qualitative description of adsorption processes. A detailed review of various methods for the investigation of adsorption of organic substances on the surface of liquid and solid electrodes is to be found in Reference 1.

In this review we shall consider the main regularities in adsorption of organic molecules on the mercury electrode surface and in the last section also the problem of the influence of the nature of the metal on adsorption of organic compounds.

I. THERMODYNAMIC METHODS OF ESTIMATION OF ORGANIC SUBSTANCE ADSORPTION FROM THE SURFACE TENSION AND DOUBLE LAYER CAPACITY MEASUREMENTS

As has been shown independently by a number of authors,[2-4] it is possible to apply to the electrode/solution interface the thermodynamic equations relating the surface tension σ to the electrode potential E, as well as to the adsorptions and activities of ions and molecules of the solution (Γ_i and a_i, respectively). At constant pressure and temperature the basic equation of electrocapillarity can be written as

$$d\sigma = -q\,dE - \sum_i \Gamma_i\,d\mu_i = -q\,dE - RT\sum_i \Gamma_i\,d\log a_i \qquad (1)$$

where μ_i is the chemical potential of the i-th component, R is the gas constant and T is the absolute temperature. In the general case, the quantity q in eqn (1) is the Gibbs adsorption of the potential-determining ions expressed in electric units. In the case of the mercury electrode, however, which can be considered with sufficient accuracy as being ideally polarizable, q coincides with the surface charge of the metal.

Equation (1) can be used for a quantitative determination of adsorption of the solution components on the mercury surface. It should be borne in mind however that the physical significance of Γ_i varies depending on the conditions of the choice of the Gibbs plane.[5] Let us consider this question for the case of a two-component system: an alcohol solution (2) in water (1). At $E =$ constant it follows from eqn (1) that

$$d\sigma = -\Gamma_1 \, d\mu_1 - \Gamma_2 \, d\mu_2 \qquad (2)^*$$

If the Gibbs plane is chosen in such a way that $\Gamma_1 = 0$, the experimental value of $\Gamma_2^0 = -d\sigma/d\mu_2$ determined by this condition is the excess of alcohol moles in a solution fraction containing unit interface over the solution fraction from homogeneous bulk containing an equal amount of water moles.

Another method for the choice of the Gibbs plane conforms to the condition

$$v_1 \Gamma_1^v + v_2 \Gamma_2^v = 0 \qquad (3)$$

where v_1 and v_2 are partial molar volumes of water and alcohol, respectively. The quantity Γ_2^v determined by this condition is the excess of alcohol moles in the solution fraction containing unit interface over the same volume fraction of homogeneous solution. This corresponds to the closest approach of the Gibbs plane to the physical boundary between the electrode and the solution.

Equations (2) and (3) in combination with the Gibbs-Duhem equation

$$N_1 \, d\mu_1 + N_2 \, d\mu_2 = 0 \qquad (4)$$

where N_1 and N_2 are molar fractions of water and alcohol in the bulk of the solution, permit to relate Γ_2^v to Γ_2^0. In fact, eliminating from these equations Γ_1^v and $d\mu_1$, we obtain

$$-\frac{d\sigma}{d\mu_2} = \Gamma_2^0 = \Gamma_2^v \left(1 + \frac{v_2 N_2}{v_1 N_1}\right) \qquad (5)$$

It follows from eqn (5) that in sufficiently diluted solutions, when the quantity $v_2 N_2/v_1 N_1$ can be neglected compared to unity, $\Gamma_2^0 \approx \Gamma_2^v$.

It follows from the above that the physical significance of Γ_2^0 and Γ_2^v does not conform to the concept of adsorption as the surface concentration of the substance being adsorbed. For the quantities Γ_1 and Γ_2 contained in eqn (2) to have the physical significance of surface concentrations it is necessary to assume that the heterogeneous region near the electrode surface is confined

---

* Equation (2) remains valid if in addition there is an indifferent electrolyte with constant activity present in the solution.

within the monolayer of the substance being adsorbed and that the Gibbs plane separates this monolayer from the homogeneous solution.[5] Let us designate the surface concentrations of water and alcohol thus determined by Γ_1^1 and Γ_2^1. Then from eqns (2) and (4) we obtain

$$-\frac{d\sigma}{d\mu_2} = \Gamma_2^0 = \Gamma_2^1 - \Gamma_1^1 \frac{N_2}{N_1} \tag{6}$$

Formula (6) is an equation in two unknowns: Γ_1^1 and Γ_2^1, which therefore cannot be determined by a strictly thermodynamic method.

In order to determine Γ_2^1 (or Γ_1^1), it is necessary to make some model assumptions regarding the structure of the interface from which it will be possible to estimate the areas per 1 mole of organic substance and 1 mole of water in the surface layer (S_2 and S_1, respectively). Then Γ_1^1 and Γ_2^1 can be found by solving the set of equations:

$$\left. \begin{aligned} \Gamma_2^0 &= \Gamma_2^1 - \Gamma_1^1 \frac{N_2}{N_1} \\ \Gamma_1^1 S_1 &+ \Gamma_2^1 S_2 = 1 \end{aligned} \right\} \tag{7}$$

A rough calculation based on the assumption that an organic substance molecule occupies on the surface an area 22 Å² and a water molecule—10 Å² leads to the conclusion that at the organic compound concentration in aqueous solutions $c_2 \leqslant 0.5\,\text{M}$ with an accuracy not less than 3 per cent, $\Gamma_2^1 \approx \Gamma_2^0$. Thus under these conditions the experimental values of Γ_2^0 can be considered as surface concentrations of organic substances. For simplicity, hence we shall write the experimental value of Γ_2^0 as Γ_2, bearing in mind however that the values thus obtained refer to the plane where $\Gamma_1 = 0$.

From electrocapillary measurements the values of Γ at a given potential can be readily determined by means of eqn (1), which gives:

$$\Gamma_i = -\frac{1}{RT} \left(\frac{\partial \sigma}{\partial \log a_i} \right)_E \approx -\frac{1}{RT} \left(\frac{\partial \sigma}{\partial \log c_i} \right)_E \tag{8}$$

where c_i is the concentration of the ith component.

In order to obtain the values for the extent of adsorption at a given electrode charge, it is necessary to introduce a new function $\xi = \sigma + qE$,[6] the total differential of which is

$$d\xi = d\sigma + q\,dE + E\,dq \tag{9}$$

It follows from eqns (1) and (9) that

$$d\xi = E\,dq - RT\sum_i \Gamma_i\,d\log a_i \tag{10}$$

whence

$$\Gamma_i = -\frac{1}{RT}\left(\frac{\partial \xi}{\partial \log a_i}\right)_q \approx -\frac{1}{RT}\left(\frac{\partial \xi}{\partial \log c_i}\right)_q \tag{11}$$

The adsorption isotherms obtained from the experimental data by means of eqns (8) and (11) are self-consistent in the sense that by calculating by means of eqn (8) the Γ,c-curve at $E = $ constant and knowing the dependence of q on E at various concentrations of the substance being adsorbed, it is possible to plot the Γ,c-curve at $q = $ constant, coinciding exactly with the adsorption isotherm calculated by means of eqn (11). Thus with a strictly thermodynamic approach to the investigation of the adsorption isotherms, the choice of the electric variable is of no fundamental significance and is determined by considerations of convenience.

Apart from electrocapillary measurements, the data on adsorption of organic compounds at the electrode/solution interface can be obtained also from differential capacity (C) measurements. The thermodynamic method of the application of the capacity measurements is based on double integration of the C,E-curve, which in accordance with the Lippmann equation gives the dependence of the surface tension on the electrode potential

$$q = \int_{E_{q=0}}^{E} C \, dE \tag{12}$$

$$\sigma = \sigma_{q=0} - \int_{E_{q=0}}^{E} q \, dE \tag{13}$$

Then the σ,E-curves (or the ξ,q-curves) thus obtained are used to estimate adsorption of the components, as is described above.

This method requires two integration constants to be known. In eqn (12) such a constant is the potential of zero charge $E_{q=0}$, and in eqn (13) $-\sigma_{q=0}$, i.e. the value of σ at the electrocapillary curve maximum.

A review of various methods for the determination of the zero charge potentials as well as the most reliable values of $E_{q=0}$ are to be found.[7] As regards the second integration constant, its determination involves electrocapillary measurements near $E_{q=0}$.

The capacity method is much simplified when it is used to study adsorption of an organic substance which is added in small concentrations to an indifferent electrolyte solution (the supporting electrolyte) and desorbed at sufficiently negative electrode potentials, so that the differential capacity curves measured in the presence of organic substance additions coincide with the C,E-curve of the supporting electrolyte. The double layer structure being determined under such conditions only by the indifferent electrolyte composition, it is quite natural to assume that at such negative potential (say, E_1) not only the differential capacity values should coincide in a pure electrolyte solution and in solutions with organic substance additions but also those of q and σ. Thus in this case the integration constant in eqn (12) is the electrode charge in the supporting electrolyte solution at $E = E_1$, and the integration constant in eqn (13) the value of σ in the supporting electrolyte solution at $E = E_1$. Under these conditions integration is performed from negative (beginning with E_1) to positive potentials (the 'backward integration' method).[8] Since in further calculations of adsorption by means of eqn (8) it is necessary to know only the change in the surface tension with increasing organic substance concentration rather than the absolute values of σ, in calculating the electrocapillary curve of the supporting electrolyte it is possible to put the constant $\sigma_{q=0}$ equal to zero. Thus in calculating adsorption from differential capacity measurements, the only necessary constant under given conditions is the value of $E_{q=0}$ in the supporting electrolyte solution.

The drawback of this method is that at small organic substance concentrations there is not enough time for the equilibrium in the double layer to be established within the half-period of the alternating current used to perform the differential capacity measurements and it becomes necessary to extrapolate the non-equilibrium capacity values measured to zero frequency.[9–12] This method was used for the investigation of thiourea adsorption on mercury[13] and, along with others for the adsorption of aniline, phenol and pentafluorophenol.[14–16]

II. BASIC QUALITATIVE REGULARITIES IN ADSORPTION OF ORGANIC SUBSTANCES ON THE MERCURY ELECTRODE

Beginning with the works of Gouy,[17] the method of electrocapillary curve measurements has been widely used for the investigation of adsorption of a great variety of organic compounds.[14–16,18–60] As is clear from Fig. 1, adsorption of an organic substance (in this case n-butanol) markedly lowers the surface tension of mercury, the decrease of σ being maximal near $E_{q=0}$ of the supporting electrolyte and becoming less pronounced with increasing negative or positive surface charge. In the case of aliphatic compounds, at large enough $|q|$, the electrocapillary curve measured in the presence of organic substance additions coincides with the σ,E-curve of the supporting electrolyte (Fig. 1). According to eqn (8), this means that at large charges organic molecules are desorbed from the electrode surface.

The adsorption of an organic substance on the electrode surface increases with increase of its bulk concentration (Fig. 1) and of the indifferent electrolyte concentration (the salting out effect), as well as with increasing length of the hydrocarbon chain in molecules of organic substances of a homologous series. A detailed study of the last-mentioned effect for the case of adsorption of normal acids, alcohols and amines of the fatty series[50-51] showed the Traube rule to be valid under such conditions.

The electrocapillary curves measured in the presence of organic compounds often exhibit a shift of the electrocapillary maximum $\Delta E_{q=0}$.

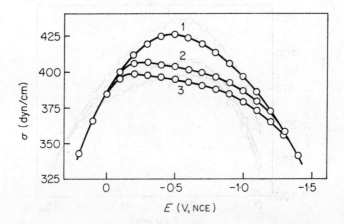

Fig. 1. Electrocapillary curves of a mercury electrode in 1 N Na_2SO_4 solution (1) and with additions of n-butyl alcohol in concentrations: 2 — 0·1 M, 3 — 0·2 M.

The comparison of the values of $\Delta E_{q=0}$ and the adsorption shifts of the potential ΔE at the solution/gas interface in the same solutions shows that the adsorption of molecules of aliphatic oxygen compounds at both interfaces involves identical in sign and nearly equal potential drops.[19-22] This result can be explained as being due to the substitution of the water dipoles oriented with their negative end towards air or mercury by the organic substance molecules oriented with their hydrocarbon chain towards the interface, so that the C—O bond sets up a positive potential difference between the outer phase and the bulk of the solution. However, Devanathan,[61] as well as Bockris, Devanathan and Müller,[62] associated the shift of $E_{q=0}$ in the presence of aliphatic compound only with the expulsion of water dipoles from the double layer, the polar group of the organic molecule located outside the double layer having, in their opinion, no effect on the value of $\Delta E_{q=0}$ being measured. These concepts are difficult to accept.[64,65] It should be noted that a similar concept for the case of the solution/air interface advanced by Kamiensky[63] leads to impossible values of chemical hydration energies of cations and anions.[66]

The comparison of the adsorption potential drops at the two interfaces in the case of aromatic compounds shows considerable discrepancies between $\Delta E_{q=0}$ and ΔE to the extent of their having different signs in some cases.[20,21] Another anomaly is that the electrocapillary curves measured in the presence of some aromatic compounds do not coincide with the σ,E-curve of the supporting electrolyte even when the most positive values of q are reached. The σ,E-curves measured in the presence of aniline, which are shown in Fig. 2

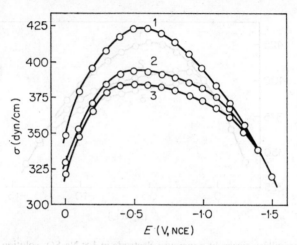

Fig. 2. Electrocapillary curves of a mercury electrode in 1 N KCl solution (1) and with additions of aniline in concentrations: 2 — 0·05 M, 3 — 0·1 M.

are an example of this case. The reasons for these anomalies were first established by Gerovich and collaborators.[28-31] The investigation of adsorption on mercury of benzene, naphthalene, anthracene, phenanthrene and chrysene as well as the comparison of the electrocapillary behaviours of aromatic and of the corresponding hydroaromatic hydrocarbons showed that the shift of the point of zero charge in the negative direction (or a decrease in the positive value of $E_{q=0}$) and the adsorption of aromatic compounds at larger positive q are associated with the effect of the interaction between π-electrons of the aromatic ring and the charges of the mercury surface. Later these conclusions were experimentally confirmed in the case of adsorption of other aromatic and heterocyclic compounds.[35-46] The substitution of fluorine for hydrogen atoms in aromatic compounds leads to the depletion of the aromatic ring π-electron density owing to the strong electron affinity of fluorine atoms. As a result, the π-electron interaction hardly affects the adsorption on mercury of the molecules of pentafluoroaniline, pentafluorobenzoic acid and pentafluorophenol.[16,52]

According to Blomgren, Bockris, and Jesch,[37] in the case of adsorption on mercury of molecules both of aliphatic and aromatic compounds the role of the functional group is determined mainly by its influence on the solubility. The comparison of the adsorption behaviour of organic molecules at the solution/mercury and solution/gas interfaces shows these relations to be somewhat more complicated.[54] In fact, it follows[50-51] that the adsorption behaviour of aliphatic alcohols, acids and amines does not differ much at the solution/mercury and solution/gas interfaces. However, in the case of acids and especially of alcohols, the surface activity at the interface with mercury is somewhat lower than at the interface with air, whereas the opposite is true for amylamine and partly for butylamine. Therefore the nature of the polar group, in spite of its being turned towards the solution and thus removed from the mercury surface, affects somewhat the change in the adsorption energy when passing from the free surface of the solution to the interface with mercury.

The slight difference in the adsorptivities of normal aliphatic compounds of the fatty series at the interfaces with mercury and with air indicates that the gain in free energy upon contact of mercury with hydrocarbon tails of organic molecules is approximately compensated by its consumption in the removal of adsorbed water molecules. Such compensation however cannot be general, as is evident from the consideration of adsorption of organic compounds with horizontal orientation at the interface. The horizontal orientation is to be expected in the case of hydrocarbons without polar groups, at any rate at small surface coverages. Under these conditions the adsorptivity of n-hexane decreases greatly when passing to the interface with mercury.[53] In this case the decrease in the free energy gain when a hydrocarbon molecule is transferred from the bulk of the solution to the surface is due to the fact that with horizontal orientation of hydrocarbon molecules water is displaced from the mercury surface mainly by $>CH_2$ rather than by $-CH_3$ groups. This phenomenon is also observed in the case of adsorption of perfluorinated compounds where it is even more pronounced.[49] Hence the gain in free energy upon wetting of mercury by the $-CF_3$ and $>CF_2$ groups is much less than upon its wetting by water.

In addition to hydrocarbons, the horizontal orientation is also characteristic of aliphatic compounds with two or more functional groups, e.g. glycols, glycerine, or ethylenediamine. Under these conditions, the interaction of the polar group with the mercury surface results in the adsorptivity of the organic compound being appreciably higher at the solution/mercury than at the solution/gas interface.[21]

A sharp increase in the surface activity when passing to the solution/mercury interface is observed in the case of tetrabutyl- and tetrapropylammonium cations.[67] This effect can be accounted for by electrostatic interaction of the charge of R_4N^+ cation with the induced charge in the metal phase. However in the case of propyl-, butyl-, and amylammonium cations, the mirror image effect is much weaker due to strong hydration of the functional group in compounds of the RNH_3^+ type.[67]

As has been already pointed out above, the main difference in the adsorption behaviours of aromatic and heterocyclic compounds at the interface with mercury and at the solution/gas interface is due to the π-electronic interaction. However of essential importance for the increase of adsorptivity of these compounds when passing to the interface with mercury is the interaction of mercury with the polar group as well. The π-electronic interaction favours a more horizontal orientation of the molecules of aromatic substances at the solution/mercury interface and this, in its turn, facilitates the interaction of the polar group with the metal. Thus the introduction of the polar group into the aromatic compound molecule along with decreasing the surface activity at the solution/gas interface (due to increasing solubility) can lead to an increase in the surface activity at the interface with mercury. Such an effect is observed in the case of a transition from benzene to aniline.[54]

When two surface-active substances are simultaneously present in the solution their adsorption is largely determined by the interaction of the molecules of these substances within the adsorption layer.[24,60] It is of interest to note that the regularities of the simultaneous adsorption of n-butyl alcohol and aniline are essentially different for positive and negative charges of the mercury electrode.[60] Thus at $q < 0$ the simultaneous presence of these two substances in the solution can increase the adsorption and the electrocapillary curve of the solution of the aniline-butanol mixture lies below the σ,E-curves for the solutions of individual substances. At positive electrode charges with increasing q, the σ,E-curve in the presence of two additions approaches the electrocapillary curve for the aniline solution. Butanol is desorbed from the electrode surface covered by aniline molecules, as it would be desorbed in the absence of aniline. Evidently, the difference in the behaviour of the aniline—n-butanol mixture on different branches of the electrocapillary curve is accounted for by different orientations of adsorbed aniline molecules at $q > 0$ and at $q < 0$.[14]

The differential capacity being the second derivative of the surface tension with respect to the potential, the shape of the C,E-curve is much more sensitive to any changes in the double layer structure than that of the electrocapillary curve. High sensitivity of the double layer capacity to traces of organic impurities in the solution was the reason why many investigators failed to verify by this method the Lippmann equation. After Proskurnin and Frumkin's work[68] the method of differential capacity measurements come into general use for investigations of the electric double layer structure and adsorption of various organic substances.

However, the differential capacity of the double layer is a much more complex function of adsorption of organic molecules than surface tension, so that a strictly thermodynamic approach to the estimation of adsorption from the C,E-curves measurements is not very extensively used. In quantitative investigations the C,E-curves measurements are as a rule supplemented by electrocapillary measurements or some model representations of the systems being investigated.[10-16,46,56-61,69-82]

In most cases the region of organic substance adsorption on the C,E-curves corresponds to a region of low capacity values limited on both sides by the adsorption—desorption peaks (Fig. 3). At large enough positive and negative electrode charges, when organic molecules are practically completely desorbed from the surface, the C,E-curves measured in solutions with organic substance additions coincide with the C,E-curve of the supporting electrolyte. As is clear from Fig. 3, with increasing organic substance concentration, the

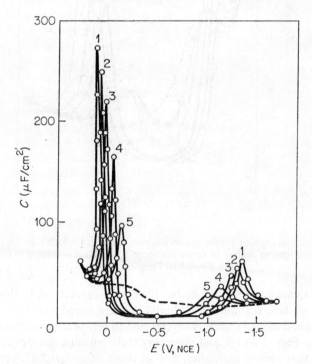

Fig. 3. Differential capacity curves of a mercury electrode in 0·1 N Na_2SO_4 solution (dashed line) and with additions of n-butyl alcohol in concentrations: $1 - 0·8$, $2 - 0·6$, $3 - 0·4$, $4 - 0·2$, and $5 - 0·1$ M. Frequency 400 cps.

adsorption region widens, the adsorption peaks become higher and the capacity at the minimum of the C,E-curve tends to approach a limiting value of the order of 4–5 μF/cm^2. These regularities make it possible to use the method of differential capacity curves measurements for a qualitative determination of adsorptivity of various organic substances at the electrode/solution

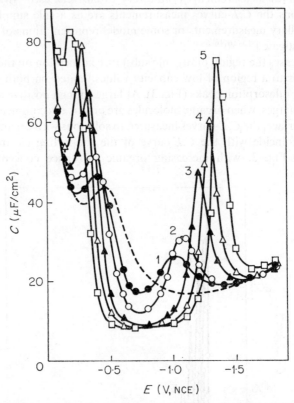

Fig. 4. Differential capacity curves of a mercury electrode in 1 N KCl solution (dashed line) and with additions of aniline in concentrations: $1 - 0.01$, $2 - 0.02$, $3 - 0.05$, $4 - 0.1$, and $5 - 0.2$ M. Frequency 400 cps.

interface. Quite a number of studies have been carried out in this direction (for a list of these see[1] and[64]). It should be emphasized however that the use of this method alone in some cases can lead to erroneous conclusions. In fact comparing Figs. 3 and 4, and observing their outward similarity, one could conclude that aniline molecules, just as those of n-C$_4$H$_9$OH, are desorbed from the mercury surface at a large enough positive surface charge. That this conclusion is erroneous is evident from σ,E-curves in aniline solutions

(Fig. 2). Thus in respect of adsorption the system KCl + aniline proves to be more complex than Na_2SO_4 + n-butanol. While in the presence of aliphatic compounds both maxima on the C,E-curves are connected with the adsorption-desorption processes, only the cathodic maxima correspond to these processes in the case of aniline, whereas anodic maxima are associated with the process of re-orientation of adsorbed aniline molecules: vertical orientation of molecules characteristic of $q < 0$ being replaced by a horizontal one which favours the π-electronic interaction of the benzene ring with the positive charges of the mercury surface.[14]

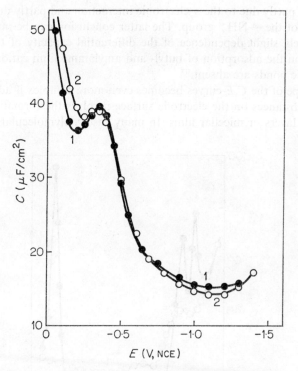

Fig. 5. Differential capacity curves of a mercury electrode in solutions: $1 - 0.1$ N HCl, $2 - 0.1$ N HCl + 0.1 M aniline. Frequency 400 cps.

An even more interesting example presents the C,E-curves in acid solutions of aniline,[83] pyridine[46] and phenylenediamine, which, as can be seen from Fig. 5, hardly differ from the C,E-curve in a pure supporting electrolyte solution. In this case, however, a possible conclusion about the absence of adsorption of organic substance over the whole potential range under investigation would be at variance with the results of electrocapillary measurements.[31,36,39]

The double layer capacity being independent of the organic substance adsorption on a positively charged electrode seems to be due to a strong π-electronic interaction, in the presence of which aniline molecules and anilinium cations horizontally oriented on the electrode surface form, as it were, an extension of the metal surface in the direction of the solution, so that the electrode and the solution sides of the double layer do not move apart and hence its capacity changes insignificantly, if at all. At $q < 0$ the slight influence of the anilinium cations adsorption on the double layer capacity is partly due to the π-electronic interaction and partly due to strong hydration of the $-NH_3^+$ group. The latter conclusion can be inferred from the relatively slight dependence of the differential capacity of the mercury electrode on the adsorption of butyl- and amylammonium cations in which π-electronic bonds are absent.[67]

The shape of the C,E-curves becomes even more complex if adsorption of organic substances on the electrode surface involves the formation of poly-molecular layers or micellar films. In many cases polymolecular adsorption

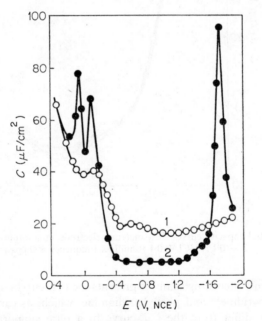

Fig. 6. Differential capacity curves of a mercury electrode in solutions: $1 - 0.05$ N Na_2SO_4, $2 - 0.05$ N $Na_2SO_4 + 0.008$ N $C_6H_{13}OSO_3Na + 0.001$ N $[(C_4H_9)_4N]_2SO_4$. Frequency 400 cps.

leads to a capacity decrease in a certain potential range up to $0.9 \, \mu F/cm^2$.[84] The processes of formation and destruction of micellar films on the C,E-curves correspond to the appearance of additional peaks,[85,86] the frequency dependence of which differs sharply from that of the adsorption-desorption peaks.[86]

If two organic compounds are present together in the solution, their desorption from the electrode surface as a rule occurs simultaneously and involves the formation of one common peak on the C,E-curve. The position of this peak depends on the adsorption characteristics of each of these substances, on their concentration ratio as well as on the interaction between molecules of these compounds within the adsorption layer.[60,82] In rare instances, the adsorption peaks of two substances become separated. This case is illustrated in Fig. 6 which shows the C,E-curve for co-adsorption of $[(C_4H_9)_4N]^+$ cations and $C_6H_{13}OSO_3^-$ anions. The comparison with the electrocapillary curves for the same system and for solutions with additions of one or the other of these surface-active substances shows that the right hand anodic peak on the C,E-curve corresponds to desorption of $[(C_4H_9)_4N]^+$ cations and the left hand one to desorption of $C_6H_{13}OSO_3^-$ anions.

III. THEORY OF THE EFFECT OF ELECTRIC FIELD ON ADSORPTION OF ORGANIC MOLECULES

This theory was first developed by one of the authors in 1926.[87] Since it is described in detail in our review,[64] we shall dwell here only on the analysis of the basic assumptions and report some new results. Two following assumptions form the basis of this theory.

1. The First Assumption

The electric double layer in the presence of an organic substance whose molecules do not change their orientation on the electrode surface can be represented as two capacitors connected in parallel, with only adsorbate molecules contained between the plates of one of them and only water molecules between those of the other. This model conforms to the equation:

$$q = q_0(1 - \theta) + q_1\theta \tag{14}$$

where θ is the surface coverage with organic substance; q_0 and q_1 the values of q at $\theta = 0$ and $\theta = 1$, respectively.

The quantity q_0 is determined using the experimental C,E-curve in the supporting electrolyte solution, so that

$$q_0 = \int_0^E C_0 \, dE \tag{15}$$

where C_0 is the value of C at $\theta = 0$ and the potential E (henceforth as well) is read from the point of zero charge in the pure supporting electrolyte solution.

The quantity q_1 is determined by means of an approximate formula

$$q_1 = C_1(E - E_N) \tag{16}$$

where C_1 and E_N are constants: C_1 is the value of C at $\theta = 1$ and E_N is the shift of the point of zero charge when passing from $\theta = 0$ to $\theta = 1$. In the case of adsorption of organic substances for which $C_1 \ll C_0$ (e.g. aliphatic compounds with one polar group) the errors in the estimation of q_1 by means of eqn (16) are relatively insignificant, the change of the charge in the process of adsorption for high values of the charge being primarily determined by the term $q_0(1 - \theta)$ in eqn (14). In principle, however, it would be possible to take into consideration the dependence of C_1 on E as well.

2. The Second Assumption

Adsorption of organic substance on the electrode surface can be described by Langmuir's equation corrected for intermolecular interaction of adsorbed particles. It was admitted that this interaction can be taken into account by introducing the term $a\theta^2$ into the equation of state of the adsorbed layer by analogy with the three-dimensional van der Waals equation of state.[88] This gives the equation

$$Bc = \frac{\theta}{1 - \theta} \exp(-2a\theta) \tag{17}$$

in which B is the adsorption equilibrium constant, c is the organic substance concentration and a is the attraction constant: if $a > 0$ it is the attraction forces between adsorbed molecules that predominate, while if $a < 0$ the repulsion forces are stronger. The quantity a, increasing with the length of hydrocarbon chain, takes account of the increase of the gain in free energy in adsorption (linear with θ), which is associated with two factors:

(a) dispersion interaction between parallel oriented hydrocarbon chains;

(b) expulsion of hydrocarbon chains from the bulk of the solution, which also makes the drawing together of nonpolar parts of molecules more favourable.

The application of eqn (17) to non-localized adsorption from solution is justified since in this case adsorption of the solute involves the displacement of the solvent.[89] The importance of the last point was emphasized by Bockris et al.[62] When deriving eqn (17) it is necessary to assume however that the solvent and adsorbate molecules occupy equal areas on the surface. If the adsorbate molecules occupy an n times larger area on the surface than the solvent molecules, two different equations can be obtained, depending on the assumptions made in the derivation:[90,110,142]

$$Bc = \frac{\theta}{n(1 - \theta)^n} \exp(-2a\theta) \tag{18}$$

and

$$Bc = \frac{\theta}{n(1 - \theta)^n} \left(1 - \theta + \frac{\theta}{n}\right) \exp(-2a\theta) \tag{19}$$

It can be readily seen that at $n = 1$ both the equations reduce to the isotherm eqn (17).

It should be noted that the basic assumptions of the theory in eqns (14) and (17) proved to be quite sufficient for expressing quantitatively, in conjunction with strictly thermodynamic relations, the shape of the electrocapillary curves in the presence of tert. amyl alcohol.[64,87] However, the verification of the theory under more rigid conditions, which involved the comparison of calculated and experimentally measured C,E-curves,[69,76-80] as well as the discussion of the main statements of the theory made a more detailed analysis of assumptions one and two necessary.

The analysis of the first assumption by means of Gibbs thermodynamics shows[95,98] that eqn (14) is strictly equivalent to the condition of the congruence of the adsorption isotherm with respect to the electrode potential. This condition implies that the adsorption isotherms measured at different E are similar and can be written in the form

$$B(E) \cdot c = f(\theta) \tag{20}$$

where $B(E)$ is a function of the electrode potential and $f(\theta)$ is a function of the coverage independent of E.

An alternative to condition eqn (20) is the assumption about the congruence of the adsorption isotherm with respect to the electrode charge, which can be written as:

$$B(q) \cdot c = f(\theta) \tag{21}$$

where $B(q)$ is a function of the electrode charge. The analysis of eqn (21) shows[95,98] it to be equivalent to the relation

$$E = E_0(1 - \theta) + E_1\theta \tag{22}$$

where the potentials E, E_0, and E_1 correspond to a certain $q = $ constant at the surface coverages with organic substance, $\theta = 0$ and $\theta = 1$, respectively.

The question of the choice of one of the models of the double layer in the presence of organic substance leads to a discussion on the choice of the electric variable.[91-94,97] With such a statement of the problem, a number of arguments were adduced in favour of eqn (21) to the effect that the choice of the electrode charge as an independent electric variable would be more convenient and suitable (see e.g.[94,99]). In reality, however, as we have pointed out earlier, with a strictly thermodynamic approach to the investigation of adsorption on electrodes, the choice of the electric variable is of no fundamental importance. But when the thermodynamic approach to investigation of adsorption on electrodes is supplemented by some model assumptions, e.g. by those of eqns (20) or (21), the choice of one of them should be dictated by the agreement of the relevant models with the experimental data. It should be noted that under the condition $C_0 = C_1$, which to the first approximation is realized in the case of adsorption on mercury of thiourea molecules[13] or those of benzene m-disulphonate ions,[56] eqns (20) and (21) are compatible. The difference between them can be experimentally established only under the condition $C_0 \gg C_1$, e.g. in the case of adsorption on mercury of aliphatic compounds with one polar group. Under these conditions the verification of the model of two parallel capacitors was carried out by various methods[69,71,95,96,98] showing this model to ensure better agreement with experimental data than the model based on eqn (21). The strongest argument in favour of the model of two parallel capacitors seems to be the dependence of the adsorption potential drop $E_{q=0}$ on the extent of adsorption (or on $\theta = \Gamma/\Gamma_m$). It follows from eqns (14)–(16) under the assumption that $C_0 = $ constant and $q = 0$ that:

$$E_{q=0} = \frac{E_N \theta}{(C_0/C_1)(1 - \theta) + \theta} \tag{23}$$

i.e. according to the two parallel capacitors model the dependence of $E_{q=0}$ on θ (or on Γ) should deviate from linearity, the stronger the more the ratio C_0/C_1 differs from unity. From eqn (22) at $q = 0$, when $E_0 = 0$ and $E_1 = E_N$, we obtain a different relation

$$E_{q=0} = E_N \theta \tag{24}$$

The use of the two parallel capacitors model is justified if the condition of equipotentiality is valid both on the metal surface and at the boundary between the inner part of the double layer and the bulk of the solution.[95,96] In the former case this condition is trivial, whereas in the latter it is only approximately valid due to the high value of the dielectric constant outside the inner layer.[100] But on the free surface of the solution one of the equipotential surfaces (i.e. the metal surface) is absent. Therefore, in this case, in the absence of effects caused by changes in the orientation of adsorbed molecules, in accordance with eqn (24) an additive summation of dipole effects is to be expected. Figure 7 shows the experimental dependences of $E_{q=0}$ on Γ for the case of adsorption at the two interfaces of n-propyl alcohol and n-caproic acid obtained by different methods. As is clear from the figure, in the case of the solution/mercury interface these dependences deviate markedly from the straight line and agree well with eqn (23). The same result

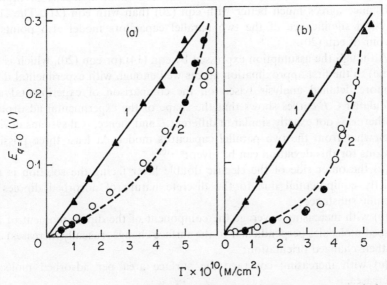

Fig. 7. Dependence of the adsorption potential drop on adsorption of n-propyl alcohol (a) and caproic acid (b) at the solution/gas (1) and solution/mercury (2) interfaces. Data obtained: from the maximum on σ, E curves (open circles) and from the minimum on C,E-curves in dilute supporting electrolyte solutions (full circles). Dashed curves calculated by means of eqn (23) under the condition that: $E_N = 0.31$ V, $C_0/C_1 = 3.5$, $\Gamma_m = 6 \times 10^{-10}$ mol/cm² (a) and $E_N = 0.29$ V, $C_0/C_1 = 7$, $\Gamma_m = 5 \times 10^{-10}$ mol/cm² (b). Supporting electrolyte -0.01 N H_2SO_4.

was obtained[95] for the case of adsorption on mercury of n-valeric acid and n-amylamine. It should be noted that similar forms of the curves of the dependence of $E_{q=0}$ on Γ were found for camphor,[34] n-butylamine[42] and some other aliphatic compounds,[40] although they were explained by a change in the orientation of adsorbed dipoles. The dependence of the adsorption potential drop on θ being approximately linear in these systems at the solution/gas interface (see Fig. 7), this explanation cannot be considered correct.

It follows from eqn (23) that the dependence of $E_{q=0}$ on log c can be linear only with $C_0 \approx C_1$ and a logarithmic adsorption isotherm, which is obtained

2

from isotherm eqn (17) at $a < 0$. These conditions are realized in the case of adsorption on mercury of organic ions and thiourea molecules. The Esin–Markov effect observed in their presence (see[40,101]) can serve as an additional proof of the validity of eqn (23).

Good agreement between the $E_{q=0}$–Γ-curves with eqn (23), as well as some other considerations,[98] shows that the structure of the interface between the electrode and an aqueous solution containing small additions of organic substance agrees much better with eqn (20) than with eqn (21). The clear physical significance of the two parallel capacitors model also points in favour of eqn (20).

Although the assumption expressed by eqn (14) (or eqn (20), which is the same) to the first approximation agrees well enough with experimental data, a more detailed analysis based on the comparison of experimental and calculated C,E-curves shows that the shape of the experimental adsorption isotherms is not exactly similar at different E and, hence, real systems deviate somewhat from the two parallel capacitors model. At least three possible reasons for this deviation can be given:[87,102]

(a) the outer side of the electric double layer facing the solution is not strictly equipotential due to the discrete nature of adsorbed dipoles of organic substance;

(b) with increasing coverage the component of the dipole moment of adsorbed molecules perpendicular to the surface changes (usually increases) due to their changed orientation;

(c) with increasing coverage the surface area per adsorbed molecule decreases.

The shape of the adsorption isotherm described by eqn (17) being determined by the value of the attraction constant a, in using this isotherm the deviation of the system from the two parallel capacitors model can be formally represented in terms of the dependence of a on E. An analysis shows that the first two reasons for the deviations of real systems from this model correspond to a linear dependence of a on E:[102,150]

$$a = a_0 + kE \qquad (25)$$

in case 'a', $k > 0$ (adsorption on mercury of alcohols and amines of the fatty series[76,79,80]) and in case 'b', $k < 0$ (adsorption on mercury of phenol[15] and pyridine[46,77]). The third reason for the deviation of real systems from the two parallel capacitors model given above corresponds to a parabolic dependence of a on E.[64,87] Such a dependence was observed in the case of the system $Na_2SO_4 + [(C_4H_9)_4N]_2SO_4$.[103]

It is evident that in the general case any deviation of the system from the two parallel capacitors model can be formally expressed by means of a dependence of the attraction constant contained in eqn (17) on the electrode potential: $a = a(E)$. The introduction of this dependence appears justified, since, as has been pointed out above, it is only a slight correction to the first basic assumption of the theory.

It remains now to substantiate the second basic assumption of the theory, viz. the applicability of eqn (17) for the description of real systems. As was shown by one of the authors,[104] a convenient criterion in the choice of the adsorption isotherm is the dependence of $(\partial \log c/\partial \theta)_E$ on θ, which at the potential of maximum adsorption $E = E_m$, when $d\theta/dE = 0$, can be readily determined experimentally from the experimental C,E-curves by means of the relation

$$\theta = \frac{C_0 - C}{C_0 - C_1} \tag{26}$$

It can be shown that the $(\partial \log c/\partial \theta)_E$ vs. θ curve always passes through a minimum at a certain value of $\theta = \theta^*$, which, being a function of

$$n = S_{\text{org}}/S_{\text{H}_2\text{O}},$$

is at the same time independent of the value of the attraction constant. It can be also readily shown that in the case of eqn (18)

$$\theta^* = \frac{1}{1 + \sqrt{n}} \tag{27}$$

and in the case of eqn (19)

$$\theta^* = \frac{2n - 1 - \sqrt{(n^2 - n + 1)}}{3(n - 1)} \tag{28}$$

The dependence of θ^* on n calculated by means of these equations is shown in Fig. 8. It was experimentally established[104] from the dependence of $(\partial \log c/\partial \theta)_E$

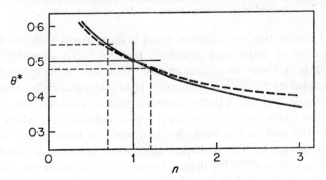

Fig. 8. Dependence of θ^* on n calculated by means of eqn (27)—solid line and eqn (27)—dashed line.

on θ for some aliphatic compounds (alcohols, acids, amines, tetrabutyl-ammonium cation) as well as for aniline, that the values of θ^* lie within the range 0·48–0·55 (see Fig. 9). As follows from Fig. 8, this range corresponds to the values of n from 0·7–1·2. Considering the rather small sensitivity of the shape of the isotherm to the change of n within this range, we can set $n = 1$ to a good enough approximation.

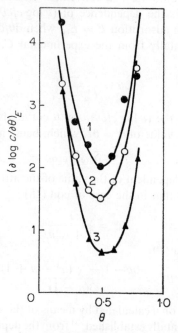

Fig. 9. Dependence of $(\partial \log c/\partial \theta)_E$ on θ obtained from experimental adsorption isotherms: 1—aniline, 2—n-amylamine and 3—tert-amyl alcohol.

At first glance, this experimental result is inconsistent with the difference in dimensions of water and adsorbate molecules. To eliminate this discrepancy it is necessary to assume an adsorption site on the mercury surface to be occupied by a group of associated H_2O molecules rather than by one water molecule.[105] From a thermodynamic point of view, this can mean that the work due to the transfer of such a group of molecules as a whole from the surface to the bulk is less than the total work of transfer of each of the molecules separately. The S-shaped adsorption isotherm of water vapour on mercury obtained,[106–108] demonstrating a considerable attractive interaction between adsorbed H_2O molecules, supports this conclusion.

The concept of the association of water molecules adsorbed on the mercury surface theoretically justifies the use of eqn (17) as a semi-empirical basis for the consideration of adsorption of organic molecules at the mercury/ electrolyte interface.

The two basic assumptions of the theory with account taken of possible deviations of real systems from the two parallel capacitors model (by introducing the dependence $a = a(E)$) lead to the following formula for the adsorption equilibrium constant B, the differential capacity C and the decrease of the surface tension $\Delta\sigma$ due to organic substance adsorption:[109]

$$B = B_0 \exp\left[-\frac{\int_0^E q_0\, \mathrm{d}E + C_1 E\left(E_N - \frac{E}{2}\right)}{A} \right] \exp(a_0 - a) \qquad (29)$$

$$C = C_0(1 - \theta) + C_1\theta - A\frac{\mathrm{d}^2 a}{\mathrm{d}E^2}\theta(1 - \theta)$$

$$+ \frac{1}{A}\left[q_0 + C_1(E_N - E) + A\frac{\mathrm{d}a}{\mathrm{d}E}(1 - 2\theta) \right]^2 \frac{\theta(1 - \theta)}{1 - 2a\theta(1 - \theta)} \qquad (30)$$

$$\Delta\sigma = -A[\log(1 - \theta) + a\theta^2] \qquad (31)$$

in which $A = RT\Gamma_m$, B_0 and a_0 are the values of the functions $B(E)$ and $a(E)$ at $E = 0$, respectively. In the case of a linear dependence of a on E (see eqn (25)) or at $a = $ constant, eqns (29) and (30) are simplified accordingly.

The theory based on the above assumptions is applicable in the first instance to adsorption on mercury of aliphatic compounds, in the presence of which, with account taken for the small linear dependence of a on E, there is good agreement between calculated and experimentally measured differential capacity curves. This is illustrated in Fig. 10 for the case of adsorption on mercury of iso-amylamine.[79] Good agreement between calculated and experimental C,E-curves is observed in the case of adsorption on mercury of $n\text{-}C_3H_7OH$, $n\text{-}C_4H_9OH$, $n\text{-}C_5H_{11}OH$, iso-$C_5H_{11}OH$, tert-$C_5H_{11}OH$, $n\text{-}C_4H_9NH_2$, $n\text{-}C_5H_{11}NH_2$, $n\text{-}C_4H_9COOH$, and $C_2H_5COC_2H_5$.[76,78-80]

The picture is more complicated in the case of aromatic and heterocyclic compounds, which can be adsorbed in two different positions: horizontal and vertical. The number of molecules adsorbed in either position depends, in its turn, both on the organic substance concentration and on the electrode potential. For these systems the two parallel capacitors model cannot be applied over the whole range of adsorption potentials.[14] It must be substituted by the model of three parallel capacitors, with water molecules contained between the plates of one of them, organic substance molecules in the first, say vertical, position between the plates of the second, and organic substance molecules in the second, say horizontal, position between the plates of the third capacitor.[59] This model corresponds to

$$q = q_0(1 - \theta_1 - \theta_2) + q_1\theta_1 + q_2\theta_2 \qquad (32)$$

where θ_1 and θ_2 are the coverages for vertical and horizontal orientations of adsorbed molecules, respectively; q_1 is the value of q at $\theta_1 = 1$ and $\theta_2 = 0$; q_2 is the value of q at $\theta_2 = 1$ and $\theta_1 = 0$.

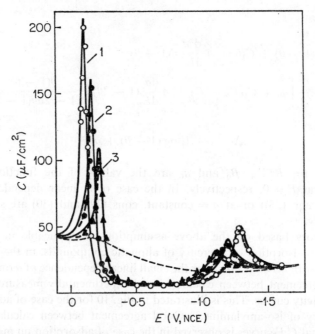

Fig. 10. Differential capacity curves of a mercury electrode in 1 N Na₂SO₄ solution (dashed line) and with additions of isoamylamine in concentrations: 1—0·04, 2—0·02, and 3—0·01 M. Solid lines—calculated curves (different symbols) and dashed lines—experimental data at the frequency 400 cps.

In using eqn (32), however, it is necessary to choose in addition two adsorption isotherms relating the quantities θ_1 and θ_2 with the bulk concentration of organic substance. It should be stressed that the choice of these isotherms cannot be arbitrary since they are related to eqn (32) by the basic equation of electrocapillarity eqn (1). It can be shown that in the general case the following isotherms are consistent with the three parallel capacitors model

$$B_1 c = \frac{\theta_1}{n_1(1 - \theta_1 - \theta_2)^{n_1}} \exp\left(-2a_{11}n_1\theta_1 - 2a_{12}n_1\theta_2\right) \\ B_2 c = \frac{\theta_2}{n_2(1 - \theta_1 - \theta_2)^{n_2}} \exp\left(-2a_{22}n_2\theta_2 - 2a_{12}n_2\theta_1\right) \tag{33}$$

in which the adsorption equilibrium constants B_1 and B_2 are some functions of the electrode potential; the attraction constant a_{11} takes account of the interaction between vertically adsorbed molecules, a_{22} between horizontally adsorbed molecules and a_{12} between the adsorbed molecules in different positions; n_1 is the ratio of the area occupied by a vertically oriented organic substance molecule to the area per an associate of adsorbed water molecules; n_2 is the similar ratio for the case of horizontal orientation of adsorbate molecules.* By means of eqn (32) it is possible to find the explicit dependence of B_1 and B_2 on E. In fact, if we assume

$$q_1 = C_1(E - E_{N1})$$

and

$$q_2 = C_2(E - E_{N2}),$$

we obtain for B_1 and B_2:

$$B_1 = B_{01} \exp\left[-\frac{\int_0^E q_0 \, dE + C_1 E\left(E_{N1} - \frac{E}{2}\right)}{RT\Gamma_{m1}}\right] \tag{34a}$$

$$B_2 = B_{02} \exp\left[-\frac{\int_0^E q_0 \, dE + C_2 E\left(E_{N2} - \frac{E}{2}\right)}{RT\Gamma_{m2}}\right] \tag{34b}$$

The set of eqn (33) can be exactly solved only with the help of an electronic computer. There is an additional difficulty due to the fact that in eqns (33) and (34), apart from organic substance concentration, it is possible to vary relatively arbitrarily 12 parameters: B_{01}, B_{02}, C_1, C_2, E_{N1}, E_{N2}, n_1, n_2, a_{11},

---

* Parry and Parsons[57] also used the set of eqn (33) to describe such systems; they assume however the quantities B_1 and B_2 to be some functions of the electrode charge rather than of its potential. The analysis of this assumption shows that it is equivalent to the model of three capacitors connected in series and filled with water molecules and adsorbate molecules in two different positions, respectively. Apparently, it is difficult to give a physical interpretation to such model, although it is not inconsistent with the experimental dependence of θ on q in sodium p-toluenesulphonate solutions.

a_{22}, a_{12}, and $A = n_1 RT\Gamma_{m1} = n_2 RT\Gamma_{m2}$. In order to avoid these difficulties, we tried to use for the evaluation of the three parallel capacitors model a first rough approximation, according to which $a_{11} = a_{12} = a_{22} = 0$. Since under these conditions the comparison of theory with experiment can be only semi-quantitative, it appeared justified to make the following additional assumptions: $C_0 = C_2 =$ constant, $n_1 = 1$, and $n_2 = 2$. Even with all these simplifications, the expression for the differential capacity of the double layer in the presence of an organic substance adsorbed in two different positions is rather cumbersome:

$$C = C_0 - 2A\alpha\theta_1 + \frac{A}{2(1 + \theta_2)} \{8\alpha^2(E - E_m)^2\theta_1(1 - \theta_1)$$
$$+ \beta^2\theta_2(1 - \theta_2) + 8\theta_1\theta_2[\alpha^2(E - E_m)^2 + \alpha\beta(E - E_m)]\} \quad (35)$$

where $\alpha = (C_0 - C_1)/2A$; $E_m = -C_1 E_{N1}/(C_0 - C_1)$; $\beta = -2C_0 E_{N2}/A$; $A = RT\Gamma_{m1} = 2RT\Gamma_{m2}$.

Under the same conditions the decrease in surface tension due to organic substance adsorption can be written as

$$\Delta\sigma = -A \left[\log(1 - \theta_1 - \theta_2) + \frac{\theta_2}{2} \right] \quad (36)$$

In Figs. 11–13 calculations by means of these equations are compared with the experimental C,E- and σ,E-curves for the mercury electrode in the

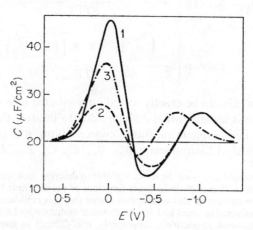

Fig. 11. Theoretically calculated differential capacity curves: 1—$E_{N1} = 0.5$ V and $E_{N2} = -0.5$ V; 2—$E_{N1} = 0.5$ V and $E_{N2} = 0.25$ V; 3—$E_{N1} = 0$ and $E_{N2} = -0.5$ V.

presence of ortho- and para-phenylenediamines.[59] It was assumed in the calculations that: $C_0 = 20\ \mu F/cm^2$; $C_1 = 7\ \mu F/cm^2$; $A = 1\cdot6\ \mu F/cm^2$; $B_{01} = 100\ l./mole$; $B_{02} = 500\ l./mole$; $c = 0\cdot02\ M$; $E_{N1} = +0\cdot5\ V$, and 0; $E_{N2} = -0\cdot25$, and $-0\cdot5\ V$. As is clear from Fig. 12, when passing from

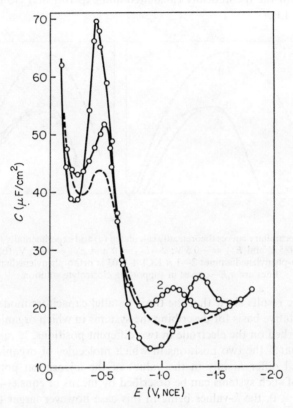

Fig. 12. Differential capacity curves of a mercury electrode in 1 N KCl solution (dashed line) and with 0·02 M addition of: 1—para-phenylenediamine, 2—ortho-phenylenediamine. Frequency 400 cps.

ortho- to para-phenylenediamine, the anodic peak grows appreciably, the decrease of capacity in the middle part of the C,E-curve compared to the curve of the supporting electrolyte becomes greater and the cathodic peak shifts in the direction of more negative potentials. All these phenomena can be observed in Fig. 11 when passing from curve 3 calculated for $E_{N1} = 0$ to curve 1, which was calculated assuming $E_{N1} = +0\cdot5\ V$. Thus the difference in the adsorption behaviour of ortho- and para-phenylenediamines can be connected with the change in the dipole moment component perpendicular to the surface observed in the case of their vertical orientation, whereas the quantity E_{N2} due to π-electronic interaction of these molecules with the mercury surface remains unchanged. This conclusion is supported by the comparison of the theoretically calculated and experimental electrocapillary curves (Fig. 13).

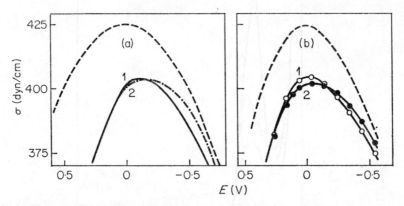

Fig. 13. Electrocapillary curves theoretically calculated (a) and experimentally measured (b): (a) 1—$E_{N1} = 0.5$ V and $E_{N2} = -0.5$ V; 2—$E_{N1} = 0$ and $E_{N2} = -0.5$ V. (b) 1—1 N KCl + 0.03 M para-phenylenediamine; 2—1 N KCl + 0.03 M ortho-phenylenediamine. Dashed lines are σ,E—curves in supporting electrolyte solution.

The above results show that the three parallel capacitors model is a good semi-quantitative basis for describing the systems in which organic substance can be adsorbed on the electrode in two different positions. It can be readily shown[150] that if the two positions in which molecules of organic substance are adsorbed differ mainly in the values of the adsorption potentials, the behaviour of such systems can be described by means of eqns (14), (17), and (25) with $k < 0$, the k-values being in this case however larger than in the case of aliphatic substances adsorption. This gives a theoretical basis for the application of the two parallel capacitors model to the description of adsorption on mercury of phenol[15] and pyridine.[46,77]

The three parallel capacitors model in conjunction with the set of isotherms (eqn (33)) can be also used as a basis in describing the simultaneous adsorption on the electrode of two surface-active organic substances. Only in that case in the first of the equations (eqn (33)) the total concentration c should be substituted by c_1 (the concentration of the first substance), and in the second equation by c_2 (the bulk concentration of the second organic substance). The physical significance of other quantities appearing in eqns (32) and (33) will be changed accordingly: all the quantities with the subscript 1 refer to the first substance and with the subscript 2 to the second.

An accurate solution of the set of eqn (33) in the case of the simultaneous adsorption on the electrode of two organic substances presents the same difficulties as in the case of one substance adsorbed in two different positions. For some particular cases where these difficulties could be avoided the three parallel capacitors model was successfully used by Tedoradze. Arakeljan and Belokolos.[82] An essentially similar theoretical approach to the description of such systems was developed also by Kastening and Holleck.[110]

IV. PROBLEMS OF THEORY OF MOLECULAR ADSORPTION OF ORGANIC SUBSTANCES

In two recent papers,[62,99] the second of which is further development of the first, Bockris and collaborators advanced a molecular theory of adsorption of organic substances and of its dependence on the electrode charge. Bockris and collaborators contrast their theory as a molecular one with our concepts, which are of a thermodynamic nature and involve finding the necessary values of the constants from experiments.

There is no doubt that a theory entirely based on molecular data is a considerable advance over a thermodynamic theory supplemented with a macromodel (the two or three parallel capacitors models). It seems to us however that in developing such a theory one should determine first from a thermodynamic analysis which quantities appearing in mathematical expressions should be treated from the molecular point of view. In point of fact, Butler's theory[23] can be considered as an example of an approach of that kind, since Butler's relations can be obtained from the equations of Frumkin's initial theory[87] if the double layer capacity is calculated as the capacity of a flat capacitor of constant thickness from molecular polarizabilities, the shift of the point of zero charge—from dipole moments of adsorbed molecules, and the adsorption isotherm is approximated by means of the Henry equation.

We would like to discuss here the possibilities, as it appears to us, of further development of the theory of adsorption of organic substances on the basis of thermodynamic relations, which would bring us to an accurate molecular theory, at least for simplest aliphatic compounds.

It follows from eqn (29) at $a = $ constant that the dependence of adsorption on the potential is determined by the expression:

$$B = B_0 \exp\left[-\frac{\int_0^E q_0\, dE + C_1 E\left(E_N - \dfrac{E}{2}\right)}{RTT_m}\right] \quad (37)$$

This expression takes full account of the effect of the free energy stored in the electric double layer undisturbed by the adsorption process since this energy is determined by numerical integration of the experimental C,E-curves in a pure supporting electrolyte solution. As is evident from eqn (37), the adsorption is determined by the following parameters: C_1, E_N, Γ_m, and B_0, which depend on the nature of the adsorbate. In addition, a relation must be established between the adsorbate nature and the value of the attraction constant a contained in the adsorption isotherm eqn (17). Thus in order to pass from the thermodynamic to the molecular theory it is necessary to interpret the experimental dependence of C_0 on E, to estimate the parameters C_1, Γ_m. E_N and B_0 as well as to interpret the intermolecular interaction of adsorbed particles. Let us consider each of these problems in more detail.

1. The Dependence of C_0 on E

A number of investigators aimed at solving this problem completely or partially.[62,111–116] Bockris, Gileadi, and Müller[99] did not take up the question of the adequacy of the double layer model used by them when treating the adsorption of organics, for the solution of the above problem, although earlier[62] they pointed to the necessity of resorting to more complex concepts for explaining the dependence of C_0 on E at $q \geqslant 0$. In principle we can suppose that it will be possible to develop an adequate molecular theory of the C,E-curves in not too distant future.

2. The Quantity C_1

This quantity depends primarily on the thickness of the hydrocarbon layer between the double layer ions and metal and on its dielectric constant. Difficulties arise because this layer can contain water molecules as well.[80] The physical significance of the parameters contained in the quantity C_1 is quite clear, but their quantitative determination requires further experimental work on the dependence of C_1 on the hydrocarbon chain length.

3. The Quantity Γ_m

The determination of this quantity involves establishing the geometric dimensions and orientation of adsorbed organic molecules. In calculating Γ_m however, it is necessary to take into consideration the possibility of embedding of water molecules between adsorbate molecules as well, since adsorption saturation is usually established before water molecules are completely expelled from the mono-molecular layer at the interface.[117] Nevertheless, the calculation of Γ_m can be considered as being the simplest problem in the molecular adsorption theory of organic compounds.

4. The Quantity E_N

For better understanding of the physical significance of this quantity it can be conveniently represented as the difference of Galvani potentials at the point of zero charge at the metal/solution interface $\varphi_{q=0}$ after its coverage with adsorbed organic molecules and in a pure solvent, respectively:

$$E_N = (\varphi_{q=0})_{org} - (\varphi_{q=0})_{H_2O} \tag{38}$$

The effect of adsorbed substance upon the change of $\varphi_{q=0}$ is usually associated with the change in the dipole effect upon substitution of adsorbate molecules for those of the solvent. This approach is to a certain degree justified due to the existence of an approximate parallelism between E_N and the adsorption potential at the solution/gas interface at $\theta = 1$ in the case of adsorption of fatty acids, alcohols and amines.[21] It should be borne in mind however that along with this dipole effect the quantity $\varphi_{q=0}$ can also vary owing to the change in the electron density distribution in the metal surface layer depending on the medium the metal is in contact with. Judging by the influence of adsorption of noble gases on metal electron work functions,[118] this effect, about which little is known, can become quite appreciable.

The attempts to calculate the dipole effect, to the consideration of which we have to confine ourselves at present, meet difficulties owing to the following two circumstances:

(a) at present there is no general agreement about the sign of the dipole component of the quantity $(\varphi_{q=0})_{H_2O}$,[119,120] although many workers, Bockris et al.[62] included, suppose that at $q = 0$ there is a certain preferred orientation of water dipoles with their negative ends turned towards mercury, as was suggested by Frumkin.[21] There is no doubt however that the potential drop caused by this orientation is not large in absolute value (of the order of 0·1 V);

(b) a quantitative calculation of the potential drop in the adsorbed dipole layer from the dipole moments by means of the well-known elementary formula is possible under exceptional conditions, as follows from the surface potential measurements of films of long chain ω-bromosubstituted fatty acids and alcohols.[121] The reasons for these difficulties have already been discussed.[22] However, attempts[122,123] to interpret semi-quantitatively the surface potentials of monolayers of long-chain substances at the solution/gas interface on the basis of dipole moments of individual bonds lead to useful results.

It should be noted that in a model of the n-butanol molecule directed normal to the surface (as suggested[99]) the dipole moment is oriented almost parallel to the surface and cannot affect the potential drop in the adsorption layer. It is known, however, that the presence on the surface of monolayers of aliphatic alcohols, e.g. hexadecyl alcohol, causes the surface potential to shift by $\sim$400 mV in the positive direction. In these condensed layers the molecules are oriented normal to the surface and the dipole structure of the polar group should not differ from that in the n-butanol molecule. The polar group being far removed from the interface in the case of solution/mercury and solution/gas interfaces, there should be no difference in its dipole structure. Thus if we adopted the model used,[99] we should have to assume the potential drop between the gas phase and water to be about -400 mV. But such high negative values of this potential difference lead to improbable relations between chemical hydration energies of cations and anions.[66]

5. The Quantity B_0

This quantity is a measure of the free energy of adsorption at the uncharged mercury/solution interface $-\Delta G_A^0$, since in sufficiently dilute organic substance solutions, where the mole fraction of the solvent can be equated with unity,

$$B_0 = \frac{1}{55 \cdot 5} \cdot \exp\left(-\frac{\Delta G_A^0}{RT}\right) \qquad (39)*$$

As is well known, calculations of the adsorption energy of simple molecules on various adsorbents from the gas phase have been carried out on the basis of molecular data. In principle such a calculation is possible for the case under consideration. For this purpose, however, in addition to taking account of the image work and the dispersion interaction at the interface with metal both for water and adsorbate, and in the case of unsaturated and aromatic compounds also of the π-electronic interaction between metal and adsorbed particle, it is necessary to give a molecular interpretation of the interaction between solvent and solute in the bulk of the solution, i.e. to calculate from molecular quantities the standard free energy of evaporation and the free energy of hydration.

Apparently, we are far as yet from solving these problems. Bockris et al.[99] introduce an empirical component of the free adsorption energy ('chemical' term), which limits the 'molecular' nature of their theory.

6. The Adsorption Isotherm

As shown above, at a given number of constants eqn (7) fits best the experimental data, though the interaction between adsorbed molecules is

---

* As has been many times rightly emphasized by Bockris et al.,[37,62,99] the quantity $\triangle G_A^0$ appearing in eqn (39) is strictly speaking the difference between the standard free energies of adsorption of organic substance and of water molecules. This fact is of essential importance for understanding the comparative behaviours of organic substances at the solution/mercury and solution/gas interfaces.

allowed for in this isotherm in a semi-empirical manner. A molecular interpretation requires more rigid account to be taken of the dependence of the interaction between adsorbed particles on the coverage and hence will lead to a more complex expression for the adsorption isotherm, which will make more difficult its use with a simultaneous allowance for the field effect.

Bockris et al.[99] used the isotherm expressed by eqn (19) at $a = 0$, and treated the intermolecular interaction within the adsorption layer entirely as an interaction between oriented water dipoles. Taking into consideration the interaction between these dipoles seems in a certain sense to be equivalent to our conclusion about the association of water molecules adsorbed on mercury, since it gives an approximately Langmuir dependence of θ on the adsorbate concentration at the values of n corresponding to the relation between the geometric dimensions of water and organic substance molecules. However, by ignoring the direct interaction between adsorbed organic substance molecules as Bockris et al. do, it is impossible to explain the dependence of the shape of the isotherm on the chain length (the S-shape becoming more pronounced with increasing chain length), which is observed in the case of adsorption both at the solution/mercury and solution/gas interfaces, and with a sufficient chain length leads to characteristic condensation phenomena.

In conclusion, we would like to make the following remark regarding the level of experimental verification attempted in the case of both thermodynamic and molecular theories. In developing the theory the basic assumptions of which were considered in the previous section we sought to obtain exact agreement between calculated and experimental quantities under the conditions of a most rigorous verification, viz. a confrontation of calculated and experimentally measured C,E-curves. Therefore, the theory was made more complicated (by taking account of the dependence of a on E), which would be quite unnecessary if we considered as it was done by Bockris et al. as sufficient the statement of some similarity between the calculated and experimentally found dependence of adsorption on potential.

V. INFLUENCE OF THE NATURE OF THE METAL ON ADSORPTION OF ORGANIC MOLECULES

The dependence of adsorption of organic substances on the nature of the metal can be determined by the following factors: change of the difference between the energies of interaction of adsorbate and water with the metal surface, heterogeneity of the solid electrode surface, adsorption of hydrogen and oxygen atoms on the metal surface and finally by chemisorption processes leading to deep changes in the structure of adsorbate molecules (e.g. to the breaking of the C—H and even C—C bonds). Let us discuss briefly each of these factors.

1. The Change in the Energy of Interaction of Adsorbate and Water Molecules with Metal Surface

If, when passing from one metal to another, only this factor comes into play, the main regularities in adsorption of organic molecules established for the mercury electrode (sections 2 and 3) should be valid for other metals as well. In fact, the shape of the electrocapillary curves on 57·5 per cent indium amalgam and liquid gallium electrode in the presence of iso-amyl alcohol[124,125] is similar to corresponding σ,E-curves of the mercury electrode (Fig. 14),

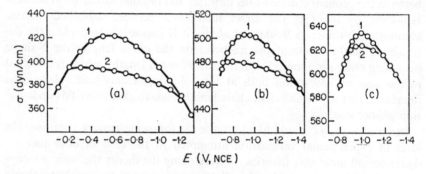

Fig. 14. Electrocapillary curves on various metals. (a) On mercury: 1—1 N KCl + 0·1 N HCl; 2—1 N KCl + 0·1 N HCl + 0·1 M iso-$C_5H_{11}OH$. (b) On 57·5 per cent indium amalgam: 1—1 N Na_2SO_4 + 0·01 N H_2SO_4; 2—1 N Na_2SO_4 + 0·01 N H_2SO_4 + 0·14 M iso-$C_5H_{11}OH$. (c) On gallium: 1—1 N KCl + 0·1 N HCl; 2—1 N KCl + 0·1 N HCl + 0·1 M iso-$C_5H_{11}OH$.

and the differential capacity curves of the solid bismuth electrode obtained recently by Palm, Past and Pullerits[132] (Fig. 15) are quite similar to corresponding C,E-curves on mercury (cf. Fig. 3). Characteristic shapes of the C,E-curves with decreasing capacity near $E_{q=0}$ and adsorption-desorption peaks in the presence of various organic substances were observed also on electrodes from thallium amalgam, gallium, lead, tin, silver, and antimony.[125–132]

Of course, when passing from one metal to another, the change in the interaction of the electrode both with adsorbate and water molecules involves a change in the parameters contained in eqn (37), but a quantitative comparison is yet possible, owing to the scarcity of experimental data, in few

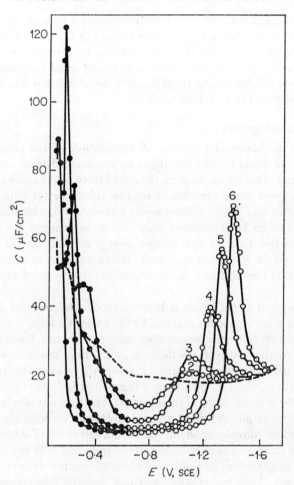

Fig. 15. Differential capacity curves of a bismuth electrode in n-amyl alcohol solutions of varying concentration: 1—0, 2—0·01, 3—0·02, 4—0·05, 5—0·1 and 6—0·2 M. Supporting electrolytes—0·1 N K_2SO_4 (open circles) and H_2SO_4 (full circles). Frequency 1000 cps.

cases only.[132] As is clear from Fig. 14, the adsorptivity of iso-$C_5H_{11}OH$ on indium amalgam and especially on gallium is markedly less than on mercury. In the latter case this can be accounted for by a stronger adsorption of water molecules on gallium surface at $q \geqslant 0$.[125] It should be pointed out, however, that at large enough negative surface charges, where this kind of adsorption of water on gallium is not observed, the differential capacity values in the presence of n-C_3H_7OH and tert-$C_5H_{11}OH$ at equal q are practically the same on mercury and gallium.[131] On passing to the iron electrode, the interaction of water molecules with the metal surface increases to such an extent that hexyl alcohol, which has a high surface activity in the case of Hg, Ga, Pb, Bi, Sn, Tl, Ag, and Sb, is practically not adsorbed on iron.[133] No adsorption of aliphatic compounds (cyclohexane, fatty acids) could be observed on the gold electrode.[134] According to Hillson,[135] hexyl alcohol has a very low surface activity on the nickel electrode.

2. Surface Heterogeneity

In itself the change in the state of aggregation of the electrode upon transition from liquid to solid metal has no pronounced effect on its adsorption properties. This is evident from the coincidence of the cathodic branches of the C,E-curves in the presence of organic substance for liquid and solid mercury electrodes.[129,136] Unlike a liquid electrode, however, the surface of a solid one can be heterogeneous with respect to adsorption energy. This is due to the fact that the free surface energy depends on the orientation of metal crystals and increases with increasing concentration of flaws at the interface (dislocations, vacancies, microdistortions of crystal lattice, etc.).

The adsorption isotherm on a heterogeneous surface must be different from that on a homogeneous surface. As shown by Temkin,[137] in the case of uniform heterogeneity the adsorption energy decreases linearly with the coverage (Temkin isotherm). In the case of a formal use of eqn (17) under these conditions, an increase in surface heterogeneity should lead to a decrease of the positive value of a down to zero (Langmuir isotherm) and then to an increase in the negative value of this constant. This explains why in some cases the dependence of adsorption of organic substances on solid electrodes with a heterogeneous surface upon the bulk concentration of adsorbate obeys formally the Langmuir isotherm. As follows from eqn (30), the decrease of a should lead to a lowering and widening of the adsorption-desorption peaks on the C,E-curves. Thus the experimental fact that the adsorption-desorption peaks on the C,E-curves are much less pronounced in the case of some solid electrodes, e.g. Ag and Sb, than on mercury[128,129] can be accounted for by heterogeneity of the surface of these metals.

The surface heterogeneity can change significantly as the result of annealing or mechanical treatment of the metal. Consequently, the pre-treatment of solid electrodes can be expected to affect their adsorption properties. Figure 16 shows the adsorption isotherms of allylphenylthiourea on a high purity iron electrode calculated for $E = -0.35$ V (NHE) by means of eqn (26) from the experimental C,E-curves.[138] It is clear from the figure that the surface activity of allylphenylthiourea decreases sharply upon transition from iron annealed at 600°C to iron annealed at 750°C.

3. Effect of Adsorption of Hydrogen and Oxygen Atoms

Let us consider this problem assuming that the adsorption of hydrogen and organic substance can be treated as a reversible process.[139,140] Under these conditions the basic equation of electrocapillarity eqn (1) can be written as:

$$d\sigma = -\Gamma_H \, d\mu_H - \Gamma_{org} \, d\mu_{org} \tag{40}$$

Let us denote by A_H the amount of hydrogen adsorbed per unit surface. The quantity A_H is not identical with Γ_H, since a fraction of hydrogen disappearing upon formation of unit surface is ionized and expended in its charging. A_H and Γ_H are related by the following expression:

$$\Gamma_H = A_H - \frac{q}{F} \tag{41}$$

In a solution of a constant composition the equilibrium potential of the electrode is determined by the quantity μ_H, and hence

$$d\mu_H = -F \, dE \tag{42}$$

Substituting expression eqns (41) and (42) into eqn (40), we obtain

$$d\sigma = (FA_H - q) \, dE - \Gamma_{org} \, d\mu_{org} \tag{43}$$

It follows from eqn (43) that

$$F\left(\frac{\partial A_H}{\partial \mu_{org}}\right)_E - \left(\frac{\partial q}{\partial \mu_{org}}\right)_E = -\left(\frac{\partial \Gamma_{org}}{\partial E}\right)_{\mu_{org}} \tag{44}$$

The quantity Γ_{org} being a function of μ_{org} and E, we can write the following general equation:

$$\left(\frac{\partial \mu_{org}}{\partial E}\right)_{\Gamma_{org}} = -\left(\frac{\partial \mu_{org}}{\partial \Gamma_{org}}\right)_E \left(\frac{\partial \Gamma_{org}}{\partial E}\right)_{\mu_{org}} \tag{45}$$

It follows from eqns (44) and (45) that

$$\left(\frac{\partial \mu_{org}}{\partial E}\right)_{\Gamma_{org}} = -\left(\frac{\partial q}{\partial \Gamma_{org}}\right)_E + F\left(\frac{\partial A_H}{\partial \Gamma_{org}}\right)_E \tag{46}$$

which can be re-written as

$$\left[\frac{\partial(\Delta G_A^0)}{\partial E}\right]_{\Gamma_{org}} = -\left(\frac{\partial q}{\partial \Gamma_{org}}\right)_E + F\left(\frac{\partial A_H}{\partial \Gamma_{org}}\right)_E \tag{47}$$

where ΔG_A^0 is the standard free adsorption energy of organic substance.

In the case of the mercury electrode the change in adsorbability of organic substance with potential is determined by the first term in the right hand side of eqn (47). In the case of hydrogen-adsorbing metals it is also necessary to take into account the second term $F(\partial A_H/\partial\Gamma_{org})_E$.

In order to compare the first and the second terms let us assume that eqn (14) is valid on the electrode under investigation and that it is possible to write an analogous expression for A_H:

$$A_H = A_H^0(1 - \theta) + A_H'\theta \tag{48}$$

Taking into consideration eqns (14) and (48), it is possible to write eqn (47) as:

$$\left[\frac{\partial(\Delta G_A^0)}{\partial E}\right]_{\Gamma_{org}} = \frac{q_0 - q_1}{\Gamma_m} - F\frac{A_H^0 - A_H'}{\Gamma_m} \tag{49}$$

For many organic substances over a wide potential range $|q_1|$ in the case of the mercury electrode is known to be several times as little as $|q_0|$, which determines the effect of the electric field. Judging by the few experimental data available,[141] there probably exists a similar relation between A_H' and A_H^0. Thus the relative value of the first and the second terms in the right hand side of eqn (49) as factors determining the value of $[\partial(\Delta G_A^0)/\partial E]_{\Gamma_{org}}$ depends on the relation between FA_H^0 and $|q_0|$.

The maximal value of FA_H^0 in the hydrogen region being $\sim 2 \times 10^{-4}$ C/cm^2 and the maximal value of q_0 for the platinum electrode (e.g. in acid sulphate solutions)—a quantity of the order of 10^{-5} C/cm^2, it is evident that the second term in the right hand side of eqn (49) is of decisive importance. Even at the positive boundary of the hydrogen region and within the double layer region, in which in the case of solutions containing no surface-active anions the amounts of H_{ads} and O_{ads}, if small, are not equal to zero, $|q_0|$ and FA_H^0 can be of the same order of magnitude and the presence of the term $F(\partial A_H/\partial\Gamma_{org})_E$ must be allowed for in the determination of the dependence of ΔG_A^0 on E.

It follows from the above that the relations derived for the mercury electrode are not applicable for the determination of the position of the adsorbability maximum of neutral molecules relative to $E_{q=0}$ in the case of the platinum electrode and apparently also for other hydrogen-adsorbing metals. Whereas $E_{q=0}$ for platinum lies in the hydrogen region,[140] the potential of maximum adsorption, due to the desorbing action of hydrogen must be shifted into the double layer region (see the data on naphthalene adsorption on platinum[142]). The magnitude of this shift is limited by the appearance of the surface not only of positive charges, but also of adsorbed oxygen, there being no doubt that the presence of oxygen also reduces the adsorbability of organic molecules.

4. Dissociative Adsorption

Until the beginning of the sixties, in most studies on adsorption of organic substances on platinum metals, a review of which was given by Sokolsky,[143] organic molecules were assumed to be physically adsorbed or reversibly chemisorbed on the electrode surface. In his review Sokolsky first suggested that adsorption of organic substances on platinum metals can involve chemical reactions. In fact, beginning with 1964, various investigators[144-148] established by independent methods that adsorption on platinum of aliphatic

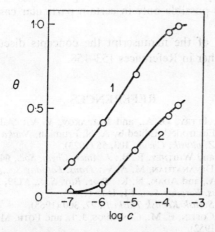

Fig. 16. Adsorption isotherms of allylphenylthiourea at $E = -0.35$ V (NHE) on an iron electrode annealed at: 1—600°C and 2—at 750°C.

alcohols and hydrocarbons involves the dissociation of organic molecules with the breaking of the C—H and even C—C bonds. Thus, in the case of methanol adsorption on platinum three hydrogen atoms break away from the CH_3OH molecule, so that the composition of the chemisorbed particle corresponds to the formula HCO.[145*] In recent years dissociative adsorption of organic molecules has been experimentally established also for other platinum metals: palladium, ruthenium, rhodium and their alloys. Moreover, research has been considerably extended to include many organic substances belonging to different classes.

The data on the adsorption of organic substances from electrolyte solutions obtained so far do not permit us to give a full description of the adsorbed layer structure and in most cases lead only to definite conclusions regarding the stoichiometry of the adsorbed layer (relative amounts of carbon, oxygen,

* According to Breiter $C_2H_2O_3$.[151]

and hydrogen on the surface). A more detailed discussion of the regularities in dissociative adsorption on the basis of the experimental evidence available now can be found in Reference 1. As follows from direct measurements of ion adsorption, carried out with radioactive tracers, the adsorption of organics on platinum exerts a very strong influence on the composition of the ionic side of the double layer at the platinum/solution interface.[152]

To conclude our discussion of the effect of the nature of the metal on adsorption of organic molecules, we would like to emphasize that the application of experimental data on adsorption of organic compounds obtained for mercury to other metals (see e.g.[35-37,149]), as follows from the above discussion, is possible only in certain particular cases and even then with great caution.

After completion of the manuscript the concepts discussed above have been developed further in References 153–158.

REFERENCES

1. DAMASKIN, B. B., PETRY, O. A., and BATRAKOV, V. V.: 'Adsorption of Organic Compounds at Electrodes', Edited by A. N. Frumkin, *Nauka* (Moscow, 1968).
2. FRUMKIN, A. N.: *Z. physik. Chem.*, **103**, 55 (1923).
3. GRAHAME, D. C., and WHITNEY, R. B.: *J. Amer. Chem. Soc.*, **64**, 1548 (1942).
4. PARSONS, R., and DEVANATHAN, M. A. V.: *Trans. Faraday Soc.*, **49**, 404 (1953).
5. GUGGENHEIM, E. A., and ADAM, N. K.: *Proc. Roy. Soc.*, **A139**, 218 (1933).
6. PARSONS, R.: *Trans. Faraday Soc.*, **51**, 1518 (1955).
7. FRUMKIN, A. N.: *Svensk Kemisk Tidskrift*, **77**, 300 (1965).
8. GRAHAME, D. C., COFFIN, E. M., CUMMINGS, J. I., and POTH, M. A.: *J. Amer. Chem. Soc.*, **74**, 1207 (1952).
9. FRUMKIN, A. N., and MELIK-GAIKAZJAN, V. I.: *Dokl. Akad. Nauk. SSSR*, **77**, 855 (1951).
10. MELIK-GAIKAZJAN, V. I.: *Zhur. Fiz. Khim.*, **26**, 560 (1952).
11. LORENZ, W., and MÖCKEL, F.: *Z. Elektrochem.*, **60**, 507 (1956).
12. TEDORADZE, G. A., and ARAKELJAN, R. A.: *Dokl. Akad. Nauk. SSSR*, **156**, 1170 (1964).
13. SCHAPINK, F. W., OUDEMAN, M., LEU, R. W., and HELLE, J. N.: *Trans. Faraday Soc.*, **56**, 415 (1960).
14. DAMASKIN, B. B., MISHUTUSHKINA, I. P., GEROVICH, V. M., and KAGANOVICH, R. I.: *Zhur. Fiz. Khim.*, **38**, 1797 (1964).
15. DAMASKIN, B. B., GEROVICH, V. M., GLADKIKH, I. P., and KAGANOVICH, R. I.: *Zhur. Fiz. Khim.*, **38**, 2495 (1964).
16. DAMASKIN, B. B., ANDRUSEV, M. M., GEROVICH, V. M., and KAGANOVICH, R. I.: *Elektrokhimiya*, **3**, 667 (1967).
17. GOUY, G.: *Ann. Chim. Phys.* (8), **8**, 291 (1906); (8), **9**, 75 (1906).
18. FRUMKIN, A. N.: 'Elektrokapilljarnye javlenija i elektrodnye potentialy', Odessa, 1919.
19. FRUMKIN, A. N.: *Z. physik. Chem.*, **111**, 190 (1924).
20. FRUMKIN, A. N., DONDE, A. A., and KULVARSKAJA, R. M.: *Z. physik. Chem.*, **123**, 321 (1926).
21. FRUMKIN, A. N.: *Ergebn. exakt. Naturwiss.*, **7**, 235 (1928); *Coll. Symp. Ann.*, **7**, 89 (1930).
22. FRUMKIN, A. N., and WILLIAMS, J. W.: *Proc. Nat. Acad. Sci. U.S.*, **15**, 400 (1929).

23. BUTLER, J. A. V.: *Proc. Roy. Soc.*, **A122**, 399 (1929).
24. BUTLER, J. A. V., and OCKRENT, C.: *J. Phys. Chem.*, **34**, 2286, 2297, 2841 (1930).
25. BUTLER, J. A. V., and WIGHTMAN, A.: *J. Phys. Chem.*, **35**, 3293 (1931).
26. FRUMKIN, A. N., GORODETSKAJA, A. V., and CHUGUNOV, P. S.: *Acta physicochim. URSS*, **1**, 12 (1934).
27. KRJUKOVA, T. A., and FRUMKIN, A. N.: *Zhur. Fiz. Khim.*, **23**, 819 (1949).
28. GEROVICH, M. A., and OLMAN, O. G.: *Zhur. Fiz. Khim.*, **28**, 19 (1954).
29. GEROVICH, M. A.: *Dokl. Akad. Nauk. SSSR*, **96**, 543 (1954); **105**, 1278 (1955).
30. GEROVICH, M. A., and RYBALCHENKO, G. F.: *Zhur. Fiz. Khim.*, **32**, 109 (1958).
31. GEROVICH, M. A., and POLJANOVSKAJA, N. S.: *Nauchn. Dokl. Vysshei Shkoly, Khim. i Khim. Tekhn.*, **N4**, 651 (1958).
32. GRAND, R.: *Ann. Physique*, **10**, 738 (1955); *J. chim. phys.*, **60**, 1315 (1963).
33. CONWAY, B. E., BOCKRIS, J. O'M., and LOVREČEK, B.: *CITCE, Proc.*, **6**, 207 (1955).
34. STROMBERG, A. G., and ZAGAINOVA, L. S.: *Zhur. Fiz. Khim.*, **31**, 1042 (1957).
35. ANTROPOV, L. I., and BENERGEE, S. N.: *J. Indian Chem. Soc.*, **35**, 531 (1958); **36**, 451 (1959).
36. BLOMGREN, E., and BOCKRIS, J. O'M.: *J. Phys. Chem.*, **63**, 1475 (1959).
37. BLOMGREN, E., BOCKRIS, J. O'M., and JESCH, C.: *J. Phys. Chem.*, **65**, 2000 (1961).
38. GIERST, L.: *Trans. Symp. Electrode Processes, Philadelphia*, p. 294 (J. Wiley and Sons N.Y., 1959).
39. CONWAY, B. E., and BARRADAS, R. G.: *Electrochim. Acta*, **5**, 319, 349 (1961).
40. BARRADAS, R. G., HAMILTON, P. G., and CONWAY, B. E.: *J. Phys. Chem.*, **69**, 3411 (1965).
41. CONWAY, B. E., BARRADAS, R. G., HAMILTON, P. G., and PARRY, J. M.: *J. Electroanalyt. Chem.*, **10**, 485 (1965).
42. ZWIERZYKOWSKA, I.: *Roczn. chem.*, **38**, 663, 1169, 1195, 1367 (1964); **39**, 101 (1965).
43. NÜRNBERG, H. W., and WOLFF, G.: *Coll. Czech. Chem. Comm.*, **30**, 3997 (1965).
44. GIERST, L., and HERMAN, P.: *Z. analyt. Chem.*, **216**, 238 (1966).
45. PARTRIDGE, L. K., TANSLEY, A. C., and PORTER, A. S.: *Electrochim. Acta*, **11**, 517 (1966).
46. DAMASKIN, B. B., SURVILA, A. A., VASINA, S. JA., and FEDOROVA, A. I.: *Elektrokhimiya*, **3**, 825 (1967).
47. DEVANATHAN, M. A. V., and FERNANDO, N. J.: *Trans. Faraday Soc.*, **58**, 368 (1961).
48. FRUMKIN, A. N., KAGANOVICH, R. I., and BIT-POPOVA, E. S.: *Dokl. Akad. Nauk SSSR*, **141**, 670 (1961).
49. FRUMKIN, A. N., KUZNETSOV, V. A., and KAGANOVICH, R. I.: *Dokl. Akad. Nauk SSSR*, **155**, 175 (1964).
50. KAGANOVICH, R. I., GEROVICH, V. M., and OSOTOVA, T. G.: *Dokl. Akad. Nauk SSSR*, **155**, 893 (1964).
51. KAGANOVICH, R. I., and GEROVICH, V. M.: *Elektrokhimiya*, **2**, 977 (1966).
52. KUZNETSOV, V. A., and DAMASKIN, B. B.: *Elektrokhimiya*, **1**, 1153 (1965).
53. KAGANOVICH, R. I., GEROVICH, V. M., and GUSAKOVA, O. JU.: *Elektrokhimiya*, **3**, 946 (1967).
54. FRUMKIN, A. N., and DAMASKIN, B. B.: *J. Pure a. Appl. Chem.*, **15**, 263 (1967).
55. ANDRUSEV, M. M., AJUPOVA, N. KH., and DAMASKIN, B. B.: *Elektrokhimiya*, **2**, 1480 (1966).
56. PARRY, J. M., and PARSONS, R.: *Trans. Faraday Soc.*, **59**, 241 (1963).
57. PARRY, J. M., and PARSONS, R.: *J. Electrochem. Soc.*, **113**, 992 (1966).
58. PARSONS, R., and ZOBEL, F. G. R.: *Trans. Faraday Soc.*, **62**, 3511 (1966).
59. DAMASKIN, B. B., FRUMKIN, A. N., and DJATKINA, S. L.: *Izv. Akad. Nauk SSSR, Khim. Ser.*, 2171 (1967).
60. ARAKELJAN, R. A., and TEDORADZE, G. A.: *Elektrokhimiya*, **4**, 144 (1968).
61. DEVANATHAN, M. A. V.: *Proc. Roy. Soc.*, **A267**, 256 (1962).
62. BOCKRIS, J. O'M., DEVANATHAN, M. A. V., and MÜLLER, K.: *Proc. Roy. Soc.*, **A274**, 55 (1963).

63. KAMIENSKI, B.: *Electrochim. Acta*, **1**, 272 (1959).
64. FRUMKIN, A. N., and DAMASKIN, B. B.: 'Modern Aspects of Electrochemistry', Edited by J. O'M. Bockris and B. E. Conway, Vol. 3, p. 149 (Butterworths, London, 1964).
65. FRUMKIN, A. N., DAMASKIN, B. B., and CHIZMADZHEV, JU. A.: *Elektrokhimiya*, **2**, 875 (1966).
66. FRUMKIN, A. N.: *Electrochim. Acta*, **2**, 351 (1960).
67. DAMASKIN, B. B., KAGANOVICH, R. I., GEROVICH, V. M., and DJATKINA, S. L.: *Elektrokhimiya*, **5**, 507 (1969).
68. PROSKURNIN, M. A., and FRUMKIN, A. N.: *Trans. Faraday Soc.*, **31**, 110 (1935).
69. HANSEN, R. S., MINTURN, R. E., and HICKSON, D. A.: *J. Phys. Chem.*, **60**, 1185 (1956); **61**, 953 (1957).
70. HANSEN, R. S., KELSH, D. J., and GRANTHAM, D. H.: *J. Phys. Chem.*, **67**, 2316 (1963).
71. BREITER, M., and DELAHAY, P.: *J. Amer. Chem. Soc.*, **81**, 2938 (1959).
72. LORENZ, W.: *Z. Elektrochem.*, **62**, 192 (1958).
73. LORENZ, W., MÖCHEL, F., and MÜLLER, W.: *Z. physik. Chem. (N.F.)*, **25**, 145 (1960).
74. LORENZ, W., and MÜLLER, W.: *Z. physik. Chem. (N.F.)*, **25**, 161 (1960).
75. FRUMKIN, A. N., and DAMASKIN, B. B.: *Dokl. Akad. Nauk SSSR*, **129**, 862 (1959).
76. DAMASKIN, B. B., and GRIGORJEV, N. B.: *Dokl. Akad. Nauk SSSR*, **147**, 135 (1962).
77. KLJUKINA, L. D., and DAMASKIN, B. B.: *Izv. Akad. Nauk SSSR, Khim. Ser.*, 1022 (1963).
78. DAMASKIN, B. B.: *Electrochim. Acta*, **9**, 231 (1964).
79. LERKH, R., and DAMASKIN, B. B.: *Zhur. Fiz. Khim.*, **38**, 1154 (1964); **39**, 211, 495 (1965).
80. DAMASKIN, B. B., SURVILA, A. A., and RYBALKA, L. E.: *Elektrokhimiya*, **3**, 146, 927, 1138 (1967).
81. ZOLOTOVITSKY, JA. M., and TEDORADZE, G. A.: *Izv. Akad. Nauk SSSR, Khim. Ser.*, 2133 (1964); *Elektrokhimiya*, **1**, 1339 (1965).
82. TEDORADZE, G. A., ARAKELJAN, R. A., and BELOKOLOS, E. D.: *Elektrokhimiya*, **2**, 563 (1966).
83. DJATKINA, S. L., and DAMASKIN, B. B.: *Elektrokhimiya*, **2**, 1340 (1966).
84. MELIK-GAIKAZJAN, V. I.: *Zhur. Fiz. Khim.*, **26**, 1184 (1952).
85. EDA, K.: *J. Chem. Soc. Japan*, **80**, 349, 461, 465, 708 (1959); **81**, 689 (1960).
86. DAMASKIN, B. B., NIKOLAEVA-FEDOROVICH, N. V., and IVANOVA, R. V.: *Zhur. Fiz. Khim.*, **34**, 894 (1960).
87. FRUMKIN, A. N.: *Z. Physik.*, **35**, 792 (1926).
88. FRUMKIN, A. N.: *Z. physik. Chem.*, **116**, 466 (1925).
89. FRUMKIN, A. N.: *J. Electroanalyt. Chem.*, **7**, 152 (1964).
90. DAMASKIN, B. B.: *Elektrokhimiya*, **3**, 1390 (1967).
91. PARSONS, R.: *Trans. Faraday Soc.*, **55**, 999 (1959).
92. PARSONS, R.: *J. Electroanalyt. Chem.*, **7**, 136 (1964).
93. DAMASKIN, B. B.: *J. Electroanalyt. Chem.*, **7**, 155 (1964).
94. PARSONS, R.: *J. Electroanalyt. Chem.*, **8**, 93 (1964).
95. FRUMKIN, A. N., DAMASKIN, B. B., GEROVICH, V. M., and KAGANOVICH, R. I.: *Dokl. Akad. Nauk SSSR*, **158**, 706 (1964).
96. FRUMKIN, A. N., DAMASKIN, B. B., and SURVILA, A. A.: *Elektrokhimiya*, **1**, 738 (1965).
97. DUTKIEVICZ, E., GARNISH, J. D., and PARSONS, R.: *J. Electroanalyt. Chem.*, **16**, 505 (1968).
98. FRUMKIN, A. N., DAMASKIN, B. B., and SURVILA, A. A.: *J. Electroanalyt. Chem.*, **16**, 493 (1968).
99. BOCKRIS, J. O'M., GILEADI, E., and MÜLLER, K.: *Electrochim. Acta*, **12**, 1301 (1967).
100. MOTT, N. F., PARSONS, R., and WATTS-TOBIN, R. J.: *Phil. Mag.*, **7**, 483 (1962).
101. GRIGORJEV, N. B., and KRYLOV, V. S.: *Elektrokhimiya*, **4**, 763 (1968).
102. DAMASKIN, B. B.: *Elektrokhimiya*, **1**, 1123 (1965).

103. DAMASKIN, B. B., VAVŘIČKA, S., and GRIGORJEV, N. B.: Zhur. Fiz. Khim., 36, 2530 (1962).
104. DAMASKIN, B. B.: Dokl. Akad. Nauk SSSR, 156, 128 (1964).
105. DAMASKIN, B. B.: Elektrokhimiya, 1, 63 (1965).
106. CASSEL, H. M., and SALDITT, F.: Z. physik. Chem., A155, 321 (1931).
107. BERING, B. P., and IOILEVA, K. A.: Izv. Akad. Nauk SSSR, Khim. Ser., 9 (1955).
108. NICHOLAS, M. E., JOYNER, P. A., TESSEM, B. M., and OLSON, M. D.: J. Phys. Chem., 65, 1373 (1961).
109. DAMASKIN, B. B.: Uspekhi Khim., 34, 1764 (1965).
110. KASTENING, B., and HOLLECK, L.: Talanta, 12, 1259 (1965).
111. MACDONALD, J. R.: J. Chem. Phys., 22, 1857 (1954).
112. MACDONALD, J. R., and BARLOW, C. A.: J. Chem. Phys., 36, 3062 (1962).
113. WATTS-TOBIN, R. J.: Phil. Mag., 6, 133 (1961).
114. MOTT, N. F., and WATTS-TOBIN, R. J.: Electrochim. Acta, 4, 79 (1961).
115. SCHWARZ, E., DAMASKIN, B. B., and FRUMKIN, A. N.: Zhur. Fiz. Khim., 36, 2419 (1962).
116. DAMASKIN, B. B.: Elektrokhimiya, 1, 1258 (1965); 2, 828 (1966).
117. KIPLING, J. J.: J. Colloid. Sci., 18, 502 (1963).
118. MIGNOLET, J. C. P.: Rec. trav. chim. P.B., 74, 685 (1955).
119. FRUMKIN, A. N., JOFA, Z. A., and GEROVICH, M. A.: Zhur. Fiz. Khim., 30, 1455 (1956).
120. CASE, B., and PARSONS, R.: Trans. Faraday Soc., 63, 1224 (1967).
121. GEROVICH, M. A., FRUMKIN, A. N., and VARGIN, D.: J. Chem. Phys., 6, 906 (1938).
122. ADAM, N. K.: 'Physics and Chemistry of Surfaces' (Oxford University Press, London, 1941).
123. DAVIES, J. T., and RIDEAL, E. K.: 'Interfacial Phenomena', 2nd edition (Academic Press, N.Y., 1963).
124. POLJANOVSKAJA, N. S., and FRUMKIN, A. N.: Elektrokhimiya, 1, 538 (1965).
125. FRUMKIN, A. N., POLJANOVSKAJA, N. S., GRIGORJEV, N. B., and BAGOTSKAJA, I. A.: Electrochim. Acta, 10, 793 (1965).
126. LOSHKAREV, M. A., KRIVTSOV, A., and KRJUKOVA, A. A.: Zhur. Fiz. Khim., 23, 221 (1949).
127. BORISOVA, T. I., ERSHLER, B. V., and FRUMKIN, A. N.: Zhur. Fiz. Khim., 22, 925 (1948); 24, 337 (1950).
128. LEIKIS, D. I.: Dokl. Akad. Nauk SSSR, 135, 1429 (1960).
129. LEIKIS, D. I., and SEVASTJANOV, E. S.: Dokl. Akad. Nauk SSSR, 144, 1320 (1962); Izv. Akad. Nauk, Khim. Ser., 1964, 450.
130. FRUMKIN, A. N., PETRY, O. A., and NIKOLAEVA-FEDOROVICH, N. V.: Dokl. Akad. Nauk SSSR, 147, 878 (1962).
131. GRIGORJEV, N. B., and BAGOTSKAJA, I. A.: Elektrokhimiya, 2, 1449 (1966).
132. PALM, U. V., PAST, V. E., and PULLERITS, R. JA.: Elektrokhimiya, 2, 604 (1966); 3, 376 (1967); 5, 886, 1009 (1969).
133. JOFA, Z. A., BATRAKOV, V. V., and CHO NGOK BA: Electrochim. Acta, 9, 1645 (1964).
134. DAHMS, H., and GREEN, M.: J. Electrochem. Soc., 110, 1075 (1963).
135. HILLSON, P. J.: J. Chim. phys., 49, 88 (1952); Trans. Faraday Soc., 48, 462 (1952).
136. GORODETSKAJA, A. V., and PROSKURNIN, M. A.: Zhur. Fiz. Khim., 12, 411 (1938).
137. TEMKIN, M. I.: Zhur. Fiz. Khim., 15, 296 (1941).
138. JOFA, Z. A., BATRAKOV, V. V., and NIKIFOROVA, YU. A.: Extend. Abstr. 17 Meet. CITCE, p. 60 (Tokyo, 1966).
139. FRUMKIN, A. N.: Dokl. Akad. Nauk SSSR, 154, 1432 (1964).
140. FRUMKIN, A. N., BALASHOVA, N. A., and KAZARINOV, V. E.: J. Electrochem. Soc., 113, 1011 (1966).
141. BREITER, M., and GILMAN, S.: J. Electrochem. Soc., 109, 622 (1962).
142. BOCKRIS, J. O'M., GREEN, M., and SWINKELS, D. A. J.: J. Electrochem. Soc., 111, 743 (1964); J. O'M. Bockris, and D. Swinkels, ibid., 111, 736 (1964).

143. SOKOLSKY, D. V.: 'Hydrogenization in Solutions', *Akad. Nauk Kazakh. SSR*, Chapter II (Alma-Ata, 1962).
144. NIEDRACH, L. W., *J. Electrochem. Soc.*, **111**, 1309 (1964).
145. PODLOVCHENKO, B. I., and GORGONOVA, E. P.: *Dokl. Akad. Nauk SSSR*, **156**, 673 (1964).
146. PETRY, O. A., PODLOVCHENKO, B. I., FRUMKIN, A. N., and LAL, HIRA: *J. Electroanalyt. Chem.*, **10**, 253 (1965); **11**, 12 (1966).
147. GILMAN, S.: 'Hydrocarbon Fuel Cell Technology', Edited by S. Baker, p. 349 (Academic Press, N.Y.-London, 1965).
148. BAGOTSKY, V. S., and VASILJEV, YU. V.: *Electrochim. Acta*, **11**, 1439 (1966).
149. FISCHER, H., and SEILER, W.: *Corros. Sci.*, **6**, 159 (1966).
150. DAMASKIN, B. B.: *Elektrokhimiya*, **4**, 675 (1968).
151. BREITER, M.: *J. Electroanalyt. Chem.*, **14**, 407 (1967); **15**, 221 (1967).
152. KAZARINOV, V. E., and MANSUROV, G. N.: *Elektrokhimiya*, **2**, 1338 (1966).
153. DAMASKIN, B. B.: *J. Electroanalyt. Chem.*, **21**, 149 (1969); **23**, 431 (1969).
154. DAMASKIN, B. B.: *Elektrokhimiya*, **5**, 606, 771 (1969).
155. KIRJANOV, V. A., KRYLOV, V. S., DAMASKIN, B. B., and CHIZHOV, A. V.: *Elektrokhimiya*, **6**, 533, 1020, 1518 (1970).
156. DAMASKIN, B. B., DJATKINA, S. L., and BOROVAYA, N. A.: *Elektrokhimiya*, **6**, 712 (1970).
157. FRUMKIN, A. N., PETRY, O. A., and DAMASKIN, B. B.: *J. Electroanalyt. Chem.*, **27**, 81 (1970).
158. DAMASKIN, B. B., FRUMKIN, A. N., and CHIZHOV, A. V.: *J. Electroanalyt. Chem.* (in press).

ION SOLVATION

Brian Case

I. INTRODUCTION

The stability of ions in solution, whether they be charged atoms or molecules, depends primarily on interaction with the solvent broadly described as solvation. This is obvious from the fact that, apart from the simple monatomic ions, most are unstable in the gaseous phase. Stabilization of ions by solvation plays an important role in electron transfer reactions both when they occur in the bulk of the solution and at electrode surfaces.[1,2] Examples of electrode reactions in which solvation effects are evident are provided by studies of metal deposition[3] and the kinetics of reactions in mixed aqueous-organic solvents.[4]

Ion solvation is determined by the dielectric properties and structure of the solvent and is characterized quantitatively by the free energy of solvation which is just the difference between the energies of the ion in solution and gas phases. Electrochemical reactions, especially those of organic species, are being increasingly studied in media other than water since organic ions are frequently unstable in that solvent. Absolute assignment of thermodynamic properties of ions in different media can only be arrived at by non-thermodynamic methods dependent for their validity on sound physical understanding of factors governing ion stabilization. Correlation of known thermodynamic data with solvent properties is very relevant to this end.

In the following discussion, structural and dielectric properties of water and the interpretations of ion hydration are first reviewed, although water and aqueous solutions are probably a good deal more complex than some of the organic solvent systems which are only now becoming of general interest. Subsequently, ion solvation in various non-aqueous solvents is described, mainly in the light of what has been learned about aqueous solutions.

II. WATER AND AQUEOUS SOLUTIONS

1. The Molecular Structure of Water

Liquid water is structurally very complex. An understanding of the mutual interaction of the solvent molecules naturally precedes interpretation of the influence of ions on the water structure.

A number of theories, the earliest of which was the classical paper of Bernal and Fowler[5] based on X-ray diffraction studies,[6] treat water as a mixture of two or more polymerically bonded structures. Alternatively the liquid may be regarded as composed of gas-like monomeric molecules occupying cavities in an almost completely hydrogen-bonded framework.

One of the central problems in understanding the nature of water is whether it is a mixture of discrete distinguishable molecular aggregates or whether there is a continuous distribution of fluctuating states of intermolecular bonding. Theories involving discrete structures imply that hydrogen bonding in water is predominantly covalent and therefore directional. Evidence for non-linear hydrogen bonds has been given[7] and except for certain specific cases the degree of covalency of this bond is usually considered to be small.[8]

A. Spectroscopic Studies of Water. Raman and infra-red spectra provide information on the vibrational modes of water molecules. The spectra thus give some insight into the short term structure of the liquid since both the intramolecular and the intermolecular vibrations are influenced by the immediate environment of the molecules. Intramolecular vibrations of water are characterized by three bands, the main feature near 3490 cm^{-1} in the infra-red or 3440 cm^{-1} in the Raman spectrum being due to the O—H stretching mode. A second band near 1645 cm^{-1} is associated with bending of the molecule. Lower frequency intermolecular motions, hindered rotation or libration, are represented by the broad band in the infra-red spectrum centred around 700 cm^{-1} and by bands in the Raman spectra near 450 and 550 cm^{-1}. Other bands at still lower frequencies are due to hindered translational modes.

Spectra of Water Stretching Modes. Analysis of these spectra is complicated by the breadth and asymmetry of the stretching bands and interpretation in terms of molecular motions and interactions is not at all simple.[9] Coupling of vibrations to those of neighbouring molecules and Fermi resonance with overtones introduce further complications. The simplest spectra to interpret are those of dilute solutions of HDO in H_2O or D_2O since the stretching vibrations are essentially uncoupled in this case.

Falk and Ford[10] recorded the infra-red stretching bands for isotopically substituted water up to 130°C and Franck and Roth[11] have extended the observation of the OD stretching vibration up to a temperature of 400°C and pressure of 4000 bars. At densities of 0.9 g cm^{-3} and higher the shape of the bands is almost gaussian at low temperature. Above 200°C the shape becomes increasingly asymmetric but is perfectly smooth with no shoulder down to a density of 0.1 g cm^{-3}.

Lack of structure in the uncoupled OH and OD vibrations is not consistent with a model of water as a mixture of a small number of different types of molecule and favours a description of the liquid in terms of a wide distribution of hydrogen bond energies. Bands which are superficially simple may, however, be the resultant of two or more component bands[12] and therefore these spectra are not proof of a continuum state for water.

Raman studies of the uncoupled OH and OD vibrations have received rather different interpretations by Wall and Hornig[9] and by Walrafen.[13] The former authors, using an empirical correlation between the stretching frequency and the O . . . O hydrogen bonded distance converted the band shape to an intermolecular distance distribution which may be compared with the most recent X-ray results.[14] They concluded that the breadth of the uncoupled bands reflects a continuous variation of intermolecular distance and there is little hydrogen bond breakage in liquid water at low temperature.

The OD stretching band is asymmetric with a shoulder on the high frequency side. As the temperature rises from 32–92°C the peak intensity decreases but the intensity at 2570 cm^{-1} is independent of temperature. An isosbestic point may indicate two component bands due to chemical species which are in equilibrium with one another. Infra-red spectra for HOD in H_2O—D_2O solutions obtained by Senior[15] show almost identical behaviour with the Raman bands over this temperature range. Assuming that the components are gaussian in form, the band may be reduced to two components, one being ascribed to non-hydrogen bonded molecules and the other being due to bonded species. Intramolecular Raman intensities of the decoupled OD vibration are in agreement with stepwise hydrogen bond breakage.

Walrafen[16] suggests that the intermolecular spectra of water arise solely from completely hydrogen-bonded units of five molecules in a tetrahedral structure, the breaking of one O—H . . . O unit producing complete loss of intensity. Thus, as far as these bands are concerned Walrafen's model of water is that of a distribution of particles of ice dispersed in a solvent of non-bonded water molecules.

Apart from the fact that Walrafen's resolution of the bands may not be unique, they are not necessarily gaussian in shape.[17] Objections to his interpretation on the basis that a sharp bond stretch band near 3700 cm^{-1}, due to gas-like monomers in the liquid, is not observed is less tenable, since the environment of an unbonded water molecule in water will be very different from that in the gas or as a dilute solution in another solvent. Stevenson[18] quotes other independent experimental observations which suggest that the monomer concentration is less than 1 per cent between 0 and 100°C. For example, the ratio of the absorptivities of liquid water and water vapour at the same ultra-violet wavelength between 1800 and 1910 Å is constant at a given temperature and corresponds to a concentration of monomer 100 times less than previously supposed.

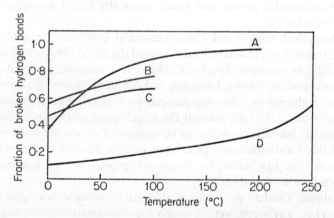

Fig. 1. Estimates of the fraction of broken hydrogen bonds in water as a function of temperature, after (A) Walrafen,[13] (B) Grjotheim and Krogh-Moe,[45] (C) Némethy and Scheraga,[36] and (D) Luck.[20]

The prediction of Walrafen's two-state theory that at 90°C nearly all the hydrogen bonds have been broken is improbable. Nearly all the observed changes in the uncoupled stretching band above 100°C and anomalous properties[19] including the high critical temperature of water must then be due to forces other than hydrogen bonding. Luck's work on the infra-red spectra of hydroxylic liquids[20] indicates that extensive hydrogen bond breakage in water only commences above 200°C.

Combination bands in the infra-red spectra of water and D_2O in the region of 8000 cm^{-1} have been resolved into three components, the intensities of which vary with temperature. Buijs and Choppin[21] assigned the ice-like band at 1.25μ to molecules with two hydrogen bonds, the second to molecules with one bonded OH group and the third to monomers. Assuming an equilibrium between bound and unbonded water the relative contribution of each species was calculated from the spectroscopic data. The temperature variation of the fraction of broken hydrogen bonds which they arrived at is illustrated in Fig. 1. This work, which implies that D_2O contains more unbonded molecules than H_2O at the same temperature, contradicts the usual opinion that D_2O is the more structured solvent. Luck[20] has considered the Buijs and Choppin analysis in more detail deriving from the spectra possible concentrations of free molecules, free OH groups in singly bonded molecules, energetically unfavourable H bonds and linear H bonds.

Since the fundamental stretching bands fail to show the structure which they should display if three distinct types of water molecule were present[22] a more likely explanation of these spectra is that temperature alters the frequencies of the fundamental modes and hence alters the Fermi resonances in the combination bands.

B. Theoretical Models and Physico-chemical Properties. (*a*) *Mixture and Cavity Models*. Early models of water treated the liquid as a mixture of several distinguishable polymeric species. Eucken,[23] for example, assumed water to contain monomers, dimers, tetramers and octamers. When the very extensive nature of hydrogen bonding was realized such descriptions were superseded by others which take into account the largely associated state of the liquid. Most models still assume water to be composed of a small number of distinguishable local structures. Interstitial models describe water in terms of monomeric gas like molecules in an otherwise more or less completely hydrogen-bonded framework.

Bernal and Fowler[5] gave one of the earliest realistic descriptions of the liquid. They envisaged very extensive polymerization with each water molecule tetrahedrally bound to four others. Equilibrium between pseudo-crystalline tridymite and quartz-like structures was postulated with the more open tridymite lattice breaking down to the denser form as the temperature was raised. They noted that the dielectric constant of water and ice and its variation with temperature and frequency are consistent with very few molecules undergoing free rotation.

Interstitial models are typified by Pauling's theory[24] in which water is likened to the clathrate compound chlorine hydrate. With single molecules enclosed within cages of pentagonal dodecahedra, water would be a clathrate of itself.[25] Such a model is difficult to reconcile with the low viscosity of water and von Stackelberg has pointed out[26] that a molecule of even moderate polarity would interact with the surrounding water dipoles to prevent the formation of a polyhedral cage. Narten, Danford, and Levy have interpreted recent X-ray data by assuming that water molecules may occupy cavities in an anisotropically expanded ice-I structure.[14]

Most theories involving mixed species in the liquid imply that there is an appreciable fraction of broken hydrogen bonds even near the melting point although this appears unlikely. Davis and Litovitz[27] avoided this inconsistency with a two-state theory of water. Breaking down of an open ice-like structure of layers of puckered hexagonal rings[28] occurs by rupture of bonds between the rings followed by rotation of these rings through 60° into a body-centred cubic close packed relation with those in layers above and below. The mixture of double rings in open ice-like and close-packed form explains the demonstration in X-ray spectra of near neighbours at a distance between that of first and second nearest neighbour distances in ice. This model predicts that only about 18 per cent of the hydrogen bonds are broken at 0°C and gives a satisfactory account of the thermal properties of water.

A less rigid formulation due to Frank and Quist[29] of water as 'any quasi-crystalline framework with single molecules occupying interstitial sites and making no contribution to the total volume' bears resemblance to the Narten model but is perhaps more appropriate. It neatly accounts for the maximum density of water by permitting molecules to fill up holes in the expanded framework and theoretical prediction of the experimental pressure–volume–temperature relations is good. However, it provides no adequate explanation of the thermodynamics of solution of non-polar substances.

The behaviour of non-polar solutes which dissolve evolving heat and producing a large decrease in entropy was interpreted by Frank and Evans with the suggestion that these substances promote the formation of ice-like structure in their environment.[30] This leads to a dynamic picture of water[31] as an assembly of flickering clusters of molecules with half-lives of 10^{-10}–10^{-11} sec. The small spread of dielectric relaxation times[32] precludes the existence of agglomerates for periods longer than this. A quantitative treatment of this model was developed by Némethy and Scheraga.[33,34] They assume that equilibrium exists between clusters and non-bonded monomeric water. Molecules in the cluster cores are completely bonded while those on the periphery may be joined by from one to four hydrogen bonds to their neighbours. Statistical mechanical calculations of the thermodynamic properties of the liquid were performed using the simplifying assumption that the energy levels of the five distinguishable forms of water molecule were equally spaced. The derivation of the combinatorial factor in their calculation is erroneous, however.[38] A closely related paper[34] describes the mechanism of solvation of non-polar substances.

Vand and Senior[35] noted that although the Némethy and Scheraga model gives satisfactory agreement with experimental values of entropy and free energy it fails to predict the behaviour of the specific heat of water. It is also much less successful when applied to D_2O.[36] These authors replace the five energy levels of the Némethy and Scheraga treatment with three levels derived from the spectroscopic observations of Choppin and Buijs.[21] Objections to this assignment have been mentioned previously. Marchi and Eyring[37] have also calculated the thermodynamic properties of water on a similar model. The two main differences between the assumed structures are that in Marchi

3

and Eyring's theory the water molecules are either completely bonded or entirely free and that the monomeric species are gas-like with complete rotational freedom rather than being restrained to librate about two axes as in Némethy and Scheraga's picture. Models which require an appreciable fraction of the molecules to be in an unbonded monomeric state have been criticized by Stevenson[18] as indicated above.

According to Vand and Senior it is not possible to reconcile the thermodynamic data for liquid water using a model based on three sharp energy levels. A close fit of calculated and experimental heat capacities has been obtained by taking into account a distribution of energy states for the three types of water molecule. Levine and Perram,[38] employing a lattice model in which the energy of the hydrogen bond depends on the number of bonds made by the water molecule, show that the number of monomers is much smaller than that calculated by Némethy and Scheraga. Their theory is readily modified to include an arbitrary distribution of bond energy levels making the hydrogen-bond energy temperature dependent. Other approaches which assume a spectrum of energy states have been made by Conway,[39] who used an electrostatic model, and by Cohen and co-workers.[40]

(b) Distorted Bond Models. In mixture models the hydrogen bond is usually regarded as either intact or broken and hence primarily covalent. An alternative though less popular description in terms of bent or strained hydrogen bonds was proposed by Pople.[41] High pressure ices, and salicylic acid, for example, are now known to contain distorted hydrogen bonds.[42] Bernal[43] has developed a very similar theory, on a random network basis, which may correlate slightly better with the observed radial distribution function for water.

Lennard-Jones and Pople applied their concept of association in liquids arising from lone-pair electron interactions to the case of water.[44] In this model very few hydrogen bonds are presumed to be broken at the melting point. Pople considered that the effect of increased temperature would be to magnify the amplitude of the hydrogen bond bending motion, that is, a relaxation of the hindered rotation of the molecules rather than complete rupture of the individual bonds.[41] This view implies that the hydrogen bond is predominantly electrostatic. There is a sharp distinction between models exemplified by those of Pople and of Frank and Wen.[31] In the former the degree of covalency of the H bond must be minor to allow long range structure together with bending of the bond. In the latter the bonding is regarded as primarily covalent and hence 'all or nothing'. Uncertainty still exists as to the magnitude of the bond energy. Estimates range from 1·3–6·8 kcal.[45,46] Pauling's value of 4·5–5·0 kcal[47] is widely quoted and is comparable to recent theoretical calculations.[48]

Distorted bond models would seem to be more in accord with spectroscopic evidence than most of the mixture models. Success in predicting the properties of the substance is necessary, though by no means sufficient, as is witnessed by the wide variety of models which satisfactorily account for many of the liquid's peculiarities. No model involving rigidity of structure, an appreciable concentration of monomeric molecules, a small number of discrete species, or in which a range of energy states of the hydrogen bonded molecules is not considered can claim to be a very realistic representation. Many of the proposed descriptions must contain elements of the truth, but a definitive theory has not yet emerged.

2. The Ionic Hydration Sphere

Early work on ion transport and viscosities of ionic solutions showed that ions profoundly alter the structure of the solvent in their vicinity. Two effects were distinguished; association of water with the ion and disruption or enhancement of structure in the surrounding medium. Frank and Wen[31] represented aquo-ions as being surrounded by an inner shell of oriented water encompassed by a secondary shell of modified structure which in turn was surrounded by water with the bulk properties. Ions which increased the viscosity of water were said to be 'structure making'; those which decreased it being 'structure breaking'. Numerous papers and reviews have been devoted to the question of hydration numbers, the definition of which is usually determined by the technique used to measure them. There is no 'chemical' bond in the usual sense between simple ions and water molecules—merely orientation of the molecular dipoles into positions of minimum energy in the electrostatic field of the ion. Kinetic interchange between molecules in the inner and outer spheres is usually very rapid. The various physical properties of ionic solutions (conductivity,[49] viscosity,[50] compressibility[51]) probably reflect the influence of ions on the first and second hydration sheaths to different extents.

Transition-metal ions, on the other hand, are frequently more specifically 'bonded' to immediately adjacent water molecules. This is indicated by the slower exchange rate of labelled water in the inner co-ordination sphere[52] especially with ions having high crystal field stabilization, for example Fe^{3+} and Cr^{3+}. Octahedral co-ordination about the aquated chromic ion has been definitely established by an isotope dilution technique using $H_2^{18}O$.[53] Such ions also have unexpectedly high energies of solvation.[54]

When the residence time of a water molecule in the primary hydration sphere of an ion exceeds 10^{-4} sec two separate oxygen 17 NMR signals may be observed, corresponding to molecules in the bulk and in the solvation sheath respectively. Integration of the absorption curves leads to deduction of primary hydration numbers of six for aluminium III and four for beryllium II.[55,56] No definite conclusion could be reached on the hydration number when the technique was applied to the paramagnetic nickelous ion.[57] By comparing line widths for solutions containing alkaline earth cations with that of an aluminium III solution and taking the co-ordination number for this ion as six the primary hydration numbers shown in Table 1 were obtained. It may be noted that hydration numbers of these divalent ions are similar for ions of similar size. The interpretation of non-integral hydration numbers as reflecting an equilibrium between at least two hydrated forms is novel.[58]

Relative rates of exchange of hydrate water have been calculated from NMR spectra[59,60] and the number of 'sites' near an ion which are occupied by water were also computed in this work. Indeed, Samoilov[61] has proposed that ion hydration should be characterized by the measured rate of interchange of water in the primary hydration shell.

For simple ions the rate of exchange of inner hydration sphere water is very rapid and the number of nearest neighbour water molecules must be deduced by rather less direct methods than NMR spectroscopy. In order to compute definite hydration numbers it is frequently necessary to postulate that the water molecules hydrating the ion suffer a more or less complete change in the property concerned. For example the permittivities of electrolyte solutions measured by microwave techniques may be treated by assuming that the inner sphere water is immobilized and has the limiting high frequency permittivity of 5·5.[70]

The wavelengths of compressional waves in solutions may be measured (using ultrasonic interferometry) and the compressibilities of the solutions calculated. Again, if the molecules in the solvate sphere are considered incompressible, hydration numbers can be obtained from these data.[71] Measurements of activity coefficients,[66] transport of water with ions[72] and diffusion coefficients[64] have also led to estimates of specific hydration numbers.

It is probably unrealistic to expect complete immobilization, for example, of molecules in the ion's solvation sphere, especially for simple non-transition metals. A more satisfactory procedure is to characterize hydration numbers by a parameter related to the *partial* loss or change of some particular property. A relevant paper by Glueckauf[73] on dielectric properties follows this approach.

Water Structure in the Ion's Environment. Much recent experimental work has been directed towards determining the orientation of water molecules in the primary solvation sheath. The only definite conclusion is that Eu^{3+} and Er^{3+} occupy interstitial positions in an ice-like structure. The smallest cations (Li^+ and Be^{2+}) have the largest peripheral electrostatic fields. In these

Table 1. Ion Hydration Numbers

Technique	Ref.	Li⁺	Na⁺	K⁺	Rb⁺	Cs⁺	Ag⁺	F⁻	Cl⁻	Br⁻	I⁻
Permittivity	62	6	4	4	3·5	—	—	—	0	0	0
Proton Resonance	63	4	3·1	2·1	1·6	1·0	—	1·6	0	0	0
Diffusion											
Coefficients	64	2·3–2·8	0·7–1·2	0·4–0·9	—	0	—	1·8	0–0·5	0·2–0·7	0·7–1·2
Activity Coefficients	65	3·4	2·0	0·6	—	1·4	—	—	0·9	0·9	0·9
	66	4·2–4·7	2·7–3·3	1·7–1·8	1·5	—	—	—	—	—	—
Hydration Entropies	67	5	4	3	3	0	4	—	—	—	—
Ion Exchange	68	3·3	1·5	0·6	—	0	—	—	2	—	0·7
Various	69	4	3	2	1	—	—	3	—	—	—

Technique	Ref.	Be²⁺	Mg²⁺	Ca²⁺	Sr²⁺	Ba²⁺	Zn²⁺	Cd²⁺	Hg²⁺	Pb²⁺	Al³⁺	Cr³⁺
Proton Resonance	58	4	3·8	4·3	5·0	5·7	3·9	4·6	4·9	5·7	6	6

spherically symmetrical fields water molecules probably adopt the angled dipole lone-pair orientation.[74] Normal dipole orientation is more likely around the larger ions. About the largest cations the peripheral field is very weak and only partial orientation of water molecules in the inner sphere will be possible. Failure of these partially orientated molecules to fit into the normal water-structure could cause the structure-breaking effect of these ions.

Hydrogen bonding with water protons is possible for anions but not for cations. It has frequently been suggested that the positive end of the water dipole can approach anions more closely than the negative end can approach cations.[75] Buckingham's treatment of the enthalpy of hydration of ions[76] considers the interactions of the ion's electrostatic field with the dipoles and quadrupoles of surrounding water molecules and takes into account the simultaneous water-water interactions. Due to an inversion of ion-quadrupole interaction energy for ions of opposite charge the model predicts a difference in hydration enthalpy for anions and cations of the same (crystal) radius and co-ordination number. This conclusion is, however, based upon the assumption of normal water orientation and that the centre of the H_2O molecule is the multipole origin. Blandamer and Symons[77] contend that the interactions of anions and the cations with water molecules are essentially symmetrical if one considers the molecule to consist of two (tetrahedral) lobes of fractional positive charge (protons) and two of fractional negative charge (lone-pair electrons).[78] In this context, it has been suggested[79] that the oxygen orbitals of inner-sphere water molecules may undergo trigonal splitting. Recent calculations of the electron density distribution in the water molecule show that the sp^3 hybridization is less pronounced than was formerly supposed.[80]

3. The Surface Phase of Water and Aqueous Solutions

Although ideas on the structure of water in the bulk are still somewhat speculative, even less is known about the air-water surface phase. It is almost certain that there will be some degree of preferential orientation due to asymmetry of the field at the liquid-gas boundary. Fletcher[81] argued that surface molecules will orient themselves with their most polar vertices directed into the liquid and concluded that near the freezing point molecules at the surface are completely orientated with the protons towards the gas phase. A variety of evidence including the measured free energies of hydration of ions, which are discussed below, indicates that the potential drop at the water surface due to dipole orientation is less than 0·2 V.[82] A highly ordered surface layer such as suggested by Fletcher would produce a surface potential jump of several volts. The surface potential of water has also been computed by Stillinger and Ben-Naim[83] using an idealized representation of the water molecule as a point dipole plus point quadrupole encased in a spherical exclusion envelope. The permanent quadrupole moment distorts the dipolar symmetry of the electrostatic field and leads to a surface dipole layer with an associated potential. This model leads to a much more realistic value of about 30 mV for the surface potential at 25°C with the surface water molecules tending to orient their protons away from the vapour and into the liquid.

Evidence from surface tension studies indicates that the surface phase of water is more disordered than the bulk. The surface tension of most liquids decreases with increasing temperature which indicates a positive surface excess entropy. This excess entropy for water at 25°C[84]

$$S^0 = - \left(\frac{\partial \gamma}{\partial T}\right) = 0 \cdot 157 \text{ erg cm}^{-2} \text{ deg}^{-1} \qquad (1)$$

if due to a single molecular layer would be equivalent to $2 \cdot 18$ cal deg^{-1} mole^{-1}, that is, an entropy increase equal to that of bulk water for a temperature rise from 25–63°C. The ellipticity of light reflected from the surface of a liquid can provide information on the thickness of the surface layer and the change of refractive index within it. From such studies it has been concluded[85] that the transition layer probably extends only one or two molecular diameters into the bulk.

The surface tension of most aqueous inorganic salt solutions is greater than that of pure water and therefore there is a deficiency of solute in the surface layer. A solute-free region approximately 4 Å thick was inferred by Langmuir[86] on application of the Gibbs adsorption isotherm to data for KCl and other salts. More refined calculations[87] give values of about 5 Å for 0·1 M alkali chlorides decreasing to 3 Å in 1–2 M solutions. The surface tensions of many salt solutions are approximately proportional to the concentration and, at least at low concentration, there is little specific effect of ions of the same valency.[88] A coulombic theory of this phenomenon, based on repulsion of ions from the interface into the medium of higher dielectric constant by image forces screened by the ionic atmosphere is in approximate agreement for dilute solutions of 1:1 electrolytes.[89] The image forces were interpreted by Schäfer[90] on a more structural model as a loss of hydration energy of the ions in the surface region. Use of an effective dielectric constant close to the ion as given in Webb's theory[91] and an adjustable value for the distance of closest approach of the ion to the surface detract from the applicability of this description. Schmutzer[92] introduced the idea of representing forces repelling ions from the surface by an infinite potential barrier at a distance δ from the interface. He obtained good agreement with the surface tension data of salts like KCl, NaCl, and LiF by the choice of an appropriate value of δ which for KCl is 3·2 Å. The surface tension increment caused by the ion-free layer is:

$$\Delta\gamma = \frac{mvRT\delta\phi}{1000} \tag{2}$$

where m is the solute molality, v the number of ions per molecule of solute, and ϕ the osmotic coefficient of the solution. This is much the largest term in Schmutzer's expression for the surface tension. Plots of $\Delta\gamma$ against $mv\phi$ for the concentration range 1–4 M have slopes giving values of δ which correspond to monovalent cations retaining one, and di- and trivalent ions having at least two, water molecules between the ion and the gas phase. It would seem that cations approach the solution-gas interphase with their primary hydration sheath intact.

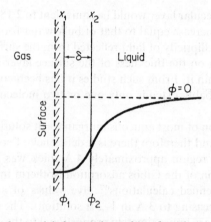

Fig. 2. Variation of potential close to the surface of a solution (after Randles[94]).

Certain salts, notably KCNS and KClO$_3$ have an equivalent solvent-free thickness of only one third of a molecular layer. Obviously some anions are not subject to the repulsive forces to nearly the same extent as the simple alkali cations or even the chloride, hydroxide, and sulphate anions.

Frumkin[93] has measured the surface potentials of many aqueous electrolyte solutions and has shown that all the well dissociated salts of alkali cations with Cl$^-$, Br$^-$, I$^-$, NO$_3^-$, ClO$_3^-$, and ClO$_4^-$ have more positive surface potentials than pure water. A surface potential drop can be produced by some degree of preferred solvent dipole orientation or by the production of an effective dipole at the surface by charge separation if either anions or cations are able to approach the surface more closely than ions of the opposite charge. A positive surface potential indicates orientation of the positive end of the dipole towards the bulk solution. Differences in surface potentials $\chi(\text{soln}) - \chi(\text{water})$ for $1m$ solutions of KF, KCl, KI, NaClO$_4$, and KCNS are -2, $+2$, $+27$, $+48$, and $+53$ mV respectively. Since the surface potential of water itself is only about 100–200 mV the observed changes are almost certainly mainly due to the solute. The lyotropic series of anions, that of decreasing increment of surface tension at the same concentration:

$$F^- < Cl^- < Br^- < NO_3^- < I^- < ClO_4^- < CNS^- < PF_6^-$$

applies equally to the increasing positive increment of surface potential. These results confirm the idea that anions are less strongly desorbed from the surface than are cations.

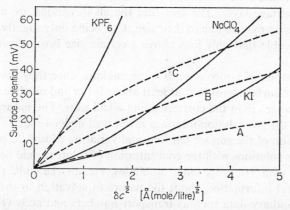

Fig. 3. Experimental surface potentials as a function of concentration, and values calculated assuming that anions are subject to adsorptive chemical potentials of (A) zero, (B) $1\cdot38kT$, and (C) $2\cdot3kT$ ($\delta = 4$ Å in each case).[94]

Randles[94] has adapted the simple theory of the metal-solution double layer to apply to the surface phase of aqueous solutions. Cations are restricted to a plane of closest approach at a distance x_2 from the surface as in Fig. 2 while anions may penetrate the interphase further, to a distance x_1 from the surface. The potential ϕ, at plane 1, relative to the solution bulk is equal to the contribution of the ions to the surface potential (with sign reversed) and was calculated by Randles using this model. A comparison of the calculated values with the observed surface potential changes is shown in Fig. 3, curve A. It is obvious that the simple theory is inadequate. Only when allowance is made for an *attractive* force by reducing the chemical potential of the ions within the planes 1 and 2, can any measure of agreement between theory and experiment be obtained, curve B, and then only in the region of low salt concentration. In fact, the surface potential change in solutions of KPF$_6$ is so large and the lowering of the surface tension of water by this salt so unusual that the PF_6^- ion must be substantially adsorbed. This holds to a lesser extent for CNS^- and ClO_4^- also.

An explanation of these observations which fits in well with knowledge of properties of ions in water was proposed. Cations simply retain their tightly bound inner-sphere water molecules. The lyotropic anions series also applies to the decrease in hydration energy and the larger ions are those which cause structure-breaking in the bulk aqueous solutions. Gurney[95] has suggested that there may be non-coulombic mutual attraction between ions which are alike in disrupting the water structure and also between ions which promote ordering of the water in their environment due to interaction of their hydration shells. This could well explain why the large anions are attracted into the disordered surface layer. The fact that the alkali cations are all desorbed from this region may be due to their size, Cs^+ being only slightly larger than F^-. It is possible that only ions above a certain size break down the water structure.

The behaviour of aqueous acids is anomalous since they cause negative increments of surface tension,[96] at least above $0·1$ m and much larger surface potential changes than the corresponding alkali salts. The nature of proton solvation at the gas-solution interface is not well understood.

Investigation of the surface tensions and variation of surface potentials of non-aqueous solutions with the concentration of solute should be rewarding. Correlation with viscosity, ionic mobilities, etc. in the bulk phase could provide useful information about the nature of solvation in these solvents. Lack of subsidiary data such as transport numbers and activity coefficients often means, however, that such measurements usually cannot be rigorously interpreted at the present time.

4. Thermodynamics of Ion Hydration

A. Determination of Electrolyte Hydration Energies. The free energies of formation of hydrated salts are known in many cases with considerable precision. Experimentally they may be determined from the standard electrode potentials of the elements forming the anion and cation, for example:

$$-FE^0_{Na^+/Na} = \Delta G^0_{form}H^+(aq) - \Delta G^0_{form}Na^+(aq) \qquad (3)$$

and

$$-FE^0_{Cl^-/Cl_2} = \Delta G^0_{form}H^+(aq) + \Delta G^0_{form}Cl^-(aq) \qquad (4)$$

when the potentials are referred to the standard hydrogen electrode. Consequently, from a knowledge of the free energy of formation of the elements in the gas phase and their ionization potentials and electron affinities respectively, the total hydration energy of the gaseous ions may be obtained from the cycle:

$$M(g) \xrightarrow{I_M} M^+(g) \qquad X^-(g) \xleftarrow{A_X} X(g)$$

$$\Delta G_{subl} \uparrow \qquad \downarrow \qquad \Delta G_{hyd}M^+X^- \downarrow \qquad \uparrow \Delta G_{form}X(g)$$

$$M \qquad M^+(aq) \qquad X^-(aq) \qquad X$$

$$\Delta G_{form}M^+(aq) \qquad \qquad \Delta G_{form}X^-(aq)$$

An alternative route is the measurement of the saturation solubility of the crystalline salt and the activity of the electrolyte in the saturated solution. These, together with the lattice energy of the crystal give the required $\Delta G_{hyd}M^+X^-$.

Conventional single ion hydration energies are referred either to the gas-phase proton[97] e.g.

$$\Delta G_{hyd,conv}M^{z+}(g) = \Delta G_{hyd}M^{z+}(g) - z\Delta G_{hyd}H^+(g) \qquad (5)$$

$$\Delta G_{hyd,conv}X^{z-}(g) = \Delta G_{hyd}X^{z-}(g) + z\Delta G_{hyd}H^+(g) \qquad (6)$$

or to hydrogen gas[98]

$$\Delta'G_{hyd,conv}M^{z+}(g) = \Delta G_{hyd}M^{z+}(g) - z\Delta G_{form}H^+(aq) \qquad (7)$$

$$\Delta'G_{hyd,conv}X^{z-}(g) = \Delta G_{hyd}X^{z-}(g) + z\Delta G_{form}H^+(aq) \qquad (8)$$

Enthalpies of hydration are usually obtained calorimetrically and hydration entropies may be calculated from the temperature variation of the standard potentials of the elements. Gold[99] has compiled self-consistent conventional values of thermodynamic hydration data and more recently Halliwell and Nyburg have thoroughly re-examined the literature on hydration enthalpies.[97]

B. Real Hydration Energies of Ions. Although measurement of electrode potentials and salt solubilities can only lead to hydration energies relative to that of the proton, little attention has been paid to a method which gives the hydration energies of single ions. The 'real' hydration energy, α, is defined for the processes

$$M^{z+}(g) \xrightarrow{\alpha_{M^{z+}}} M^{z+}(aq) \qquad (9)$$

$$X^{z-}(g) \xrightarrow{\alpha_{X^{z-}}} X^{z-}(aq) \qquad (10)$$

The 'real' hydration energy differs from the usual 'chemical' hydration energy μ which is the energy of interaction of the ion with the bulk solution, by the energy required for the ion to traverse the liquid/gas interphase which is structurally different from the bulk, i.e.

$$\alpha_{M^{z+}} = \mu_{M^{z+}} + zF\chi \tag{11}$$

$$\alpha_{X^{z-}} = \mu_{X^{z-}} - zF\chi \tag{12}$$

χ, the surface potential, is the electrostatic potential drop across this interphase. The salt hydration energy is the sum of the 'chemical' hydration energies of the ions.

$$\Delta G_{hyd}M^{z+}X^{z-} = \mu_{M^{z+}}(hyd) + \mu_{X^{z-}}(hyd) = \alpha_{M^{z+}} + \alpha_{X^{z-}} \tag{13}$$

There is no *thermodynamic* way in which the chemical hydration energies of individual ions can be determined.

Real ion hydration energies are obtained by the following route. The standard potential of a metal M_1 is related to the overall energy change in the reactions:

$$\tfrac{1}{2}H_2 \rightarrow H^+(aq) + e \tag{14}$$

$$e + M_1^+(aq) \rightarrow M_1 \tag{15}$$

(For simplicity we consider a univalent cation.) From the definition of the real free energy of hydration we then have:

$$-FE^0_{M_1^+/M_1} = \Delta G^0_{form}H^+(g) + \alpha_{H^+}(aq) - \Delta G^0_{form}M_1^+(g) - \alpha_{M_1^+}(aq) \tag{16}$$

Now consider the (hypothetical) process by which the metal is sublimed, ionized and then the ion and the ejected electron are returned to the elemental metal:

$$
\begin{array}{ccc}
M_1(g) & \rightarrow M_1^+(g) & + \quad e \\
\uparrow & \downarrow & \downarrow \\
M_1 & M_1^+(M_1) & + e(M_1)
\end{array}
$$

The overall free energy change is zero, and so:

$$\Delta G^0_{form}M_1^+(g) + \alpha_{M_1^+}(M_1) + \alpha_e(M_1) = 0 \tag{17}$$

Combining eqns (16) and (17) we have

$$-FE^0_{M_1^+/M_1} = \Delta G^0_{form}H^+(g) + \alpha_{H^+}(aq)$$
$$+ \alpha_{M_1^+}(M_1) - \alpha_{M_1^+}(aq) + \alpha_e(M_1) \tag{18}$$

From eqn (16) the real hydration energy of the cation of another metal, M_2, is:

$$\alpha_{M_2^+}(aq) = +FE^0_{M_2^+/M_2} + \Delta G^0_{form}H^+(g)$$
$$- \Delta G^0_{form}M_2^+(g) + \alpha_{H^+}(aq) \tag{19}$$

and combination of this with eqn (18) gives

$$\alpha_{M_2^+}(aq) = F[E^0_{M_2^+/M_2} - E^0_{M_1^+/M_1}] - \Delta G^0_{form} M_2^+(g)$$
$$+ \alpha_{M_1^+}(aq) - \alpha_{M_1^+}(M_1) - \alpha_e(M_1) \quad (20)$$

Randles[100] has measured the difference between the real free energy of solution of the cation M_1^+ in water and in the metal M_1, i.e.

$$\alpha_{M_1^+}(aq) - \alpha_{M_1^+}(M_1),$$

in the case of mercury. The potential of the cell

$$Hg|Hg^{2+}(aq)|N_2|Hg(Jet) \quad (21)$$

is given by:

$$-2FE = [\alpha_{Hg_2^{2+}}(H_2O) - \alpha_{Hg_2^{2+}}(Hg)] + {}^{H_2O}\Delta^{Hg}\psi \quad (22)$$

${}^{H_2O}\Delta^{Hg}\psi$ being the volta potential difference between a point just outside the surface of the mercury jet and one just outside the surface of the aqueous solution of mercurous ions. The impressed potential for which ${}^{H_2O}\Delta^{Hg}\psi$ is zero can be determined and is known as the compensation potential E_C. Referred to unit activity of mercurous ion in solution:

$$-2FE^0_C = \alpha^0_{Hg_2^{2+}}(H_2O) - \alpha^0_{Hg_2^{2+}}(Hg) \quad (23)$$

$\alpha_e(M)$ is the negative of the electronic work function, W_{Hg}, of mercury and the real free energy of hydration of any cation can be calculated from eqns (20) and (23) if the relevant free energy data and electrode potentials are known.

$$\alpha^0_{M_1^+}(aq) = F[E^0_{M_1^+/M} - E^0_{Hg_2^{2+}/Hg} - E^0_C] - \Delta G^0_{form} M_1^+(g) + W_{Hg} \quad (24)$$

Similarly

$$\alpha^0_{X^-}(aq) = F[-E^0_{X^-/X_2} + E^0_{Hg_2^{2+}/Hg} + E^0_C] - \Delta G^0_{form} X^-(g) - W_{Hg} \quad (25)$$

Values of the real hydration energies of the alkali and halide ions are given in Table 2 together with values of single ion chemical hydration energies obtained by a theoretical division of the hydration energies of the salts. The close agreement between the two indicates that the surface potential is small $\sim \pm 0.1$ V as has been known for some time.

C. Theoretical Calculations of Ion Hydration Energies. The problem of computing individual hydration energies has been tackled many times. As the surface potential of water probably lies between 100–200 mV it is unlikely that these calculations will ever give a precision significantly greater than the ± 3 kcal by which the experimental real hydration energies differ from the chemical hydration energies. Apart from the comparative ease with which a model of hydration can be formulated to account for the observed energies

Table 2. 'Real' and 'Chemical' Hydration Energies
(kcal/mole)

		H^+	Li^+	Na^+	K^+	Rb^+	Cs^+	Tl^+	Ag^+	Ca^{2+}	Zn^{2+}	Cd^{2+}	F^-	Cl^-	Br^-	I^-
$\alpha_i^0(H_2O)$	(a)	260·5	122·1	98·2	80·6	75·5	67·8	82·0	114·4	380·8	484·6	430·5	103·8	75·8	72·5	61·4
	(b)	256·5	117·0	94·0	77·0	72·5	63·0	—	110·5	371·0	492·5	428·0	—	74·5	69·0	60·5
$\Delta G_i^0(H_2O)$	(c)	260·2	121·8	98·0	80·3	75·2	67·5	81·7	114·2	380·1	484·0	429·8	87·2	73·8	66·9	58·0

References:: (a) 94; (b) 101; (c) 98.

within $\sim$2 kcal/mole as is evident from the diversity of models, such calculations do help to clarify and emphasize various factors contributing to the hydration of ions.

One of the first attempts to explain the free energy change accompanying the dissolution of a salt was that of Born.[102] He treated the ions in solution as rigid spheres of radius r and charge ze immersed in a continuous medium of dielectric constant ε. The solvation energy was taken to be the difference between the electrostatic energy of the ion in the solution and that of the ion in a vacuum. This very simple model gives the familiar relation:

$$\Delta G = - \frac{N(ze)^2}{2r} \left[1 - \frac{1}{\varepsilon} \right] \tag{26}$$

for the solvation energy and correctly predicts the order of magnitude of free energy and enthalpy of hydration of many salts. More elaborate theories are usually concerned with either determining a more realistic function to represent the dielectric constant in the vicinity of the ion rather than that of the bulk solvent or with calculating the free energy of solvation on the basis of a detailed examination of the electrical interactions of the ion with the solvent molecules in its environment.

Of the early papers following Born's that of Bernal and Fowler[5] is the most penetrating. An analysis by Eley and Evans[103] of structural effects influencing the entropy of hydration has scarcely been bettered. Early work has been admirably summarized by Conway.[67] Only the salient features will be described below.

(a) Continuous Dielectric Models. Webb[91] calculated the energy of charging an ion, represented by a point charge surrounded by a spherical cavity of radius r_0 by employing an equation giving the energy per unit volume of solvent as a function of the field strength E due to the ion. The relation between the field strength and distance r from the charge is:

$$E + 4\pi P = \frac{e}{r^2} \tag{27}$$

From the theory due to Langevin, P, the polarization per unit volume of solvent may be expressed in terms of the number of water molecules per cm^3, the optical polarizability of the water molecule α, and μ, its dipole moment. An expression for the electrostatic energy of a spherical shell of solvent of thickness dr was obtained and the solvation energy could then be evaluated by integrating this expression between the limits $r = r_0$ and $r = \infty$.

A similar approach was employed by Grahame[104] who deduced a semi-empirical relation for the variation of the dielectric constant ε with field. He assumed that ε tended to 80, the value of the bulk dielectric constant of water, at low field strength and to a limiting value of three immediately adjacent to an ion. Various parameters modifying the variation of ε with field strength were tried. The equation for the free energy of solvation was solved for various values of the effective radius of the ion. Although adequate agreement with experimental data was achieved, Grahame concluded that the effect of dielectric saturation near an ion was slight.

Debye's original theory[105] of dielectrics which had been employed by Webb was extended by Onsager[106] and amplified by Kirkwood[107] who took into account short range interactions which hinder the rotation of molecular dipoles. Booth[108] applied a modification of Kirkwood's theory of polar dielectrics to obtain a better representation of the dielectric constant in the ion's field:

$$\varepsilon = n^2 + \frac{\alpha \pi N_0 [n^2 + 2]\mu}{E} \cdot L \left[\frac{\beta(n^2 + 2)\mu E}{2kT} \right] \qquad (28)$$

where n is the refractive index of water; N_0 is the number of molecules per unit volume; μ is the dipole moment; E is the field strength; α and β are numerical factors; L is the Langevin function.

The theory was applied in detail to water assuming that ordering of the solvent is complete for nearest neighbours of the water molecule but vanishes outside this region. It was found that the reduction of ε due to dielectric saturation would only be significant in fields greater than 10^6 V cm^{-1}.

Employing the dependence of ε on field strength as given by Booth, Glueckauf[109] re-examined the calculation of hydration energies on the continuous dielectric model. Basing his choice on comparison between calculation of the bulk dielectric constants of aqueous electrolyte solutions and the experimental results of Hasted, Collie, and Ritson[62] Glueckauf[73] chose a value of $\beta = 2 \cdot 38 \times 10^{-4}$ rather than a higher value also quoted by Booth or one measured experimentally by Malsch.[110] The novelty of his approach is that allowance was made for empty space round an ion and between neighbouring water molecules in which the dielectric constant falls to unity. This was postulated as a consequence of the mode of packing six co-ordinated water molecules round an ion. In the mathematical treatment the empty space was averaged into a spherical shell of thickness Δ, and use of the differential form of the dielectric constant permits elegant solution of the integrals involved. A value of $\Delta = 0 \cdot 55$ Å was chosen to give the best correlation of the enthalpies of hydration with the measured values for the alkali chlorides.[98] The same value of Δ was employed by Latimer, Pitzer, and Slansky[111] as an empirical correction to the ionic radius which would make experimental hydration energies conform to the Born equation. When considering ionic partial molar volumes[112] Glueckauf also notes that ions will be surrounded by a dead space corresponding to a hollow sphere of thickness 0·55 Å if it is assumed that isomorphic replacement occurs between water molecules and ions of the same effective radius, 1·38 Å.

In a further discussion of his model Glueckauf[113] has determined those values of β and Δ which give the closest fit with the free energies, enthalpies and entropies of hydration quoted by Noyes.[98] Good agreement is obtained for the alkali halides when the high frequency value of $n^2 = 1·76$ is used in eqn (28). Adequate representation for these quantities for polyvalent ions is only achieved for $n^2 = 5·5$ which is the long wave value for the refractive index and the more appropriate value for use with Booth's equation. Unfortunately with $n^2 = 5·5$ the alkali halide hydration energies are discrepant. Glueckauf attributed this lack of consistency in the continuum model to the fact that water molecules close to high-field ions show structural discontinuities.

Conway and Desnoyers[114] have argued that the integrated rather than the differential form of the dielectric constant must be employed in three-dimensional calculations. A treatment of the free energy of hydration by Laidler and Pegis[115] which considers ions to be charged spheres with a dielectric constant of about two and uses the integral dielectric constant has been shown to give an unsatisfactory value for the enthalpy of proton hydration.[97] The reason for this is presumed to be the neglect of specific orientations of solvent molecules around the ions. In a revised version of the model,[117] Laidler and Muirhead-Gould[119] have included an 'orientation term' derived on the basis of Halliwell and Nyburg's adaptation of Buckingham's calculations. A rather arbitrary addition of 25 per cent to the crystal values of the ionic radii detracts from the conviction of this analysis.

Several pertinent objections have been directed at continuum models.[118] The main point is that application of the Booth equation, derived from a macroscopic theory of dielectric saturation, is questionable. A uniform structure is implicit in Booth's calculation whilst in fact the first layer of water molecules round an ion is almost certainly disrupted. It is also debatable whether the experimental value of β determined by Malsch is relevant since in his experiments the dielectric saturation caused by the impressed field must be much less than is experienced by inner layer water molecules.

(b) *Structural Theories.* Considering the limitations of the dielectric continuum model it might seem that examination of the structural effects of the electrical interactions of ions and solvent molecules would be more profitable.

The earliest calculations of this type were made by Bernal and Fowler[5] who considered a detailed model of the charge distribution in the water molecules which were assumed to be tetrahedrally deployed about the ions. The coordinated water molecules round both anions and cations were also assumed to bind two other water molecules each. These authors represented the enthalpy of hydration as the sum of three terms, the first giving the ion-dipole interaction for the ion and its four nearest neighbours, the second the 'Born' energy of solvation of the ion plus its water shell and the third giving the energy of the water molecule which had been replaced by the ion. When both the energy of interaction of the ion with its solvation sheath and that required to break water-water bonds to form this configuration are considered it is found that for maximum stabilization, the orientations of water molecules about an anion and a cation are different (see Fig. 4).

A more thorough investigation of this model was made by Eley and Evans.[103] They retained the four co-ordination and water orientation aspects of Bernal and Fowler's treatment but the assumed charge distribution in the water molecule was simplified so that the negative end of the dipole coincided with the centre of the oxygen atom. The energetic contributions of the following processes were evaluated and the sum equated to the ion hydration energy:

(a) a tetrahedral group of five water molecules is removed from the solution and transferred to the gas phase;

(b) this group is dissociated into single molecules;

(c) four water molecules are co-ordinated about a gaseous ion;

(d) the tetrahedral complex is placed in the cavity left by the water molecules which were removed;

(e) the water molecule remaining in the gas phase is condensed to the bulk of the solution.

Verwey[46] considered a model in which the water molecules co-ordinated round the cation were not constrained to lie in a plane passing through the centre of the ion. He showed that orientations out of this plane (see Fig. 4) could provide extra stabilization by allowing water molecules in the primary hydration sphere to bond with three other solvent molecules in the outer solvation sphere. He also noted that orientation of a water molecule adjacent to an anion as in the Bernal and Fowler model would allow further bonding (via the oxygen atom lone pair electrons and the H atom remote from the ion) to *three* other water molecules rather than two. In contrast to the assumption made by Bernal and Fowler, Verwey calculated that, according to his model, six co-ordination was preferred for all alkali halide ions except Li^+. Instead of almost identical hydration energies for anions and cations of the same size predicted by other models Verwey predicted more negative hydration energies for the anions. He maintained that this was due to neglect by Eley and Evans of the reorientation of water molecules outside the primary hydration shell.

A rather different interpretation was arrived at by Buckingham[76] who considered the detailed interactions between an ion and its nearest neighbour solvent molecules arranged in tetrahedral or octahedral configurations. The solvent outside the first shell was treated as a continuum with a field-dependent dielectric constant. Water molecules were taken to be polarizable spheres

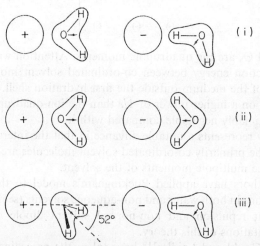

Fig. 4. The disposition of water molecules about simple ions, after (i) Bernal and Fowler,[5] (ii) Buckingham,[76] and (iii) Verwey.[46]

with permanent multiple moments at their centres. Several types of ion-solvent interaction not previously considered could therefore be investigated. The interaction energy was expressed as

$$U = U_e + U_i + U_d + U_r \qquad (29)$$

where U_e is the electrostatic energy of the permanent multipole moments of water in the field due to the ion; U_i is the energy due to induced multipole moments; U_d is the contribution of London dispersion forces; and U_r a repulsion energy due to overlap of electron clouds.

Treating the system as spherically symmetrical about the ion centre, Buckingham computed each of the components of the interaction energy as a power function of r, the distance between the centre of the ion and that of the solvent molecule for a normal orientation of the solvent dipoles.

It was shown that although the ion-dipole interaction energy remains unchanged on reversing the sign of the ion, the quadrupole moments are unaffected by inversion and the ion-quadrupole energy term is thus changed in sign. This results in a difference of solvation energies for anions and cations of the same size, viz.:

$$U_e^+ - U_e^- = \frac{|z|e(\Theta_a + \Theta_b)}{r^3} \tag{30}$$

$$\left.\begin{array}{l} \Theta_a = \Theta_{zz} - \Theta_{xx} \\[2mm] \Theta_b = \Theta_{zz} - \Theta_{yy} \end{array}\right\} \tag{31}$$

where Θ_a and Θ_b are the quadrupole moments. Attention was also given to lateral interaction energy between co-ordinated solvent molecules and the polarization of the medium outside the first hydration shell. All these other terms depend on a higher power of $1/r$ than the ion-quadrupole energy and thus are presumably negligible compared with it.

This theory represents a notable advance in that the interactions between the ion and the primarily co-ordinated solvent molecules are given explicitly in terms of the multipole moments of the solvent.

Several authors have applied Buckingham's model to the energetics of solvation of ions in both water and non-aqueous solvents (see below). Neglect of ion-solvent repulsion and non-normal solvent dipole orientation are, however, limitations of this theory.

Muirhead-Gould and Laidler[119] have taken the preceding treatment one step further. Besides terms already considered, the repulsive forces were calculated on the basis of a Lennard-Jones potential with the constant evaluated so as to agree with the experimental ion-water internuclear separation. Both tetrahedral and octahedral hydration were considered, again with normal orientation of the water molecules. Calculated free energies were minimized with respect to rotation of the water molecules about their axes of symmetry.

These calculations are consistent with tetrahedral hydration for the smaller ions and octahedral hydration for the larger ones. It is found that the values are seriously in error for all the univalent anions except fluoride, suggesting that the orientation of water next to anions should be that suggested by Eley and Evans[103] as in Fig. 4. Complexity of the calculations rules out an analysis of this model and for the same reason representation of the water molecule as three point charges at the atom centres plus a polarizable dipole is necessarily grossly simplified. Laidler's model also includes several other sources of indeterminate error. Firstly, the polarizability of the water molecule is assumed to be independent of field strength. Secondly, there is considerable disagreement about the appropriate values of the ionic radii. This important point, which has been the subject of much discussion, is considered below. Thirdly, the value of the hydrogen bond energy is rather uncertain. On the Laidler model change of this energy by 1 kcal/mole would alter the hydration energies by 8–12 kcal/mole.

(c) *Ionic Radii*. Innumerable attempts have been made to correlate not only solvation energies but also partial molal volumes, mobility, etc. with ion size. It is relevant therefore to consider just what the ion size parameter (radius) should be. It has usually been implicitly assumed that the effective radius of an ion in solution is equal to its crystal radius. Benson and Copeland[120] have estimated the magnitude of forces in a crystal and in an ionic solution and concluded that there is no basis for believing that ions in solution are under less electrostatic pressure than are ions in a solid crystal. Ion-water distances in crystal hydrates are close to ion-water distances in solution and the enthalpies of hydration of most salts are small, implying that the lattice and solvation energies must be nearly equal. They conclude that the radius of an ion in either solid or solution should be within 0·02 Å of its 'hard sphere' value.

Thus the use of crystal radii seems justified for ions *in solution*. There are pitfalls however in employing Born-type models which have only recently been recognized. Firstly the radius of an ion, or even an uncharged species, will change on transfer from gas to liquid phase where it will be subject to compressional forces. Secondly the radius of the dissolved 'species' will depend on its surface charge density so that models of solvation using a continuous charging process to evaluate the electrostatic energy must use a variable parameter for the ion size. Finally, the radius will depend on the degree of dielectric saturation induced by the ion in the solvent.

Two devices have been employed to produce agreement with the Born equation using crystal radii. Certain authors[98] have attributed values as low as 1·8 to the effective dielectric constant next to the ion which is much lower than the value of $\varepsilon = 5$ expected for irrotationally bound water.[62] Alternatively, some empirical correction is made to the crystal radius either by multiplication by,[115] or more usually by addition of, a constant.[111] Glueckauf[112] has explained this correction as accounting for the 'void space' about ions in which the dielectric constant is unity but the existence of a cavity surrounding the ions is difficult to reconcile with the small or negative apparent molar volumes of the smaller cations.

Stokes has pointed out that the self energy of an ion in the gas phase is many times greater than its self energy in solution.[121] He computed the solvation energy as the difference of the self energy of the ion in solution and gas plus a small term representing the solvation energy of the neutral species

$$\Delta\mu_{hyd} = \mu_{aq}(\text{ion}) - \mu_{vac}(\text{ion}) + \Delta\mu(\text{neut}) \tag{32}$$

where

$$- \mu_{aq}(\text{ion}) = \frac{Nz^2e^2}{2} \left[\frac{2nr_w}{r_c(r_c + 2nr_w)\varepsilon'_{aq}} + \frac{1}{\varepsilon_{aq}(r_c + 2nr_w)} \right] \quad (33)$$

$$- \mu_{vac}(\text{ion}) = \frac{Nz^2e^2}{2} \cdot \frac{1}{r_{vac}\varepsilon_{vac}} \quad (34)$$

where ε'_{aq} is the effective dielectric constant of bound water; $2nr_w$ is the thickness of n layers of water molecules around the ion, taken to be zero for univalent anions and 2·8 Å for univalent cations; and $\varepsilon_{vac} = 1$.

The most important point made by Stokes is that the relevant parameter for the gas phase ion is the van der Waals radius r_{vac} rather than the crystal radius r_c. Using the van der Waals radius reasonable agreement was found between the calculated single ion hydration energies and those quoted by Noyes[98] when the effective dielectric constant of the n layers of co-ordinated water molecules was taken to be

$$1/\varepsilon'_{aq} = \tfrac{1}{2}[\tfrac{1}{5} + \tfrac{1}{78}] = 1/9 \quad (35)$$

The question of the exact values of the crystal radii has been discussed recently.[122] Two distinct sets are in general use at present, one due to Pauling[25] and the other based on a value of 1·33 Å for the fluoride ion and which has been extended and modified by Goldschmidt,[123] Zachariansen[124] and Ahrens.[125] Waddington[126] has critically reassessed the data for the alkali metal halides and alkaline earth chalogenides using the method of the undetermined parameter. Essentially this means finding an additive set of radii in the form

$$r_+ = a - \delta \quad (36)$$

$$r_- = b + \delta \quad (37)$$

so that the sums $a + b$ represent the observed internuclear distances with the minimum deviation. Values of r_+ and r_- determined by Waddington for the alkali halides have an average deviation, α, from additivity of 0·021 Å while on the Goldschmidt scale $\alpha = 0·035$ Å and Pauling's values give $\alpha = 0·069$ Å. Similarly for the alkaline earth chalcogenides all deviations are less than 2 per cent.

Five different methods were used to evaluate the undetermined parameter δ:

(a) the Landé method, based upon the assumption that in lithium halides the anions are in contact;

(b) the Pauling method in which the radii in crystals containing isoelectronic ions (e.g. NaF, KCl) are assumed to be in the inverse ratio of their effective nuclear charges z'_e. z'_e is the charge on the ion minus the screening effect S_e of the electrons in the ion;

(c) from ionic susceptibilities, assuming that r^2 is proportional to χ according to the Lamor equation;

(d) from molar refractions by taking R to be proportional to r^4;

(e) from minima in electron density distributions.

Good agreement was found between methods (a)–(d) for alkali halides and the value of $\delta = 1.852$ Å so derived was used to calculate the ionic radii given in column 4 of Table 3. As was pointed out by Gourary and Adrian,[127] the ionic radii derived by assigning values to Cl⁻ and Na⁺ on the basis of an accurate determination of the electron density distribution in the NaCl and LiF crystals due to Witte and Wolfel[128] and given in column 5 of Table 3

Table 3. *Ion Radii in* Å

Ion	Pauling (a)	Goldschmidt (b)	Waddington (c)	Gourary and Adrian (d)	Stokes (e)
Li⁺	0·60	0·78	0·73₉	0·94	
Na⁺	0·95	0·98	1·00₉	1·17	1·35₂
K⁺	1·33	1·33	1·32₀	1·49	1·67₁
Rb⁺	1·48	1·49	1·46₀	1·63	1·80₁
Cs⁺	1·65	1·65	1·71₈	1·86	1·99₇
F⁻	1·36	1·33	1·32₂	1·16	1·90₉
Cl⁻	1·81	1·81	1·82₂	1·64	2·25₂
Br⁻	1·95	1·96	1·98₃	1·80	2·29₈
I⁻	2·16	2·20	2·24₁	2·05	2·54₈

References: (a) 129; (b) 123; (c) 126; (d) 127; (e) 121.

differ considerably from those indicated by the other four methods. Ideally this method should be applied to a crystal in which the anion and cation are isoelectronic. It may also weight the outermost regions of the electron distribution too heavily.

Agreement between the values of the undetermined parameter for the alkali earth chalcogenides, obtained by various methods, is poor.

Blandamer and Symons[130] note that several properties of ions in solution appear to correlate more closely with ion size when compared on the new 'electron density' scale rather than Pauling's. The partial entropies of the alkali halide ions, corrected to Gurney's scale, lie on a single line when plotted against the reciprocal of the ion radii rather than on two distinct lines as formerly. The increase of limiting equivalent conductivities of ions in water varies more regularly with ion size using the Gourary and Adrian radii. Jain has recomputed single ion hydration energies by Stokes method but using the 'electron density' crystal radii rather than Pauling's values.[131] He apparently achieves a much better agreement with Noyes 'experimental' single-ion hydration energies than did Stokes but this may in part be due to Jain's use of a smaller effective dielectric constant $\varepsilon'_{aq} = 2 \cdot 7$ and the assumption that $n = 1$ rather than 0 for the halide ions. It may be mentioned in passing that it is inconsistent to compare the calculated energies with Noyes empirical set of values which were derived using a different set of ionic radii.

Morris[122] has redetermined the conventional enthalpy of hydration of the proton by the method of Halliwell and Nyburg (discussed in the next section) using the Gourary and Adrian ionic radii. His finding that the modified extrapolation yields a greater value for the proton enthalpy is contrary to the conclusion reached by Case.[268] In any event, general criticisms of the method still apply.[133]

Free energy changes associated with hydration reactions of alkali halide ions in the gas phase, determined by mass spectrometry, correlate well with the new experimental ionic radii but not with Pauling's values.[132]

(d) *Recent Estimates of Single Ion Hydration Energies.* A comparison of models of hydration to decide which best represents experimental results is hampered by the lack of a sound independent method of assigning individual ion hydration energies. The desirability of correlating emf and pH scales in different solvents poses this problem for other media also.

Once the solvation energy for any ion has been established, that for any other ion can be found using the total solvation energies of salts containing the ions in question. Thus, most efforts have been directed at finding the solvation energy for the proton. Methods (usually semi-empirical) employed to estimate the enthalpy of hydration of H^+ have given results varying as much as 50 kcal mole^{-1}. Work up to about 1960 has been reviewed by Conway[133] and by Vasil'ev and colleagues.[116] Recently extrapolation procedures with reasonable theoretical foundation have been proposed and the latest values of ΔH_H^+ and ΔG_H^+ are probably not in error by more than ± 5 kcal mole^{-1}.

Izmailov[134] has determined single ion solvation energies for water, alcohols and several other solvents. The data were obtained from emf measurements or when these were not available, from solubility measurements. Free energies of solution obtained from salt solubilities are usually less accurate than potentiometric values. Data for butanol, isoamyl alcohol and acetone are also considered less reliable because of incomplete dissociation of salts in these solvents. The difference between the standard potentials of metal electrodes, that is, the emf of the hypothetical cell:

$$M_1 | \underset{a_1=1}{M_1^+(S)} \; \Big| \; \underset{a_2=1}{M_2^+(S)} | M_2$$

is directly related to the *difference* in solvation energy of the two cations:

$$FE^0 = F(E_2^0 - E_1^0) = -[\Delta G^0_{solv}M_2^+(g) - \Delta G^0_{solv}M_1^+(g)]$$
$$- [\Delta G^0_{form}M_2^+(g) - \Delta G^0_{form}M_1^+(g)] \quad (38)$$

Similarly the *sum* of solvation energies of a cation and an anion may be obtained by measuring the standard emf of a cell of the type:

$$M_1 | M_1^+X^- \; \Big| \; M_2^+X^- | M_2X | M_2$$

from which

$$FE^0 = [\Delta G^0_{solv}M_1^+ + \Delta G^0_{solv}X^-]$$
$$- [\Delta G^0_{form}M_2^+(g) + \Delta G^0_{form}M_1^+(g)] - \Delta G^0_{latt}M_2X \quad (39)$$

ΔG^0_{latt} being the lattice energy of the insoluble salt M_2X.

To find individual ion solvation energies Izmailov plotted the quantities

$$[\Delta G^0_{solv}M_1^+ - \Delta G^0_{solv}M_2^+] = \alpha \quad (40)$$

and

$$[\Delta G^0_{solv}M_1^+ + \Delta G^0_{solv}X^-] = \beta \quad (41)$$

against the reciprocal of the co-ion radius for various cations M_2^+ and anions X^-. The sum of these quantities was taken to extrapolate to a value of $2\Delta G^0_{solv}M_1^+$ as $1/r \to 0$ since when r becomes large the energies of anions and cations of the same size should tend to the same value. Izmailov seems to have neglected the fact, as did Verwey,[135] that for self consistency the quantities α and β should follow the Born relation when r is large and slopes of plots of these quantities should be equal in magnitude but opposite in sign in that region.

Subsequently[101] Izmailov favoured a different method of extrapolation. It was assumed that the free energies of solvation of isoelectronic alkali and halide ions approach equality with increasing size. By plotting

$$\Delta G^0 H^+ + \tfrac{1}{2}[\Delta G^0 X^- - \Delta G^0 M^+]$$

against $1/n^2$, where n is the principal quantum number of the molecular orbitals which are formed by bonding the lone electron pairs of the solvent to the ions, and extrapolating to $1/n^2 = 0$, $\Delta G^0 H^+$ is obtained. Although this method does not rely on the Born relation for its validity and employs a more definite parameter than the ion radius, a long non-linear extrapolation of experimental data is involved. Izmailov's single ion solvation energies are not very reliable for calculating the free energies of transfer of ions between different solvents.

Noyes[54] has used an empirical extrapolation to obtain individual ion hydration energies which have been regarded as absolute 'experimental' values by several authors. The conventional free energies of hydration, referring to the processes:

$$M^+(g) + e(g) + H^+(H_2O) \rightarrow M^+(H_2O) + \tfrac{1}{2}H_2(g) \tag{42}$$

$$X^-(g) + \tfrac{1}{2}H_2(g) \rightarrow X^-(H_2O) + H^+(g) + e(g) \tag{43}$$

were expressed as the sum of a neutral and an electronic part, the neutral contribution being the value for an uncharged species of the same size and electron distribution as the ion, that is:

$$\Delta G^0_{\text{conv hyd}} = \Delta G^0_{\text{el}} + \Delta G^0_{\text{neut}} \tag{44}$$

The electronic part of the solvation energy is

$$\Delta G^0_{\text{el}} = \sum_{i=1}^{n} \frac{c_i}{r^i} \tag{45}$$

and thus

$$\Delta G^0_{\text{neut}} - \Delta G^0_{\text{conv hyd}} M^+ = \Delta G^0_{\text{form}} H^+(\text{aq}) - \sum \frac{c_i}{r^i} \tag{46}$$

and

$$\Delta G^0_{\text{conv hyd}} X^- - \Delta G^0_{\text{neut}} = \Delta G^0_{\text{form}} H^+(\text{aq}) + \sum \frac{c_i}{r^i} \tag{47}$$

Values of the left hand sides of eqns (46) and (47) were plotted against $1/r$ for these ions, and the Born value was used for c_1. By varying c_2 and ΔG^0_{neut} empirically a good fit to experimental data was obtained with limiting slopes at $r = \infty$ conforming to the Born equation. Use of the hydration energies of the noble gases as ΔG^0_{neut} did not prove satisfactory. The best agreement was found for

$$\Delta G^0_{\text{neut}} = 1 \cdot 58 r^2 + 1 \cdot 325 \text{ kcal/mole} \tag{48}$$

It is reasonable to assume that the hydration energy should be proportional to the surface area of the neutral species and the constant $1 \cdot 325$ merely relates to the change in standard state in the transfer from gas to solution. The value of $\Delta G^0_{\text{form}} H^+(\text{aq})$ so found, $103 \cdot 8$ kcal/mole when combined with the free energy of formation of the gaseous proton, $+364\cdot0$ kcal/mole,[136] gives a value of

$$\Delta G^0_{hyd}H^+(g) = -260\cdot2 \text{ kcal/mole}$$

compared with $-256\cdot5$ kcal proposed by Izmailov.

An important paper on the enthalpy of hydration of the proton has been published by Halliwell and Nyburg.[97] The method is based on obtaining $\Delta H^0_{hyd}H^+(g)$ from differences of conventional hydration enthalpies of pairs of ions with the same radii but of opposite charge. They define conventional enthalpies of hydration by the processes:

$$Mz^+(g) + zH^+(H_2O) \rightarrow Mz^+(H_2O) + zH^+(g) \tag{49}$$

and

$$Xz^-(g) + zH^+(g) \rightarrow Xz^-(H_2O) + zH^+(H_2O) \tag{50}$$

so that:

$$\Delta \bar{H}^0_{hyd}Mz^+(g) = \Delta H^0_{hyd}Mz^+(g) - z\Delta H^0_{hyd}H^+(g) \tag{51}$$

and

$$\Delta \bar{H}^0_{hyd}Xz^-(g) = \Delta H^0_{hyd}Xz^-(g) + z\Delta H^0_{hyd}H^+(g) \tag{52}$$

The difference between conventional thermodynamic quantities as used by Noyes and by Halliwell and Nyburg, for example, should be noted. They are related by the expression:

$$\Delta \bar{Y}^0_{hyd} = \Delta Y^0_{conv\,hyd} \pm z\Delta Y^0_{form}H^+(g) \tag{53}$$

where Y can be the free energy, enthalpy or entropy of hydration and $\Delta Y^0_{form}H^+(g)$ relates to the reaction

$$\tfrac{1}{2}H_2(g) \rightarrow H^+(g) + e(g) \tag{54}$$

It is readily seen that for the simple case of uni-uni valent ions

$$\Delta \bar{H}^0_{hyd}M^+(g) - \Delta \bar{H}^0_{hyd}X^-(g)$$
$$= \Delta H^0_{hyd}M^+(g) - \Delta H^0_{hyd}X^-(g) - 2\Delta H^0_{hyd}H^+(g) \tag{55}$$

Halliwell and Nyburg plotted experimental values of conventional hydration enthalpies against a function of the ionic radius and obtained a smooth curve for cations and a second curve for anions. From these curves it is possible to interpolate enthalpies at any convenient value of the radius and hence to obtain the left hand side of eqn (55) for (hypothetical) pairs of anions and cations of the same radius. A plot of this enthalpy difference against a reciprocal function of the radius may be extrapolated to $r_i = \infty$ where the intercept is equal to $-2\Delta H^0_{hyd}H^+(g)$.

The difficulty in this method lies in choosing the right function of the radius against which to plot the experimental quantities. Halliwell and Nyburg based their choice on Buckingham's model[76] of hydration for which the ion-dipole energy is proportional to $(r_i + r_w)^{-2}$, the ion-quadrupole energy proportional to $(r_i + r_w)^{-3}$ and the Born charging energy for the medium outside the inner hydration sphere is proportional to $(r_i + 2r_w)^{-1}$. The radius of a co-ordinated water molecule, r_w, if equated to the average effective radii

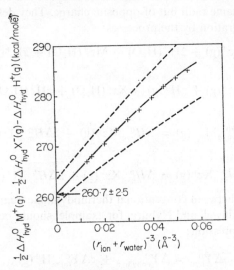

Fig. 5. Interpolated values of the difference between the conventional enthalpies of hydration of cations and anions plotted against a function of the ionic radii (after Halliwell and Nyburg[97]).

of water molecules in crystalline hydrates, is 1·38 Å. It was justifiably concluded that in the plot of $^{M+}\Delta^{X-}\Delta\bar{H}_{hyd}^{0}$ the ion-dipole and Born charging energies would cancel and hence the correct radius function is $(r_i + r_w)^{-3}$ since the ion-quadrupole energy dominates. There is, however, no such justification for plotting the *separate* anion and cation conventional hydration energies against this function of the radius as was pointed out by Conway,[133] though it is probable that the interpolated values do not depend critically on the exact form of the curve. A more pertinent criticism of the method is the assumption that the water molecules are oriented normally to the ions. If this is *not* so the ion-dipole energies need not be equal for anions and cations of equal sizes and the extrapolation function should include terms in both $(r_i + r_w)^{-2}$ and $(r_i + r_w)^{-3}$. Conway has calculated the various electrostatic ion-solvent interaction enthalpies and has shown that the ion-dipole contribution decreases less rapidly with ion size than the ion quadrupole term. Consequently the slope of the $\Delta\Delta\bar{H}^\circ$ plot should probably be less than that given by Halliwell and Nyburg with a correspondingly greater intercept than in Fig. 5. Conway proposes a value of $\sim -267 \pm 7$ kcal mole^{-1} for the enthalpy of hydration of the proton compared with the original estimate of $-260\cdot5 \pm 2\cdot5$ kcal mole^{-1}.

A recent experimental determination of the quadrupole moments of the water molecules gives a value of $-0\cdot13 \times 10^{-26}$ esu cm^2 for the component Θ_{zz} in the direction of the dipole axis.[137] This is less than one tenth of the value previously assumed and suggests that the effect of the quadrupole moment of the water molecule on the solvation energy of ions is very much less than supposed by Halliwell and Nyburg.

D. The Hydrated Electron and Potential Scales. Solvated electrons have been known for many years in solutions of alkali metals in ammonia,[138] although the existence of hydrated electrons was not seriously considered before the early 1950's[139,140] since the products of chemical reactions of hydrogen atoms produced by radiolysis of solutions are identical with those produced by solvated electrons. Many of the rates of such reactions are anomalously high[141] however and this suggested that a second reducing species, more powerful than the H atom, was also present. With the advent of high electron current pulse accelerators and fast reaction techniques[142] the absorption spectrum of the hydrated electron has been positively identified[143] and the rate constants for reaction with a great variety of ions and both inorganic and organic compounds have been determined.[144]

Whilst the hydrated electron is obviously not an 'ion' in the usual sense it is useful to consider its properties from this point of view. The high energy electrons initially produced by radiolysis of water become thermalized in $\sim 10^{-11}$ sec. The speed of thermalized electrons is such that they can induce electronic but not orientational polarization in the surrounding water. Rotation of molecules in low temperature glasses at 77°K will be extremely slow if possible at all. It is found, however, that the electron adsorption band in pulse irradiated 15 M sodium hydroxide[145] is identical with that of 6 M sodium hydroxide aqueous glass at -196°C except for the expected shift due to temperature change. Exactly the same thing is found for liquid and glassy methanol.[146]

The inference that there is identical orientation of solute around the electron in both fluid and rigid phases suggests that electrons become trapped (solvated) when an instantaneous solvent orientation distribution around them produces the necessary potential well, the electron then stabilizing this formation for the period of its lifetime which is $t_{1/2} \sim 800$ μsec in alkaline solution.

Hydration Energy of the Electron. The free energy of hydration

$$e^-(g) \rightarrow e_{aq}^- \tag{a}$$

has been calculated by Baxendale[147] from thermodynamic data for the reactions:

$$\Delta G \text{ kcal/mole}$$

$$e_{aq}^- + H_2O \rightleftharpoons H + OH_{aq}^- \qquad +8{\cdot}4 \tag{b}$$

$$H \rightarrow \tfrac{1}{2}H_2 \qquad -48{\cdot}6 \tag{c}$$

$$H_{aq}^+ + OH_{aq}^- \rightarrow H_2O \qquad -19{\cdot}3 \tag{d}$$

giving

$$e_{aq}^- + H_{aq}^+ \rightarrow \tfrac{1}{2}H_2 \qquad -59{\cdot}5 \tag{e}$$

The free energy change for reaction (a) was derived from the equilibrium constant obtained from the rates of the forward and backward reactions. Baxendale's calculation has been adjusted to a more recent value of k_b.[148]

Taking reaction (e) together with the reactions

$$\tfrac{1}{2}H_2 \rightarrow H^+(g) + e^-(g) \tag{f}$$

and

$$H^+(g) \rightarrow H_{aq}^+ \tag{g}$$

the free energy change for the hydration step (a) may be calculated. The hydration energy of the proton $\sim -257 \pm 7$ kcal/mole can be derived from the estimate of the hydration enthalpy of the proton[97] together with its entropy in the gas phase, $+27{\cdot}4$ cal/°C/mole[136] and in solution, ~ -5 cal/°C/mole[95] and agrees closely with the value quoted by Noyes.[98] The electron hydration energy so calculated is approximately -40 kcal/mole and is certainly smaller than that of the iodide ion which is about 57 kcal/mole. Theoretical calculations[149] of the mean radius of charge distribution in the ground state of the hydrated electron give a value of $2{\cdot}5$–$3{\cdot}0$ Å which is in accord with the greater effective radius of this species than that of the iodide ion, $2{\cdot}2$ Å, as indicated by the relative hydration energies. Further evidence is provided by a calculation[150] of 5 Å for the effective diameter of e_{aq}^- in the reaction

$$e_{aq}^- + e_{aq}^- \rightarrow H_2 + 2OH_{aq}^- \tag{h}$$

The hydration energy of -40 kcal or $1{\cdot}75$ eV agrees closely with the absorption maximum at 7200 Å ($1{\cdot}72$ eV). There is evidence that the absorption spectrum corresponds to a charge-transfer-to-solvent process. A linear relation has been noted between the energy of λ_{max} for the charge-transfer spectra of iodide ions and that for electrons solvated in those media[151] as was predicted by Platzman and Franck.[152]

The activation energies[153] of a variety of reactions of the hydrated electron with widely different rates are approximately equal to the activation energy of diffusion controlled reactions in aqueous solutions,[154] about 3 kcal/mole. It is suggested that the activation energy is that necessary to reorganize the hydration sphere around the electron as in the case of other aqueous ions. A review of the kinetics of the solvated electron and atomic hydrogen in water has been given by Stein.[155]

Baxendale has made the interesting observation that a new absolute scale of electrode potentials may now be adopted based on the measured free energy of

$$\tfrac{1}{2}H_2 \rightarrow H^+_{aq} + e^-_{aq} \qquad (i)$$

rather than arbitrarily assigning to it a value of zero. Absolute aqueous electrode potentials would then all be 2·58 V more negative than conventional values. Measurements of the free energy of the above reaction in other solvents may permit fixing absolute potential scales for these media also and should lead to absolute values for thermodynamic quantities of ion transolvation and hence to deeper understanding of solvation in non aqueous solvents. Several other very much more indirect methods have been applied to this problem and are described in Sections III.3 and V.4.

E. The Free Energies of Transfer of Ions between Light and Heavy Water. There is evidence from many sources that deuterium oxide is a more structured solvent than light water at the same temperature.[36,156,157] The dielectric constants of the liquids have been measured over a range of temperatures by Wyman and Ingalls.[158] At 25°C the difference in the dielectric constants is only 0·4 per cent. Computing the difference of enthalpy and entropy of solvation of ions in the two solvents using the Born equation with a radius equal to that of the ion plus the diameter of a water molecule leads to values of only about 2 and 7 cal mole^{-1} respectively for alkali halide ions.[159] The experimental enthalpies of solution are approximately 100 times greater than this.[160,161] It is evident that the mode of co-ordination of solvent molecule in the inner solvation sphere and the influence of the ion on the liquid structure around it must be rather different for the two solvents.

In that they reflect the structural differences between light and heavy water the thermodynamic quantities characterizing transfer of salts from one liquid to the other are of great interest. Calorimetric values for heats of transfer of alkali halides have recently been supplemented by the work of Wu and Friedman.[162]

Free energy changes for the process $MX(H_2O) \rightarrow MX(D_2O)$ where M^+ is an alkali cation and X^- is a halide ion have been determined from equilibrium emf measurements. La Mer and Noonan[163,164] studied the cell:

$$\text{Ag, AgCl} | \text{KCl}(D_2O) | \text{KCl}(H_2O) | \text{AgCl, Ag}$$

for which the net process was assumed to be the transfer of t^+ equivalents of potassium chloride from light to heavy water, and calculated the free energy of transfer on the assumption that the energy change due to intermixing of the two liquids at the boundary was zero. The latter ambiguity was avoided by Salomaa and Aalto[165] who employed an amalgam electrode reversible to the cation in the cell.

$$\text{M(Hg)} | \text{MCl}(D_2O) | \text{AgCl, Ag} - \text{Ag, AgCl} | \text{MCl}(H_2O) | \text{M(Hg)}$$

the potential of which is directly proportional to the free energy of transport of the salt. Values determined in this way are given in Table 4. The free energy of transfer of potassium chloride is very close to that quoted by La Mer and Noonan and gives credence to their view that the effect of solvent interchange is negligible. These results are however different from those of Greyson[166] who used the cell

$$\text{Ag, AgCl} | \text{MCl}(D_2O) | ^+ | \text{MCl}(H_2O) | \text{AgCl, Ag}$$

with a cation exchange resin $| ^+ |$ between the light and heavy water solutions which suggests that there is a contribution due to a potential difference arising from the transport involved in the passage of current through the membrane. An independent determination of the free energy of transfer of NaCl and KCl was made by Kerwin[167] from the solubilities of these salts. This allowed Greyson to estimate the correction due to solvent transport in the membrane. The corrected energies are still only about half as large as those calculated by Salomaa and Aalto (see Table 4). The difficulty of satisfactorily estimating the activities of saturated solutions of highly soluble salts has been mentioned.[159] Thus some doubt remains concerning the adequacy of the solvent correction used by Greyson. This should also be borne in mind when considering the entropies of transport of alkali chlorides which have been derived from the temperature variation of the emf's of membrane cells.

Swain and Bader[159] consider that the enthalpy of transfer is primarily determined by the difference in the frequency of libration of solvent molecules in the inner solvent sphere of the ion. Assignment of individual ionic heats of transfer on the basis of a statistical mechanical calculation of this quantity for the fluoride ion was made. The ion was assumed to be four co-ordinated and was chosen because it has a clear maximum at 698 cm^{-1}

Table 4. *Thermodynamic Functions for Transfer of Alkali Halides From H_2O to D_2O (cal/mole)*

	ΔH_t^0			ΔG_t^0				$T\Delta S_t^0$	
	(a)	(b)	(c)	(d)	(c, e)	(f)	(g)	(h)	(i)	(g)
KF	60	60	—			−3 {92}	34		63 {90}	26
LiCl	400	420	423	202 {23}	111 {50}	−32 {92}	247	221	390 {90}	153
NaCl	560	510	545	212 {9}	116	110 {12}	346	333	402 {12}	215
KCl	615	560	—	219 {9}	121	130 {12}	377	396	432 {12}	239
NaBr	[695]	650	—			170 {12}	424		483 {12}	272
KBr	750	695	—			180 {12}	455		519 {12}	296
RbBr	790	—	—				482			310
CsBr	815	—	—				496			321
NaI	[842]	810	830		0{100}	230 {12}	524		585 {12}	320
KI	890	865				240 {12}	555		627 {12}	342

Values in brackets [] are interpolated.
Values in braces {} are uncertainties.

References: (a) 160; (b) 161; (c) 162; (d) 165; (e) 167; (f) 166; (g) 159; (h) Based on data from references 161, 162, and 165; (i) Reference 166 based on data from reference 161.

4

for the librational band in both sodium fluoride and potassium fluoride. When the reverse procedure is employed for other alkali halide ions the calculated librational frequencies agree well with those actually observed. The iodide ion appears to be six co-ordinated unlike all the other ions which were taken to have four nearest neighbours.

While the change in librational frequency of co-ordinated water molecules seems to account reasonably well for the difference in enthalpies, the free energies of transfer calculated by Swain and Bader differ substantially from those obtained experimentally. This is doubtless due to the effects of the ions on the structure of water outside the primary hydration shell. These effects, reflected in the entropy of solution, were completely ignored in Swain and Bader's treatment. While Swain and Bader have attributed all the thermo-dynamic differences to changes in librational frequency at a single wavelength there is evidence of lower frequency bands corresponding to other modes of rotation.[168] Recalculation of the free energies and enthalpies of transfer on Swain and Bader's model but omitting their assumption of three-fold degeneracy of the librational mode can lead to lower values for ΔG which are more consistent with those of other authors.[169]

Several phenomena indicate that anions interact more strongly with water than cations of the same size. Noyes[54] has suggested that while cations fit into the water structure relatively easily, the configurations of the hydrogen-bonded water molecules round an anion are more dependent on the nature of the anion. Similarly, anions have the greatest effect on the librational bands in the infra red spectra. Fluoride ion increases the frequency at which maximum absorption occurs while chloride bromide and iodide ions effect a shift to lower frequencies.[159] The alkali cations absorb at slightly lower frequencies, the lower the larger the ion; lithium ion has no effect. Lithium fluoride is the sole alkali halide which is more soluble in D_2O than H_2O. Googin and Smith[170] showed that most of the effect of dissolved electrolytes on the isotopic fractionation of hydrogen between aqueous electrolyte solutions and the vapours was due to the anions.

These facts indicate that the free energy of transfer of alkali halides will be dominated by the anion hydration energy. Indeed, this is the interpretation placed by Salomaa on the fact that the electrochemically measured free energies of transfer of LiCl, NaCl, and KCl are virtually the same. Considering similar effects found for the transfer of electrolytes from water to methanol-water mixtures (see Section III.3) it is plausible that the cation solvation energies in the two types of water are very nearly equal and the value of 210 cal/mole which was also found for cadmium chloride is approximately the transfer energy of the chloride ion alone.

Hepler[169] has satisfactorily accounted for the magnitude of the free energies of transfer of univalent halides between H_2O and D_2O. These all lie between 100–250 cal/mole. A similar value for silver bromate, 190 cal/mole is obtained from the ratio of the solubilities of this salt in the two liquids whereas employing the Born equation with the macroscopic dielectric constants $\varepsilon_0 = 78 \cdot 30$ for H_2O and $\varepsilon_0' = 77 \cdot 94$ for D_2O gives a transfer energy of only 5–20 cal for ions between 2 and 0·5 Å.

Hepler noted that the curves of dielectric constant against radius which were used by Ritson and Hasted[56] to explain their measurements of the dielectric constants of aqueous electrolyte solutions could be approximated by

$$\varepsilon_{sat} = 5 \cdot 00$$

for $r < 1 \cdot 5$ Å

$$\varepsilon = \frac{\varepsilon_0 - \varepsilon_{sat}}{2 \cdot 5} (r - 1 \cdot 5) + \varepsilon_{sat} \tag{56}$$

for $4 \cdot 0 > r > 1 \cdot 5$ and

$$\varepsilon_0 = 78 \cdot 3$$

for $r > 4 \cdot 0$ Å, where ε_{sat} is the dielectric constant under dielectric saturation. Thus by splitting up the free energy expression

$$G = k \int_{r_i}^{\infty} \frac{1}{\varepsilon r^2} \, dr \tag{57}$$

into integrals for the three regions specified and evaluating each for the appropriate variation of dielectric constant with radius he obtained:

$$
{}_{H_2O}\Delta G^{D_2O} = k \left(\frac{1 \cdot 5 - r_i}{1 \cdot 5 r_i} \right) \left(\frac{\varepsilon_{sat} - \varepsilon_{sat}'}{\varepsilon_{sat} \cdot \varepsilon_{sat}'} \right)
$$
$$
+ k \left[0 \cdot 417 \frac{Y - Y'}{YY'} + \frac{X'}{(Y')^2} \ln \left(\frac{0 \cdot 375 \varepsilon_0'}{\varepsilon_{sat}'} \right) - \frac{X}{Y^2} \ln \left(\frac{0 \cdot 375 \varepsilon_0}{\varepsilon_{sat}} \right) \right]
$$
$$
+ 0 \cdot 25 k \left[\frac{\varepsilon_0 - \varepsilon_0'}{\varepsilon_0 \varepsilon_0'} \right] \tag{58}
$$

where $X = (\varepsilon_0 - \varepsilon_{sat})/2 \cdot 5$ and $Y = 1 \cdot 5 X - \varepsilon_{sat}$, the primed symbols referring to D_2O. The ratio of the high field dielectric constants was assumed to be equal to the ratio of refractive indices of the liquids.

Results of this calculation are summarized in Fig. 6 curve A. Curve B illustrates the behaviour if the dielectric is saturated in the first layer of water molecules round an ion, i.e. using ε_{sat} for $r \leqslant r_i + 2r_w$ and normal outside this region. It would seem from these calculations that some degree of dielectric saturation must be invoked to explain the experimental transfer energies on the continuous dielectric model. The contribution to the free energy made by solvent outside the region where the dielectric constant is diminished is, from eqn (58), only 2 cal mole^{-1}.

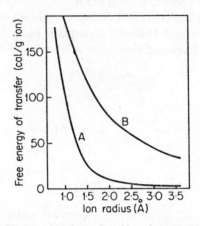

Fig. 6. Calculated molar free energies of transfer of ions from H_2O to D_2O (after Hepler[169]).

F. Isotope Fractionation and the Chloride Ion Solvation Energy. The real free energy of transfer of chloride ion between light and heavy water has been determined[173] from the compensation potential of the cell:

$$Ag|AgCl|NaCl \text{ in } H_2O|Air|NaCl \text{ in } D_2O|AgCl|Ag$$

Here:

$$E^0 = - (\alpha_{Cl}^{D_2O} - \alpha_{Cl}^{H_2O})/F \qquad (59)$$

and

$$^{H_2O}\Delta^{D_2O}\alpha_{Cl^-}^0 = 80 \pm 50 \text{ cal/mole}$$

Solomaa's argument[165] that the chemical transfer energies of alkali chlorides between the two solvents is virtually equal to that of the chloride ion alone is supported by the value of $^{H_2O}\Delta^{D_2O}\mu_{Cl^-}$ calculated by Swain and Bader.[159] If it is accepted that

$$^{H_2O}\Delta^{D_2O}\mu_{Cl^-} \simeq 210 \text{ cal/mole}$$

then from the measured real transfer energy of chloride ion

$$^{H_2O}\Delta^{D_2O}\chi \simeq 6 \text{ mV}$$

It is generally observed that D_2O behaves like H_2O at a lower temperature, the inference being that heavy water is the more structured solvent due to its smaller librational energy. If this applies to both the surface phases then the calculated value of $\Delta\chi$ suggests a small positive value for the surface potential of water.

The equilibrium constant L for the reaction

$$2D^+ \cdot nD_2O + [2n + 1]H_2O = 2H^+ \cdot nH_2O + [2n + 1]D_2O \quad (60)$$

is of interest in the interpretation of the effect of deuterium in the aqueous solvent on rate constants of acid catalysed reactions and on the ionization constants of weak acids. Normally the hydrated proton (or deuteron) is formulated as H_3O^+ with $n = 1$ but the reaction as written above preserves generality. The equilibrium constant L is independent of the value of n. Comparison of the emf's of pairs of cells with the electrolyte dissolved in light and heavy water respectively,[163] e.g.

$$Pt(D_2)|DCl \text{ in } D_2O|AgCl|Ag$$

$$Pt(H_2)|HCl \text{ in } H_2O|AgCl|Ag$$

gives the free energy changes involved in the process

$$H_2(g) + 2D^+ \cdot nD_2O + 2Cl^-(D_2O) + 2nH_2O$$
$$= D_2(g) + 2H^+ \cdot nH_2O + 2Cl^-(H_2O) + 2nD_2O \quad (61)$$

Combination of this with the free energy for the reaction:[171,172]

$$D_2(g) + H_2O = H_2(g) + D_2O \quad (62)$$

which is known from thermochemical and spectroscopic data, gives the free energy of the equilibrium

$$2D^+ \cdot nD_2O + Cl^-(D_2O) + [2n + 1]H_2O$$
$$\rightarrow 2H^+ \cdot nH_2O + Cl^-(H_2O) + [2n + 1]D_2O \quad (63)$$

for which the equilibrium constant L' is 18·2 at 25°C.

It is evident that

$$\frac{L}{L'} = \frac{[Cl^-(D_2O)]}{[Cl^-(H_2O)]} \quad (64)$$

and consequently

$$-RT \ln (L/L') = 2^{H_2O}\Delta^{D_2O}\alpha^0_{Cl^-} \quad (65)$$

The preliminary experimental value of $^{H_2O}\Delta^{D_2O}\alpha_{Cl^-} = 80 \pm 50$ cal mole^{-1} corresponds to $L = 13·9 \pm 2·7$. In former estimates of L from emf data

$_{H_2O}\Delta^{D_2O}\alpha_{Cl^-}$ was either assumed to be very small or it was effectively equated to the *chemical* free energy of transfer and calculated theoretically from spectroscopic data of uncertain reliability. Independent estimates of L by NMR and fractionation between solution and vapour give rather lower values of L in the range 8·0–11·4.

III. ALCOHOLS AND ALCOHOL-WATER MIXTURES

1. Potentiometric Determination of Solvation Energies

Compared with water, thermodynamic data for electrolytes in non-aqueous solvents are very sparse. This is in part due to neglect of the electrochemistry of ionic solutions in these media and in part a result of the nature of the solvents themselves. No other solvent is as universal as water even though several of the more common e.g. N methyl formamide and propylene carbonate have very high dielectric constants. The chief difficulty, apart from the low solubilities of many salts, is the lack of suitable reversible electrodes. It is frequently found that the silver halide and mercurous salt electrodes which work well in aqueous media are very much less reliable in other solvents. The usual reason is that in, for example, amides and nitriles silver halides are very soluble in solutions containing a moderate concentration of halide ion 0·01–0·1 M, say, due to the formation of complex silver halides. Mercurous salt electrodes are frequently rendered useless by an increase in the stability of the mercuric state with decomposition or disproportionation of the salt and sometimes oxidation of the solvent also. In certain studies useful results can be obtained by limiting the solution concentration to that below which complex formation is negligible. For results of the highest precision variable potentials may have to be extrapolated back to the time of immersion of the electrodes in the solution.[174] When there is no alternative, reference electrodes which have a fixed potential but lack thermodynamic reversibility are often used. They are usually of the type

$$Ag|AgX(sat)Et_4NX|$$

and contain a reproducible, though unknown, liquid junction potential. An even more common procedure is to use an aqueous calomel reference electrode. Of course no meaningful thermodynamic calculations can be made from potentials measured in this way owing to the indeterminate liquid junction potential between aqueous and non aqueous solvent but useful comparison of half-wave potentials for example may be made in this way provided that care is taken that the boundary potential is constant.

More is known about the thermodynamics of ion solvation in methanol and ethanol than in any other solvents and the methanol + water system has been extensively studied. The data have been collated and discussed by several authors.[175,176,177] Free energies of transfer of the hydrogen halides from water to alcohol or mixed solvents are obtained directly from the difference of standard potentials of the corresponding silver halide electrodes in the two liquids. The standard potential of amalgam cells such as[178,179]

$$Ag|AgCl|MCl(S)|M(Hg)|MCl(H_2O)|AgCl|Ag$$

is directly proportional to the free energy of the transfer reaction

$$MCl(H_2O) \rightarrow MCl(S)$$

Schug and Dadgar[180] have measured the energy of the alkali halide ions relative to that of the sodium ion by measuring the partition equilibrium between an amalgam of two different alkali metals and the corresponding alkali halide solution in aqueous and methanolic solution, that is

$$Na(Hg) + \frac{1}{z} M^{z+}(S) = Na^+(S) + \frac{1}{z} M(Hg) \tag{66}$$

Time and amalgam flow-rate dependence of the potentials of amalgam cells has frequently been noted and is presumably due to dissolution of the amalgam. The effect is worse in methanol than in water because of the lower hydrogen over voltage in this solvent. Although a method for correcting for amalgam corrosion has been suggested,[181,182] some of the published data for such cells are unreliable.

An alternative method of determining the free energy of transfer is to measure the saturation solubility of a salt in both solvents. Provided that the dissolved electrolyte is in equilibrium with the solid salts and not crystalline solvates like NaI . 2H$_2$O the free energy of transfer is given by:

$$\Delta G_t^0 MX = 2RT \ln \left(\frac{a_{sat}(H_2O)}{a_{sat}(CH_3OH)} \right) \tag{67}$$

The method is only of value when the activity coefficients of the salt in the saturated solutions can be measured or estimated with reasonable accuracy.

2. Real Solvation Energies of Ions in Methanol-Water Mixtures

The real solvation energy of an ion in a non-aqueous solvent may be measured in exactly the same way as are hydration energies described in Section II.4. In practice, however, it is much easier to determine the real free energy of transfer from water to the other solvent. Rybkin,[183] and Case and Parsons[173] have independently measured the transfer energies of chloride

Table 5. *The Molal Real Free Energies of Transfer of Ions from Water to Methanol-Water Mixtures in kcal/mole*[173]

Weight % of Methanol	0	10	20	30	40	43·12	50	60	70	80	87·68	90
H^+	0	4·64	6·50	7·26	7·76	7·90	8·17	8·46	8·60	8·61	8·43	8·29
Li^+	0	4·48	6·16	6·73	7·03	7·08	7·21	7·30	7·30	7·21	7·09	7·07
Na^+	0	4·37	5·92	6·36	6·53	6·56	6·57	6·50	6·37	6·05	5·81	5·72
K^+	0	4·33	5·86	6·26	6·36	6·37	6·38	6·11	5·75	5·23	4·68	4·51
Rb^+	0					6·50					5·54	
Cs^+	0					6·52					5·55	
Cl^-	0	−4·80	−6·80	−7·70	−8·35	−8·52	−8·90	−9·40	−9·85	−10·30	−10·63	−10·75
Br^-	0	−4·74				−8·25				−10·15		
I^-	0	−4·67				−7·82				−9·16		

ion from water to mixtures of methanol and water, ethanol and water, and various other solvents. The cell used was

$$Ag|AgCl|MCl(H_2O)|N_2|MCl(S)|AgCl|Ag$$

set up according to Randles'[100] modification of the Kenrick apparatus. The measured compensation potentials, when corrected to unit activity of chloride ion in both solvents gives the standard free energy of transfer of the ion as:

$$FE = [\alpha^0_{Cl^-}(S) + RT \ln a_{Cl^-}(S)] - [\alpha^0_{Cl^-}(W) + RT \ln a_{Cl^-}(W)] \quad (68)$$

Since the activity coefficient of a single ion is not a measurable quantity a slight ambiguity is introduced by the necessary use of the mean ionic activity coefficient. Provided the solutions are dilute, however, and particularly in the Debye-Hückel region, the error should be very slight.

Real transolvation energies for chloride ion are listed in Reference 173. Values for any other ion are readily calculated from the chemical free energies of transfer of electrolytes derived from potentiometric data. This is readily apparent from the definition of the real free energy of an ion

$$\alpha_i = \mu_i + z_i e \chi$$

The free energy of transfer of a salt MX is

$$^W\Delta^S G^0_t(MX) = (\mu^0_{M^+}(S) + \mu^0_{X^-}(S)) - (\mu^0_{M^+}(W) + \mu^0_{X^-}(W)) \quad (69)$$

thus

$$^W\Delta^S G^0_t(MX) = (\alpha^0_{M^+}(S) + \alpha^0_{X^-}(S)) - (\alpha^0_{M^+}(W) + \alpha^0_{X^-}(W)) \quad (70)$$

and the terms involving the difference in the surface potentials of the solvents cancel out. On the basis of the values of $\alpha^0_{Cl^-}(S) - \alpha^0_{Cl^-}(W)$,[173] real free energies of transfer of several ions have been calculated as a function of the composition of the methanol-water mixtures and are given in Table 5. Several sets of data are available for the free energy of transfer to pure methanol and these are included in Table 6. Those in columns 2 and 3 are based on amalgam potential measurements, those in the fourth column on amalgam partition equilibrium, and the fifth gives values derived from salt solubilities.[184] There is close agreement between the value for lithium found by Hartley and Macfarlane with the result expected from extrapolation of Åkerlof's data which extends up to 90 per cent methanol. Basing the relative solvation energies of the alkali ions given by Schug and Dadgar on the value for Na^+ determined by Hartley produces satisfactory agreement with the

Table 6. *Molal Real Free Energies of Transfer of Ions from Water to Methanol*

		$-^W\Delta^M\alpha_i^0$ kcal/mole			
Ion	(a)		(e)	(f)	(g)
H^+	5·84	—	—	—	—
Li^+	—	7·00 (b)	—	—	—
Na^+	—	6·17 (c)	5·82	(6·17)	6·30
K^+	—	5·73 (d)	4·92	5·77	5·88
Rb^+	—	5·54 (d)	—	5·76	5·75
Cs^+	—	—	—	5·75	5·73
Cl^-	−11·20	—	—	—	—
Br^-	−10·68	—	—	—	−10·45
I^-	−9·70	—	—	—	−9·6

All values are based on $^W\Delta^M\alpha_{Cl^-}^0 = -11\cdot20$ kcal/mole.
References: (a) 173; (b) 185; (c) 186; (d) 182; (e) 179; (f) 180; (g) 184.

other data. It seems that errors due to amalgam dissolution have not been eliminated in Gladden and Fanning's work but the average values from electrode potential measurements are close to those derived from solubilities so that one can have reasonable confidence in them. The *absolute* real solvation energies of these ions in the methanol-water mixtures are easily obtained from the absolute real hydration energies given by Randles.[100]

3. Individual Ion Solvation Energies in the Methanol-Water System

A. Treatments Based on the Born Model. Although real solvation energies of single ions are readily measured there is no thermodynamic method of obtaining individual 'chemical solvation' energies as was noted above for hydration energies. Just as for the aqueous system various attempts have been made to divide the salt solvation energies on the basis of some model of the ion in solution.[187]

The free energies of alkali chlorides and hydrogen halides are all higher in methanol-water mixtures than in water as predicted by the Born equation. Feakins and Watson[175] showed that the transfer energies of the hydrogen halides and the alkali chlorides were related in a reasonably linear manner to the inverse of the anion and cation radii respectively:

$$\Delta G_t^0(H^+X^-) = \Delta G_t^0(H^+) + ar_a^{-1}$$
$$\Delta G_t^0(M^+Cl^-) = \Delta G_t^0(Cl^-) + br_c^{-1}$$

The slopes of the lines are many times greater, and for the alkali chlorides are opposite in sign to those given by the Born relation. Data representative of the latest emf measurements[188] are shown in Fig. 7. By extrapolation of the hydrogen halide lines to $r_a^{-1} = 0$ values of $\Delta G_t^0 H^+$ are obtained. Independent proton transfer energies may be derived from the alkali chloride plots which give the chloride ion transfer energies at different solvent compositions. The two sets of values, Fig. 8, both show the proton to be preferentially solvated in the methanol-water mixtures and both show clear

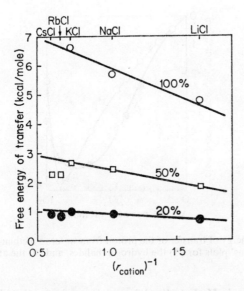

Fig. 7. Molar free energies of transfer of alkali chlorides from water to methanol-water mixtures (weight per cent) as a function of the reciprocal cation radii.

minima at high methanol concentration. Qualitatively this accords with the conclusions of de Ligny and Alfenaar[194] in their more detailed method of estimating single ion transolvation energies described below. This method also makes some allowance for the deviation of the values for rubidium and caesium chlorides from the simple Feakins' plot. Further discussions on the influence of solvent structure on the free energies enthalpies and entropies of transfer of ions from water to methanol-water and other solvent-water mixtures are available.[189,190]

Using the same method of extrapolation as described above for aqueous solutions Izmailov[134] finds the free energies of both anions and cations to be more positive in methanol than in water.

For the proton

$$^W\Delta^M G_t \mathrm{H}^+ = +4 \text{ kcal/mole}$$

It should be noted that Izmailov worked with the full solvation energy so that his estimates of transfer energies, being the difference of two values of about 250 kcal, may not be very reliable.

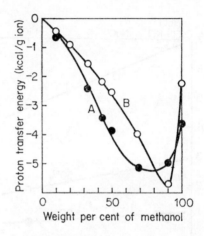

Fig. 8. Molar free energy of transfer of the proton from water to methanol-water mixtures, derived from Feakins' plots for (A) the hydrogen halides, and (B) the alkali chlorides.[188]

B. Organic Ions in Hydroxylic Solvents. Support for the idea that cations may be rather more strongly solvated and have lower free energies in organic solvents containing an electronegative oxygen atom than they are in water comes from Grunwald's work on 50 weight per cent dioxane-water mixtures.[191] The change of the standard free energy of electrolytes dissolved in this solvent when the mole fraction of water x_w was changed was determined from vapour pressure measurements. Division into contributions of the single ions can be made on the basis of the dG/dx_w values for the model ions $\mathrm{Ph_4P^+}$ and $\mathrm{Ph_4B^-}$. It was assumed that the solvation energy of these large ions is equal to that of an uncharged molecule of the same size plus the electrostatic energy of charging this molecule. Thus

$$\frac{dG}{dx_w} = \frac{dG_{\text{neut}}}{dx_w} - \frac{Ne_0^2}{2r\varepsilon}\left(\frac{d\ln\varepsilon}{dx_w}\right) \tag{73}$$

After inserting the appropriate experimental values for $d \ln \varepsilon / dx_w$ and estimating the radius r as 4·2 Å it is found that

$$\frac{dG_{Ph_4P^+Ph_4B^-}}{dx_w} = 2 \frac{dG_{neut}}{dx_w} - 9 \cdot 6 \text{ kcal} \tag{74}$$

$$= 29 \cdot 4 \text{ kcal}$$

Thus

$$\frac{dG_{neut}}{dx_w} = 19 \cdot 5 \text{ kcal}$$

which is in very good agreement with the value of 20·4 kcal for $dGPh_4C/dx_w$ and may therefore be considered to justify the original assumption. Accordingly, by equating dG/dx_w for the Ph_4P^+ and Ph_4B^- ions values for a variety of other ions can be calculated.

For cations the slopes of free energy versus concentration of water are positive indicating a lower free energy state in dioxane than in water. The anions on the other hand lose the stabilization by hydrogen bonding which is possible in water but not in dioxane.

Following Feakin's[175] suggestion that the discrepancy in his results might be due to neglect of the contribution of the energy of transfer of the 'uncharged' ion, de Ligny and Alfenaar[192] considered this point in greater detail. Employing the same type of extrapolation they subtracted from the free energy of transfer a 'neutral' contribution which was quite reasonably assumed to be proportional to the area of the ion; that is

$$\Delta G_t(\text{neut}) = kr^2 \tag{75}$$

The proportionality constant k was taken from data on the solubilities of methane and the noble gases in methanol-water mixtures. Subsequently[193] even better estimates for $\Delta G_t(\text{neut})$ were made using the free energies of transfer of the approximately spherical molecules carbon tetrachloride, ferrocene and tin tetramethyl. The values of $\Delta G_t^0(H^+) - \Delta G_{t,el}^0(M^+)$ and $\Delta G_t^0(H^+) + \Delta G_{t,el}^0(X^-)$ so obtained are plotted against the reciprocal ionic radii. The extrapolated curves meet at $1/r = 0$ with the limiting slopes required by the Born equation. Using data for the alkali halides alone the extrapolation is rather long with consequent uncertainty in the intercept value of $\Delta G_t^0(H^+)$. Inclusion of $\Delta G_t^0(H^+) - \Delta G_{FIC}^0 + \Delta G_{FOC}^0$ where FOC = Ferrocene, FIC = Ferrocinium ion, determined by measuring the standard potential of the ferrocene-ferrocinium ion couple in methanol-water mixtures, makes the method rather more convincing. The large radius of the ferrocinium ion means that a data point is obtained much nearer the $1/r$ origin than the points for the alkali halide ions, permitting a more accurate

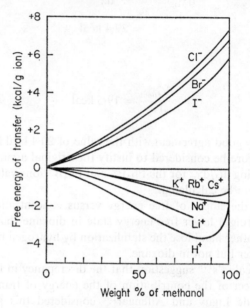

Fig. 9. Molal free energies of transfer of alkali halide ions from water to methanol-water mixtures (after Alfenaar and de Ligny[193]).

extrapolation. The transfer energies of the alkali halide ions derived in this way are shown in Fig. 9. Monotonically increasing free energies are found for the halide ions and the reverse for cations except the proton which is most stable in an 80 mole percent alcohol solution.

Table 7. *Molal Standard Free Energies of Transfer of Single Ions from Water to Methanol at 25°C in kcal/mole*

Ion	H^+	Na^+	K^+	Rb^+	Cs^+	Cl^-	Br^-	I^-
$^W\Delta^M G_t$ (a)	−0·1	−0·55	−0·17	−0·08	+0·01	+5·46	+4·81	+4·03
(b)	−2·53	−2·39	−1·78	−1·77	−1·84	+7·33	+6·82	+5·79

References: (a) 184; (b) 193.

From their work on the solvation of ferrocene de Ligny, Alfenaar, and van der Veen[194] predict a negative value of -0.3 V for the surface potential of water. They suggest that this is compatible with the negative value of the temperature coefficient of the surface potential. The increase in the number of unbonded water molecules at the surface with increasing temperature must more than offset the disordering effect of the rise in temperature. This conclusion is opposite to that of Randles and Schiffrin[195] and the apparent corroboration from calculation of the number of unbonded molecules according to the Némethy and Scheraga model is suspect due to error in that treatment.[38]

Whilst the work of de Ligny and Alfenaar provides the most plausible and detailed attempt to obtain single ion values it still involves a non-linear extrapolation and the electrostatic field of the ions will almost certainly alter the solvent structure in a different way from uncharged atoms and molecules so that the correction for the 'neutral' contribution can only be an approximation.

Since the free energies of transfer of sodium and potassium chlorides change almost linearly with the composition of the methanol-water solvent, at least up to about 90 per cent of methanol, it seems likely that this could be so for the chloride ion alone. In this case, if the chloride ion transfer energy were about $+3$ kcal/mole the proton would be more stable in the mixed solvent than in water up to approximately 80 weight per cent of alcohol followed by a rapid decrease in solvation in the alcohol rich solvents and a higher energy in methanol than in water. This reversal is not unlikely when the depolymerizing effect of methanol on water is considered.[176]

4. Correlation with Surface Properties

It is fortunate that extensive surface tension measurements which help to define the nature of the solution-gas surface layer have been made on methanol-water mixtures.[196] Knowledge of the composition of the surface phase helps in the interpretation of the experimental real free energies of ions since, as shown by Mackor,[197] it may lead to a separation of bulk and surface contributions and may permit a reasonable estimate of the variation of the surface potential drop to be made.

Kipling[198] has reanalysed the surface tension data of alcohol-water mixtures and has shown that the results are consistent with a simple monolayer model. Adsorption isotherms of methanol and water against the concentration of alcohol in the bulk solution are shown in Fig. 10. There is a marked similarity between the form of the methanol isotherm and the variation of the real free energy of transfer of the chloride ion. The surface concentration of methanol rises rapidly to about 80 per cent of a complete monolayer as the bulk mole fraction increases to 0·3. Further increase in the bulk mole fraction causes a relatively minor change in surface composition.

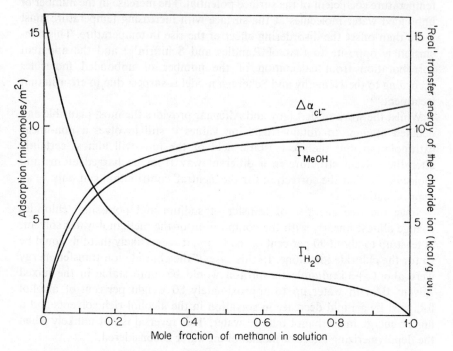

Fig. 10. Surface composition of methanol-water mixtures, and the 'real' free energy of transfer of the chloride ion from water to these solutions.

We may suppose that in the region of bulk composition where the mole fraction is between about 0·4 and 0·9 the surface composition scarcely changes. If the surface potential is similarly invariant in this range the change in $\alpha^0_{Cl^-}$ will equal the change in $\mu^0_{Cl^-}$. It has been noted above that the variation of $\mu^0_{Cl^-}$ is probably an almost linear function of the molar composition. From the slope of the linear region of the real free energy plot

$$^W\Delta^M\mu^0_{Cl^-} = + 4 \text{ kcal mole}^{-1}$$

or

$$^W\Delta^M\mu^0_{H^+} = + 1·4 \text{ kcal mole}^{-1}$$

This value for the transolvation energy of the proton is approximately the mean of the estimates of Izmailov and of de Ligny and Alfenaar. Cation free energies of transfer on this scale are small, ranging from 0·9 kcal for Na^+ to 1·5 kcal for Cs^+ and are not markedly of opposite sign to that given by the Born equation.[199] Values proposed by Feakins and by de Ligny would imply a minimum in the surface potential of water-methanol mixtures which seems unlikely.

Differential real free energies of transfer (see Section V.4) of the ferrocinium ion between water and methanol-water mixtures have been calculated

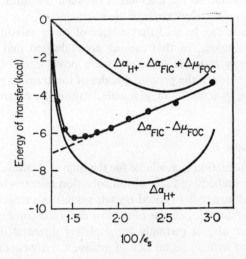

Fig. 11. Energy of transfer of the ferrocene-ferrocinium ion couple from water to methanol-water mixtures as a function of the dielectric constant of the solvent.

by Parsons[200] from values of $(\Delta\mu_{H^+} - \Delta\mu_{FIC} + \Delta\mu_{FOC}) = \beta$ measured by Alfenaar:

$$\delta_{FIC} = \Delta\alpha_{FIC} - \Delta\mu_{FOC} = \Delta\alpha_{H^+} - \beta \tag{76}$$

δ is the sum of the electrostatic and surface contributions to the real solvation energy. By analogy with the behaviour of the chloride ion it might be expected that the variation of δ would be approximately linear above $x(MeOH) = 0·4$ and that the change in the electrostatic solvation energy of this large ion approximates to the Born solvation energy. When δ is plotted against $1/\varepsilon_s$, the reciprocal of the solvent dielectric constant, a linear relation with a slope of $+3·5$ kcal mole^{-1} is indeed observed for the alcohol-rich mixtures.

Since $^W\Delta^M\alpha_{FIC} = -3.5$ kcal/mole a value of $F^W\Delta^M\chi = 7.0$ kcal/mole is found, in exact agreement with the estimate from the real transfer energies of the chloride ion. However the slope of the δ_{FIC} curve in Fig. 11 is some four times larger than that predicted by the Born equation if a value of 3·8 Å[201] is taken for the radius of the ferrocinium ion. Furthermore, even if the assumption of a region of constant surface composition is valid it is possible that χ will change due to inductive effects and the effect of chain formation on the charge distribution in the methanol molecules in the surface layer.[202] Though the composition of the surface layer may not change the bulk composition does and so the interaction between the surface molecules and the solvent molecules below changes also.

In conclusion it can be said that methanol water mixtures appear to be more basic than water, in that cations are stabilized and anions are in a higher free energy state. It does not seem possible at present to choose unambiguously between the proposed scales of ion transfer energies and consequently unified potential and pH scales cannot be regarded as definitely established.

5. Higher Alcohols

Much less information is available for the higher alcohols. Izmailov[101] has published a comprehensive table of ion solvation energies in alcohols up to $C_5H_{11}OH$, which are mainly based on salt solvation energies determined by solubility measurements. Because of the low dielectric constants of the higher alcohols salts are almost certainly incompletely dissociated in these media and this, together with a general lack of precise activity coefficient data limits the accuracy of the solvation energies. Case and Parsons[173] have measured the real free energies of transfer of chloride ion from water to ethanol-water mixtures and to butanol. Real free energies of solvation of a variety of ions based on these measurements are given in Table 8 together with values for methanol and water for comparison. The trend of real free energy of transfer with composition of the ethanol-water mixtures is very similar to that of the methanol system. According to Kipling's calculations for ethanol[198] the surface composition is approximately constant from 20–60 mole per cent of alcohol. On the basis of similar assumptions to those used for methanol it is estimated that

$$\mu^0_{H^+}(EtOH) - \mu^0_{H^+}(H_2O) = 1.6 \text{ kcal mole}^{-1}$$

and

$$\chi(EtOH) - \chi(H_2O) = -0.38 \text{ V}$$

Popovych and Dill[203] have assigned single ion transfer energies for these solvents on the assumption, which is basically that due to Grunwald,[191] that a reference anion and a reference cation have equal transolvation energies. The reference electrolyte chosen in this case was tri isoamyl-n-butylammonium tetraphenylborate. A remarkably similar trend to that observed with methanol-water mixtures was deduced with the solvation energy of the proton being lower than in the pure solvents and passing through a minimum at about 65 per cent of ethanol.

The similarity in the change of real free energy of solvation of the chloride ion in the higher alcohols compared with water, together with their similar dielectric constants suggests that the values of the proton transfer energy will be the same for propanol and n-butanol as for ethanol within the accuracy of present measurements.

Table 8. *Molal Real Free Energies of Solvation of Ions at* 25°C
$-\alpha_i^0$ kcal/mole

	H_2O (a)	MeOH (b)	EtOH (b)	n-BuOH (b)	HCOOH (b)	F (b)	N.M.F. (c)	MeCN (b)
H^+	260·5	266·3	265·9	264·5	253·8	264·0	—	257·8
Li^+	122·1	128·8	127·1	127·5	124·3	—	129·5	121·1
Na^+	98·2	104·1	102·0	101·5	102·3	—	105·8	98·8
K^+	80·6	86·1	86·7	82·0	83·8	83·2	88·3	83·0
Rb^+	75·5	80·8	80·7	73·0	80·8	77·9	—	78·2
Cs^+	67·8	76·5	72·7	70·0	71·8	—	84·8	70·3
F^-	103·8	—	—	—	—	—	—	—
Cl^-	75·8	64·6	63·4	63·6	73·3	71·9	65·2	60·8
Br^-	72·5	61·9	60·7	63·1	67·8	—	63·2	61·8
I^-	61·4	51·7	50·9	49·2	54·9	—	—	53·7
Tl^+	82·0	87·1	87·3	—	—	85·8	—	—
Ag^+	114·5	120·9	120·8	—	112·9	—	—	124·6
Cu^+	136·2	—	—	—	—	—	—	154·0
Zn^{2+}	484·6	494·8	489·2	—	495·6	491·6	—	477·5
Cd^{2+}	430·5	442·8	439·7	—	434·6	438·0	—	427·6
Cu^{2+}	498·7	507·3	513·3	—	—	508·7	—	519·6
Pb^{2+}	357·8	372·4	369·3	—	—	—	368·2	351·6
Ca^{2+}	380·8	—	—	—	368·8	—	—	368·8

F = Formamide.
N.M.F. = N-Methyl Formamide.
References: (a) 204; (b) 173, 174; (c) 205.

IV. AMIDES

1. Solvent Structures

The simple aliphatic amides formamide (F), N-methyl formamide (NMF) N-N-dimethyl formamide (DMF) and the corresponding acetamides are good solvents for both inorganic salts and a wide range of organic solutes. The high dipole moments of the amides, Table 9, are a consequence of structural properties which are also primarily responsible for the specific solvation of electrolytes.

Table 9. *Dielectric Constants and Dipole Moments of some Solvents at 25°C*

	ε	Ref.	μ (Debye)	Ref.
Water	78·3	—	1·69	213
Methanol	32·6	—	1·84	213
Formamide	109	214	3·71	215
N-Methyl Formamide	171–182	214, 216	3·82	217
Dimethyl Formamide	37	214	3·80	217
Acetamide	(74)	216	3·75	217
N-Methyl Acetamide	(86)	216	3·71	217
Dimethyl Acetamide	38	216	3·80	217
Acetonitrile	36·2	218	3·37	219
Dimethyl Sulphoxide	46·6	220	3·9	221
Sulpholane	44	222	—	—
Propylene Carbonate	65·2	224	—	—
2,4 Dimethylsulpholane	29·5	223	—	—

Values in parentheses are extrapolated.

Solid formamide is composed of layers of molecules held together by van der Waals forces.[206] Each layer consists of polymeric hydrogen-bonded chains cross linked by weaker H bonds between the chains. In the liquid phase the layered structure is broken down but the extensive chains are not disrupted. Acetamide also forms a three-dimensional network but with only one type of H bond. Six-membered rings are the basic component of this structure.[207] N-methyl acetamide has a layered structure with hydrogen bonding between the amine hydrogen atoms and the carbonyl oxygen atoms of adjacent layers.[208]

Infra-red studies of the monosubstituted amides support a predominance of the trans configuration in the liquid chains.[209] This is also the conclusion drawn from NMR investigations.[210] Although no evidence was found for cis forms in the acetamides it appears that the fraction of cis isomers increases with the size of the alkyl substituent in the formamides. H bonding with the carbonyl oxygen will produce linear polymers in the case of trans disposition (a) and dimers when in the cis orientation (b)

The fact that the apparent dipole moment of N-methyl formamide in an inert solvent decreases on dilution shows that the linear form is preferred.[211] A dimeric form as in (b) would presumably cancel out the molecular moment; increasing dilution would lead to an increased proportion of monomers and hence an increased apparent dipole moment. Mitzushima[212] has demonstrated from the effect of temperature on the Raman spectrum of NMA that there is no rotation about the C—N bond and therefore the existence of the dimeric form is ruled out. The extremely high dielectric constants of N-methyl formamide and acetamide must be due to the extensive polymerization in which the molecular dipoles are aligned virtually parallel to the chains.

The relatively high values of the dielectric constants of dimethyl formamide and dimethyl acetamide, which are not hydrogen bonded, are produced by the high intrinsic dipole moments of the solvent molecules. Dimethyl sulphoxide is another example of a non H-bonded solvent with a high dielectric constant $D \simeq 47$.

2. Factors Influencing Solvation

Besides the ion-dipole interaction between solute and solvent H bonding promotes anion solvation and π complex formation favours cation solvation.

The electron-donating ability of the amide carbonyl oxygen is increased by delocalization of the carbonyl π bond charge with a concomitant increase in the C—N bond order. Effectively the oxygen receives a fractional negative charge and the nitrogen becomes slightly positive

$$\begin{array}{c} \diagdown \quad \quad \diagup \\ \overset{+}{N}\!=\!\!\!=\!C \\ \diagup \quad \quad \diagdown\!\!\diagdown \\ \quad \quad \quad \quad O— \end{array}$$

This donation of the nitrogen lone pair electrons produces a high π-bond moment and is further increased by alkyl substituents on the nitrogen atoms. Similarly acetamide should have a higher oxygen atom electron density than formamide. These considerations show why cations are extensively solvated by the amides.[225] Although the basicities of the solvents, which have been determined from the phenol-adduct frequency shifts,[226] parallel the order expected from the inductive effect of the N and C substituents, that is:

$$DMA > DMF, \quad NMA > A, \quad NMF > F$$

steric repulsion between the bulky methyl groups is large for the acetamides. Consequently when the co-ordinating ability of the amides with the Ni^{2+} and Cr^{2+} ions in the octahedral complexes $[M(amide)_6](ClO_4)_2$ is considered the order is[196]

$$DMF > NMF > A \gg DMA > NMA$$

rather different from that shown above.

Anion solvation is enhanced by hydrogen bonding. Thus in protic solvents anions are progressively less solvated in the order

$$OH^- > F^- \gg Cl^- > Br^- > N_3^- > I^- > SCN$$

and in water the anion hydration enthalpies are greater than for cations of the same size. In *aprotic* solvents the determining factor is interaction due to mutual polarizability of solvent and anion. The larger the ion the larger is its polarizability and the halide ion energies will then be in the reverse order to that in hydrogen bonding liquids. This is reflected in the relative solubilities of alkali halides in methanol and DMF. Hydrogen bonding causes KCl to be more soluble in methanol than in DMF while KI is more readily dissolved in the amide.[228] Similarly, hydroxides are only slightly soluble in DMF and DMA due to the absence of H bonding with the solvent.

The factors which influence electrolyte solubility also determine the degree of dissociation of dissolved salts. Dissociation is brought about by solvent molecules interposing themselves between anion and cation and thereby shielding their electric fields. Thus the higher the dielectric constant of a solvent the greater will tend to be its dissociating power. Ions, however, frequently modify solvent structure in their vicinity causing a change in effective dielectric constant. Therefore the solvent bulk dielectric constant is not an absolute indicator of its dissociating power. Specific ion-solvent interactions may play an equally important part in enabling solvent molecules to squeeze between the ions in an ion-pair and the extent of electrolyte dissociation may therefore give an indication of the degree of this specific solvation.

Formamide and NMF are both what Janz and Danyluk[229] class as levelling solvents and electrolytes tend to be completely dissociated in them. Ion pairing is indicated in some instances, as for example, in the variation of the heat of solution of lithium chloride as discussed below. Owing to its much higher dielectric constant some salts, magnesium sulphate for example,[230] are completely dissociated in NMF while slight association is evident in formamide. On the other hand DMF is a differentiating solvent in which hydrogen halides,[231] and some bromides, iodides and tetra alkyl ammonium salts are incompletely dissociated.[232] Acetamide is also a levelling solvent in which most simple salts with the exception of zinc, cadmium and mercury halides, which probably form complex ions, are completely dissociated. The behaviour of NMA and DMA parallels that of the corresponding formamides and many salts are incompletely dissociated in the aprotic solvent.

3. Thermodynamics of Solvation

After the alcohols, amides are the most extensively investigated class of electrolyte solvents. Strehlow and Pavlopoulos[181] studied the hydrogen, cadmium-cadmium chloride and several metal electrodes in formamide and calculated the difference of energies of formation of the solvated proton and solvated metal cations. Standard potentials of the silver-silver chloride electrode have been determined in formamide[233] and N-methyl acetamide[234] and activity coefficients and thermodynamic properties of HCl in formamide are available over a range of temperatures.

Amalgam cells of the type:

$$Ag|AgCl|MCl(m)|M(Hg)|MCl(m = 0.1)|AgCl|Ag$$

and

$$M(Hg)|MCl(m)|AgCl|Ag$$

where M = Na or Cs were employed in formamide and NMF by Povarov[235] for the determination of activity coefficients of NaCl and CsCl. Results are close to values obtained by Vasenko[236] from freezing point measurements.

Kessler[237] has used this information to calculate standard free energies of solvation of alkali halides in these solvents. Values for formamide derived from emf measurements of Strehlow and those for NMF from the results of Luksha and Criss[205] on the potentials of metal amalgams against silver chloride and bromide electrodes are given in Table 10. The real free energies

Table 10. *Free Energies of Solvation of Salts in Water, Formamide and N-methyl Formamide*
$$M^{z+}(g) + zX^-(g) \to M^{z+}(soln) + zX^-(soln)$$

$-\Delta G_{solv}$ (kcal mole^{-1})

Salt	Water (a)	Formamide (b)	N-Methyl Formamide (c)
HCl	336·3	335·9	
LiCl	197·9		194·7
NaCl	174·0		171·0
KCl	156·4	154·9	153·5
RbCl	151·3	149·8	
CsCl	143·6		149·9
TiCl	157·8	157·7	
NaBr	170·7		169·0
ZnCl₂	636·2	635·4	
CdCl₂	582·1	581·8	
CuCl₂	650·3	652·5	
PbCl₂	509·4	512·0	

(a) From Rosseinsky.[204]
(b) Pavlopoulos and Strehlow.[181]
(c) From values of $\Delta G_{form} MX_z$ (soln) given by Latimer[280] and Luksha and Criss.[205]

of transfer of chloride ion from water to formamide and N-methyl formamide have been measured by Case and Parsons.[173] In Table 8 values of the real solvation energies of several ions in these solvents are given. Use of revised data for the electron affinities for the halogen atoms[238] causes these values to differ slightly from those quoted previously. Figure 12 shows the real free energies of transfer of the alkali halide ions between water and N-methyl formamide. Although the *absolute* values of the real transfer energies differ from the chemical transolvation energies by a constant (unknown) amount for each ion, this being proportional to the difference of the surface potentials of the two solvents, the trend of the energies is instructive. Cations transfer from water to formamide with a small free energy change which is practically independent of the ion. This is an indication of the similarity in solvent behaviour of the two liquids. Other similarities have been noted above. In contrast, although anion solvation energies in general decrease with increasing size of the ion, in these amides the stabilization of the chloride ion relative to the iodide ion is less than the relative stabilization which occurs in water. Parallel behaviour is observed for anions in methanol and acetonitrile.

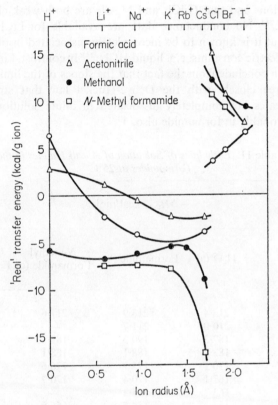

Fig. 12. Real free energies of transfer of ions from water to some non-aqueous solvents at 25°C.

Heats of solution of a variety of salts in formamide, N-methyl formamide, dimethyl formamide and N-methyl acetamide have been determined. Standard enthalpies of solution calculated from solubility data[239] agree well with those obtained by extrapolating to infinite dilution the measured integral heats of dilute solutions except for salts with high solubility like sodium bromide.

Although it might be anticipated that dilute solutions of alkali halides in solvents with high dielectric constants ($D_{NMF} = 182$, $D_{DMF} = 36.7$ at 25°C) would be completely dissociated, ion pairing does seem to occur in the formamides. Limiting slopes of plots of heat of solution against $m^{1/2}$ in agreement with the Debye–Hückel theory is found for KCl and CsCl only in N-methyl formamide.[240] Sodium chloride, bromide, and iodide have slopes between 7 and 10 times the theoretical value while the limiting slope for LiCl is about 280 times as great. Conductance data indicate that ion pairing occurs in LiCl solutions and that $GdCl_3$ and $MgCl_2$ are both weak electrolytes in this medium.[241] No conductance data are available for LiCl in N-methyl formamide but it is known to be incompletely dissociated in other solvents with high dielectric constants, e.g. liquid HCN.[242] In contrast, Finch, Gardner and Steadman conclude from the fact that the slopes of the limiting heats of solutions agree closely with the Debye-Hückel law that strontium and barium chlorates are completely dissociated in dilute solutions in NMF, DMF and probably in formamide also.[243]

Table 11. *Enthalpies of Solvation of Alkali Halides in the Formamides at 25°C*

	$-\Delta H_{solv}$ kcal/mole			
	H_2O (a)	Formamide (b)	N-methyl Formamide (c)	N-N-dimethyl Formamide (d)
LiCl	213·4	213·9	217·6	219·0
LiBr	210·0	211·7	—	219·6
NaCl	187·3	190·3	189·4	—
NaBr	183·9	188·1	188·1	191·1
NaI	170·1	175·7	176·6	182·3
KCl	167·1	170·4	170·9	—
KBr	163·7	168·3	—	—
CsCl	153·3	156·5	156·5	—
CsI	136·1	141·8	—	148·3

References:
(a) 204.
(b) From ΔH_{soln} (F) (Reference 244), ΔH_{soln} (H_2O) (Reference 136), and ΔH_{solv} (H_2O) (Reference 204).
(c) From ΔH_{soln} (NMF) (Reference 241).
(d) From ΔH_{soln} (DMF) (Reference 240).

Some general solvation effects are evident from Fig. 13 where the transolvation enthalpies of the alkali halides have been separated into single ion values on the arbitrary basis of $^W\Delta^{DMF}H^0_{Cs^+} = 0$. Both free energy and enthalpy data show a *lesser* decrease in solvation in the usual sequence

$$F^- > Cl^- > Br^- > I^-$$

in the amide solvents than in water. Such behaviour is due to the exceptionally strong stabilization of the smaller anions in water by hydrogen-bonding.

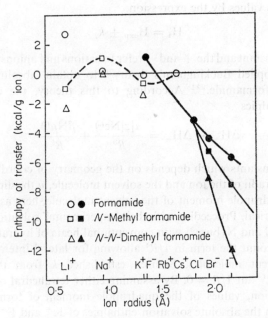

Fig. 13. Standard enthalpies of transfer of ions from water to the formamides; values based on $^W\Delta^{DMF}H_{trans}Cs^+ = 0$.

The relative increase in stabilization of the larger anions increases from formamide to dimethyl formamide. This is the order of decreasing hydrogen bonding ability of the solvents and forces due to mutual polarizability of ion and solvent in the order

$$F^- < Cl^- < Br^- < I^-$$

therefore play an increasing part in stabilization of the anions. Anion solvation enthalpies are linearly related to the reciprocal anion radii for alkali halides in formamide[244] as in water[111] and methanol.[175] This relation with ion size does not hold for the cations. Hydrogen-bonding is the most import-ant stabilizing effect for anions in each of these solvents and causes the sol-vation energy to vary in the same way as the Born electrostatic term. An 'experimental' scale of single ion enthalpies can be established for formamide by plotting the enthalpies of solvation of a series of halide salts with the same cation against the reciprocal of the anion radii. Just as in water the inter-action energy of an anion with the solvent is greater than for a cation of the same size.[46]

Assuming that the experimental solvation enthalpies so derived are related to the absolute values by the expression

$$H_i = H_{exp} \pm k \qquad (77)$$

where k is a constant and the $+$ and $-$ refer to cations and anions respectively, Somsen has applied Buckingham's theory[76] to obtain 'absolute' solvation enthalpies in formamide.[244] According to this theory, the difference in solvation enthalpies

$$\Delta H_+ - \Delta H_- = \frac{\alpha|z|Ne\Theta}{R^3} + \frac{\beta N\mu\Theta}{R^4} \qquad (78)$$

α and β are constants which depends on the geometry of coordination, R is the sum of the radii of the ion and the solvent molecule, μ the dipole moment and Θ the quadrupole moment of the solvent molecule, here assumed to be axially symmetrical. Proceeding with their 'experimental' solvation enthalpies as did Halliwell and Nyburg[97] with conventional heats of hydration but also taking into account the term in $1/R^4$ allowing for lateral interaction of co-ordinated solvent molecules, Somsen established k from the value of $H_{exp}^{(+)} - H_{exp}^{(-)}$ at $1/R = 0$. By assuming either tetrahedral or octahedral primary solvation, values of the quadrupole moment of formamide were calculated from the absolute solvation enthalpies of K^+ and F^- which both have a radius of 1·33 Å. When repulsive forces as well as those considered by Buckingham are considered ions appear to be tetrahedrally coordinated in water but octahedrally solvated in formamide.

Criticisms of this method[133] apply equally well to formamide as to water. It is certainly unlikely that formamide behaves as an axially symmetrical molecule. Moreover, part at least of the difference of solvation energies and enthalpies of equal sized anions and cations may be due to the surface potential of the solvent. This term cancels out in Halliwell and Nyburg's type of extrapolation to find the proton hydration energy but not in Somsen's method. Calculations of quadrupole moments made in this way therefore seem doubtful.

From the limited data available[243] the solvation enthalpy of chlorate ion appears to be less than that of the larger bromate ion in both water and formamide. Presumably the solvent-solvent hydrogen bonding energy is greater than the solvent-ion interaction for these large anions. The usual decrease of solvation energy with ion size is reversed, the larger more polarizable ion which causes less disturbance to the hydrogen bonded solvent structure being preferentially solvated.

An interesting theory by van Eck[245] helps to explain the observed enthalpies of solvation of the alkali cations. The total enthalpy of solvation of an ion is regarded as being the sum of three terms (a) an electrostatic term dependent on the charge ΔH_{el}, (b) a 'neutral' term, ΔH_{neut}, and (c) the enthalpy required to make a hole in the solvent to accommodate the ion, ΔH_h.

$$\Delta H_{solv} = \Delta H_{el} + \Delta H_{neut} + \Delta H_h \tag{79}$$

For inert gas type cations $-\Delta H_{el}$ can be equated to the sum of the ionization potential and the electron affinity of the corresponding metal atoms

$$-\Delta H_{el}(M^+) = I_M + A_M \tag{80}$$

Thus the enthalpy of transfer of an alkali cation between two solvents A and B should be merely equal to the difference between the neutral and cavity formation terms

$$^A\Delta^B H_t(M^+) = {}^A\Delta^B[\Delta H_{neut} + \Delta H_h] \tag{81}$$

van Eck has shown that $[\Delta H_{neut} + \Delta H_h]$ is constant for different ions in water. These terms are, in any case, small compared with the total enthalpy of solvation. From the near constancy of the enthalpies of transfer of ions Na^+, K^+, Rb^+, and Cs^+ from water to F, NMF, and DMF in Fig. 13 this would seem to be true for the formamides also. Lithium differs from the other cations probably because of different coordination in water and the amides. This would mean that the electrostatic contribution to the solvation enthalpy would no longer be the same in each solvent. The enthalpy of transfer of lithium ion between NMF and DMF is, however, equal within experimental error to that of the other alkali cations.

While the absolute solvation enthalpies of the alkaline earth cations decrease with ion size in the formamides they become *relatively* more stable in that order compared with the same ions in water. In fact they behave similarly to the halide ions. The alkaline earth chlorate solvation enthalpies increase uniformly from water to DMF, this being the order of decreasing solvent structure. A neat explanation is provided by the model of van Eck. Solvation of these cations may be related to the ease with which the solvent can accommodate itself to the dissolved ions and the energy necessary to form a cavity will be smallest for the least structured medium.

4. Ion Mobilities

A considerable body of information has been built up on the conductivity of electrolytes in the amides. From these data limiting ion mobilities at infinite dilution may be derived. In that they reflect both the effective radii of ion plus 'primary' solvating solvent molecules and the viscosity of the medium which the ion sees in its immediate environment, these ionic mobilities provide an insight into the correlation between solvation and the effect of ions on the solvent structure.

When the ratio of crystallographic to Stokes' Law radius is plotted against this radius for tetra-alkylammonium ions the curves exhibit maxima for solutions in F, and NMA just as in water.[246] It has been suggested that ions of a certain size may slip between solvent molecules so that the viscous retardation force is less than would be calculated from Stokes Law using the bulk viscosity but the higher charge density on smaller ions results in increased ion-solvent attraction.[247] Walden products of ions in the amides have also been interpreted in terms of solvation and the effect of the ions on solvent structure. High values of the Walden product such as are shown by the tetra-alkylammonium ions in formamide[246] indicate that these large organic ions cause substantial disruption of the solvent structure with a resultant lowering of viscosity in their neighbourhood. Other cations have lower Walden products in the amide than in water. This is probably due to the larger size of the solvent molecule which means that the effective radius of the solvodynamic species will be larger assuming the same degree of solvation.[248] N-methyl formamide appears similar to formamide in that Walden products for simple alkali metal salts are low[249] while those for Et_4NBr are higher than the corresponding values for water. Walden products for the alkali halide ions in DMF are lower than for the other formamides.[250] Although this could be accounted for by assuming the solvated ions to be even larger than in the other solvents it is possible that the aqueous values are abnormally high due to structure-breaking effects of the ions in water. The mobility of iodide is greater than that of the hydrogen-bonded chloride ion in water but the reverse is true for DMF[251] which illustrates the greater solvent affinity of the more highly polarizable ion in that solvent. Prue and Sherrington[250] have calculated Stokes Law radii of ions in DMF and concluded that, by this criterion, anions were unsolvated while solvation of the cations decreased from Li^+ to Cs^+ as in water. Any arguments of this nature must, however, be regarded as tentative since changes in local viscosity due to the influence of the ions field on the solvent structure cannot be distinguished from change in the effective radii of the ion.

Close similarity is found between Walden products for NMA, DMA, and

DMF. Sodium ion diffuses more slowly than the solvent molecules themselves in NMA and at a rate consistent with solvation by three or four molecules of the amide.[252] Water has little effect on ion mobilities in NMA, suggesting that ions are preferentially solvated by the amide.[253] Walden products of electrolytes in this solvent increase with temperature. A possible reason is that larger solvodynamic units are formed at the higher temperatures when structural breakdown liberates more free molecules to solvate the ions.

V. ACETONITRILE AND OTHER APROTIC SOLVENTS

1. Solvent Properties

The most commonly studied aprotic solvents other than amides are acetonitrile, dimethyl sulphoxide, sulpholane (a) and propylene carbonate (b)

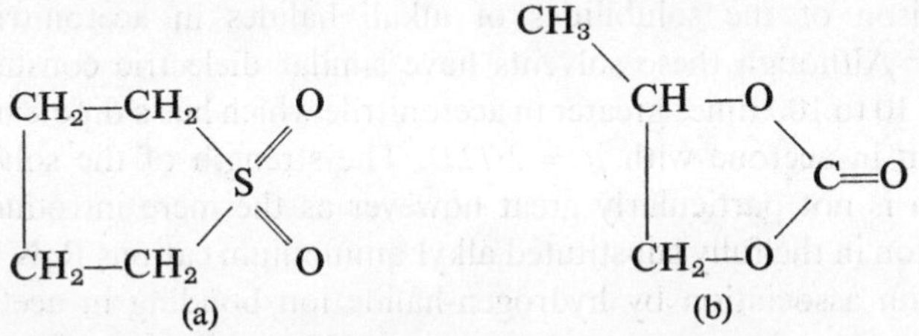

These compounds have high dielectric constants and dipole moments, Table 9. Dimethyl sulphoxide and sulpholane are highly associated liquids with temperature dependent structures.[254,255] It is possible that in dimethyl sulphoxide the oxygen atom bonds with hydrogen atoms of the methyl groups. The structure may include chains of molecules:

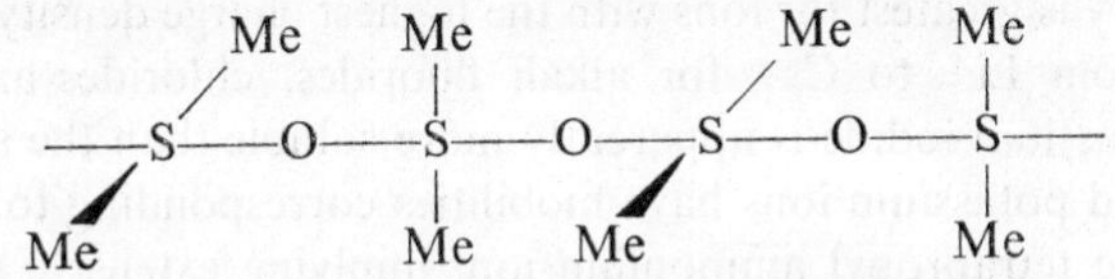

since this solvent has a high entropy of fusion and dilute solutions of dimethyl sulphoxide in benzene do seem to contain such chains.[256]

The high dielectric constant of propylene carbonate (69·0 at 23°) and other physical properties such as its high viscosity (1·67 centistokes at 38°)[257] suggest that this solvent also has a very ordered structure.

2. Electrolytes in Aprotic Media

Acetonitrile is a relatively poor solvent for electrolytes.[260] The cyanide group does not produce strong interaction with the alkali cations and according to Parker[228] its cation solvating ability is amongst the least of the common solvents where Me_2SO, $Me_2NAc > Me_2NCHO$, $H_2O > Me_2CO$, sulpholane $> MeOH > MeCN > PhCN$, $PhNO_2$.

Silver salts are exceptional in being usually more soluble in acetonitrile than in water.[261] Silver and cuprous ions probably form more specific complexes with the solvent than is implied in the usual term 'solvation'. The solubilities of those salts which do dissolve increase with increasing polarizability of the anion

$$SCN^- > I^- > N_3^- > Br^- > Cl^- > F^- > OH^-$$

as is common in aprotic solvents. Although the solubility of most salts is low due to poor solvation a large number of electrolytes are dissociated in solution since the solvent does have a relatively high dielectric constant. The influence of ion-dipole forces on solvation when other effects are absent is evident from a comparison of the solubilities of alkali halides in acetonitrile[259] and acetone.[262] Although these solvents have similar dielectric constants, solubilities are 10 to 100 times greater in acetonitrile which has a dipole moment of $3\cdot37D$ than in acetone with $\mu = 2\cdot72D$. The strength of the solvent-anion interaction is not particularly great however as the mere introduction of a single proton in the fully substituted alkyl ammonium cations R_4N^+ serves to promote ion association by hydrogen-halide ion bonding in acetonitrile[263] and acetone.[264]

Dimethyl sulphoxide is a much better solvent than acetonitrile,[258] mainly as a consequence of cation affinity with the electronegative oxygen atom.[254] Water dissolves exothermically with a substantial increase in viscosity of the solvent and appears therefore to enhance the structure by hydrogen bonding. As is evident from the high viscosity of solutions of lithium salts cation-solvent affinity is greatest for ions with the highest charge density. A decrease is evident from Li^+ to Cs^+ for alkali fluorides, chlorides and bromides although potassium iodide is apparently more soluble than the sodium salt.

Sodium and potassium ions have mobilities corresponding to a size equal to that of the tetrapropyl ammonium ion, implying extensive solvation.[220] Like acetonitrile, dimethyl sulphoxide apparently acts as a Lewis base complexing with certain transition metal ions, in chloride solutions of Cu^{2+}, Cr^{3+}, and Ni^{2+} for example.[265] Polar gases, carbonyl compounds and heterocyclics are soluble in this solvent but non-polar aliphatic compounds are not.[228] Thus it is not surprising that the same trend of anion solvation is found as for acetonitrile and that hydroxides, for example, are virtually insoluble in both solvents. As solvation increases with increasing anion polarizability in these aprotic solvents, chlorides and nitrates are frequently incompletely dissociated in acetonitrile, dimethyl sulphoxide and sulpholane, whereas iodides and picrates are much stronger electrolytes. In dimethyl

Table 12. *Solubilities of Alkali Halides in Some Organic Solvents at 25°C*

Solvent	Methanol (a)	Formic Acid (a)	Dimethyl Sulphoxide (b)	Acetonitrile (a)
Dielectric Constant	31·5	57·0	46·6	36·2
LiF	—	—	0·54	—
LiCl	41·0	27·5	9·9	$1·4 \times 10^{-1}$
LiBr	140·0	80·9	31·4	8·8
LiI	171·0	146	17·9	154
NaF	0·03	—	—	$3·0 \times 10^{-3}$
NaCl	1·4	5·2	0·5	$2·5 \times 10^{-4}$
NaBr	17·4	19·4	6·2	$4·0 \times 10^{-2}$
NaI	83·0	61·8	16·4	24·9
KF	10·2	—	—	$2·4 \times 10^{-3}$
KCl	0·53	19·2	0·18	$2·4 \times 10^{-3}$
KBr	2·10	22·7	6·5	$2·4 \times 10^{-2}$
KI	17·0	35·3	45·4	2·1
RbCl	1·34	56·9	0·47	$3·6 \times 10^{-3}$
RbBr	2·52	50·6	—	$4·7 \times 10^{-2}$
RbI	10·1	47·4	—	1·7
CsCl	3·01	130·5	1·18	$8·4 \times 10^{-3}$
CsBr	2·25	71·7	—	$1·4 \times 10^{-1}$
CsI	3·79	29·5	—	$9·9 \times 10^{-1}$

Solubilities in grams of salt per 100 g of solvent.
References: (a) 259; (b) 258.

sulphoxide potassium picrate is stronger than potassium chloride whereas
the reverse is true for methanol. The large variation of cation mobilities
compared to those of anions[220] in dimethyl sulphoxide suggests that cation
solvation is much the stronger. This may also be reflected in the degree of
dissociation since LiCl appears to be completely dissociated up to 0·3 m or
so while there is evidence for ion pairing in solutions of CsI and RbI above
about 10^{-2} m.[266,267]

In spite of its high dielectric constant propylene carbonate is a poor
solvent of inorganic salts. It is a typical aprotic solvent in which the solubilities of the alkali halides decrease in the order

$$Li^+ > Na^+ > K^+ > Cs^+$$

and

$$I^- > Br^- > Cl^- > I^-$$

Salts with oxy anions like nitrate, vanadate, and sulphate have very low
solubilities.[268] Complex ions are readily formed. Silver iodide, for example,

5

which has a very low solubility in the pure solvent readily dissolves in sodium iodide solution presumably by the formation of a highly polarizable complex cation. Other silver and mercuric compounds decompose in this solvent.

Structural effects appear very important in determining solvation in this medium. Iodide solutions are very viscous and it is noteworthy that salts having ions which are 'surface active' in water are soluble in propylene carbonate. Thus perchlorates, iodides and KF_6 are soluble as are tetra alkyl ammonium halides, solvation increasing with increasing size:

$$Bu_4N^+ > PrN^+ > Et_4N^+$$

It is to be expected that those salts which do dissolve will be highly dissociated though little detailed work has yet been done on the properties of electrolyte solutions in this intriguing solvent.[269]

3. Enthalpies of Transfer of Electrolytes from Water to Aprotic Solvents

There are few data at present on the standard potentials of elements in propylene carbonate and dimethyl sulphoxide from which free energies of solvation can be computed. Propylene carbonate is resistant to oxidation by the alkali metals so that the determination of their standard potential should not prove difficult.

Heats of solution of a number of salts in dimethyl sulphoxide and propylene carbonate have been measured.[270,271] Substantial ion association is evident from conductivity measurements on lithium trifluoroacetate in propylene carbonate and the sodium salt is only partially dissociated above 10^{-3} molar. Some ion-pairing appears to occur with lithium perchlorate since the heat of solution of this salt depends markedly on concentration. These observations are somewhat unexpected in a solvent with a dielectric constant approaching that of water. Heats of transfer of the salts from water to propylene carbonate have been determined and salt solvation energies for alkali halides derived from their hydration energies.

Wu and Friedman have proposed that, without knowledge of dielectric and structural effects, a radius increment R is useful for characterizing the solvating ability of various media[270] (see Section 4B). A good linear relation holds between the alkali cation solvation energies in propylene carbonate, relative to the sodium ion, and $(r_+ + R_+)$ according to the modified Born expression.

$$\Delta H_{solv} = - \frac{N(ze)^2}{2(r_+ + R_+)} \left(\frac{1 - \varepsilon}{\varepsilon} \right) \times$$
$$\left[1 - \frac{T}{(\varepsilon - 1)} \left(\frac{\partial \ln \varepsilon}{\partial T} \right) + \frac{T}{(r_+ + R_+)} \left(\frac{\partial R_+}{\partial T} \right) \right] \quad (82)$$

with $R_+(PC) = 0.80$ Å and $R_+(W) = 0.70$ Å.

The relative solvation enthalpies of the halide ions are also adequately represented with $R_-(PC) = 0.28$ Å and $R_-(W) = -0.20$ Å. This analysis is obviously inapplicable, however, since the linear anion and cation plots lead to two very different values for the heat of solvation of the sodium ion. A negative value of R_- for water cannot be explained on this model and the fact that $R_+(PC) > R_+(W)$ would mean that all the ion enthalpies are lower in propylene carbonate whereas in fact the enthalpies of transfer of many salts are negative.

A more informative view of these data has been suggested.[272] Ionic contributions to the enthalpies of transfer of salts from water to propylene carbonate are assigned on the arbitrary assumption of a value of zero for Cs^+ and plotted against the crystal radii of the ions as illustrated in Fig. 14.

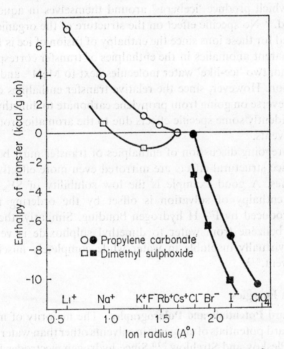

Fig. 14. Standard enthalpies of transfer of ions from water to propylene carbonate and dimethyl sulphoxide; values based on $^W\Delta^S H_{trans}Cs^+ = 0$.

Should the separation be made in a different way the curves will only be moved relative to each other; the trend of the transolvation energies will not be changed. The branch of the curve for the alkali cations shows that, insofar as this is reflected in the enthalpies, the lithium ion is less stabilized with

respect to the caesium ion in propylene carbonate than in water. Consideration of the values for these ions in dimethyl sulphoxide[271] reveals that here the
enthalpy contribution shows a similar relative stabilization of the alkali ions
in this solvent and in water. On these criteria the relative basicities of the
solvents are:

$$DMSO > H_2O > PC$$

In both non-aqueous solvents the chloride ion is relatively less solvated than
iodide which is just what is expected from the different order of stabilization
by hydrogen bonding and by the less effective ion dipole interactions.

The transfer energies of alkyl ammonium cations are several kilocalories
higher than would be anticipated from extrapolation of the data for other
cations, but this is not unexpected when the structure-promoting effect of
these ions which produce 'icebergs' around themselves in aqueous solution
is considered.[31] No specific effect on the structure of the organic media need
be postulated for these ions since the enthalpy of fusion of ice is 1·4 kcal/mole
and the apparent anomalies in the enthalpies of transfer correspond to those
expected from two 'ice-like' water molecules next to MeN^+ and four around
the Et_4N^+ ion. However, since the relative transfer enthalpies of the I^- and
BPh_4^- ions reverse on going from propylene carbonate to dimethyl sulphoxide
there are evidently some specific effects due to the aromatic group in the non-
aqueous solvents.

All the foregoing discussion of enthalpies of transfer must be regarded as
tentative since structural effects are mirrored even more effectively in solvation entropies. A good example is the low solubility of SF_6 in water. A
favourable enthalpy of solvation is offset by the ordering in the water
structure produced by F····H hydrogen bonding. Similarly, the enthalpy of
transfer of benzene from water to dimethyl sulphoxide is very small but
benzene is virtually insoluble in water while completely miscible with the
organic solvent.[271]

4. Solvation Energies

A. **Standard Potentials and Polarography.** The majority of measurements
of the standard potentials of elements in solvents other than water and alcohols
are due to Pleskov and Strehlow.[273] Since hydrogen electrodes frequently do
not work well in non-aqueous media it is often necessary to use a different
reference electrode, a common one being the half cell:

$$Ag|0·01 \text{ M } AgNO_3$$

originally used by Pleskov. The ferrocene-ferrocinium ion potential in
acetonitrile-water mixtures[274] and the standard potentials of rubidium and

thallium[275] in acetonitrile have been determined in this way. Relative potentials of lithium and thallium amalgam electrodes have been reported for dimethyl sulphoxide[176] but other direct measurements in non-aqueous media are very rare. The real free energy of transfer of chloride ion from water to acetonitrile has been determined[173] and the corresponding real ion solvation energies in this solvent are given in Table 8.

Determinations of polarographic half-wave potentials are much more common[277,278,279] since the difficulties produced by corrosion reactions mentioned before in the methanol-water systems are obviated by this technique. In media such as formic acid it is doubtful whether amalgam dissolution could ever be reduced sufficiently to give accurate standard potentials and even the standard potential of lithium in water is uncertain.[280] Provided that electrolyte activities can be reasonably estimated and the electrolyte is not associated, polarographic half-wave potentials of ions undergoing reversible reaction will generally not differ very much from the corresponding standard electrode potentials when referred to the same reference electrode.

The fact that standard potentials in any solvent are defined relative to a value of zero for the standard potential of the hydrogen electrode in that solvent shows that comparison of standard potentials of an element in different solvents means very little. Polarographic half-wave potentials are frequently measured against an *aqueous* calomel electrode and thus often include an indeterminate liquid junction potential. Comparisons between potentials in the same solvent against the same reference electrode are informative, however, as can be seen from the data in Table 13. The tendency of dimethyl sulphoxide to solvate cations strongly in the order

$$Li^+ > Na^+ > K^+ > Rb^+ > Cs^+$$

Table 13. *Polarographic Half-Wave Potentials*
(Volts against the Aqueous Sat. Calomel Electrode)

	Water (a)	Dimethyl Sulphoxide (b)	Acetonitrile (a)
Li^+/Li	-2.33	-2.45	-1.95
Na^+/Na	-2.12	-2.07	-1.85
K^+/K	-2.14	-2.11	-1.96
Rb^+/Rb	-2.13	-2.06	-1.98
Cs^+/Cs	-2.09	-2.03	-1.97

References: (a) 277; (b) 281.

is reflected in the decreasing negative values of $E_{1/2}$. Poor solvation of cations in acetonitrile is apparent from the similarity of the $E_{1/2}$ values and the slight trend towards easier reduction of the smaller ions.

The trend of polarographic half-wave potentials for the alkali metals in sulpholane indicates that the alkali metal ions are solvated more weakly by sulpholane than by water. The stabilization of the lithium ion relative to the caesium ion is greater in water than in sulpholane. Exactly the same behaviour is evident in the variation of enthalpies of solvation of these ions in propylene carbonate.[272]

More definite conclusions about ion solvation in different solvents can only be drawn from data referred to an absolute scale of potentials and solvation energies.

B. Voltage Series in Non-Aqueous Media. In order to define an absolute scale of potentials it is only necessary to know the free energy of transfer of a single ion between each of the solvents. Although potential scales are for convenience based on the hydrogen electrode, theoretically this is a bad choice since the proton is the least likely ion to be free from specific solvation effects. The ideal ion on which to base a potential scale would have the same solvation energy in every solvent or, failing that, should have a small energy of transfer between any pair of solvents which can be readily and accurately estimated.

(a) *The Pleskov Scale.* Simple electrostatic considerations suggest that a large ion with low charge and polarizability should have a small solvation energy. Pleskov therefore proposed a scale based on the rubidium ion which was then the largest monovalent cation whose standard potential had been accurately measured in a number of solvents.[282] That this proposal is a reasonable first approximation is evident from Table 14, where the standard

Table 14. *Standard Molal Electrode Potentials at 25°C Relative to $E_m^0 Rb/Rb^+ = 0$ (a)*

Solvent	H_2O	MeOH	MeCN	HCOOH	$HCONH_2$	NH_3
Electrode						
H_2/H^+	+2·925	+2·912	+3·17	+3·45	+2·855	+1·930
Li/Li^+	−0·120	−0·183	−0·06	−0·03	—	−0·31
Na/Na^+	+0·211	+0·184	+0·30	+0·03	—	+0·08
K/K^+	0	−0·009	+0·001	+0·09	−0·017	−0·05
Rb/Rb^+	0	0	0	0	0	0
Cs/Cs^+	+0·002	—	+0·001	+0·001	—	−0·02

Reference: (a) 273.

potentials of the alkali cations in a number of solvents are given relative to the rubidium/rubidium ion couple. The variation of proton stability is what would be intuitively expected, this ion being more reducible in acidic and neutral media than in water and less so in basic solvents. The rubidium ion is not, however, particularly large and transfer energies of ~ 2 kcal/mole would be expected between solvents of high and low dielectric constants.

The transolvation energy of Rb^+ may be estimated by calculating individual ion solvation energies by a method proposed by Latimer, Pitzer, and Slansky.[111] These authors showed that the hydration energies of alkali halides can be split into separate values for the anions and cations such that:

$$-\Delta G_{\text{hyd}} = \frac{Ne^2}{2} \left(1 - \frac{1}{\varepsilon_w}\right) \left[\frac{1}{r_+ + R_+} + \frac{1}{r_- + R_-}\right] \quad (83)$$

By adding $R_+ = 0.85$ Å and $R_- = 0.25$ Å to the Pauling cation and anion radii respectively eqn (83) represents the single ion hydration energies to within 1–2 kcal/mole. An empirical addition of radii increments effectively combines dielectric saturation, modification of solvent structure, electrostriction, etc. into a single parameter. The free energies of solution of salts in several non-aqueous solvents can also be represented by this equation provided that the correct values of R_+ and R_- are chosen.[274] To obtain these parameters the difference between the free energy of transfer of the caesium ion and the other alkali ions is written in the form

$$\Delta G(M^+) - \Delta G(Cs^+) = \frac{Ne^2}{2} \left[\left(1 - \frac{1}{\varepsilon_w}\right) \left(\frac{1}{r_+ + 0.85} - \frac{1}{r_{Cs} + 0.85}\right)\right.$$
$$\left. - \left(1 - \frac{1}{\varepsilon_s}\right) \left(\frac{1}{r_+ + 0.85 + \Delta R_+} - \frac{1}{r_{Cs} + 0.85 + \Delta R_+}\right)\right] \quad (84)$$

where ΔR_+ is equal to $R_+(s) - R_+(w)$.

Taking ΔR_+ to be small compared with the radius increment for either solvent, the above expression simplifies to:

$$\Delta G(M^+) - \Delta G(Cs^+) - \left\{\left(\frac{1}{\varepsilon_s} - \frac{1}{\varepsilon_w}\right) \left[\frac{1}{r_+ + 0.85} - \frac{1}{r_{Cs} + 0.85}\right]\right\} \cdot \frac{Ne^2}{2}$$
$$= \frac{Ne^2}{2} \left(1 - \frac{1}{\varepsilon_s}\right) \left[\frac{1}{(r_+ + 0.85)^2} - \frac{1}{(r_{Cs} + 0.85)^2}\right] \Delta R_+ \quad (85)$$

$\Delta G(M^+) - \Delta G(Cs^+)$ values are known from experiment, and thus by plotting the left hand side of eqn (85) against

$$\frac{1}{(r_+ + 0.85)^2} - \frac{1}{(r_{Cs} + 0.85)^2}$$

ΔR_+ may be determined from the slope of the line. An exactly analogous expression applies for the difference in solvation energies of the iodide and other halide ions and hence ΔR_- can be found.

Single ion transfer energies from water to methanol calculated in this way from eqn (85) are given in Table 7 and certainly agree more closely with de Ligny and Alfenaar's values[194] than do Feakins estimates.[190] It is possible that by restricting this method to free energy *differences* between water and non aqueous solvent and taking *relative* values for the ions many of the limitations of the Born treatment are eliminated. There is, however, a discrepancy of ~0.2 V between the standard potential of rubidium in acetonitrile, relative to the hydrogen electrode in water, calculated by Strehlow and by Coetzee and Campion[275] using essentially the same method. (Part of this may in fact be due to a redetermination of the rubidium standard potential in acetonitrile by Coetzee.) Transfer energies of the alkali metal ions between water and sulpholane have also been estimated by this method.[255]

(b) *Organic Reference Ions*. Since there is some doubt concerning the validity of Strehlow's empirical method other independent ways of obtaining individual ion solvation energies based on values for much larger ions than the alkali cations have been proposed. Large ions are necessarily either complex or organic. Requisites of such ions are that they should be of low charge, spherical and that their redox potential for the reaction

$$R^{n+} \rightleftharpoons O^{(n+1)+} + e^-$$

should be well below the oxidation potential of the solvent.

Ferrocene and cobalticene fulfil these criteria well and cobalticinium ion is almost reversibly reduced in most solvents.[279] Moreover, errors due to amalgam dissolution which restrict the accuracy of the measurement of oxidation potentials of alkali metals are obviated in this method. The redox potential of ferrocene (Foc) against the hydrogen electrode corresponds to:

$$-FE = \Delta G_{\text{form}}H^+(S) + \Delta G_{\text{soln}}Foc(S) - \Delta G_{\text{form}}Fic(S) \qquad (86)$$

As a first approximation, the difference of the redox potentials in two different solvents may be equated to the transolvation energy of the proton. Results of de Ligny and Alfenaar show that this is a good approximation for proton transfer from water to mixtures of methanol and water up to pure methanol.[193] The difference in redox potentials of ferrocene and cobalticene is 1.33 V, independent of the solvent and the difference

$$E^0_{\text{Foc/Fic}} - E^0_{\text{Rb/Rb}^+}$$

is very similar in different solvents. Estimation of the correction term

$$^w\Delta^s G_t Foc - {}^w\Delta^s G_t Fic$$

which is the difference in transolvation energy of ferrocene and ferrocinium ion should be much more reliable than an estimation of the rubidium ion transfer energy. The greater size of the ferrocinium ion and the fact that 'non-electrostatic' interactions are subtracted out means that the Born equation should give this term quite accurately.

An application which supports the validity of the ferrocene method is provided by the measurement of the free energy of transfer of silver ion from water to acetonitrile-water mixtures from the emf of the cell[274]

$$\text{Pt}|\text{Foc/Fic}(\text{H}_2\text{O} - \text{CH}_3\text{CN})|\underset{0\cdot01\,\text{M}}{\text{AgNO}_3}|\text{Ag}$$

A smooth increase in stability of Ag^+ in the mixed solvent over water is observed with

$$\mu_{\text{Ag}^+}(\text{AN}) - \mu_{\text{Ag}^+}(\text{W}) = -9\cdot3 \text{ kcal/mole}$$

Of course this is what would be expected on the basis of strong co-ordination of the ion by the cyanide group of acetonitrile but is the opposite of behaviour predicted from the dielectric properties of the solvents. A similar value for $^{\text{W}}\Delta^{\text{AN}}G_t\text{Ag}^+$ follows from the measured value of the real transfer energy[173]

$$\alpha_{\text{Ag}^+}(\text{AN}) - \alpha_{\text{Ag}^+}(\text{W}) = -10\cdot1 \text{ kcal/mole}$$

and the estimate (Section (c)) of the difference of surface potentials of water and acetonitrile

$$^{\text{W}}\Delta^{\text{AN}}\chi = -0\cdot14 \text{ V}$$

whence

$$^{\text{W}}\Delta^{\text{AN}}\mu_{\text{Ag}^+} = {}^{\text{W}}\Delta^{\text{AN}}\alpha_{\text{Ag}^+} - \text{F}^{\text{W}}\Delta^{\text{AN}}\chi = -6\cdot9 \text{ kcal/mole}$$

Thus there is reason to believe that the 'ferrocene' scale of electrode potentials in non aqueous solvents relative to the hydrogen electrode in water is more trustworthy than the 'rubidium' scale.

Two further proposals for the basis of a universal potential scale also seem less sound than the ferrocene method. The first is based on the redox behaviour of the complex of ferrous ion with 4,7-dimethyl-1,10-phenanthroline[283]

The apparent advantage of using this very large ion is offset by the unfavourable high charge of the redox system $+2/+3$ since the electrostatic energy increases with the square of the charge. An alternative possibility, which is attractive because of its experimental simplicity, is to determine the transfer energy of the proton via the measurement of the pK values of a suitable acid-base pair.

$$H^+ + B \leftrightarrows HB^+$$

or

$$H^+ + A^- \leftrightarrows HA$$

in the two solvents. This treatment, which has been thoroughly discussed by Bates,[284] relies upon the assumption that

$$\Delta G_t BH^+ = \Delta G_t B$$

or

$$\Delta G_t HA = \Delta G_t A^-$$

Unfortunately in most suitable acid-base pairs the charge is localized in a substituent group rather than effectively spread over a large molecule as in the ferrocinium ion so the above assumptions do not usually hold at all well. The method works better for organic bases like substituted anilines than for substituted carboxylic acids, at least in water-alcohol mixtures. Presumably this is in part due to a similarity in solvent orientation about the free base and its protonated conjugate acid in contrast to a different ordering of solvent round the carboxyl anion and the carboxylic acid.[285]

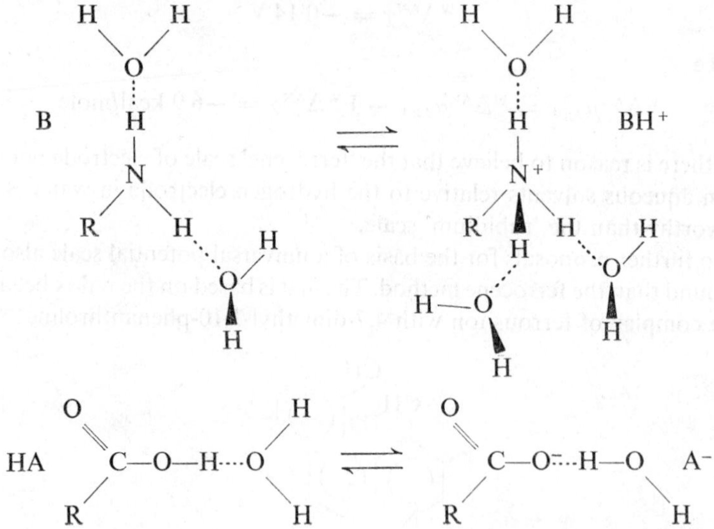

Parker and Alexander have summarized many of the proposals which have been employed to rationalize solvation energies, medium effects and voltage scales in a range of organic solvents.[187]

(c) *Differential Solvation Energies of Aromatic Ions.* In order to calculate the absolute real solvation energy of an ion one must know the free energy of formation of that ion in the gas phase. Such calculations are usually restricted to simple monotomic ions since in general it is not possible to measure ionization potentials or electron affinities of unstable gaseous polyatomic ions. Normally the only thermodynamic data that can be derived for polynuclear species are the free energies of transfer from one solvent to another, or in the case of 'organic' ions like cobalticinium, the difference in this quantity for the ion and the neutral molecule.

Quite recently pulse sampling techniques have been developed which permit the determination of electron affinities A of aromatic hydrocarbons.[286] Electron affinities are derived from the temperature variation of experimental electron capture coefficient k for the reaction:

$$R(g) + e \rightleftharpoons R^-(g)$$

R being an aromatic molecule. Ionization potentials of these compounds may be determined by photoionization or electron impact.[287,288]

Energy changes associated with the formation of a solvated anion from a solution of the hydrocarbon in a solvent s are represented by the cycle:

$$
\begin{array}{c}
A_R \\
R(g) \rightarrow R^-(g) \\
-\alpha_{R^0} \uparrow \qquad \downarrow \alpha_{R^-} \\
R(s) \qquad R^-(s)
\end{array}
$$

where α_{R^0} is the free energy of solution of the gaseous hydrocarbon; A_R is its electron affinity; and α_{R^-} is the real solvation energy of the R^- anion.

The free energy of formation is therefore given by

$$\Delta G_{\text{form}} R^-(s) = -\alpha_{R^0} + \alpha_{R^-} - A_R \qquad (87)$$

and similarly for the cation

$$\Delta G_{\text{form}} R^+(s) = -\alpha_{R^0} + \alpha_{R^+} - I_R \qquad (88)$$

Polarographic techniques can be used to obtain half-wave potentials which are close to the standard potentials corresponding to the reduction process given above since many aromatic anions are stable in aprotic solvents in the absence of water and oxidizing substances. The cations are much more chemically reactive, however, and fast cyclic sweep methods with potential variation up to 500 V/sec must be employed to observe the uncomplicated one electron transfer reaction.

Both oxidation and reduction potentials have been determined for certain alternant hydrocarbon molecules in acetonitrile.[289] This is one of the few non-aqueous solvents whose oxidation potential at a platinum surface is sufficiently more positive than that of the hydrocarbons of interest to enable their oxidation potentials to be measured. With a reference electrode reversible to the silver ion the overall reactions in the two cases are:

$$R + Ag^+ \rightarrow R^+ + Ag$$
$$R + Ag \rightarrow R^- + Ag^+$$

for which

$$-FE^0_{ox} = \Delta G^0_{form} R^+(AN) - \Delta G^0_{form} Ag^+(AN) \tag{89}$$

$$-FE^0_{red} = \Delta G^0_{form} R^-(AN) + \Delta G^0_{form} Ag^+(AN) \tag{90}$$

If we define the difference between the energy of solution of the neutral molecule and that of its gaseous ion as

$$\delta_{R^+} = \alpha_{R^0} - \alpha_{R^+} \tag{91}$$

with an analogous expression for the anion, then eqns (89) and (90) give:

$$-FE^0_{ox} = \delta_{R^+} - \delta_{Ag^+} - I_R + I_{Ag} \tag{92}$$

and

$$-FE^0_{red} = \delta_{R^-} + \delta_{Ag^+} - A_R - I_{Ag} \tag{93}$$

$\delta_{Ag^+}(AN)$ has been measured by combining the real free energy of transfer of Ag^+ between water and acetonitrile with the absolute real free energy of hydration. Referred to the activity of silver ion in 0·001 M $AgClO_4$ + 0·1 M Et_4NClO_4

$$\delta_{Ag^+} = 4·70 \text{ eV}$$

Differential real solvation energies of alternant hydrocarbon ions may be calculated from the data in Tables 15 and 16 and are given in columns 1–4 of Table 16. To obtain absolute values of the real solvation energies this quantity must be known for the hydrocarbons themselves for which determination of the solubilities and activities of the compounds in acetonitrile is required.

The particular interest of differential real free energies of solvation is that specific interaction of the ion with the solvent is effectively subtracted out. With the delocalization of charge over the large aromatic ions considered here the peripheral field will not be too different from that of the neutral molecule.

For alternant ions in which the charge distributions in anion and cation are identical the 'Born' interactions with the solvent should also be equal. The energies of interaction of the charge with multipoles of the solvent molecules decrease rapidly with the effective size of the ion and should be very small for

ION SOLVATION 127

Table 15. *Oxidation and Reduction Potentials, Electron Affinities (A) and Ionization Potentials (I) for Hydrocarbons in Acetonitrile* + 0·1 M Et$_4$NClO$_4$ *at 25°C (a)*

	$E^{0,red}$ (V)	$E^{0,ox}$ (V)	A (eV) (b)	I (eV) (c)	(d)
Anthracene	−2·37	0·99	0·56	7·55	7·66
Pyrene	−2·49	1·12	0·59	—	7·72
Phenanthrene	−2·88	1·45	0·31	8·03	8·06
Chrysene	−2·73	1·29	0·40	8·01	7·90†
1-2 benzanthracene	−2·40	1·16	0·63 (0·70)	7·54	7·67†
Triphenylene	−2·87	1·50	0·29	8·19	8·01†

The reference electrode is: Ag | 0·001 M AgClO$_4$, 0·1 M Et$_4$NClO$_4$ (Acetonitrile). References: (a) 289; (b) 286; (c) 287, 288; (d) 290. †Calculated values assuming $I + A = 8·3$ eV together with values of A from the Table.

Table 16. *Differential Real Free Energies of Solvation of the Anions and Cations of Aromatic Hydrocarbons*

	$-\delta^-$		$-\delta^+$		$1/F(\delta^- - \delta^+)$	
	(kcal/mole)				(Volts)	
Anthracene	40·8		42·9	45·4	0·09	0·20
Pyrene	37·4		—	43·8	—	0·28
Phenanthrene	34·8		43·3	43·9	0·37	0·40
Chrysene	35·3		46·5	43·9	0·49	0·38
1-2 benzanthracene	38·5 (36·9)		38·7	41·7	0·01 (0·08)	0·14
Triphenylene	35·5		45·8	41·7	0·45	0·27
					0·28 ± 0·20	0·28 ± 0·08

these large ions. It would thus be expected that the difference in the chemical solvation energies of the cation and anion of the same large polynuclear ion should be close to zero. The difference of real solvation energy will then merely be due to the difference in the energy required to transfer the ion across the gas-solution interface which is dependent on the sign of the ion's charge, that is

$$\alpha_{R^+} - \alpha_{R^-} = \mu_{R^+}(S) + F\chi(S) - (\mu_{R^-}(S) - F\chi(S)) \qquad (94)$$

$$\alpha_{R^+} - \alpha_{R^-} \rightarrow 2F\chi(S) \qquad (95)$$

or

$$\delta_{R^+} - \delta_{R^-} \rightarrow 2F\chi(S) \qquad (96)$$

Values of $(\delta_{R^+} - \delta_{R^-})/F$ are listed in Table 16. No definite trend with size is observed for these ions and the near constancy of the values confirms that to a good approximation

$$^W\Delta^{AN}\mu_{R^+} = {^W\Delta^{AN}}\mu_{R^-} \tag{97}$$

and that the average difference may be interpreted as being due to the surface potential of acetonitrile.

$$\chi(\text{AN}) = -0.14 \text{ V}$$

This value is of the same order as estimated of χ for water and for methanol. The sign of χ corresponds to a preferred surface orientation with the positive end of the acetonitrile dipole directed towards the gas phase:

$$
\begin{array}{c}
\text{CH}_3 \quad \text{gas} \\
| \\
-\text{C}^+- \\
||| \\
\text{N}_- \quad \text{liquid}
\end{array}
$$

assuming the polarity is as shown in the diagram.

It has been suggested that the analysis above may only apply to ions in a certain size-range. For small ions different interactions with the solvent and different orientation of solvent molecules around the ions will mean a difference in solvation energy of anions and cations of equal size. Very large ions may, however, merely appear as 'holes' or 'bubbles' in the liquid. Should the solvent orientation at the surface of a very large ion revert to that at the solvent-gas interface the real solvation energy of any such ion will be zero as the surface energy term will be cancelled out.

Measurements of the change of surface potential with increase in the salt concentration for solutions of sodium perchlorate in water, acetonitrile and methanol show an increasing positive $\Delta\chi$ for water and negative values for the other solvents.[291] The sign of $\Delta\chi$ therefore seems to be the same as the probable sign of the surface potential of each of the pure solvents. It is possible that disruption of the solvent structure by the salt at the gas-solution interface augments the effective number of relatively free solvent dipoles and hence enhances the natural surface potential.

REFERENCES

1. HUSH, N. S.: *Trans. Faraday Soc.*, **57**, 557 (1961).
2. MARCUS, R. A.: *Ann. Rev. Phys. Chem.*, **15**, 155 (1964).
3. BOCKRIS, J. O'M., and DAMJANOVIC, A.: *Modern Aspects of Electrochemistry*, Vol. 3, Edited by J. O'M. Bockris and B. E. Conway, Chapter 4 (Butterworths, London, 1964).

4. PARSONS, R., and BIEGLER, T.: Private communication.
5. BERNAL, J. D., and FOWLER, R. H.: *J. Chem. Phys.*, **1**, 575 (1933).
6. STEWART, G. W.: *Phys. Rev.*, **37**, 9 (1931).
7. BAUER, W. H.: *Acta Cryst.*, **19**, 901 (1965).
8. DAVIES, C. M., and LITOVITZ, T. A.: *J. Chem. Phys.*, **42**, 2563 (1965).
9. WALL, T. T., and HORNIG, D. F.: *J. Chem. Phys.*, **43**, 2079 (1965).
10. FALK, M., and FORD, T. A.: *Canad. J. Chem.*, **44**, 1699 (1966).
11. FRANCK, E. U., and ROTH, K.: *Disc. Faraday Soc.*, **43**, 108 (1967).
12. FRANK, H. S.: *Disc. Faraday Soc.*, **43**, 137 (1967).
13. WALRAFEN, G. E.: *J. Chem. Phys.*, **44**, 1546 (1966).
14. NARTEN, A. H., DANFORD, M. D., and LEVY, H. A.: *Disc. Faraday Soc.*, **43**, 97 (1967).
15. SENIOR, W. A.: *Disc. Faraday Soc.*, **43**, 131 (1967).
16. WALRAFEN, G. E.: *Hydrogen Bonded Solvent Systems*, Edited by A. K. Covington and P. Jones, pp. 9, 137 (Taylor and Francis Ltd., London, 1968).
17. EISENBERG, D., and KAUZMANN, W.: *The Structure and Properties of Water*, p. 239 (The Clarendon Press, Oxford, 1969).
18. STEVENSON, D. P.: *J. Phys. Chem.*, **69**, 2145 (1965).
19. IVES, D. J. G.: 'Some Reflections on Water', Inaugural Lecture, Birkbeck College, J. W. Ruddock (London, 1963).
20. LUCK, W. A. P.: *Physico Chemical Processes in Mixed Aqueous Solvents*, p. 11, Edited by F. Franks (Heinemann, London, 1967).
21. CHOPPIN, G. R., and BUIJS, K.: *J. Chem. Phys.*, **39**, 2035, 2042 (1963).
22. HORNIG, D.: *J. Chem. Phys.*, **40**, 3119 (1964).
23. EUKEN, A.: *Z. Elektrochem.*, **52**, 264 (1948).
24. PAULING, L.: *Hydrogen Bonding*, Edited by D. Hadzi, p. 1 (Pergamon Press, London, 1959).
25. PAULING, L.: *The Nature of the Chemical Bond*, 3rd edition, Chapter 12, p. 469 (Cornell University Press, Ithaca, N.Y., 1960).
26. VON STACKELBERG, M., and MÜLLER, H. R.: *Z. Elektrochem.*, **58**, 25 (1954).
27. DAVIS, C. M., and LITOVITZ, T. A.: *J. Chem. Phys.*, **42**, 2563 (1965).
28. LONSDALE, K.: *Proc. Roy. Soc.*, *London*, **A247**, 424 (1958).
29. FRANK, H. S., and QUIST, A. S.: *J. Chem. Phys.*, **34**, 604 (1961).
30. FRANK, H. S., and EVANS, M.: *J. Chem. Phys.*, **13**, 507 (1945).
31. FRANK, H. S., and WEN, W. Y.: *Disc. Faraday Soc.*, **24**, 133 (1957).
32. QUIST, A. S., and MARSHALL, W. L.: *J. Phys. Chem.*, **69**, 3165 (1965).
33. NÉMETHY, G., and SCHERAGA, H. A.: *J. Chem. Phys.*, **36**, 3382 (1962).
34. NÉMETHY, G., and SCHERAGA, H. A.: *J. Chem. Phys.*, **36**, 3401 (1962).
35. VAND, V., and SENIOR, W. A.: *J. Chem. Phys.*, **43**, 6 (1965).
36. NÉMETHY, G., and SCHERAGA, H. A.: *J. Chem. Phys.*, **41**, 680 (1964).
37. MARCHI, R. P., and EYRING, H.: *J. Phys. Chem.*, **68**, 221 (1964).
38. LEVINE, S., and PERRAM, J. W.: In Reference 16, p. 115.
39. CONWAY, B. E.: *Canad. J. Chem.*, **37**, 178 (1959).
40. COHEN, N. V., COTTI, M., IRIBARNE, J. V., and WEISSMANN, M.: *Trans. Faraday Soc.*, **58**, 490 (1962).
41. POPLE, J. A.: *Proc. Roy. Soc.*, **A205**, 163 (1951).
42. HAMILTON, W. C., and IBERS, J. A.: *Hydrogen Bonding in Solids* (Benjamin, N.Y., 1968).
43. BERNAL, J. D.: *Proc. Roy. Soc.*, **A280**, 299 (1964).
44. LENNARD-JONES, J., and POPLE, J. A.: *Proc. Roy. Soc.*, **A202**, 166, 323 (1950).
45. GRJOTHEIM, K., and KROGH-MOE, J.: *Acta Chem. Scand.*, **8**, 1193 (1954).
46. VERWEY, E. J.: *Rec. Trav. Chim.*, **60**, 887 (1941).
47. PAULING, L.: *The Nature of the Chemical Bond*, 3rd edition, p. 449 (Cornell University Press, Ithaca, N.Y., 1960).
48. EISENBERG, D. S.: Thesis (Oxford, England, 1964), quoted by: M. Orentlicher and P. O. Vogelhut, *J. Chem. Phys.*, **45**, 4719 (1966).
49. NIGHTINGALE, E. R.: *J. Phys. Chem.*, **63**, 1381 (1959).

50. NIGHTINGALE, E. R.: *Chemical Physics of Ionic Solutions*, p. 87, Edited by B. E. Conway and R. G. Barradas (Wiley, N.Y., 1966).
51. GOTO, S., and ISEMURA, T.: *Bull. Chem. Soc., Japan*, 37, 1693, 1697 (1964).
52. HUNT, J. P., and TAUBE, H.: *J. Chem. Phys.*, 19, 602 (1951).
53. PLANE, R. A., and HUNT, J. P.: *J. Amer. Chem. Soc.*, 79, 3343 (1957).
54. NOYES, R. M.: *J. Amer. Chem. Soc.*, 84, 513 (1962).
55. CONNICK, R. E., and FIAT, D. N.: *J. Chem. Phys.*, 39, 1349 (1963).
56. ALEI, M., and JACKSON, J. A.: *J. Chem. Phys.*, 41, 3402 (1964).
57. CONNICK, R. E., and FIAT, D. N.: *J. Chem. Phys.*, 44, 4103 (1966).
58. SWIFT, T. J., and SAYRE, W. G.: *J. Chem. Phys.*, 44, 3567 (1966).
59. MCDONALD, C. L., and PHILLIPS, W. D.: *J. Amer. Chem. Soc.*, 85, 3736 (1963).
60. LUZ, A., and SCHULMAN, R. G.: *J. Chem. Phys.*, 43, 3750 (1965).
61. SAMOILOV, O. YA.: *Disc. Faraday Soc.*, 24, 141 (1957).
62. HASTED, J. B., RITSON, D. M., and COLLIE, C. H.: *J. Chem. Phys.*, 16, 1 (1948).
63. HINDMAN, J. C.: *J. Chem. Phys.*, 36, 1000 (1962).
64. ROBINSON, R. A., and STOKES, R. H.: *Electrolyte Solutions*, 2nd edition, p. 331 (Butterworths, London, 1959).
65. GLUECKAUF, E.: *Trans. Faraday Soc.*, 51, 1235 (1955).
66. GLUECKAUF, E., *The Structure of Electrolyte Solutions*, Chapter 7 (Wiley, N.Y., 1959).
67. CONWAY, B. E., and BOCKRIS, J. O'M.: *Modern Aspects of Electrochemistry*, Vol. 1, Chapter 2 (Butterworths, London, 1954).
68. GLUECKAUF, E., and KITT, G. P.: *Proc. Roy. Soc. (London)* A228, 322 (1955).
69. BOCKRIS, J. O'M.: *Quart. Rev. (London)*, 3, 173 (1949).
70. HAGGIS, G. H., HASTED, J. B., and BUCHANAN, T. J.: *J. Chem. Phys.*, 20, 1452 (1952).
71. ALLAM, D. S., and LEE, W. H.: *J. Chem. Soc.* A, 426 (1966).
72. RUTGERS, A. J., and HENDRIKX, Y.: *Trans. Faraday Soc.*, 58, 2184 (1962).
73. GLUECKAUF, E.: *Trans. Faraday Soc.*, 60, 1637 (1964).
74. VASLOW, F.: *J. Phys. Chem.*, 67, 2773 (1963).
75. BATSONOV, S.: *J. Struct. Chem., U.S.S.R.*, 4, 158 (1963).
76. BUCKINGHAM, A. D.: *Disc. Faraday Soc.*, 24, 151 (1957).
77. BLANDAMER, M. J., and SYMONS, M. C. R.: *J. Phys. Chem.*, 67, 1304 (1963).
78. BURNELLE, L., and COULSON, C. A.: *Trans. Faraday Soc.*, 53, 403 (1957).
79. JØRGENSEN, C. K.: *Inorganic Complexes*, p. 123 (Academic Press Inc., London, 1963).
80. BADER, R. F., and JONES, G.: *Canad. J. Chem.*, 41, 586 (1963).
81. FLETCHER, N. H.: *Phil. Mag.*, 7, 255 (1962).
82. FRUMKIN, A. N., IOFA, Z. A., and GEROVICH, M. A.: *J. Phys. Chem. U.S.S.R.*, 30, 1455 (1956).
83. STILLINGER, F. H., JR., and BEN-NAIM, A.: *J. Chem. Phys.*, 47, 4431 (1967).
84. CLAUSSEN, W. F.: *Science*, 156, 1226 (1967).
85. MCBAIN, J. W., BACON, R. C., and BRUCE, H. D.: *J. Chem. Phys.*, 7, 818 (1939).
86. LANGMUIR, I.: *J. Amer. Chem. Soc.*, 39, 1897 (1917).
87. HARKINS, W. D., and MCLAUGHLIN, H. M.: *J. Amer. Chem. Soc.*, 47, 2083 (1925).
88. SCHWENKER, G.: *Ann. Physik.*, 11, 525 (1931).
89. ONSAGER, L., and SAMARAS, N. N. T.: *J. Chem. Phys.*, 2, 528 (1934).
90. SCHÄFER, K. L., PEREZ MASIÀ, A., and JÜNTGEN, H.: *Z. Elektrochem.*, 59, 425 (1955).
91. WEBB, T. J.: *J. Amer. Chem. Soc.*, 48, 2589 (1926).
92. SCHMUTZER, E.: *Z. Phys. Chem.*, Leipzig, 204, 131 (1955).
93. FRUMKIN, A. N.: *Z. Phys. Chem.*, 109, 34 (1924).
94. RANDLES, J. E. B.: *Disc. Faraday Soc.*, 24, 194 (1957).
95. GURNEY, R. W.: *Ionic Processes in Solution*, Chapter 16 (McGraw-Hill, London 1953).
96. LORENZ, P. B.: *J. Phys. and Colloid Chem.*, 54, 685 (1950).
97. HALLIWELL, H. F., and NYBURG, S. C.: *Trans. Faraday Soc.*, 59, 1126 (1963).
98. NOYES, R. M.: *J. Amer. Chem. Soc.*, 86, 971 (1964).
99. BENJAMIN, L., and GOLD, V.: *Trans. Faraday Soc.*, 50, 797 (1954).

100. RANDLES, J. E. B.: *Trans. Faraday Soc.*, **52**, 1573 (1956).
101. IZMAILOV, N. A.: *Dokl. Akad. Nauk SSSR*, **149**, 288, 320, 348 (1963).
102. BORN, M.: *Physik. Z.*, **1**, 45 (1920).
103. ELEY, D. D., and EVANS, M. G.: *Trans. Faraday Soc.*, **34**, 1093 (1938).
104. GRAHAME, D. C.: *J. Chem. Phys.*, **18**, 903 (1950).
105. DEBYE, P.: *Polar Molecules*, p. 110 (Chem. Catalog Co. Inc., N.Y., 1929).
106. ONSAGER, L.: *J. Amer. Chem. Soc.*, **58**, 1486 (1936).
107. KIRKWOOD, J. G.: *J. Chem. Phys.*, **7**, 911 (1939).
108. BOOTH, F.: *J. Chem. Phys.*, **19**, 391, 1327, 1615 (1951).
109. GLUECKAUF, E.: *Trans. Faraday Soc.*, **60**, 572 (1964).
110. MALSCH, J.: *Phys. Zeit*, **29**, 770 (1928); **30**, 837 (1929).
111. LATIMER, W. M., PITZER, K. S., and SLANSKY, C. M.: *J. Chem. Phys.*, **7**, 108 (1939).
112. GLUECKAUF, E.: *Trans. Faraday Soc.*, **61**, 914 (1965).
113. GLUECKAUF, E.: Reference 50, p. 67.
114. CONWAY, B. E., DESNOYERS, J. E., and SMITH, A. C.: *Phil. Trans.*, **A256**, 389 (1964).
115. LAIDLER, K. J., and PEGIS, C.: *Proc. Roy. Soc.*, **A241**, 80 (1957).
116. VASIL'EV, V. P., ZOLOTAREV, E. K., KAPUSTINSKII, A. F., MISHCHENKO, K. P., POD-
    GORNAYA, E. A., and YATSIMIRSKII, K. B.: *Zhur. Fiz. Khim.*, **34**, 1763 (1960).
117. LAIDLER, K. J.: Reference 50, p. 75.
118. LEVINE, S.: Reference 50, p. 106.
119. MUIRHEAD-GOULD, J. S., and LAIDLER, K. J.: *Trans. Faraday Soc.*, **63**, 944, 953 (1967).
120. BENSON, S. W., and COPELAND, C. S.: *J. Phys. Chem.*, **67**, 1194 (1963).
121. STOKES, R. H.: *J. Amer. Chem. Soc.*, **86**, 979 (1964).
122. MORRIS, D. F. C.: *Structure and Bonding*, **4**, 63 (1968).
123. GOLDSCHMIDT, V. M.: *Skrifter Norske Videnskaps-Akad Oslo Mat. Naturv. Kl.*, **8**,
    1 (1926).
124. ZACHARIASEN, W. H.: *Z. Krist.*, **80**, 13 (1931).
125. AHRENS, L. H.: *Geochim. Cosmochim. Acta*, **2**, 155 (1952).
126. WADDINGTON, T. C.: *Trans. Faraday Soc.*, **61**, 1482 (1965).
127. GOURARY, B. S., and ADRIAN, F. J.: *Solid State Phys.*, Vol. 10, 127 (Academic Press
    Inc., New York, 1960).
128. WITTE, H., and WOLFEL, E.: *Z. Physik. Chem. (Frankfurt)*, **3**, 296 (1955).
129. PAULING, L.: Reference 47, p. 514.
130. BLANDAMER, M. J., and SYMONS, M. C. R.: *J. Phys. Chem.*, **67**, 1304 (1963).
131. JAIN, D. V. S.: *Ind. J. Chem.*, **3**, 466 (1965).
132. KEBARLE, P., ARSHADI, M., and SCARBOROUGH, J.: *J. Chem. Phys.*, **50**, 1049 (1969).
133. CONWAY, B. E.: *Modern Aspects of Electrochemistry*, Vol. 3, Chapter 2, Edited by
    J. O'M. Bockris and B. E. Conway (Butterworths, London, 1964).
134. IZMAILOV, N. A.: *Zhur. Fiz. Khim.*, **34**, 2414 (1960).
135. VERWEY, E. J.: *Chem. Weekblad*, **37**, 530 (1940).
136. ROSSINI, F. D., WAGMAN, D. D., EVANS, W. H., LEVINE, S., and JAFFE, I.: 'Selected
    Values of Chemical Thermodynamic Properties', N.B.S. Circular 500 (1952).
137. VERHOEVEN, J., and DYMANUS, A.: *J. Chem. Phys.*, **52**, 3222 (1970).
138. SYMONS, M. C. R.: *Quart. Rev.*, **13**, 99 (1959).
139. STEIN, G.: *Disc. Faraday Soc.*, **12**, 227 (1952).
140. PLATZMAN, R. L.: *Natl. Acad. Sci. Natl. Res. Council Pubn.*, **305**, 34 (1953).
141. BARR, N. F., and ALLEN, A. O.: *J. Phys. Chem.*, **63**, 928 (1959).
142. DORFMAN, L. M.: *Science*, **141**, 493 (1963).
143. HART, E. J., and BOAG, J. W.: *J. Amer. Chem. Soc.*, **84**, 4090 (1962).
144. ANBAR, M.: 'Solvated Electron', *Adv. in Chem. Ser. 50*, p. 55 (Amer. Chem. Soc.,
    Washington D.C., 1965).
145. ANBAR, M., and HART, E. J.: *J. Phys. Chem.*, **69**, 1244 (1965).
146. DAINTON, F. S., KENNE, J. P., KEMP, T. J., SALMAN, G. A., and TEPLY, J.: *Proc. Chem.
    Soc.*, 265 (1964).
147. BAXENDALE, J. H.: *Radiation Research*, Suppl. 4, 139 (1964).

148. HART, E. J., GORDON, S., and FIELDEN, F. M.: *J. Phys. Chem.*, **70**, 150 (1966).
149. JORTNER, J.: Reference 147, p. 24.
150. ANBAR, M., and NETA, P.: *Intern. J. Appl. Radiation*, **16**, 227 (1965).
151. WALKER, D. C.: *Quart. Rev.*, **21**, 79 (1967).
152. PLATZMAN, R. L., and FRANCK, J.: *Z. Physik*, **138**, 411 (1954).
153. ANBAR, M., and HART, E. J.: *J. Phys. Chem.*, **71**, 3700 (1967).
154. EIGEN, M., KRASE, W., MAASS, G., and DE MAEYER, L.: *Progr. Chem. Kinetics*, **2**, 287 (1963).
155. STEIN, G.: Reference 16, p. 87.
156. BERNAL, J. D., and TAMM, G.: *Nature*, **135**, 229 (1935).
157. BEN-NAIM, A.: *J. Chem. Phys.*, **42**, 1512 (1965).
158. WYMAN, J., and INGALLS, E. N.: *J. Amer. Chem. Soc.*, **60**, 1182 (1938).
159. SWAIN, C. G., and BADER, R. F. W.: *Tetrahedron*, **10**, 182 (1960).
160. LANGE, H. E., and MARTIN, W.: *Z. Physik. Chem.*, **A180**, 233 (1937).
161. LANGE, H. E., and MARTIN, W.: *Z. Elektrochem.*, **42**, 662 (1936).
162. WU, Y. C., and FRIEDMAN, H. L.: *J. Phys. Chem.*, **70**, 166 (1966).
163. LA MER, V. K., and NOONAN, E.: *J. Amer. Chem. Soc.*, **61**, 1487 (1939).
164. LA MER, V. K., and NOONAN, E.: *J. Phys. Chem.*, **43**, 247 (1939).
165. SALOMAA, P., and AALTO, V.: *Acta. Chem. Scand.*, **20**, 2035 (1966).
166. GREYSON, J.: *J. Phys. Chem.*, **66**, 2218 (1962); **71**, 259, 2210 (1967).
167. KERWIN, R. E.: Thesis, University of Pittsburg (1964).
168. GIGUÈRE, P. A., and HARVEY, K. B.: *Can. J. Chem.*, **34**, 798 (1956).
169. HEPLER, L. G.: *Aust. J. Chem.*, **17**, 587 (1964).
170. GOOGIN, J. M., and SMITH, H. A.: *J. Phys. Chem.*, **61**, 345 (1957).
171. ROSSINI, F. D., KNOWLTON, J. W., and JOHNSON, H. L.: *J. Res. Natl. Bur. Stand.*, **24**, 369 (1940).
172. UREY, H. C.: *J. Chem. Soc.*, 562 (1947).
173. CASE, B., and PARSONS, R.: *Trans. Faraday Soc.*, **63**, 1224 (1967).
174. AGARWAL, R. K., and NAYAK, B.: *J. Phys. Chem.*, **70**, 2568 (1966).
175. FEAKINS, D., and WATSON, P.: *J. Chem. Soc.*, 4686, 4734 (1963).
176. FRANKS, F., and IVES, D. J. G.: *Quart. Rev.*, **20**, 1 (1966).
177. BATES, R. G., and ROBINSON, R. A.: Reference 50, p. 221.
178. ÅKERLÖF, G.: *J. Amer. Chem. Soc.*, **52**, 2353 (1930).
179. GLADDEN, J. K., and FANNING, J. C.: *J. Phys. Chem.*, **65**, 76 (1961).
180. SCHUG, K., and DADGAR, A.: *J. Phys. Chem.*, **68**, 106 (1964).
181. PAVLOPOULOS, T., and STREHLOW, H.: *Z. Physik. Chem.*, *N.F.*, **1**, 89 (1954).
182. BRÄUER, K., and STREHLOW, H.: *Z. Physik. Chem.*, *N.F.*, **17**, 346 (1958).
183. RYBKIN, YU. F.: 'Kandidat Dissertation', Gorkii State University, Kharkhov (1963).
184. STREHLOW, H.: *The Chemistry of Non-Aqueous Solvents*, p. 146, Edited by J. J. Lagowski (Academic Press, N.Y., 1966).
185. MACFARLANE, A., and HARTLEY, H.: *Phil. Mag.*, **8**, 326 (1929).
186. HARTLEY, H., and MACFARLANE, A.: *Phil. Mag.*, **10**, 611 (1935).
187. PARKER, A. J., and ALEXANDER, R.: *J. Amer. Chem. Soc.*, **90**, 3313 (1968).
188. ANDREWS, A. L., BENNETTO, H. P., FEAKINS, D., LAWRENCE, K. G., and TOMKINS, R. P. T.: *J. Chem. Soc.*, A, 1486 (1968).
189. BENNETTO, H. P., and FEAKINS, D.: Reference 16, p. 235.
190. FEAKINS, D.: *Physico-chemical Processes in Mixed Aqueous Solvents*, p. 71, Edited by F. Franks (Heinemann Educational Books Ltd., London, 1967).
191. GRUNWALD, E., BAUGHAM, G., and XOHNSTAMM, G.: *J. Amer. Chem. Soc.*, **82**, 5801 (1960).
192. DE LIGNY, C. L., and ALFENAAR, M.: *Rec. Trav. Chim.*, **84**, 81 (1965).
193. ALFENAAR, M., and DE LIGNY, C. L.: *Rec. Trav. Chim.*, **86**, 929 (1967).
194. DE LIGNY, C. L., ALFENAAR, M., and VAN DER VEEN, N. G.: *Rec. Trav. Chim.*, **87**, 585 (1968).
195. RANDLES, J. E. B., and SCHIFFRIN, D. J.: *J. Electroanal. Chem.*, **10**, 360 (1965).

196. BUTLER, J. A. V., THOMSON, D. W., and MACLELLAN, W. H.: *J. Chem. Soc.*, 674 (1933).
197. MACKOR, E. L.: *Rec. Trav. Chim.*, **70**, 747 (1951).
198. KIPLING, J. J.: *J. Coll. Sci.*, **18**, 502 (1963).
199. PARSONS, R.: Reference 16, p. 246.
200. PARSONS, R.: Private communication.
201. STRANKS, D. R.: *Disc. Faraday Soc.*, **29**, 73 (1959).
202. DE LIGNY, C. L., and ALFENAAR, M.: Personal communication.
203. POPOVYCH, O., and DILL, A. J.: *Anal. Chem.*, **41**, 456 (1969).
204. ROSSEINSKY, D. R.: *Chem. Rev.*, **65**, 467 (1965).
205. LUKSHA, E., and CRISS, C. M.: *J. Phys. Chem.*, **70**, 1496 (1966).
206. LADELL, J., and POST, B.: *Acta Cryst.*, **7**, 559 (1954).
207. SENTI, F., and HARKER, D.: *J. Amer. Chem. Soc.*, **62**, 2008 (1940).
208. KATZ, J. L., and POST, B.: *Acta Cryst.*, **13**, 624 (1960).
209. SUZHKI, I.: *Bull. Chem. Soc., Japan*, **35**, 540 (1962).
210. LaPLANCHE, L. A., and ROGERS, M. T.: *J. Amer. Chem. Soc.*, **86**, 337 (1964).
211. MIZUSHIMA, S.: *Molecular Structure*, Chapter 6 (Academic Press, N.Y., 1954).
212. MIZUSHIMA, S., SIMANOUTI, T., NAGAKURA, S., KWATANI, K., TSUBOI, M., BALOA, H., and FUJIOKA, O.: *J. Amer. Chem. Soc.*, **72**, 3490 (1950).
213. SMITH, J. W.: *Electric Dipole Moments*, p. 175 (Butterworths, London, 1955).
214. BASS, S. J., NATHAN, W. I., MEIGHAN, R. M., and COLE, R. H.: *J. Phys. Chem.*, **68**, 509 (1964).
215. KURLAND, R. J., and WILSON, E. B.: *J. Chem. Phys.*, **27**, 585 (1957).
216. LEADER, G. R., and GORMLEY, J. F.: *J. Amer. Chem. Soc.*, **73**, 5731 (1951).
217. MEIGHAN, R. M., and COLE, R. H.: *J. Phys. Chem.*, **68**, 503 (1964).
218. BILLON, J. P.: *J. Electro Anal. Chem.*, **1**, 486 (1960).
219. *Organic Solvents*, p. 224, Edited by A. Weissberger (Interscience Publ. Inc., N.Y., 1955).
220. SEARS, P. G., LESTER, G. R., and DAWSON, L. R.: *J. Phys. Chem.*, **60**, 1433 (1956).
221. COTTON, F. A., and FRANCIS, R.: *J. Amer. Chem. Soc.*, **82**, 2986 (1960).
222. FERNÁNDEZ-PRINI, R., and PRUE, J. E.: *Trans. Faraday Soc.*, **62**, 1257 (1966).
223. HIRSCH, E., and FUOSS, R. M.: *J. Amer. Chem. Soc.*, **77**, 6115 (1955).
224. WATANABE, M., and FUOSS, R. M.: *J. Amer. Chem. Soc.*, **78**, 527 (1956).
225. BUFFAGNI, S., and DUNN, T. M.: *J. Chem. Soc.*, 5105 (1961).
226. DRAGO, R. S., MEEK, D. W., JOESTEN, M. D., and LA ROCHE, L.: *Inorg. Chem.*, **2**, 124 (1963).
227. GOPAL, R., and RIZVI, S. A.: *J. Ind. Chem. Soc.*, **43**, 179 (1966).
228. PARKER, A. J.: *Quart. Rev.*, **16**, 163 (1962).
229. JANZ, G. J., and DANYLUK, S. S.: *Chem. Rev.*, **60**, 209 (1960).
230. JOHARI, G. P., and TEWARI, P. H.: *J. Phys. Chem.*, **69**, 3167 (1965).
231. SEARS, P. G., WOLFORD, R. K., and DAWSON, L. R.: *J. Electrochem. Soc.*, **103**, 633 (1956).
232. SEARS, P. G., WILHOIT, E. D., and DAWSON, L. R.: *J. Phys. Chem.*, **59**, 373 (1955).
233. AGARWAL, R. K., and NAYAK, B.: *J. Phys. Chem.*, **71**, 2062 (1967).
234. DAWSON, L. R., ZUBER, W. H., and ECKSTROM, H. C.: *J. Phys. Chem.*, **69**, 1335 (1965).
235. POVAROV, YU. M., GORBANEV, A. J., KESSLER, YU. M., and SAFONOVA, I. V.: *Dokl. Akad. Nauk SSSR*, **155**, 1411 (1964).
236. VASENKO, E. N.: *Zhur. Fiz. Khim.*, **23**, 959 (1949).
237. KESSLER, YU. M.: *Elecktro Khimya*, **2**, 1467 (1966).
238. BERRY, R. S., and REIMANN, C. W.: *J. Chem. Phys.*, **38**, 1540 (1963).
239. STRACK, G. A., SWANDA, S. K., and BAHE, L. W.: *J. Chem. Eng. Data*, **9**, 416 (1964).
240. HELD, R. P., and CRISS, C. M.: *J. Phys. Chem.*, **71**, 2487 (1967).
241. HELD, R. P., and CRISS, C. M.: *J. Phys. Chem.*, **69**, 2611 (1965).
242. COATES, J. E., and TAYLOR, E. G.: *J. Chem. Soc.*, 1245 (1936).
243. FINCH, A., GARDNER, P. J., and STEADMAN, C. J.: *J. Phys. Chem.*, **71**, 2996 (1967).
244. SOMSEN, G.: *Rec. Trav. Chim.*, **85**, 517, 526 (1966).

245. VAN PANTHALEON VAN ECK, C. L.: Thesis (Leiden, 1958).
246. NOTLEY, J. M., and SPIRO, M.: *J. Phys. Chem.*, **70**, 1502 (1966).
247. HALE, J. M., and PARSONS, R.: *Advances in Polarography*, Vol. 3, p. 829 (Pergamon Press, 1960).
248. GOPAL, R., and HUSAIN, M. M.: *J. Ind. Chem. Soc.*, **40**, 981 (1963).
249. FRENCH, C. M., and GLOVER, K. H.: *Trans. Faraday Soc.*, **51**, 1418 (1955).
250. PRUE, J. E., and SHERRINGTON, P. J.: *Trans. Faraday Soc.*, **57**, 1795 (1961).
251. SEARS, P. G., WILHOIT, E. D., and DAWSON, L. R.: *J. Chem. Phys.*, **23**, 1274 (1955).
252. WILLIAMS, M. D., ELLARD, J. A., and DAWSON, L. R.: *J. Amer. Chem. Soc.*, **79**, 4652 (1957).
253. DAWSON, L. R., LESTER, G. R., and SEARS, P. G.: *J. Amer. Chem. Soc.*, **80**, 4233 (1958).
254. SCHÄFER, H. L., and SCHAFFERNICHT, W.: *Angew. Chem.*, **72**, 618 (1960).
255. COETZEE, J. F., SIMON, J. M., and BERTOZZI, R. J.: *Anal. Chem.*. **41**, 766 (1969).
256. LINDBERG, J. J., KENTTÄMAA, J., and NISSEMA, A.: *Suomen Kem.*, **34B**, 98 (1961).
257. 'Propylene Carbonate', Technical Bulletin, Jefferson Chemical Co., P.O. Box 303, Houston, Texas, U.S.A.
258. BUTLER, J. N.: *J. Electro. Anal. Chem.*, **14**, 89 (1967).
259. PAVLOPOULOS, T., and STREHLOW, H.: *Z. physik. Chem.*, **202**, 474 (1954).
260. CRUSE, K., and HUBER, R.: *Angew. Chem.*, **66**, 632 (1954).
261. KOLTHOFF, I. M., and COETZEE, J. F.: *J. Amer. Chem. Soc.*, **79**, 1852 (1957).
262. LANNUNG, A.: *Z. physik. Chem.*, **161**, 255 (1932).
263. WALDEN, P., and BIRR, E.: *Z. physik. Chem.*, **153**, 1 (1931); **163**, 263, 281 (1933).
264. MCDOWELL, M. J., and KRAUS, C. A.: *J. Amer. Chem. Soc.*, **73**, 3293 (1951).
265. DRAGO, R. S., and MEEK, D.: *J. Phys. Chem.*, **65**, 1446 (1961).
266. DUNNETT, J. S., and GASSER, R. P. H.: *Trans. Faraday Soc.*, **61**, 922 (1965).
267. CRAWFORD, J. M., and GASSER, R. P. H.: *Trans. Faraday Soc.*, **62**, 3451 (1966); **63**, 2758 (1967).
268. CASE, B.: Thesis (Bristol University, 1965).
269. NELSON, R. F., and ADAMS, R. N.: *J. Electro Anal. Chem.*, **13**, 184 (1967).
270. WU, YUNG-CHI, and FRIEDMAN, H. L.: *J. Phys. Chem.*, **70**, 501, 2020 (1966).
271. ARNETT, E. M., and MCKELVEY, D. R.: *J. Amer. Chem. Soc.*, **87**, 1515 (1965).
272. FRIEDMAN, H. L.: *J. Phys. Chem.*, **71**, 1723 (1967).
273. STREHLOW, H.: Reference 184, p. 134.
274. KOEPP, H.-M., WENDT, H., and STREHLOW, H.: *Z. Elektrochem.*, **64**, 483 (1960).
275. COETZEE, J. F., and CAMPION, J. J.: *J. Amer. Chem. Soc.*, **89**, 2513 (1967).
276. COGLEY, D. R., and BUTLER, J. N.: *J. Electrochem. Soc.*, **113**, 1074 (1966).
277. COETZEE, J. F., MCGUIRE, D. K., and HENDRICK, J. L.: *J. Phys. Chem.*, **67**, 1814 (1963).
278. TAKAHASHI, R.: *Talanta*, **12**, 1211 (1965).
279. SCHNEIDER, H., and STREHLOW, H.: *J. Electro Anal. Chem.*, **12**, 530 (1966).
280. LATIMER, W. H.: *Oxidation Potentials*, p. 336, 2nd edition (Prentice-Hall Inc., N.Y., 1952).
281. SCHÖBER, G., and GUTMANN, V.: *Advances in Polarography*, Vol. 3, p. 940, Edited by I. S. Longmuir (Pergamon, London, 1960).
282. PLESKOV, W. A.: *Usp. Khim.*, **16**, 254 (1947).
283. NELSON, I. V., and IWAMOTO, R. T.: *Anal. Chem.*, **33**, 1795 (1961).
284. BATES, R. G.: Reference 184, Chapter 3.
285. GUTBEZAHL, B., and GRUNWALD, E.: *J. Amer. Chem. Soc.*, **75**, 559, 565 (1953).
286. BECKER, R. S., and CHEN, E.: *J. Chem. Phys.*, **45**, 2403 (1966).
287. WACKS, M. E., and DIBBLER, V. H.: *J. Chem. Phys.*, **31**, 1557 (1959).
288. WACKS, M. E.: *J. Chem. Phys.*, **41**, 1661 (1964).
289. CASE, B., HUSH, N. S., PARSONS, R., and PEOVER, M. E.: *J. Electro Anal. Chem.*, **10**, 360 (1965).
290. HEDGES, R. M., and MATSEN, F. A.: *J. Chem. Phys.*, **28**, 950 (1958).
291. MINC, S., ZAGORSKA, I., and KOCZOROWSKI, Z.: *Roczniki Chemii Ann. Soc. Chim. Polonorum*, **41**, 1983 (1967).

THEORY OF MOLECULAR ELECTRODE KINETICS

R. R. Dogonadze

I. INTRODUCTION

In the past decade considerable progress has been made in the theory of electrode processes, both in understanding the physical mechanism of an elementary act and in obtaining quantitative data in calculations of the current and other kinetic parameters. Naturally, the theoretical results obtained at present are applicable only to a limited class of electrode reactions. It should be stressed, however, that some of the conclusions of the theory are of a general nature and valid for all reactions occurring under the conditions when the slow step of the process is the discharge or ionization of ions. On the other hand, the theoretically studied class of reactions is broad enough and their consideration is of great practical interest.

This paper is concerned with the quantum theory of electrode processes for reactions of two types:

(1) redox reactions involving electron exchange between electrode and ion in solution; and

(2) hydrogen ion discharge reactions.

Now we have at our disposal theoretical prerequisites for calculating even more complex electrode reactions, but here we shall confine ourselves to the description of semi-phenomenological theory based on rather general assumption (Section II). In spite of this theory being practically free of any

135

assumptions about the nature of the model, it permits us to draw some very important conclusions regarding the nature of the course of an electrochemical reaction. On the other hand, for this semi-phenomenological theory to be complete, it is necessary to determine the parameters contained in it by comparing them with experimental data. With such a treatment, however, the true physical mechanism of the course of the electrochemical reaction remains unexplained. Therefore in two subsequent sections we shall proceed from a specific model for each component of the total system participating in the elementary act of ion discharge. Naturally, we shall thus lose some accuracy in the quantitative results of the theory, but in the author's opinion, the main object of the theory now is to elucidate the physical mechanism of the course of electrode reactions and to obtain fundamental qualitative results.

Among the assumptions about the model of the theory, the dielectric continuum model used by us to describe the solvent stands apart. As will be shown later, a polar solvent plays an important dynamic role in the charge transfer process. Apparently the importance of the medium in homogeneous electron transfer reactions in polar liquids was first stressed by Libby.[1] He pointed out that due to the fluctuation motion of the solvent atoms during electron transfer, the system has to pass through the Franck–Condon barrier. This idea was later developed and applied to various homogeneous charge transfer processes in polar liquids by Platzman and Franck,[2] who used a dielectric continuum model for the description of the solvent. It is of interest to note that it was the above authors[2] who suggested the possibility of applying the theory of radiationless transitions developed in the theory of polyatomic molecules[3] and polar crystals.[4] The radiationless transition theory was elaborated by Pekar,[4] Lax,[3] Huang, and Rhys,[5] Krivoglaz,[6] Fröhlich,[7] Kubo[8] and others. In fact analysis of different charge transfer processes in polar systems shows the physical mechanism to be quite similar in liquids and crystals. The first attempt to extend the theory to polar liquids was made by Davydov[9] and Deygen[10] in the investigation of the absorption spectra and electric conductivity of metal ammonia solutions. The most consistent theory of homogeneous redox reactions within the framework of the classical theory was given by Marcus.[11-12] Though the method used by Marcus differs somewhat from Pekar's approach based on non-equilibrium thermodynamics, the final result for potential energy of non-equilibrium polarization is the same for both theories. Later Marcus generalized his theory for the case of redox reactions on metal electrodes[13-14] (see also[15-19]). In Section III we shall discuss in detail a polar solvent model to the approximation of a dielectric continuum. The expression for the potential energy and quantum-mechanical Hamiltonian of the solvent given there is more accurate than those in the theories of Pekar and Marcus. Besides it will be shown to what approximation it is possible to derive the results of the above authors from our formulae. In our opinion, the results in Section III are better in that a more accurate expression is obtained for the basic parameter of the theory—the reorganization energy of the solvent E_s—than in earlier theories. In Pekar and Marcus's theory the quantity E_s depends only on the geometric dimensions of ions. In the formula obtained by us however, the reorganization energy proves to be dependent on such an important parameter as the space correlation of dipole moments of solvent molecules. It is to be hoped from this result that the possibilities of comparing physical results of the theory with experimental data will be better in the future.

In addition, Section III gives an adiabatic perturbation theory for the estimation of the electron transfer probability in unit time. The last Section IV deals with the calculations of electrochemical kinetics for redox and hydrogen ion discharge reactions.

In conclusion, we shall discuss two basic assumptions of the theory, which will be considered valid for all the processes treated in the present paper. The first assumption is that total current is made up of mutually independent individual ion discharge acts. According to this assumption, estimation of the current presents two independent problems:

(1) Determination of the concentration of the discharging ions and the electrical potential distribution in the region near the electrode.

(2) Calculation of the specific electrode reaction rate constant.

The first problem is of a purely statistical nature. It is one of the most important problems in modern theoretical electrochemistry. In this paper, however, we shall be only concerned with the second problem. It will be clear from what follows that calculation of the current can be formally completed, the resulting expression containing the discharging ion's concentration C_s and the electrical potential φ_s in the reaction region as given parameters.

The second assumption refers to the choice of the slow step of the total electrochemical process. Everywhere in this paper we shall consider the discharge process to be the rate determining step. This assumption can be quantitatively formulated as the condition

$$\tau_e \gg \tau_d \tag{1}$$

where τ_e is the mean time necessary for one discharge act (the reciprocal of the discharge probability in unit time) and τ_d—the mean time between diffusion jumps of a discharging ion in solution. The physical significance of condition (eqn (1)) is that before discharging ion must 'impinge' many times on the electrode. The violation of this condition means that the slow step is

the diffusion transport of ions to the electrode, for in that case each 'impingement' of the ion on the electrode should result in a discharge act (probably
with an 'adhesion' coefficient). Actually at $\tau_e \ll \tau_d$ the ion resides so long in
the reaction region that the charge exchange process can be considered as being
a semi-equilibrium one and the current will be determined mainly by the
number of 'impingements' of ions upon the electrode.

II. SEMI-PHENOMENOLOGICAL THEORY

In studying an electrode process, it is convenient to analyse it first with the
aid of a semiphenomenological theory which should be based only on
sufficiently obvious assumptions. Though the physical ideas and mathematical
methods used in this section are of a rather general nature and applicable to
many electrochemical processes involving charge transfer, for convenience
we shall make a number of particular assumptions later to obtain a closed
expression for the current and to illustrate the basic assumptions of the
semiphenomenological theory.

It is characteristic of all processes considered by us that during a charge
transfer in an elementary electrochemical act one or several electrons in the
electrode must undergo a change in quantum state. For example, if a redox
reaction occurs on the electrode, such as

$$Ox + e_f \rightarrow Red \qquad (2)$$

(electron e_f, which at first is at an arbitrary quantum energy level ε_f in the
electrode, passes to the ion in solution to the level ε_i). Similarly, during ionization of a particle adsorbed on the electrode

$$A_{ads} \rightarrow A_{sol}^+ + e_f \qquad (3)$$

the electron, which is initially at the adsorption level ε_a and ensures the
chemical bond of particle A with the electrode, may pass after ionization to
an arbitrary level ε_f in the bulk of electrode, etc. Henceforth the qualitative
nature of the energy spectrum of an electron in initial and final states will be of
importance for us. For example, if the electron is in the bulk of the electrode,
two basic cases should be distinguished when the electrode material is either
metal or insulator, intrinsic or impurity semiconductors being particular
cases of these. A qualitative type of structure of the electron energy spectrum
for metals and insulators (semiconductors) is given in Figs. 1 and 2. Continuity of the energy spectrum is characteristic of both cases, but the values of
the energy level density $\rho(\varepsilon_f)$ near the Fermi level are essentially different for

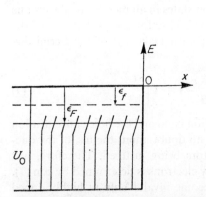

Fig. 1. Schematic diagram of the electron energy spectrum in metal; U_0 is the depth of the potential well for electrons.

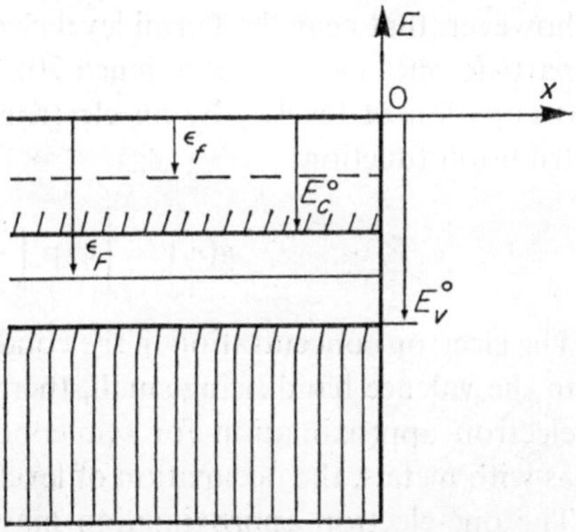

Fig. 2. Schematic diagram of the electron energy spectrum in semiconductor (insulator).

metals and insulators. For metals $\rho(\varepsilon_f)$ changes smoothly near ε_F. For example, in the free electron model

$$\rho_m(\varepsilon_f) = \frac{m_e V_e}{2\pi^2 \hbar^3} \sqrt{(2m_e(\varepsilon_f - U_0))} \tag{4}$$

where m_e is the electron mass, U_0 is the depth of a potential well in the metal, and V_e is the metal volume. As will be shown below, an essential role in the kinetics of electrode processes on metals is played only by the energy levels located near ε_F:

$$|\varepsilon_f - \varepsilon_F| \sim kT \tag{5}$$

In this energy range the relative change of $\rho(\varepsilon_f)$ is very small:

$$\frac{\Delta\rho_m}{\rho_m} \simeq \frac{kT}{2(\varepsilon_F - U_0)} < 1 \tag{6}$$

For example, at room temperature ($kT \simeq 0.025$ eV) and ($\varepsilon_F - U_0$) $\simeq 6$ eV this quantity is equal to 0.002 and at ($\varepsilon_F - U_0$) $= 3$ eV, $\Delta\rho_m/\rho_m \simeq 0.004$. This means that in electrochemical kinetics for metal electrodes we can consider $\rho(\varepsilon_f) \simeq \rho(\varepsilon_F) \equiv \rho_F$, i.e. the electron level density for a given electrode is constant.

For theoretical calculations condition eqn (5) viz. that only the electrons at energy levels near ε_F participate in the passage of the current, is of fundamental importance. This is due to the fact that owing to the large concentration of the conduction electrons it is not always the case that one particle (one electron) states can exist in the system. It has been rigorously proved, however, that near the Fermi level electron states in all metals are always one particle ones (see e.g. Reference 20). In this case for the probability of the occupation of level ε_f by an electron we must make use of the Fermi distribution function

$$n(\varepsilon_f) = \left[\exp\left(\frac{\varepsilon_f - \varepsilon_F}{kT}\right) + 1\right]^{-1} \tag{7}$$

The electron concentration in the conduction band and the hole concentration in the valence band being small, there is no doubt about the use of the one-electron approximation for semiconductors being justified. Therefore, just as with metals, the occupation of levels by electrons is determined by eqn (7). The one-electron approximation may become invalid only in highly doped semiconductors, which are not considered in this article. It is a specific feature of semiconductors that the electron level density near ε_F is zero (see Fig. 2):

$$\rho(\varepsilon_f) = 0, \quad E_v^0 \leqslant \varepsilon_f \leqslant E_c^0 \tag{8}$$

Therefore, the main contribution to the current is made by the electrons localized at the levels near the bottom of the conduction band E_c^0 and by the holes localized at the levels near the top of the valence band E_v^0. This shows that in semiconductors the dependence of the level density ρ on ε_f cannot be neglected. The concentrations of electrons in the conduction band and holes in the valence band being small, the following equations can be used:

$$\rho_e(\varepsilon_f) = \frac{m_e^* V_e}{2\pi^2 \hbar^3} \sqrt{(2m_e^*(\varepsilon_f - E_c^0))}, \quad \varepsilon_f \geqslant E_c^0; \tag{9}$$

$$\rho_p(\varepsilon_f) = \frac{m_p^* V_e}{2\pi^2 \hbar^3} \sqrt{(2m_p^*(E_v^0 - \varepsilon_f))}, \quad \varepsilon_f \leqslant E_v^0, \tag{10}$$

where m_e^* and m_p^* are effective masses of electrons and holes respectively.

In considering the electron states in an ion or adsorbed particle, we shall everywhere in this article assume the quantum energy level of the electron to be discrete and isolated. In the case of an ion, this means that the electron is localized at the level with the least possible energy ε_i and that the excited energy levels are located high enough for their occupation to be neglected. If one or more excited levels are located near ε_i, they can be taken into account by a method described later.

The spectrum of an electron in the adsorbed state is much more complicated. Let us consider this problem for metal electrodes in a strictly qualitative manner (for more detailed treatment, see References 21–23). The physical nature of adsorption will differ significantly depending on whether the electron energy in the adsorbed particle is larger ($\varepsilon_a > \varepsilon_F$) or less ($\varepsilon_a < \varepsilon_F$) than the

Fermi energy. In the former case ($\varepsilon_a > \varepsilon_F$) the electron is practically delocalized in the metal and the adsorbed particle is present on the surface as a relatively weakly bound ion. This state somewhat resembles physical adsorption. In this case, the effect of adsorbed particles on the electrode only amounts to some insignificant changes in the electron density and energy spectrum distributions. In this article for simplicity we shall consider only local adsorption levels ($\varepsilon_a < \varepsilon_F$). At $\varepsilon_a < \varepsilon_F$ the electron density distribution has a maximum near the adsorbed particle. In this sense, the corresponding energy value can be called a local level. If the adsorbed particle forms a local level, on the average, the total charge of electron and particle is zero. Owing to this sharp screening effect there is practically no interaction between adsorbed particles. Therefore, at fairly small fractional coverage of the electrode surface, the local adsorption level does not form a band structure.

For a description of the electrode process, it is necessary to know the number of elementary acts occurring per unit time, i.e. the transition probability per unit time, assuming the initial and final electron states to be given. The basic assumption of the semi-phenomenological theory is that in order to express the transition probability we assume as being valid the Arrhenius equation

$$W = \text{constant} \exp\left(-\frac{E_a}{kT}\right) \tag{11}$$

or the equation of the absolute reaction rate theory equivalent to it

$$W = \frac{kT}{h}\kappa \cdot e^{-F_a/kT} = \frac{kT}{h}\kappa \cdot e^{S_a/k} \cdot e^{-E_a/kT} \tag{12}$$

where κ is transmission coefficient, E_a, F_a, and S_a are the energy, free energy and entropy of activation respectively. Naturally, the question of the validity of eqns (11) and (12) for semi-phenomenological theory remains unsolved, since an expression for the transition probability can be obtained only from a quantum-mechanical treatment of the problem based on the use of a certain model for the whole system participating in the reaction. Unfortunately, the derivation of eqn (12) which is usually given in the theory of absolute reaction rates, is quite unsatisfactory. This is due not only to the absence of a rigorous quantum-mechanical calculation, but even to a larger degree to the lack of substantiation for the basic assumptions of the theory. The absence of a criterion of applicability of the theory of absolute reaction rates makes this method appear universal, which is not the case. This accounts for the appearance of a number of papers in which the theory of absolute reaction rates was used in calculation of such processes that cannot lead to equations of the type of (11) or (12). In particular, we shall show in Section IV that if the coordinate of the proton itself is chosen as the reaction coordinate in proton transfer processes in liquids, the transition probability is not described by eqn (12).

At present a rigorous quantum-mechanical derivation of eqn (12) is available only for charge transfer processes in condensed systems of the polar type if the system terms are assumed to have the form of parabolas (harmonic approximation). For radiationless processes in crystals this was demonstrated in References 6 and 8, and for homogeneous electron transfer reactions in polar liquids in References 24–29. The main criterion of applicability of eqn (12) for the reactions considered above can be written as

$$\hbar\omega_0 < kT \qquad (13)$$

where ω_0 is the frequency of polarization fluctuations in liquids. If condition eqn (13) is fulfilled, the transition of the system from the initial electron term to the final state term will occur at the intersection point of the potential energy curves, which corresponds to the classical path of transition of the system over the potential barrier. Initially in the theory developed in References 24–29 the intramolecular degrees of freedom of reacting ions were assumed to be 'frozen', i.e. that in a dynamical sense the total system consisted only of electron and solvent. With a view to subsequent calculations for more complex homogeneous and heterogeneous charge transfer processes in polar liquids, we have estimated the quantum-statistical transition probability for one-dimensional terms of a rather general form. The main result obtained by us can be formulated as a criterion of applicability of the classical approach to the motion of a particle:

$$\Delta E < kT \qquad (14)$$

where ΔE is the difference between the first excited state and the ground state of the system.[30] Since the equally-spaced energy levels of an oscillator differ by the quantity $\hbar\omega_0$, we see that condition eqn (13) is a particular case of condition eqn (14).

On this basis, the equation for the transition probability eqn (12) can be interpreted as follows: the classical part of the total system is responsible for the appearance of a factor exponentially dependent on the reciprocal temperature (activation factor), whereas the quantum part of the total system contributes to the transfer coefficient κ, thus establishing the adiabatic or nonadiabatic nature of the reaction course. Naturally, in this case the activation energy value will depend on the parameters of the quantum subsystem. It is important, however, that the quantum part in itself cannot lead to the activation equation. The parameters of the classical subsystem, in turn, will be contained in the transfer coefficient. Thus, the separation of the roles of classical and quantum subsystems is physical rather than mathematical. In the transfer processes of charge strongly interacting with the polar medium which are of interest to us, there must exist at least one classical component, the solvent, since eqn (13) seems to be valid for all polar solvents. For example, for water at room temperature, we have:

$$\omega_0 \simeq 10^{11} \text{ sec}^{-1} < \frac{kT}{\hbar} \simeq 4 \cdot 10^{13} \text{ sec}^{-1} \tag{15}$$

On this statement is actually based the most important assumption of the semi-phenomenological theory regarding the applicability of eqn (12) for calculation of electrode processes. In a sense, our assumption is experimentally verified by the exponential dependence of the current on the reciprocal temperature in the Tafel equation.

In addition to the classical part, the total system must contain at least one quantum subsystem—the electron. The kind of motion of other components (the intramolecular degrees of freedom of the reagents) is determined by means of the criterion eqn (14): if the corresponding degrees of freedom are 'frozen' ($\Delta E > kT$), they should be included in the quantum subsystem, whereas the degrees of freedom excited at room temperature ($\Delta E < kT$) should be considered by a classical approach. If in addition to the electron, the reacting system contains other quantum subsystems, in calculating the transition probability W it is necessary to write along with the electron states, the quantum number of these subsystems in the initial and final states, giving corresponding subscripts to the transmission coefficient and the activation free energy. As the quantum subsystem is only weakly excited at room temperature ($\Delta E > kT$), the unexcited ground state is of primary importance for electrochemical kinetics. We shall see later, however, that in certain overvoltage regions these excited states can still make a contribution.

It should be pointed out that the treatment in this section refers only to reactions in condensed phases. In the general case, it may prove inapplicable to reactions in the gas phase. To explain the essential difference between these two types of reactions, it should be recalled that the free translational motion of particles characteristic of gas reaction is practically absent in condensed phases and that in liquids particles move only by jumps (the only exception being very large ions moving almost in a classical manner, i.e. those which have shallow potential wells). Characteristic potential energy profiles for reactions in gas and condensed phases are given in Figs. 3 and 4. It should be noted that some authors sometimes use electron terms of the type shown in Fig. 3 for the treatment of reactions in liquids. In view of the foregoing consideration, such an approach to the problem is incorrect.

In estimating the total transition probability W in condensed systems, the quantum-mechanical expression for the probability is statistically averaged over the initial energies of the system, including only intrinsic degrees of freedom (vibrational and rotational). For gas reactions an analogue of W is the scattering cross section $\sigma(K_{0i}; K_{0f}; T)$, statistically averaged over initial

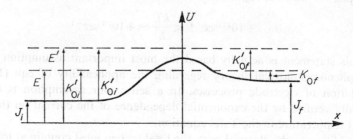

Fig. 3. Characteristic shape of the potential energy profile for reactions in the gas phase. J_i, J_f are internal energies of particles before and after reaction, E, E' are total energies of reacting particles.

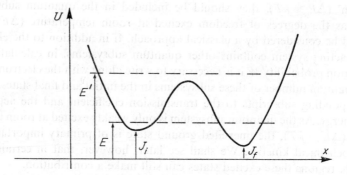

Fig. 4. Characteristic shape of the potential energy profile for reactions in the condensed phase.

energies of the intrinsic degrees of freedom of scattering particles. (K_{0i} and K_{0f} are kinetic energies of particles before and after scattering.) In order to obtain the rate constant of a gas reaction, the cross section must be statistically averaged over initial velocities of colliding particles by means of Maxwellian distribution and integrated over final velocities

$$K \sim \int \sigma(K_{0i}; K_{0f}; T)\, e^{-K_{0i}/kT}\, dK_{0i}\, dK_{0f} \tag{16}$$

It is specific to gas reactions that the Arrhenius shape of the rate constant can be due not only to the dependence of the cross section on temperature, but also to the Maxwellian averaging over velocities. This becomes evident if we consider cross sections of two types:

(1) $\sigma = 0$ at $K_{0i} < E_a$ and $\sigma = $ constant at $K_{0i} > E_a$ (threshold reaction).
(2) $\sigma \sim \delta(K_{0i} - E_a)$ (resonance reaction), where $\delta(K_{0i} - E_a)$ is the Dirac delta function and E_a is the activation energy.

Thus, in gas reactions even at 'frozen' intrinsic degrees of freedom of molecules the translational motion of particles can act as a classical subsystem.

For a general consideration of reactions on the basis of a semi-phenomenological theory, the type of dependence of the transition probability on the free electron energy in the electrode ε_f is of great importance. We shall assume everywhere in this article that the transmission coefficient in the energy range of interest to us, which is determined by eqn (5), changes very slightly. In particular, if we use a free electron model for the metal, we can write the wave function as a plane wave

$$\psi_f(X) = \frac{1}{\sqrt{(V_e)}} e^{iK_f X} \qquad (17)$$

where the wave number K_f is determined by the equality

$$K_f = \frac{1}{\hbar} \sqrt{(2m_e(\varepsilon_f - U_0))} \qquad (18)$$

The transmission coefficient is determined by the square of the exchange integral, which contains the overlap of the function ψ_f and the wave function of the electron in the final state. The latter function, in virtue of our assumption about the local character of the electron state in the ion or adsorbed particle differs from zero only in a small length a_0 (a_0 is equal to one or several Angstroms). Thus, the criterion of validity of our assumption is of the form

$$|\Delta K_f \cdot a_0|^2 \simeq \frac{m_e a_0^2 kT}{2\hbar^2} \cdot \frac{kT}{|\varepsilon_F - U_0|} < 1 \qquad (19)$$

As the actual values of $(\varepsilon_F - U_0)$ are several electron-volts, we see that at room temperature this condition is satisfied fairly well. Therefore, henceforth the transmission coefficient will be assumed to be independent of ε_f:

$$\kappa(\varepsilon_f) \simeq \kappa(\varepsilon_F) \equiv \kappa_F$$

In order to study the dependence of the free energy of activation on ε_f, let us consider a number of quasithermodynamic characteristics of the system, assuming the electron to be the only quantum particle in the whole system. Let us introduce the concept of the free energy of the system at a fixed quantum state of the electron localized in the electrode at level ε_f. Here and henceforth we denote by F the so called 'standard' free energy, i.e. the free energy calculated for one ion. In calculating $F(\varepsilon_f)$ we should remember that according to the standard procedure for quantum-mechanical calculation of the transition probability, the initial and final states of the system are determined without taking into consideration the perturbation energy, i.e. the interaction of the electron with the discharging ion. The presence of the perturbation leads only to transitions between 'pure' states of the system at the beginning and the end. Hence it follows that the electron energy, equal to $\varepsilon_f - e\varphi_m$ (where φ_m is the Galvani potential difference between metal electrode and solution), makes an additive contribution to the free energy:

$$F(\varepsilon_f) = \varepsilon_f - e\varphi_m + \text{constant} \tag{20}$$

Henceforth it will be more convenient for us to use an expression for $F(\varepsilon_f)$ written in a somewhat different form. The free energy of activation of the system in the state when $\varepsilon_f = \varepsilon_F$ and the electrode potential has an equilibrium value φ_0, can be calculated by means of a familiar equation of statistical physics:

$$F_{0F} = -kT \ln \sum_n \exp \left\{ - \frac{1}{kT} (\varepsilon_F - e\varphi_0 + Ze\varphi_{0S} + E_{00}^{(1)} + E_{0n}^{(1)}) \right\}$$

$$= \varepsilon_F - e\varphi_0 + Ze\varphi_{0S} + E_{00}^{(1)} - kT \ln \sum_n \exp \left(- \frac{E_{0n}^{(1)}}{kT} \right) \equiv J_{0F}^{(1)} + \mathscr{F}_0^{(1)} \tag{21}$$

where $\varphi_{0S}^{(1)}$ is the potential at the point from which the discharge occurs; $E_{0n}^{(1)}$ represents the energy levels of the classical subsystem in the initial state reckoned from the ground level $E_{00}^{(1)}$. The symbol $\mathscr{F}$ here and below stands for the free energy of the classical subsystem calculated assuming the energy to be reckoned from the least value of E_0. In the literature E_0 and J_{0F} sometimes refer to the energy of the system at the temperature $T \to 0$. For chemical reactions however, occurring in the liquid phase this definition is incorrect, since at $T \to 0$ all liquids (except quantum liquids) undergo a phase transition. It is clear that the energy spectrum, as well as the thermodynamic characteristics of a liquid differs essentially from the corresponding quantities for the same substance in the solid phase. The quantity J_{0F} has a particularly simple interpretation if an electron term concept is introduced for the classical subsystem. As is clear from Fig. 5 (curve a) J_{0F} practically coincides with the potential energy minimum since for the classical subsystem $\Delta E < kT$.

Along with the true thermodynamic quantity F_{0F}, we can also consider the nonequilibrium quasithermodynamic free energy of the initial state, when the electron is at an arbitrary level ε_f and the electrode potential is equal to $\varphi_m \neq \varphi_0$:

$$F^{(1)}(\varepsilon_f) = \varepsilon_f - e\varphi_m + Ze\varphi_{0S}$$
$$+ E_0^{(1)} - kT \ln \sum_n \exp\left(-\frac{E_n^{(1)}}{kT}\right) \equiv J_f^{(1)} + \mathscr{F}^{(1)}$$
$$= (\varepsilon_f - \varepsilon_F) - e\eta + Ze\eta_S + E_0^{(1)} - E_{00}^{(1)} + \mathscr{F}^{(1)} - \mathscr{F}_0^{(1)} + F_{0F} \quad (22)$$

where $\eta = \varphi_m - \varphi_0$ and $\eta_S = \varphi_S - \varphi_{0S}$ are the overvoltages on the electrode and at the point from which discharge occurs. The first three terms in the right hand side of the second equation correspond to the changes in the energies of electron and discharging ion on applying an overvoltage and excitation of the electron from level ε_F to level ε_f. The next two terms in this equation describe the change in the energy spectrum of the classical subsystem with changing potential. With reference to electron terms this means that the first three terms in $F(\varepsilon_f)$ lead to a parallel displacement of the equilibrium term vertically upwards (curve b in Fig. 5); the fourth and the fifth

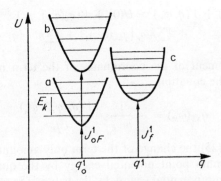

Fig. 5. Qualitative picture of the electronic terms of the classical subsystem: (a) at the equilibrium potential and $\varepsilon_f = \varepsilon_F$; (b) in the presence of overvoltage and $\varepsilon_f \neq \varepsilon_F$ (without allowing for the interaction with the electrode field); (c) at $n \neq 0$ and $\varepsilon_f \neq \varepsilon_F$ (allowing for the interaction with the electrode field).

term $F(\varepsilon_f)$ reflect the potential energy change in the classical subsystem with changing electrode potential, $E_0^{(1)} - E_{00}^{(1)}$ corresponding to the vertical displacement of the term minimum point and $\mathscr{F}^{(1)} - \mathscr{F}_0^{(1)}$ characterizing the change in the electron term shape. In order to understand the physical nature of the changes in the classical subsystem caused by overvoltage, let us consider the behaviour of the terms in the vicinity of the minimum point, assuming the deviation of normal coordinates of the classical subsystem q_K from the equilibrium values to be small. Then, the potential energy of the classical

6

subsystem, in the absence of the electrode potential, can be roughly shown as a parabola:

$$U(q) = U(0) + \tfrac{1}{2}\sum_k \hbar\omega_K q_K^2 \tag{23}$$

where ω_K are the normal frequencies of the classical subsystem (solvent and intramolecular vibrations of discharging ion). The interaction of the classical subsystem with the electric field $V(q, \varphi_m)$ can be expanded as a power series in the small deviations of q_K from $q_{K0}(\varphi_m)$, where $q_{K0}(\varphi_m)$ are the coordinates of the electron term minimum point at the electrode potential value φ_m:

$$V(q, \varphi_m) = V(q_0(\varphi_m), \varphi_m) + \sum_k \frac{\partial V(q_0(\varphi_m), \varphi_m)}{\partial q_K} (q_K - q_{K0}(\varphi_m))$$

$$+ \tfrac{1}{2}\sum_{k,k'} \frac{\partial^2 V(q_0(\varphi_m), \varphi_m)}{\partial q_K \partial q_{K'}} (q_K - q_{K0}(\varphi_m))(q_{K'} - q_{K'0}(\varphi_m)) + \ldots \tag{24}$$

If we take only the first two terms in eqn (24) the electron term can be written as

$$U(q, \varphi_m) = U(q) + V(q, \varphi_m) \simeq U(0) + V(q_0(\varphi_m), \varphi_m)$$

$$+ \tfrac{1}{2}\sum_k \hbar\omega_K[q_{K0}(\varphi_m)]^2 + \tfrac{1}{2}\sum_k \hbar\omega_K[q_K - q_{K0}(\varphi_m)]^2 \tag{25}$$

where the displacement of the coordinates of the term minimum point is determined from the equation

$$q_{K0}(\varphi_m) = -\frac{1}{\hbar\omega_K} \cdot \frac{\partial V(q_0(\varphi_m), \varphi_m)}{\partial q_K} \tag{26}$$

According to eqn (25) the change of the term only amounts to the displacement of the minimum point potential energy by the quantity $q_0(\varphi_m)$, the decrease of minimum potential energy by the value of reorganization energy of the classical subsystem $\tfrac{1}{2}\sum_k \hbar\omega_K[q_{K0}(\varphi_m)]^2$ plus the energy of interaction of the field with the classical subsystem in the new equilibrium state $V(q_0(\varphi_m), \varphi_m)$:

$$E_0^{(1)} = U(q_0(\varphi_m)) + V(q_0(\varphi_m), \varphi_m)$$

$$\simeq U(0) + V(q_0(\varphi_m), \varphi_m) + \tfrac{1}{2}\sum_k \hbar\omega_K[q_{0K}(\varphi_m)]^2 \tag{27}$$

It is important that to the first approximation the vibration frequencies (i.e. the term shape) are not affected by the field. Then the free energy of the system of classical oscillators ($\hbar\omega_K < kT$) can be written as

$$\mathscr{F}^{(1)} \simeq -kT\sum_k \ln \tfrac{1}{2} \operatorname{csch} \frac{\hbar\omega_K}{2kT} \simeq -kT\sum_k \ln \frac{kT}{\hbar\omega_K} \tag{28}$$

The shape of the potential energy surface is changed if we take into consideration the third and the subsequent terms in the energy of interaction of the classical subsystem with the electric field (see eqn (24)). In particular to the second approximation (with account taken of the third term in eqn (24)), it can be shown by means of some simple linear algebraic transformations that the displacement of the term minimum point is accompanied by a change in the frequencies of normal vibrations. The physical sense of this is that to the second approximation the elastic constants of intramolecular degrees of freedom of the classical subsystem change under the action of the electric field. The new frequencies, naturally, will depend on the field. The change of free energy and minimum potential energy upon application of overvoltage can be written as

$$\mathscr{F}^{(1)} - \mathscr{F}^{(1)}_0 \simeq kT\sum_k \ln \frac{\Omega_K(\varphi_m)}{\Omega_K(\varphi_0)}, \tag{29}$$

where $\Omega_K(\varphi)$ are the normal frequencies at Galvani potential φ. In addition, we have:

$$E^{(1)}_0 - E^{(1)}_{00} = [U(q_0(\varphi_m)) - U(q_0(\varphi_0))]$$
$$+ [V(q_0(\varphi_m), \varphi_m) - V(q_0(\varphi_0), \varphi_0)] \tag{30}$$

The difficulty of calculating these quantities for particular systems is due not only to the difficulty of determining the shape of the term $U(q)$ and the form of the interaction $V(q, \varphi_m)$, but possibly even to a larger degree to the need to know the distribution of electrical potential in the electrolyte. Therefore, to simplify the calculations we shall assume the whole potential drop to occur within the Helmholtz layer, i.e. the absence of the ψ_1-effect. As will be shown later, the basic results of the theory are all formally generalized also for the case of the presence of the ψ_1-effect. But in that case concrete expressions become very cumbersome and make quantitative estimates of the results more difficult.

In the absence of the ψ_1-effect, the quasithermodynamic nonequilibrium free energy of the initial state is

$$F^{(1)}(\varepsilon_f) = \varepsilon_f - e\varphi_m + E^{(1)} + \mathscr{F}^{(1)}$$
$$= J^{(1)}_f + \mathscr{F}^{(1)} = (\varepsilon_f - \varepsilon_F) - e\eta + F_{0F} \tag{31}$$

since in this case the electric field in that part of the space where the classical subsystem (solvent and discharging ion) is located is equal to zero. Since we have assumed that the adsorbed particle is neutral it follows that in the general case the free energy of the final state will be a constant, independent of the electrode potential:

$$F^{(2)} = \varepsilon_a + E^{(2)} + \mathscr{F}^{(2)} \equiv J^{(2)} + \mathscr{F}^{(2)} \tag{32}$$

where ε_a is the electron energy in the final state (in the adsorbed particle or reaction product ion); $E^{(2)}$ and $\mathscr{F}^{(2)}$ are the minimum potential energy and the free energy of the classical subsystem at the end of the reaction, respectively. Accordingly, the change of free energy in the course of reaction is

$$\begin{aligned}
\Delta F(\varepsilon_f) &= F^{(1)}(\varepsilon_f) - F^{(2)} \\
&= \Delta J_f + \Delta\mathscr{F} \\
&= -e\eta + (\varepsilon_f - \varepsilon_F) + (F_{0F} - F^{(2)}) \\
&= (\varepsilon_f - \varepsilon_F) - e\eta + \Delta F_{0F}
\end{aligned} \tag{33}$$

It can be readily seen that the term which is of real thermodynamic significance is ΔF_{0F}, which determines the equilibrium constant

$$K_0 = \frac{C_0^{(1)}}{C_0^{(2)}} = \exp\left(-\frac{\Delta F_{0F}}{kT}\right) \tag{34}$$

where $C_0^{(1)}$ and $C_0^{(2)}$ are surface concentrations of ions before and after discharge at the equilibrium potential. If we introduce in a strictly formal manner an equilibrium constant for the reaction involving only the electron localized at the level ε_f and leaving after the reaction a 'hole' at the same level we shall obtain for this quantity

$$K_0(\varepsilon_f) = \frac{C_0^{(1)} \cdot n(\varepsilon_f)}{C_0^{(2)}[1 - n(\varepsilon_f)]} = \exp\left(-\frac{\Delta F_0(\varepsilon_f)}{kT}\right) \tag{35}$$

where $n(\varepsilon_f)$—the Fermi distribution—appears as the electron concentration at level ε_f; similarly, $1 - n(\varepsilon_f)$ is the 'hole' concentration. Thus, each electron level in the electrode acts as an independent redox system in electrochemical kinetics, it being very important, however, that the equilibrium potential for each level has the same value. In fact, the equilibrium potential value is determined from the equation

$$\Delta F_0(\varepsilon_f) + kT \ln \frac{C_0^{(1)} \cdot n(\varepsilon_f)}{C_0^{(2)}[1 - n(\varepsilon_f)]} = 0 \tag{36}$$

After substituting for the Fermi distribution, this becomes:

$$\Delta F_0(\varepsilon_f) - (\varepsilon_f - \varepsilon_F) + kT \ln \frac{C_0^{(1)}}{C_0^{(2)}} = 0$$

which according to eqn (33) is equivalent to an equation independent of ε_f, i.e.:

$$\Delta F_{0F} + kT \ln \frac{C_0^{(1)}}{C_0^{(2)}} = 0 \tag{37}$$

In order to obtain an expression for the equilibrium potential eqns (31) and (32) should be substituted into eqn (37):

$$e\varphi_0 = (\varepsilon_F - \varepsilon_a) + (E_0^{(1)} - E^{(2)}) + (\mathscr{F}_0^{(1)} - \mathscr{F}^{(2)}) + kT \ln \frac{C_0^{(1)}}{C_0^{(2)}}$$

$$= \Delta F(\varepsilon_F, \varphi_m = 0) + kT \ln \frac{C_0^{(1)}}{C_0^{(2)}} \tag{38}$$

where $\Delta F(\varepsilon_F, \varphi_m = 0)$ is the difference of the free energies of initial and final states at $\varphi_m = 0$. Essentially eqns (36) and (37) correspond to the equilibrium condition of a chemical reaction written as an equality of overall chemical potentials, the quantity $\Delta F(\varepsilon_F, \varphi_m = 0)$ being the change of the chemical potential of the reacting system. Naturally, an expression for the equilibrium potential eqn (38) can be obtained from a consideration of standard thermodynamic cycles.

One of the most important assumptions of the semi-phenomenological theory, as well as of any other microscopic theory now available, is the assumption of the quasiequilibrium nature of the occupation of energy levels of the systems in the case of small deviations from equilibrium. Speaking more generally, this means that we shall use the Gibbs distribution for slightly nonequilibrium systems. Determination of the nonequilibrium distribution function involves a most complicated problem of solving a kinetic equation. However, according to rough estimates, for electrochemical processes under the conditions when the charge transfer is the slow step, our assumption is usually adequate. As will be shown in the next section, the quasiequilibrium Gibbs distribution was actually implicitly used by us when we chose for the transition probability the absolute reaction rate eqn (12). From the condition of quasiequilibrium in conjunction with the principle of balancing, we can find a simple relationship between the forward (W_{12}), and back (W_{21}) transition probabilities. In the general case, the forward transition probability is of the form:

$$W_{12} = \sum_{i,l} W_{12}(\varepsilon_i^{(1)} \varepsilon_l^{(2)}) g_1(\varepsilon_i^{(1)}) \, e^{-\varepsilon_i^{(1)}/kT} \left[\sum_n g_1(\varepsilon_n^{(1)}) \, e^{-\varepsilon_n^{(1)}/kT} \right]^{-1} \tag{39}$$

where $\varepsilon^{(1)}$, $\varepsilon^{(2)}$ are the energy levels of initial and final states and $g_1(\varepsilon^{(1)})$ is the statistical weight of an initial state level. The denominator in eqn (39) is a statistical sum for the initial state and can be expressed in terms of the free energy as:

$$\sum_i g_1(\varepsilon_i^{(1)}) \, e^{-\varepsilon_i^{(1)}/kT} = Q_1 = \exp\left(-\frac{F^{(1)}}{kT}\right) \tag{40}$$

In accordance with quantum mechanics, when the system passes from level $\varepsilon_i^{(1)}$ to level $\varepsilon_i^{(2)}$ under steady-state conditions, the law of conservation of energy is obeyed, i.e. $\varepsilon_i^{(1)} = \varepsilon_i^{(2)}$. Therefore, using the principle of detailed balancing:

$$W_{12}(\varepsilon_i^{(1)}, \varepsilon_i^{(2)})g_1(\varepsilon_i^{(1)}) = W_{21}(\varepsilon_i^{(2)}, \varepsilon_i^{(1)})g_2(\varepsilon_i^{(2)}) \tag{41}$$

and from eqn (39) we obtain:

$$W_{12} = [\sum_{i,l} W_{21}(\varepsilon_i^{(2)}, \varepsilon_i^{(1)})g_2(\varepsilon_i^{(2)}) \, \mathrm{e}^{-\varepsilon_i^{(2)}/kT}] \, \mathrm{e}^{F^{(1)}/kT}$$

After multiplying and dividing the right hand side by the statistical sum of final state, we finally obtain

$$W_{12} = W_{21} \, \mathrm{e}^{\Delta F/kT} \tag{42}$$

Equation (42) obtained above is of a completely general nature and applicable to any kinetic process if the condition of quasiequilibrium of initial and final states is obeyed (generalization for the case of non-steady state processes, i.e. for perturbations dependent on time is a trivial modification, see e.g.[31]). If into eqn (42) we substitute eqn (33) for $\Delta F(\varepsilon_f)$ and use eqns (34) and (35) we obtain the following relations between the forward and back electron transition probabilities in electrochemical kinetics:

$$\frac{W_{21}(\varepsilon_f)}{W_{12}(\varepsilon_f)} = \exp\left(\frac{e\eta - (\varepsilon_f - \varepsilon_F) - \Delta F_{0F}}{kT}\right) = K_0(\varepsilon_f) \, \mathrm{e}^{e\eta/kT} \tag{43}$$

Having carried out these preliminary steps we can then pass to direct estimation of the current density. For definiteness, we shall assume a cathodic process to occur on the electrode and introduce the cathodic overvoltage

$$\eta_c = \varphi_0 - \varphi_m = -\eta \tag{44}$$

The expression for the total cathodic current can be written as

$$i_c = \overrightarrow{i_c} - \overleftarrow{i_c} = eC_0^{(1)}\int np\,W_{12}\,\mathrm{d}\varepsilon_f - eC^{(2)}\int(1 - n)p\,W_{21}\,\mathrm{d}\varepsilon_f \tag{45}$$

where integration is performed over the whole electron energy spectrum in the electrode. In writing eqn (45) we have made use of the assumption about the absence of the ψ_1-effect and substituted the equilibrium concentration of discharging ions $C_0^{(1)}$. If the potential drop extends beyond the Helmholtz layer to the diffuse part of the double layer, nonequilibrium values should be written for the concentration. In this case, we are faced with the problem of choosing the distance at which the ion discharge occurs. Moreover, in the general case, we should take into consideration the possibility of an ion discharge at any distance from the electrode, i.e. calculate the integral of the type

$$\int_\delta^\infty C^{(1)}(x)W(x)\,dx \tag{46}$$

where δ is the least distance of approach of discharging ion to the electrode. However, in view of the fact, that as a rule the quantum mechanical transition probability $W(x)$ diminishes sharply at atomic distances, whereas $C^{(1)}(x)$ changes fairly smoothly, the integrand in eqn (46) has a very sharp maximum near the point $x = \delta$. This means that in the presence of the ψ_1-effect, $C_0^{(1)}$ can be substituted by the nonequilibrium ion concentration at the point $x \approx \delta$, i.e. by $C^{(1)}(\delta, \varphi_m)$.

So far we have discussed only the concentration of discharging ions. If we wish to consider the concentration of discharged ions, we have to distinguish between two cases:

(1) when these ions are not adsorbed; and
(2) when these ions are adsorbed.

In the former case, $C^{(2)}$ can be substituted by the equilibrium value $C_0^{(2)}$. In the latter case, it is necessary to consider the particular desorption mechanism in the given system and to write the steady-state condition for the electrochemical process, i.e. to equate the numbers of adsorbed and desorbed particles per unit time. The supplementary equation thus obtained allows us to express the nonequilibrium concentration of adsorbed particles in terms of the concentrations of initial and final reaction products, which in the absence of the ψ_1-effect have equilibrium values. In order not to complicate the description of general methods of semi-phenomenological theory, we shall not consider here the desorption current and shall retain the formally nonequilibrium concentration $C^{(2)}$ in the ionization current $\overleftarrow{i_c}$. An example of the procedure of taking into consideration the desorption processes in electrochemical kinetics will be given in Section IV in the treatment of the hydrogen overvoltage theory.

Using eqns (35) and (42) let us rewrite the expression for the total current in the form

$$i_c = eC_0^{(1)}\int n\rho W_{12}\left(1 - \frac{C^{(2)}}{C_0^{(2)}}\cdot\frac{C_0^{(2)}}{C_0^{(1)}}\cdot\frac{1-n}{n}\cdot\frac{W_{21}}{W_{12}}\right)d\varepsilon_f$$

$$= eC_0^{(1)}\int n\rho W_{12}\left(1 - \frac{C^{(2)}}{C_0^{(2)}}e^{-e\eta_c/kT}\right)d\varepsilon_f \tag{47}$$

The relation between forward and back currents is then:

$$\overleftarrow{i_c} = \overrightarrow{i_c} \cdot \frac{C^{(2)}}{C_0^{(2)}} \, e^{-e\eta_c/kT}; \quad i_c = \overrightarrow{i_c}\left(1 - \frac{C^{(2)}}{C_0^{(2)}} \, e^{-e\eta_c/kT}\right) \tag{48}$$

Thus the treatment can be confined to the calculation of the discharge current $\overrightarrow{i_c}$, which by means of eqn (12) can be rewritten for metal electrodes as

$$\overrightarrow{i_c} = eC_0^{(1)}\frac{kT}{h}\,\kappa_F\rho_F\int n(\varepsilon_f)\, e^{-F_a(\varepsilon_f)/kT}\, d\varepsilon_f = \int d\overrightarrow{i_c}(\varepsilon_f) \tag{49}$$

For further treatment we have to determine more accurately the nature of the dependence of the free energy of activation on ε_f and on overvoltage η_c. It will be shown in Section IV on the basis of a rigorously quantum-mechanical treatment without any assumptions about the model that the free energy of activation can contain as an argument the difference of energies of the quantum subsystem in initial and final states only in terms of the quantity ΔJ. As shown above, this is the difference of the total energies of the system in initial and final states, assuming the classical subsystem to be in the ground (unexcited) energy state. If the behaviour of the classical subsystem is described by means of a potential energy surface, ΔJ is the difference of minimum potential energies of initial and final states (see Fig. 4). Evidently, the above statement does not mean that ΔJ is the only argument of the free activation energy F_a. Moreover, an expression for F_a will certainly contain the parameters characterizing the energy spectrum and wave functions of the classical subsystem, or in terms of the potential energy surface the parameters characterizing the geometric shape and relative positions of the terms.

In the absence of the ψ_1-effect, according to eqns (31)–(33) we have:

$$F_a = F_a(\Delta J_f, \gamma) = F_a(\varepsilon_f - e\varphi_m + E^{(1)} - \varepsilon_a - E^{(2)}, \gamma) \tag{50}$$

where γ stands for the set of all parameters characterizing the geometric shape and relative position of the terms of initial and final states. Since in the absence of the ψ_1-effect, the quantities γ, $E^{(1)}$, $E^{(2)}$, and ε_a do not depend on φ_m, the whole dependence of the free energy of activation on the electrode potential φ_m is contained in the combination $\varepsilon_f - e\varphi_m$, i.e.

$$F_a = F_a(\varepsilon_f - e\varphi_m, \gamma) \tag{51}$$

However in what follows it will be sometimes more convenient to write the argument of the free energy of activation not as ΔJ_f or $\varepsilon_f - e\varphi_m$, but as $\Delta F(\varepsilon_f)$. This possibility results from the independence of $\Delta\mathscr{F}$ on ε_f and φ_m or ultimately from the constancy of the shape of the terms of initial and final states. Thus, according to eqns (33) and (44) we have

$$F_a = F_a(\Delta F(\varepsilon_f), \gamma) = F_a(\Delta J_f + \Delta\mathscr{F}, \gamma)$$
$$= F_a(\varepsilon_f - \varepsilon_F + e\eta_c + \Delta F_{0F}, \gamma) \tag{52}$$

It should be stressed that eqn (52) (as well as eqn (51)) is valid only in the case of the absence of the ψ_1-effect, or to be more precise, in the case when the coefficients determining the geometric shape of the potential energy surfaces and their relative position do not depend on the electrode potential. If this condition is not obeyed, we can use only eqn (50), taking into consideration, however, that parameters $E^{(1,2)}$ and γ may depend on the electrode potential φ_m and that change of electrostatic energy of the ion upon discharge should be added to the quantity ΔJ_f.

The problem of the dependence of the parameters on the electrode potential is of interest not only in connection with quantitative calculation of electrochemical kinetics, but has a fundamental physical significance. In fact, if γ does not depend on φ_m, from eqn (52) we can write the microscopic analogue of the Brönsted equation relating the change of the activation free energy to that of the quasithermodynamic reaction free energy $\Delta F(\varepsilon_f)$, for the elementary electrochemical discharge current involving the electron localized at an arbitrary quantum level ε_f:

$$\alpha(\varepsilon_f) = -\frac{\partial F_a(\Delta F(\varepsilon_f))}{\partial \Delta F(\varepsilon_f)} \tag{53}$$

The microscopic nature of this relation is due to its containing a parameter of a thermodynamically non-averaged subsystem, i.e. the electrode electrons. Below, however, we shall show from eqn (53) that it is possible to introduce the concept of the transfer coefficient α having a macroscopic sense. If, however, the coefficients γ depend on φ_m, eqn (50) should be used in place of eqn (52) which permits us to relate the change of the activation free energy to change of ΔJ_f:

$$\alpha(\varepsilon_f) = -\frac{\partial F_a(\Delta J_f)}{\partial \Delta J_f} \tag{54}$$

The advantage of the Brönsted equation (53) over eqn (54) is that at a known transfer coefficient $\alpha(\varepsilon_f)$ it is possible to relate the kinetic characteristic F_a directly to the quasi-thermodynamic quantity for the reaction free energy $\Delta F(\varepsilon_f)$, whereas from eqn (54) we can only relate F_a to the quantity ΔJ_f, which is not of a directly thermodynamic nature. This advantage is particularly apparent in considering the dependence of free energy of activation and the current on overvoltage. In the absence of the ψ_1-effect, it follows from eqns (52) and (60) that

$$\alpha(\varepsilon_f) = -\frac{\partial F_a(\Delta F(\varepsilon_f))}{\partial e\eta_c} = -\frac{\partial F_a(\Delta F(\varepsilon_f))}{\partial \varepsilon_f} \tag{55}$$

By means of this equation it is possible to make a rather general qualitative analysis of the form of the current-voltage dependence. For this purpose, let us consider the experimentally measured transfer coefficient, which according to eqn (49) we shall write in the form:

$$\alpha_{\exp} = kT\frac{\mathrm{d}\ln \vec{i}_c}{\mathrm{d}e\eta_c} = \frac{kT}{\vec{i}_c}\,eC_0^{(1)}\frac{kT}{h}\,\kappa_F\rho_F\!\int n\,\frac{\mathrm{d}}{\mathrm{d}e\eta_c}\,\mathrm{e}^{-F_a/kT}\,\mathrm{d}\varepsilon_f \tag{56}$$

If now we use eqn (55) we shall have:

$$\alpha_{\exp} = \frac{\int \alpha(\varepsilon_f)\,\mathrm{d}\vec{i}_c(\varepsilon_f)}{\vec{i}_c} = \langle\alpha(\varepsilon_f)\rangle_{av} \tag{57}$$

Thus, in the absence of the ψ_1-effect, the experimental value of the transfer coefficient coincides with the mean value of the microscopic transfer coefficient calculated for an individual value of ε_f. The weight of $\alpha(\varepsilon_f)$ should be equal to the contribution of the current from an individual level ε_f to the total discharge current. If we assume the main contribution to the total current to be made by a small group of electron levels in a metal electrode localized in the vicinity of a certain level ε^* dependent on overvoltage, the mean value of the transfer coefficient can be approximately substituted by

$$\alpha_{\exp} \simeq \alpha(\varepsilon^*) \equiv \alpha^*(e\eta_c) \tag{58}$$

Taking into consideration the physical significance of ε^*, it is possible to relate the macroscopic values of the free activation energy

$$F_a^* = F_a(\varepsilon^* + e\eta_c)$$

to the change of the nonequilibrium free energy of the electrode reaction at given overvoltage $\Delta F^* = \Delta F(\varepsilon_f^*)$:

$$\frac{\partial F_a^*}{\partial \Delta F^*} = \frac{\partial F_a^*}{[1 + (\partial\varepsilon^*/\partial e\eta_c)]\,\partial e\eta_c}$$

$$= \frac{1}{1 + (\partial\varepsilon^*/\partial e\eta_c)}\left(\frac{\partial F_a(\varepsilon_f + e\eta_c)}{\partial \varepsilon_f}\right)_{\varepsilon^*}\left(1 + \frac{\partial\varepsilon^*}{\partial e\eta_c}\right)$$

whence we obtain a macroscopic generalization of the Bronsted equation for eqn (53):

$$\alpha^* = - \frac{\partial F_a^*}{\partial \Delta F^*} \tag{59}$$

Another very important relation for α_{exp} can be obtained from eqn (55) if in the integral we substitute differentiation with respect to $e\eta_c$ by the derivative with respect to ε_f (see eqn (55)) and then integrate by parts. After simple transformations, we obtain:

$$\alpha_{exp} = \frac{1}{i_c} \int [1 - n(\varepsilon_f)] \, \overrightarrow{di_c}(\varepsilon_f) = 1 - \langle n(\varepsilon_f) \rangle_{av} \tag{60}$$

where averaging has the same significance as in eqn (57). Comparing eqns (57) and (60) we see that the mean value of the occupation of electron levels in a metal electrode is given by an accurate formula

$$\alpha_{exp} = \langle \alpha(\varepsilon_f) \rangle_{av} = 1 - \langle n(\varepsilon_f) \rangle_{av} \tag{61}$$

which can be written approximately as:

$$\alpha_{exp} \simeq \alpha(\varepsilon^*) \simeq 1 - n(\varepsilon^*) \tag{62}$$

In spite of the fact that eqns (61) and (62) are strictly phenomenological, we can draw from these equations some important conclusions about the nature of electrochemical kinetics for the processes under consideration. First of all, it should be pointed out that since the electron level occupation numbers in the metal $n(\varepsilon_f)$ are less than unity, according to eqn (61) we have

$$0 \leqslant \langle \alpha(\varepsilon_f) \rangle_{av} \simeq \alpha(\varepsilon^*) \leqslant 1 \tag{63}$$

The nontriviality of this assertion follows from the fact that the microscopic transfer coefficient for individual electrons $\alpha(\varepsilon_f)$ localized at arbitrary levels ε_f on the whole does not satisfy eqn (63), i.e. $\alpha(\varepsilon_f)$ can be less than zero or greater than 1. To prove this, let us make another important assumption: let us consider the entropy of activation to be practically independent of ε_f. Then in the Brönsted equation (53) (or eqn (59)) the free energy of activation can be substituted by the activation energy $E_a(\varepsilon_f)$ (or $E_a(\varepsilon^*) = E_a^*$)

$$\alpha(\varepsilon_f) = - \frac{\partial E_a(\Delta F(\varepsilon_f))}{\partial \Delta F(\varepsilon_f)} \tag{64}$$

or

$$\alpha^* = - \frac{\partial E_a^*}{\partial \Delta F^*} \tag{64a}$$

Now let us consider the three possible cases of the positions of initial terms (1) with respect to the fixed term of final state (2) shown in Fig. 6(a). As follows from a simple mathematical analysis, if the minimum of the initial term is located inside the final term (with the subscript ε_{f1} in Fig. 6(a)), the

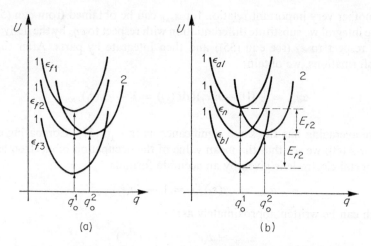

Fig. 6. Qualitative picture of the terms for electrons localized at different levels in the electrode.

transfer coefficient is negative ($\alpha(\varepsilon_{f1}) < 0$); if the final term minimum is inside the initial term, the transfer coefficient is greater than unity ($\alpha(\varepsilon_{f3}) > 1$ in Fig. 6(a)). In all the remaining cases (see e.g. the term with subscript ε_{f2}) we have $0 < \alpha < 1$. To formulate this result quantitatively, let us introduce some definitions. We shall say that the transition is of a barrierless nature if the initial state term passes through the minimum of the final term (see the term with subscript ε_{bl} in Fig. 6(b)). If at the beginning the system was at the term $U_1(q)$, the quantity

$$U_1(q_{02}) - U_1(q_{01}) = E_{r1} \tag{65}$$

will be called the reorganization energy for a forward process (q_{01} and q_{02} are the coordinates of the initial and final terms minima, respectively, see Fig. 6(b)). We are justified in giving this name to the quantity in question, because the quantity E_{r1} is equal to the work to be performed for the initial equilibrium configuration of the classical subsystem to be reorganized into the equilibrium final configuration at a fixed electron state. The reorganization energy for a back process is determined in a similar manner:

$$U_2(q_{01}) - U_2(q_{02}) = E_{r2} \tag{66}$$

where $U_2(q)$ is the final state term (see Fig. 6(b)). In the general case, the quantities E_{r1} and E_{r2} may not coincide, but if the terms of initial and final states have the same shape, i.e.

$$U_2(q - q_{02}) = U_1(q_{01} - q), \quad E_{r1} = E_{r2}$$

The process of a barrierless transition of the system from initial to final terms can be treated as follows. First, the classical subsystem becomes activated: from the equilibrium value q_{01} it passes to the intersection point q^*. Then at fixed coordinates of the classical subsystem the quantum transition of the electron from the first to the second term occurs. In this case, since for a barrierless process q^* coincides with the equilibrium configuration of the final state q_{02}, there is no need for the relaxation of the classical subsystem after the electron transition. The value of the activation energy for a barrierless process is determined by the relation (see eqn (33):

$$E_a(\varepsilon_{bl}) = \Delta J(\varepsilon_{bl}) = (\varepsilon_{bl} - \varepsilon_F) + e\eta_c + \Delta J_{0F} \tag{67}$$

The so-called activationless transition is treated in a similar manner (see the term with subscript ε_{al} in Fig. 6(b)). Here the activation energy is zero and the electron transition from initial to final terms occurs directly at the equilibrium point q_{01}, whereupon the classical subsystem relaxes to the final equilibrium configuration. A transition will be called normal if the initial state term is located between the barrierless and activationless terms (see, e.g. the term with subscript ε_n in Fig. 6(b)). In the case of a normal transition, the classical subsystem is at first activated from the equilibrium configuration q_{01} into the configuration q^*, whereupon the quantum electron transition occurs, i.e. the system passes from initial to final term, which is followed by the relaxation of the configuration of the classical subsystem from q^* to q_{02}. In terms of the reorganization energy, it is easy to formulate the criteria determining the possible values of the microscopic transfer coefficient:

$$\alpha(\varepsilon_f) \geqslant 1, \quad \Delta J = (\varepsilon_f - \varepsilon_F) + e\eta_c + \Delta J_{0F} \leqslant -E_{r1} \tag{68a}$$

$$\alpha(\varepsilon_f) \leqslant 0, \quad \Delta J = (\varepsilon_f - \varepsilon_F) + e\eta_c + \Delta J_{0F} \geqslant E_{r2} \tag{68b}$$

$$0 \leqslant \alpha(\varepsilon_f) \leqslant 1, \quad -E_{r1} \leqslant \Delta J = (\varepsilon_f - \varepsilon_F) + e\eta_c + \Delta J_{0F} \leqslant E_{r2} \tag{68c}$$

Comparing eqns (63) and (68c) we can conclude that the electron term corresponding to the level $\varepsilon^*(\eta_c)$, which makes the main contribution to the resulting discharge current is always located between the barrierless and activationless terms. For instance, if at some overvoltage $\eta_c^{(1)}$ the main contribution to the current is made by the level $\varepsilon_f^{(1)} = \varepsilon^*(\eta_c^{(1)})$ with a sufficient increase of overvoltage the term corresponding to this level can rise so high that $\Delta J(\varepsilon_f^{(1)})$ will be larger than E_{r2}. According to eqns (63) and (68), at the new value of overvoltage η_c level $\varepsilon_f^{(1)}$ will practically make no contribution to the current. Now the main contribution to the current will be made by another level ε_f situated below $\varepsilon_f^{(1)}$ so that condition eqn (68c) should be satisfied. A decrease of overvoltage, when $\Delta J(\varepsilon_f^{(1)})$ becomes less than $(-E_{r1})$, can be treated in a similar manner. In this case, the current will be determined by level ε_f situated above $\varepsilon_f^{(1)}$, so that condition eqn (68c) should be satisfied as before. Thus condition eqn (63) can be treated in a similar manner. In this case, the current will be determined by level ε_f situated above $\varepsilon_f^{(1)}$, so that condition eqn (68c) should be satisfied as before. Thus condition eqn (63) can be rewritten according to eqn (68c) in the form:

$$-E_{r1} \leqslant (\varepsilon^*(\eta_c) - \varepsilon_F) + e\eta_c + \Delta J_{0F} \leqslant E_{r2} \tag{69}$$

For a more detailed analysis of the discharge current-overvoltage curve let us introduce conditionally three regions of overvoltages:

Barrierless Region (low overvoltages). Let us consider the region of low overvoltages in which the following condition is obeyed:

$$0 < e\eta_c < -(\Delta J_{0F} + E_{r1}) \tag{70}$$

It should be noted that this overvoltage region cannot exist for all systems, since the necessary condition for eqn (70) to be valid is that

$$-\Delta J_{0F} > E_{r1} \tag{71}$$

or, since

$$\Delta J_{0F} = \Delta F_{0F} - \Delta\mathscr{F} = kT \ln \frac{C_0^{(1)}}{C_0^{(2)}}$$

that

$$kT \ln \frac{C_0^{(1)}}{C_0^{(2)}} > E_{r1} + \Delta\mathscr{F} \simeq E_{r1} \tag{72}$$

Here we make use of the fact that $\Delta\mathscr{F} \ll E_{r1}$, which can be readily shown to a harmonic approximation, for example, when $\Delta\mathscr{F} \simeq kT \ln (\omega_1/\omega_2)$, where ω_1 and ω_2 are the characteristic frequencies of initial and final terms (see derivation of eqn (28)).

The reorganization energy being as a rule much larger than kT, condition eqn (72) can be satisfied at very small concentrations of discharged ions. For example, if we take $E_{r1} \simeq 1$ eV, at room temperature condition eqn (72) gives $C_0^{(2)} \simeq 10^{-18} C_0^{(1)}$. Evidently, condition eqn (72) is never satisfied for redox systems and therefore the barrierless region can be observed only in the case of adsorption of discharged ions on the electrode and only if the adsorption energy is small enough (or the bond energy of the discharging ion in solution is very large). In other words, the barrierless discharge region can be observed only for metals with very high overvoltage. Comparing eqns (68a) and (71) we can formulate the condition of the appearance of the barrierless discharge on the electrode as the requirement that the electron localized at the Fermi level at the equilibrium potential ($\eta_c = 0$) should be characterized by a microscopic transfer coefficient greater than unity

$$\alpha(\varepsilon_F, \varphi_m = \varphi_0) > 1 \tag{73}$$

As has been already stated above, in this case, the main contribution to the current will be made by level $\varepsilon^* > \varepsilon_F$ so that

$$\alpha^* = \alpha(\varepsilon^*(\eta_c)) \lesssim 1 \tag{74}$$

or according to eqn (61)

$$n^* = n(\varepsilon^*(\eta_c)) \gtrsim 0 \tag{75}$$

It should be pointed out that the electron term corresponding to level ε^* is not exactly barrierless, for this would require $\alpha^* = 1$ and $n^* = 0$. Using the form of the Fermi distribution eqn (7) we can readily see that $n^* \to 0$ only at $(\varepsilon^* - \varepsilon_F) \to +\infty$, i.e. at $\eta_c \to -\infty$, which has no physical sense. However, α^* practically reaches the value 1 even at small deviations of ε^* from ε_F. In fact, already at $(\varepsilon^* - \varepsilon_F) \simeq 5kT$ the value of n^* is 0·007 and the transfer coefficient $\alpha^* = 0·993$, which cannot be experimentally distinguished from unity. Thus, we arrive at a very important conclusion that even at a significant decrease of overvoltage the level making the main contribution to the current ε^* deviates insignificantly from the Fermi level. Thereby, we actually prove eqn (5) for the low overvoltage region.

In the barrierless discharge region, the value of the activation energy E_a^* practically coincides with the reorganization energy of the direct process E_{r1} (see Fig. 6(b)) and has a nearly constant value independent of overvoltage:

$$(E_a^*)_{bl} \simeq E_{r1} = \text{constant} \tag{76}$$

On the other hand, by definition, this quantity is equal to the difference of minimum potential energies of the final term and the initial term corresponding to the electron level ε^*:

$$(E_a^*)_{bl} \simeq E_{r1} \simeq -\Delta J(\varepsilon^*) = (\varepsilon^* - \varepsilon_F) + e\eta_c + \Delta J_{0F} \tag{77}$$

or

$$\varepsilon^*(\eta_c) - \varepsilon_F \simeq -e\eta_c - (E_{r1} + \Delta J_{0F}) \tag{78}$$

Hence, according to eqn (78) in the barrierless discharge region ε^* decreases linearly with increasing overvoltage. This result has a simple physical sense. If at an overvoltage η_c^0, the current was determined by level $\varepsilon^*(\eta_c^0)$, over the whole overvoltage range under consideration the corresponding term occupies a barrierless configuration (see Fig. 7, the term with subscript zero). When the overvoltage is increased by a small quantity $\Delta\eta_c$, so that the system does not go out of the barrierless discharge region $(\alpha^*(\eta_c^0 + \Delta\eta_c) \simeq 1)$, the initial term $\varepsilon^*(\eta_c^0)$ rises by the quantity $\Delta\eta_c$ and the barrierless configuration is now occupied by a different term $\varepsilon^*(\eta_c^0 + \Delta\eta_c)$, which was located lower than the term $\varepsilon^*(\eta_c^0)$ by the quantity $\Delta\eta_c$ (see Fig. 7).

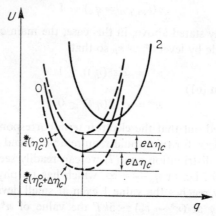

Fig. 7. Schematic representation of the terms determining the current in the barrierless region.

The experimental value of the transfer coefficient $\alpha_{exp} \simeq \alpha^*$ in the region under consideration being unity, the logarithm of the current for a barrierless discharge should depend linearly on overvoltage. On the other hand, according to eqn (76) the activation energy, and thus the transition probability does not depend on overvoltage. Hence it follows that the current-over-voltage dependence in the barrierless region should be controlled by a factor other than the transition probability. To establish this mechanism it should be pointed out that according to eqn (49) the discharge current is determined not only by the activation factor $\exp{(-E_a^*/kT)}$, but also by the electron 'concentration' factor $n^* = n(\varepsilon^*)$ (this statement will be proved more rigorously below, see eqn (85)):

$$\vec{i_c} \sim n^* \, e^{-E_a^*/kT} \simeq e^{-\varepsilon^*-\varepsilon_F/kT} \, e^{-E_{r1}/kT} = e^{e\eta_c/kT} \, e^{\Delta J_{0F}/kT} \tag{79}$$

where we used eqns (76) and (78). The equation obtained above shows that the experimental value of the discharge current activation energy in the barrierless region is determined not only by the transition probability E_a^*, but also by the occupation of the levels contributing to the current

$$\varepsilon^* - \varepsilon_F = E_a^{exp}(\eta_c) = -kT \ln n^* + E_a^*$$
$$\simeq [\varepsilon^*(\eta_c) - \varepsilon_F] + E_a^* \simeq e\eta_c + \Delta J_{0F} \tag{80}$$

Extrapolating this value of the experimental activation energy at $e\eta_c \to 0$, we obtain

$$E_a^{\text{exp}}(\eta_c \to 0) = \Delta J_{0F} \tag{81}$$

Equation (81) is very important since it allows us to determine experimentally one of the most essential parameters of the theory, ΔJ_{0F}.

The barrierless discharge region was first experimentally observed by Krishtalik, who investigated the kinetics of the hydrogen ion discharge on a mercury electrode (see, e.g. References 32–36).

Normal Region (medium overvoltages). Now we shall consider the overvoltage region determined by the relation:

$$-(\Delta J_{0F} + E_{r1}) < e\eta_c < E_{r2} - \Delta J_{0F} \tag{82}$$

To establish the nature of the current-voltage dependence in the range eqn (82), let us analyse the integrand in eqn (49) for the discharge current. In Fig. 8 the solid curve shows the qualitative form of the dependence of the integrand on the electron energy. Of great importance in this case is the assumption made by us that the dashed curve corresponding to the activation factor is bell-shaped. This assumption is readily justified in the treatment of arbitrary electron terms of initial and final terms with a sufficiently plausible shape. The activation factor reaches its maximum value at

$$\varepsilon_{al} = \varepsilon_F - e\eta_c - \Delta J_{0F} + E_{r2}$$

i.e. when the initial term has an activationless configuration (see the term with subscript ε_{al} in Fig. 6(b)). As is clear from Fig. 6, with deviations from $\varepsilon_f = \varepsilon_{al}$ in both directions ($\varepsilon_f > \varepsilon_{al}$ or $\varepsilon_f < \varepsilon_{al}$) the activation energy increases smoothly, which accounts for the bell-shape of the factor $\exp(-E_a/kT)$. Naturally, in this treatment we proceed from the assumption made by us earlier that the activation entropy depends only slightly on ε_f. Since ε_{al} depends linearly on $e\eta_c$ with changing overvoltage the dashed curve in Fig. 8

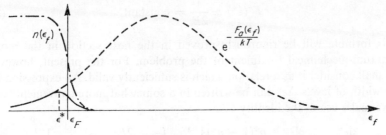

Fig. 8. Qualitative picture of the Fermi distribution function $n(\varepsilon_f)$, activation factor $\exp[-F_a(\varepsilon_f)/kT]$ and their product (solid curve).

without changing its form will shift to the right (with decreasing $e\eta_c$) or to the left (with increasing $e\eta_c$). The maximum of the whole integrand will shift in the same direction. It can readily be seen that the level at which the integrand maximum is reached coincides in its sense with the quantity ε^* introduced earlier, i.e. with the level making the main contribution to the total discharge current. This can be also directly proved mathematically if the derivative of the integrand is equated to zero,

$$\frac{d}{d\varepsilon_f} (n\, e^{-E_a/kT}) = 0 \tag{83}$$

After simple transformations we obtain the equation for determination of ε^*:

$$1 - n(\varepsilon_f) = \alpha(\varepsilon_f) \tag{84}$$

Comparing this equation with formula eqn (62), we see that at $\varepsilon_f = \varepsilon^*$ they coincide absolutely. If now we use the Laplace method (analogue of the method of steepest descent for functions of real variables), we can find an approximate expression for the discharge current

$$\vec{i}_c = eC_0^{(1)} \frac{kT}{h} \kappa_F \rho_F n(\varepsilon^*) \exp\left[-\frac{F_a^*}{kT} \right] \Delta\varepsilon^* \tag{85}$$

where

$$\Delta\varepsilon^* = \sqrt{(2\pi)kT} [n^*(1 - n^*) + kTF_a''(\varepsilon^*)]^{-1/2} \tag{86}$$

has the significance of the width of levels located near ε^* and contributing to the discharge current. In calculating the integral eqn (49) by the Laplace method, the quantity $F_a''(\varepsilon^*)$ is assumed to be practically independent of ε^*, i.e. of overvoltage. Taking this into consideration and assuming the reorganization energies of forward and back processes to be of the same order of magnitude ($E_{r1} \simeq E_{r2} = E_r$), we can estimate the second derivative of the free activation energy

$$F_a''(\varepsilon^*) \simeq \frac{1}{2E_r} = \text{constant} \tag{87}$$

This formula will be rigorously proved in the next section in the model quantum-mechanical treatment of the problem. For the present, however, we shall consider it as a relation which is sufficiently valid. An expression for the width of levels $\Delta\varepsilon^*$ can be written in a somewhat more convenient form if we make use of the identity

$$\frac{d\varepsilon^*}{de\eta_c} = -\left[1 + \frac{n^*(1 - n^*)}{kTF_a''(\varepsilon^*)} \right]^{-1} \simeq \left[1 + \frac{2E_r}{kT} n^*(1 - n^*) \right]^{-1} \tag{88}$$

Substituting eqn (88) into eqn (86) we obtain

$$\Delta\varepsilon^* = \left\{\frac{2\pi kT}{F_a''(\varepsilon^*)}\left|\frac{d\varepsilon^*}{de\eta_c}\right|\right\}^{1/2} \simeq \left\{4\pi kTE_r\left|\frac{d\varepsilon^*}{de\eta_c}\right|\right\}^{1/2} \tag{89}$$

The estimates obtained at real values of E_r (of the order of some electron-volts) show that the halfwidth of the levels ($\frac{1}{2}\Delta\varepsilon^*$) depends slightly on n^*, i.e. on potential, and amounts to several kT. For example, for $E_r = 2$ eV in the normal region ($n^* \simeq 0.5$) we have $\frac{1}{2}\Delta\varepsilon^* \simeq 2.5kT$, and in the barrierless ($n^* \simeq 0.01$) and activationless ($n^* \simeq 0.99$) regions $\frac{1}{2}\Delta\varepsilon^* \simeq 5.4kT$. Hence in calculating the transfer coefficient, the quantity $\Delta\varepsilon^*$ can be assumed to be constant; the correction for α_{exp} appearing in this case being no greater than several hundredths. Thus, the experimental activation energy will appear as the quantity (see eqn (85))

$$E_a^{exp} = -kT \ln n^* + E_a(\varepsilon^*) \tag{90}$$

whence for the experimental transfer coefficient we obtain the relation eqn (58)

$$\alpha_{exp} = \frac{d \ln \vec{i_c}}{de\eta_c/kT} = -\alpha^*\frac{d\varepsilon^*}{de\eta_c} + \alpha^*\left(1 + \frac{d\varepsilon^*}{de\eta_c}\right) = \alpha^* \tag{91}$$

It is characteristic of the normal region that at a certain overvoltage $\eta_c = \eta_c^F$ the discharge current is determined mainly by the electron localized at the Fermi level: $\varepsilon^*(\eta_c^F) = \varepsilon_F$. At small deviations of overvoltage from η_c^F, the discharge current can be written as

$$\vec{i_c} \simeq eC_0^{(1)}\frac{kT}{h}\kappa_F\rho_F\frac{1}{2}\Delta\varepsilon^*$$
$$\exp\left\{-\frac{F_a(\varepsilon_F + e\eta_c^F)}{kT} + \frac{e\eta_c - e\eta_c^F}{2kT}\left(1 - \frac{e\eta_c - e\eta_c^F}{4kT + 1/F_a''}\right)\right\}$$
$$\simeq eC_0^{(1)}\frac{kT}{h}\kappa_F\rho_F\frac{1}{2}\Delta\varepsilon^*$$
$$\exp\left\{-\frac{F_a(\varepsilon_F + e\eta_c^F)}{kT} + \frac{e\eta_c - e\eta_c^F}{2kT}\left(1 - \frac{e\eta_c - e\eta_c^F}{2E_r}\right)\right\} \tag{92}$$

Accordingly, for the transfer coefficient we have

$$\alpha_{exp} \simeq \frac{1}{2}\left(1 - \frac{e\eta_c - e\eta_c^F}{2kT + 1/F_a''}\right) \simeq \frac{1}{2} - \frac{e\eta_c - e\eta_c^F}{2E_r} \tag{93}$$

The reorganization energy being a sufficiently large quantity, we can conclude on the basis of formula eqn (93) that in the normal region over a wide enough overvoltage range the transfer coefficient is equal to $\frac{1}{2}$. For example, if $E_r = 2\,\text{eV}$ the transfer coefficient is $0\cdot5 \pm 0\cdot1$ at $\Delta e\eta_c \simeq 0\cdot8\,\text{eV}$.

Activationless Region (high overvoltages). With further increase of overvoltage, when the condition

$$e\eta_c > E_{r2} - \Delta J_{0F} \tag{94}$$

is valid, electrochemical kinetics is determined by the activationless discharge. Comparing eqns (94) and (68b) we see that the activationless region begins at overvoltages such that the microscopic transfer coefficient for the electron localized at the Fermi level becomes negative:

$$\alpha(\varepsilon_F + e\eta_c) < 0 \tag{95}$$

On the other hand, it follows from formula eqn (69) that the electron level ε^* determining the discharge current in the activationless region is located below the Fermi level, i.e. $n(\varepsilon^*) \simeq 1$ and $\alpha(\varepsilon^*) \simeq 0$. Just as in the barrierless region actual deviation of ε^* from ε_F is not large. For example, at $\varepsilon^* \simeq \varepsilon_F - 5kT$ n^* is equal to $0\cdot993$ and the transfer coefficient is practically zero: $\alpha^* = 0\cdot007$. Thus, with overvoltage changing over an extremely large range

$$\Delta e\eta_c \simeq E_{r1} + E_{r2}$$

covering the three regions of electrochemical kinetics, the level ε^*, making the chief contribution to the current, varies over a small range of energies near the Fermi level ($\pm 5kT$).

As has been already stated earlier, in the activationless region the activation energy is zero and $\Delta J(\varepsilon^*) \simeq E_{r2}$, whence we find directly the dependence of ε^* on overvoltage:

$$\varepsilon^* \simeq \varepsilon_F - e\eta_c - (\Delta J_{0F} - E_{r2}) \tag{96}$$

The linear dependence of ε^* on $e\eta_c$ in the activationless region has the same physical nature as in the barrierless region (see explanation for formula eqn (78)). Since in the overvoltage range under consideration $n^* \simeq 1$, not only E_a^* but also the experimental activation energy is zero (see eqn (90)), which results in the current having a constant value:

$$\vec{i}_c = eC_0^{(1)} \frac{kT}{h} \kappa_F \rho_F\, e^{S_a/kT} \Delta\varepsilon^* = \text{constant} \tag{97}$$

A practical realization of an activationless mechanism presents certain experimental difficulties, since increase of discharge current may lead to the violation of condition eqn (1) the physical significance of which is that diffusion control starts to operate.[46,47] In considering experimental conditions for establishing activationless or barrierless mechanisms for an electrochemical reaction, one should bear in mind that both these processes always occur simultaneously, viz. if the forward process is barrierless, the back process will be an activationless process, and vice versa.[36] This can readily be seen if according to formula eqn (47) we write the relation between direct and reverse currents for an individual electron level ε_f:

$$\overrightarrow{di_c}(\varepsilon_f) = \overleftarrow{di_c}(\varepsilon_f)\, e^{e\eta_c/kT}\, \frac{C_0^{(2)}}{C^{(2)}} \tag{98}$$

This formula shows that for each fixed overvoltage $e\eta_c$ the maximum contribution both to direct and reverse currents is made by the same level ε^* determined by eqn (84). In considering the ionization current, the transfer coefficient β can be conveniently determined using the formula

$$\beta_{\exp} = -kT\, \frac{d}{de\eta_c}\, \ln \frac{C_0^{(2)}}{C^{(2)}} \overleftarrow{i_c} \tag{99}$$

to isolate the dependence of the adsorbed particles concentration on overvoltage. If now we use formula eqn (48) we obtain from eqn (99) the usual relation between the transfer coefficients of forward and back processes:

$$\beta_{\exp} = 1 - \alpha_{\exp} \tag{100}$$

We have described qualitatively the whole discharge current-overvoltage curve and shown that, generally speaking three asymptotic overvoltage regions are possible: barrierless ($\alpha_{\exp} \simeq 1$), normal ($\alpha_{\exp} \simeq 0\cdot5$) and activationless ($\alpha_{\exp} \simeq 0$). These three asymptotic curves are shown in logarithmic scale in Fig. 9. The smooth curve is only qualitative in nature, since strictly

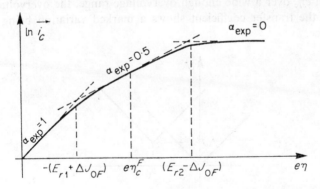

Fig. 9. Schematic representation of the dependence of the logarithm of current on overvoltage.

within the framework of the semi-phenomenological theory it is impossible to investigate the shape of the transition region, i.e. the overvoltage range near $e\eta_c = -\Delta J_{0F} \pm E_{r1,2}$. Further quantitative elaboration of the theory depends essentially on the concrete form of the dependence of the free energy of activation on ε_f. The quantity $F_a(\varepsilon_f)$ can be calculated only for a definite model; this calculation is made in subsequent sections. An attempt can be made, however, to approach the problem from a semiempirical point of view. The first approach is to try to reconstruct on the basis of the experimental data the shape of the $F_a - \varepsilon_f$ dependence. For this purpose, we shall write the $\alpha(\varepsilon_f)$ dependence as the inverse function

$$e\eta_c + kT \ln \frac{1 - n(\varepsilon_f)}{n(\varepsilon_f)} = \varepsilon_f - \varepsilon_F + e\eta_c = F(\alpha) \qquad (101)$$

where $F(\alpha)$ is the unknown function. Let us substitute in this formula $\varepsilon^*(\eta_c)$ for ε_f and replace α by $\alpha^* = 1 - n^* = \alpha_{\exp}$

$$e\eta_c = F(\alpha_{\exp}) + kT \ln \frac{1 - \alpha_{\exp}}{\alpha_{\exp}} \qquad (102)$$

The obtained relation actually solves the problem posed since it gives the form of the function F if the dependence of the transfer coefficient on overvoltage is experimentally known. With $\alpha(\varepsilon_f)$ known, it is possible to find the activation energy using formula eqn (64)

$$E_a(\varepsilon_f) = -\int \alpha(\varepsilon_f)\, d\varepsilon_f \qquad (103)$$

For practical use of the method described above, it is necessary to bear in mind that in order to find the detailed form of the functional dependence $E_a(\varepsilon_f)$ one should have at one's disposal experimental data on the dependence of $\alpha_{\exp}$ on $e\eta_c$ over a wide enough overvoltage range, the overvoltage range in which the transfer coefficient shows a marked variation being of most importance.

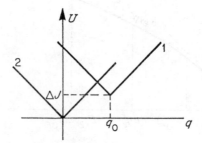

Fig. 10. Model linear terms.

Another semi-empirical approach to the problem consists in assuming the form of the dependence $E_a(\varepsilon_f)$ or the form of the electronic terms and determining the theoretical parameters from comparison with experiment. Below we shall give as an example a calculation of the current for symmetrical linear terms (see Fig. 10) which are seldom used in electrochemical literature:[37]

$$U_1(q) = (\varepsilon_f - \varepsilon_F) + e\eta_c + \Delta J_{0F} + \frac{E_r}{q_0}|q - q_0|$$

$$= \Delta J(\varepsilon_f) + \frac{E_r}{q_0}|q - q_0|,$$

$$U_2(q) = \frac{E_r}{q_0}|q| \tag{104}$$

The activation energy for linear terms is found from simple geometric considerations and is of the form

$$E_a(\varepsilon_f) = \tfrac{1}{2}(E_r - \Delta J(\varepsilon_f)), \quad |\Delta J(\varepsilon_f)| \leqslant E_r \tag{105}$$

In the case when $|\Delta J(\varepsilon_f)| > E_r$ the terms do not intersect and the activation energy can be formally considered to be equal to infinity. Accordingly, the discharge current can be written as (see eqn (49)):

$$\vec{i}_c = eC_0^{(1)} \frac{kT}{h} \kappa_F \rho_F \, e^{S_a/kT} \int_{\varepsilon_F - e\eta_c - \Delta J_{0F} - E_r}^{\varepsilon_F - e\eta_c - \Delta J_{0F} + E_r} n(\varepsilon_f) \exp\left(-\frac{E_a(\varepsilon_f)}{kT}\right) d\varepsilon_f$$

The integral in the right hand side can be calculated accurately, and finally for the discharge current we obtain the expression

$$\vec{i}_c = eC_0^{(1)} \frac{kT}{h} \kappa_F \rho_F \, e^{S_a/kT} 2kT \, e^{e\eta_c + \Delta J_{0F} - E_r/2kT} \arctg \frac{sh(E_r/2kT)}{ch(e\eta_c + \Delta J_{0F}/2kT)} \tag{106}$$

which can be asymptotically represented for the three regions as follows:
    (a) for the barrierless region:

$$\ln \vec{i}_c = \ln \Delta\varepsilon_{bl}^* A + \frac{e\eta_c + \Delta J_{0F}}{kT}; \quad \Delta\varepsilon_{bl}^* = 2kT; \quad 0 < e\eta_c < -(\Delta J_{0F} + E_r) \tag{107a}$$

    (b) for normal region:

$$\ln \vec{i}_c = \ln \Delta\varepsilon_n^* A + \tfrac{1}{2}\frac{e\eta_c + \Delta J_{0F} - E_r}{kT}; \quad \Delta\varepsilon_n^* = \pi kT;$$

$$-(\Delta J_{0F} + E_r) < e\eta_c < E_r - \Delta J_{0F} \tag{107b}$$

(c) for the activationless region:

$$\ln \vec{i}_c = \ln \Delta\varepsilon_{al}^* A; \quad \Delta\varepsilon_{al}^* = 2kT; \quad e\eta_c > E_r - \Delta J_{0F} \quad (107c)$$

where the constant A is determined from the formula

$$A = eC_0^{(1)} \frac{kT}{h} \kappa_F \rho_F \, e^{S_a/kT} \quad (108)$$

and the quantity $\Delta\varepsilon^*$ as before has the significance of the range of levels making the chief contribution to the current. The overvoltage dependence of the discharge current, corresponding to eqn (106), is plotted in Fig. 11,

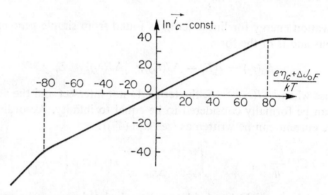

Fig. 11. Dependence of the logarithm of current on overvoltage for model linear terms; $E_r = 2$ eV, $kT = 0.025$ eV, constant $= \ln \pi A_K T - E_r/2kT$.

where the value 2 eV is chosen for the parameter E_r (reorganization energy). It can be seen from the plot that practically over the whole normal region $\alpha = 0.5$. According to the general semi-phenomenological theory, in the barrierless region $\alpha = 1$ and in the activationless region $\alpha = 0$. The transition between these three regions occurs in a very narrow overvoltage range: $\Delta e\eta \simeq 6kT \simeq 0.15$ eV. Here the transition region is that part of the curve where α varies within 0.05–0.45 (near the activationless region) or within 0.55–0.95 (near the barrierless region).

In the present section we have proceeded practically everywhere from the assumption of the electron being the only quantum subsystem in the problem. But in considering some other particular systems, we find other particles to possess quantum spectra along with the electron. The condition of a quantum treatment of the particle was formulated by us already at the beginning of this section and amounted to the requirement that the excitation energy ΔE

should be greater than the thermal energy kT. As we shall see in what follows, this is the case for example, in the treatment of the proton transfer processes. Below we shall give a method for taking into consideration other than electron quantum subsystems.

If during the ion discharge the quantum subsystem located at an initial moment at an arbitrary excited level $\varepsilon_r^{(1)}$ passes to any other excited level of the final state $\varepsilon_r^{(2)}$, we can use for the current corresponding to this process formula eqn (85), taking into consideration, however, that now the transmission coefficient κ_F will depend on r and r', since in this case the exchange integral will contain the overlapping of the wave functions of the quantum subsystem in initial and final states. Moreover, the position of the electron level ε^*, making the chief contribution to the discharge current, will depend on the subscripts r and r'. Now the difference of minimum potential energies of the classical subsystem can be conveniently written as

$$\Delta J^{rr'}(\varepsilon_f) = (\varepsilon_f - \varepsilon_F) + e\eta_c + \Delta J_{0F} + (\varepsilon_r^{(1)} - \varepsilon_{r'}^{(2)}) \qquad (109)$$

where ΔJ_{0F} comprises the difference of zero energies of the quantum subsystem, and $\varepsilon_r^{(1)}$ and $\varepsilon_r^{(2)}$ are reckoned from the energies of unexcited states. If the parameters γ determining the geometric shape and relative horizontal arrangement of the terms are assumed to be unchanged when the quantum subsystem is excited, the values ε^* for various r and r' can be readily interrelated. In fact, under this assumption, the excitation of the quantum subsystem amounts only to a vertical displacement of corresponding term, which is equivalent to a change of overvoltage by the quantity $(\varepsilon_r^{(1)} - \varepsilon_{r'}^{(2)})$. Therefore, if by $\varepsilon_0^*(\eta_c)$ we designate the electron level making the chief contribution to the current upon transition of the quantum subsystem from the ground level of the initial state to the ground level of the final state, we have

$$\varepsilon_{rr'}^*(e\eta_c) = \varepsilon_0^*(e\eta_c + \varepsilon_r^{(1)} - \varepsilon_{r'}^{(2)}) \qquad (110)$$

Now the total discharge current averaged over the initial states and summed over the final states of the quantum subsystem can be written as

$$\vec{i}_c = \sum_{r,r'} \frac{\kappa_{rr'}}{\kappa_0} \vec{i}_{c0}(e\eta_c + \varepsilon_r^{(1)} - \varepsilon_{r'}^{(2)}) e^{-\varepsilon_r^{(1)}/kT} \left[\sum_r e^{-\varepsilon_r^{(1)}/kT} \right]^{-1} \qquad (111)$$

where $\vec{i}_{c0}$ corresponds to the discharge current calculated assuming the quantum subsystem to be passing from the unexcited initial state to the unexcited final state.

It follows from the derivation of formula eqn (111) itself that the method

R. R. DOGONADZE

described above permits us to take into account also an excited state of the electron in the ion and in the adsorbed particle, for which purpose $\varepsilon_r^{(1)}$ and $\varepsilon_r^{(2)}$ should also signify excited electron levels.

III. EFFECT OF THE SOLVENT ON ELECTRODE PROCESSES

1. Solvent Model

The course of most electrochemical reactions is strongly affected by the polar solvent. Physically, this can be explained by the fact that the charge transfer process usually involves strong reorganization of polar medium. For example, if the discharge of a charged particle results in the formation of a neutral particle, the solvent near the discharging particle will be completely depolarized in the course of electrochemical reaction. Accordingly, the heat of such a reaction will comprise the ion hydration energy. In view of the strong interaction of the charge with the polar medium, in electrochemical kinetics it is impossible to neglect the effect of the solvent on the reaction course; moreover, as a rule, this effect cannot be taken into account by introducing a small correction for some basic physical effects. The mathematical sense of this is that the contribution of the polar medium cannot be estimated by the perturbation theory method and that the interaction of the charge with the solvent should be taken into consideration already in the zero approximation. There is one more reason which makes it necessary for us to exercise particular care in analysing the role of the medium in kinetics. As has been already pointed out in the previous section, according to condition eqn (15) the dynamical behaviour of the solvent is classical and therefore the activation energy of electrochemical processes should depend to a considerable extent on the properties of the polar medium.

From a theoretical point of view, in electrochemical and homogeneous kinetics it seems to be impossible at present to take into consideration accurately the properties of a polar liquid. It should be recalled that one of the vital problems of modern physics—the determination of the energy spectrum and correlation functions (space distribution of liquid atoms) has not been solved theoretically as yet. We can only hope to be able to use a polar liquid model which would be adequate enough for the problem posed. In the present and subsequent sections we shall always assume as being possible the use of the dielectric continuum model for an approximate description of the medium. As the charge interacts mainly with the dipole moments of the solvent molecules, it is sufficient to know the specific polarization for the description of the medium in a continuum model. We have already said that of essential importance in electrochemical kinetics is the dynamical behaviour of the solvent, i.e. the dependence of polarization on time. In other words, to describe the medium it is necessary to know not only the polarization value $\vec{P}(\vec{r})$ at any point of space but also the rate of the change of polarization $\dot{\vec{P}}(\vec{r})$. We shall consider the solvent state to be completely known if we know two independent function: $\vec{P}(\vec{r})$ and $\dot{\vec{P}}(\vec{r})$. Thus, $\vec{P}(\vec{r})$ and $\dot{\vec{P}}(\vec{r})$ in the continuum model play the same part as canonically conjugate variables: coordinates and velocities of particles in mechanics. From what we have just said, it is possible to draw very important physical conclusions. First of all, due to the dynamic nature of polarization, the relations of macroscopic electrodynamics, being valid only for averaged quantities, are not valid for $\vec{P}(\vec{r})$. For example, even in the absence of an external field, dynamic polarization of the solvent differs from zero. In terms of statistical physics, dynamic polarization is equivalent to polarization fluctuation. In the presence of a constant external electric field characterized by electric induction $\vec{D}$, this field is proportional only to the mean statistical (i.e. averaged over a long time interval) value of dynamic polarization:

$$\langle \vec{P} \rangle = \frac{\varepsilon_s - 1}{4\pi\varepsilon_s} \vec{D} \tag{112}$$

where ε_s is the static dielectric constant, and $\langle . . . \rangle$ denotes statistical averaging. In this case, dynamic polarization corresponds to fluctuations of $\vec{P}$ about the mean value of $\langle \vec{P} \rangle$. Physically the fluctuation nature of dynamic polarization is due to thermal vibrations of solvent atoms and molecules. Since we are interested only in polar liquids, we have to take into consideration that the solvent molecules have their own dipole moment in the absence of an external electric field. As the result, the polarization fluctuations will be caused both by the vibrations, due to the change in the relative positions of atoms inside the solvent molecule and by the orientation (libration) vibrations of the whole molecule. For example, in water vibrations of the atomic type can include those of the O—H or H—H bonds, whereas the orientation motion results from the vibration of the whole molecule about the axis of symmetry (so-called hindered rotation). The frequencies of orientation vibrations being considerably lower than the frequencies of atomic vibrations, at room temperature the orientation vibrations, are the most highly excited, and they will make the maximum contribution to the polarization fluctuation. The characteristic values of the vibration frequencies are as follows: $\omega_0 \simeq 10^{11}$ sec^{-1} for orientation vibrations and $\omega_a \simeq 10^{13} - 10^{14}$ sec^{-1} for atomic vibrations. The dynamic behaviour of polarization being chiefly due to the orientation motion of dipoles of the medium, we shall henceforth use everywhere for the frequencies of polarization vibrations the quantity ω_0 assuming atomic vibrations to be practically 'frozen'.

Naturally, along with the atomic-orientational part, polarization of the electron clouds of the atoms also makes a contribution to total polarization of the medium. In our treatment of electronic polarization we shall make use of the fact that with fixed positions of the solvent atomic nuclei and in the absence of an external field, electronic polarization cannot fluctuate at room temperature, since the characteristic frequencies for the electron are too high: $\omega_e \simeq 10^{15} - 10^{16}\ \text{sec}^{-1} \gg kT/\hbar$. In other words, electronic polarization adjusts itself instantaneously to the change in the coordinates of the solvent atomic nuclei. If we designate electronic polarization caused by the interaction of the solvent nuclei with electron clouds of atoms as $\vec{P}_{en}$, and polarization due to the dipole moment of nuclei as $\vec{P}_n$, in virtue of what has been said above, we can relate these two quantities by an equilibrium formula of macroscopic electrodynamics:

$$\vec{P}_{en} = \frac{1 - \varepsilon_0}{\varepsilon_0}\,\vec{P}_n \tag{113}$$

where ε_0 is the optical (electronic) dielectric constant. It should be stressed that, although electronic polarization itself does not fluctuate, according to formula eqn (113) fluctuation of the nuclei ($\vec{P}_n$) involves a change in the quantity $\vec{P}_{en}$ so that both $\vec{P}_n$ and $\vec{P}_{en}$ contribute to dynamic polarization. The polarization made up of these two polarizations is usually termed inertial (according to Pekar[4]) or infrared (according to Fröhlich) polarization:

$$\vec{P}_{ir} = \vec{P}_n + \vec{P}_{en} = \frac{1}{\varepsilon_0}\,\vec{P}_n \tag{114}$$

In the presence of an external electric field varying with frequency $\omega \ll \omega_e$, electronic polarization of the atoms of the medium follows instantaneously the change of external field

$$\vec{P}_{op} = \frac{\varepsilon_0 - 1}{4\pi\varepsilon_0}\,\vec{D} \tag{115}$$

where $\vec{D}$ is the electric induction of the external field and $\vec{P}_{op}$ is the electronic polarization of the medium due to this field. The quantity $\vec{P}_{op}$ is usually called inertialess (according to Pekar) or optical (according to Fröhlich) polarization. Thus, total polarization of the medium is made up of $\vec{P}_{ir}$ and $\vec{P}_{op}$:

$$\vec{P}(r) = \vec{P}_{ir}(\vec{r}) + \vec{P}_{op}(\vec{r}) \tag{116}$$

Since optical polarization does not undergo fluctuations and contribute to dynamic polarization, in what follows we shall consider only the inertial part of the total polarization $\vec{P}_{ir}$, omitting everywhere for brevity the subscript ir, $\vec{P}(\vec{r})$ being understood to mean only the infrared component.

In a theoretical treatment of polar systems it is convenient to represent specific polarization as a set of sinusoidal waves

$$\vec{P}(\vec{r}) = \frac{1}{\sqrt{V_s}} \sum_{\vec{k}} \vec{P}_{\vec{k}}\, e^{i\vec{k}\vec{r}} \tag{117}$$

where $\vec{P}_{\vec{k}}$—the Fourier amplitudes—figure as normal coordinates for the solvent and a complete set of them unambiguously describes the polarization state of the medium at any point in the space: the wave vector $\vec{k}$ determines the length of the wave and the direction of the corresponding sinusoid (harmonic); V_s is the volume of the polar medium ($V_s \to \infty$). The interaction of the charges introduced into the solvent from outside with the polar liquid can be represented as an ordinary coulombic interaction with the charges of the medium the density of which in the continuum model is of the form:

$$\rho = -\mathrm{div}\vec{P} = -i\sum_{\vec{k}}(\vec{P}_{\vec{k}}\vec{k})\, e^{i\vec{k}\vec{r}} \tag{118}$$

If polarization is divided into longitudinal ($\vec{P}_{\vec{k}}$ parallel to $\vec{k}$) and transverse ($\vec{P}_{\vec{k}}$ perpendicular to $\vec{k}$) components, it can be seen from eqn (118) that only the longitudinal polarization interacts with the external charges:

$$\vec{P}_{11}(\vec{r}) = \frac{1}{\sqrt{V_s}} \sum_{\vec{k}} \frac{\vec{k}}{k^2}(\vec{P}_{\vec{k}}\vec{k})\, e^{i\vec{k}\vec{r}} \equiv \frac{1}{\sqrt{V_s}} \sum_{\vec{k}} \frac{\vec{k}}{k} P_{\vec{k}}\, e^{i\vec{k}\vec{r}} \tag{119}$$

With certain assumptions, it is possible to show that longitudinal and transverse components do not interact; therefore we shall deal only with the longitudinal component of inertialess polarization, which for simplicity we shall write as $\vec{P}(\vec{r})$ instead of $\vec{P}_{11}(\vec{r})$.

For a quantum-mechanical treatment of the dynamic behaviour of the polar solvent, it is necessary to find first the form of the classical Hamiltonian, i.e. the energy of the medium, which depends on the functions $\vec{P}(\vec{r})$ and $\dot{\vec{P}}(\vec{r})$.

For a pure solvent, in the absence of an external field, the energy can be written as a functional expansion in small polarization fluctuations:

$$E = \tfrac{1}{2}\int\sum_{\alpha,\beta}F_{\alpha\beta}(\vec{r} - \vec{r}')\dot{P}_\alpha(\vec{r})\dot{P}_\beta(\vec{r}')\,d\vec{r}\,d\vec{r}'$$
$$+ \tfrac{1}{2}\int\sum_{\alpha\beta}\phi_{\alpha\beta}(\vec{r} - \vec{r}')P_\alpha(\vec{r})P_\beta(\vec{r}')\,d\vec{r}\,d\vec{r}' \quad (120)$$

where subscripts α, $\beta = x$, y, z determine the projections on Cartesian coordinates, the functions F and ϕ, in virtue of homogeneity of the medium, depend only on the distance between points $\vec{r}$ and $\vec{r}'$ and are basic characteristics of the polar medium in the continuum model. The physical significance of F and ϕ we shall establish later; here we shall only point out that the linear term with respect to $\vec{P}(\vec{r})$ is absent in the functional expansion owing to isotropy of the medium. The expression for the energy in eqn (120) has a standard form of the sum of kinetic (the first term) and potential (the second term) energies. In the potential energy, we have restricted ourselves to the quadratic term with respect to polarization, which is equivalent to a harmonic approximation.

Expanding the functions $F_{\alpha\beta}(\vec{r})$ and $\phi_{\alpha\beta}(\vec{r})$ into the Fourier series and substituting them into formula eqn (120) we can write for the energy

$$E = \tfrac{1}{2}\sum_{\vec{k}}\{F_{\vec{k}}|\dot{P}_{\vec{k}}|^2 + \phi_{\vec{k}}|P_{\vec{k}}|^2\} \quad (121)$$

where $F_{\vec{k}}$ and $\phi_{\vec{k}}$ are related to the Fourier applitudes $F_{\alpha\beta}(\vec{k})$ and $\phi_{\alpha\beta}(\vec{k})$ by the formulae

$$\sum_{\alpha,\beta}F_{\alpha\beta}(\vec{k})\frac{k_\alpha k_\beta}{k^2} = F_{\vec{k}}; \quad \sum_{\alpha,\beta}\phi_{\alpha\beta}(\vec{k})\frac{k_\alpha k_\beta}{k^2} = \phi_{\vec{k}} \quad (122)$$

If we introduce the frequencies for normal vibrations of longitudinal polarization according to:

$$\omega_{\vec{k}}^2 = \frac{\phi_{\vec{k}}}{F_{\vec{k}}} \quad (123)$$

the classical Hamiltonian for the solvent assumes the same form as for a set of oscillators:

$$E = \tfrac{1}{2}\sum_{\vec{k}}\{|\dot{P}_{\vec{k}}|^2 + \omega_{\vec{k}}^2|P_{\vec{k}}|^2\}F_{\vec{k}} \quad (124)$$

Now let us consider in more detail the structure of the solvent energy in formula eqn (120). Let us divide conditionally the function $\phi_{\alpha\beta}(\vec{r} - \vec{r}')$ into local and nonlocal components:

$$\phi_{\alpha\beta}(\vec{r} - \vec{r}') = \phi_0\delta_{\alpha\beta}\delta(\vec{r} - \vec{r}') + \phi_0\varphi_{\alpha\beta}(\vec{r} - \vec{r}'); \quad \phi_0 = \text{constant} \quad (125)$$

where the first term describes the part of potential energy that is chiefly due to elastic forces of a single dipole performing orientational vibrations at a limiting long-wave frequency ω_0. The second term in formula eqn (125) characterizes the interaction of polarizations at different points of the space and can be qualitatively described by the interaction of two dipoles located at points $\vec{r}$ and $\vec{r}'$. Similarly, the kinetic energy can be divided into local and nonlocal components:

$$F_{\alpha\beta}(\vec{r} - \vec{r}') = F_0 \delta_{\alpha\beta} \delta(\vec{r} - \vec{r}') + F_0 f_{\alpha\beta}(\vec{r} - \vec{r}'); \quad F_0 = \text{constant} \quad (126)$$

where the first term is the kinetic energy of the dipole and the second term corresponds to the interaction of the polarization currents arising at points $\vec{r}$ and $\vec{r}'$ with the time-dependent dipole moments. According to qualitative estimates for the problems of interest to us, the second term in formula eqn (126) is insignificant and in zero approximation can be ignored. Moreover, if in the potential energy we retain only the local interaction, we shall have for the energy an approximate expression:

$$E_{\text{loc}} = \frac{\phi_0}{2} \int \left[\frac{F_0}{\phi_0} \dot{\vec{P}}^2(\vec{r}) + \vec{P}^2(\vec{r}) \right] d\vec{r} = \frac{2\pi}{c} \int \left[\frac{1}{\omega_0^2} \dot{\vec{P}}^2(\vec{r}) + \vec{P}^2(\vec{r}) \right] d\vec{r}$$

$$(127)$$

which coincides exactly with Pekar's formula for the classical Hamiltonian of the polar medium.[4,27] From eqn (127) we obtain the values of the constants F_0 and ϕ_0 contained in eqns (125) and (126)

$$F_0 = \frac{4\pi}{c\omega_0^2}; \quad \phi_0 = \frac{4\pi}{c}; \quad c = \frac{1}{\varepsilon_0} - \frac{1}{\varepsilon_S} \quad (128)$$

The analysis carried out above gives physical premises of Pekar's theory, which are rigorously valid only in the absence of nonlocal interaction of dipole moments of the medium, or according to eqn (123) in the absence of dispersion of vibration frequencies ($\omega_{\vec{k}}^{\rightarrow} = \omega_0$), when only the limiting long-wave dipole vibration frequency ω_0 is of importance. As follows from estimates, as well as from the comparison of theory with experimental data, Pekar's Hamiltonian eqn (127) describes correctly the order of magnitude of the energy of the system. It is of interest, however, to take into account the nonlocal interaction of dipoles. This introduces into the theory, in addition to the quantities c and ω_0, some new characteristics of the medium, which enable us to investigate the effect of the substitution of one solvent for another. Below we shall show that the most important parameters of Pekar's theory (e.g. reorganization energy) of all the solvent characteristics contain only the quantity c, which changes but slightly upon substitution of one solvent for another. Assuming that in first approximation we have to take into account only the nonlocal interaction in the potential energy (i.e. $F_{\vec{k}} = F_0$) we can rewrite the frequency dispersion law eqn (123) as

$$\omega_{\vec{k}}^2 = \omega_0^2(1 + \varphi_{\vec{k}}) \tag{129}$$

where the quantity $\varphi_{\vec{k}}$ is related to the Fourier-component function $\varphi_{\alpha\beta}(\vec{r})$ by the formula

$$\varphi_{\vec{k}} = \sum_{\alpha,\beta} \frac{k_\alpha k_\beta}{k^2} \varphi_{\alpha\beta}(\vec{k}) \tag{130}$$

To relate $\varphi_{\vec{k}}$ to physically observed quantities, we shall introduce the function

$$S(\vec{r} - \vec{r}) = \langle \vec{P}(\vec{r})\vec{P}(\vec{r}')\rangle_{av} \tag{131}$$

which characterizes the space correlation of dipole moments of the particles located at points $\vec{r}$ and $\vec{r}'$ (symbol av here denotes averaging over a large time interval). The Fourier component of the correlation function is found after substitution of eqn (117) into eqn (131):

$$S_{\vec{k}} = \langle |P_{\vec{k}}|^2 \rangle_{av} \tag{132}$$

Now the quantity $S_{\vec{k}}$ can be related to the mean potential energy of the oscillator with the wave vector $\vec{k}$ (see eqn (124)). Considering that according to the virial theorem for an oscillator, the mean potential energy is equal to the mean kinetic energy, we obtain

$$S_{\vec{k}} = \frac{\langle \varepsilon_{\vec{k}} \rangle_{av}}{\phi_{\vec{k}}} = \frac{\hbar\omega_{\vec{k}}(n_{\vec{k}} + \frac{1}{2})}{\phi_0(1 + \varphi_{\vec{k}})} \simeq \frac{c}{4\pi} \cdot \frac{kT}{1 + \varphi_{\vec{k}}} \tag{133}$$

where the mean oscillator energy is expressed in terms of Bose-Einstein distribution:

$$\langle \varepsilon_{\vec{k}} \rangle_{av} = \hbar\omega_{\vec{k}}[(e^{\hbar\omega_{\vec{k}}/kT} - 1)^{-1} + \tfrac{1}{2}] = \frac{\hbar\omega_{\vec{k}}}{2}\,\mathrm{cth}\,\frac{\hbar\omega_{\vec{k}}}{2kT} \simeq kT \tag{134}$$

and the classical nature of the motion of the medium dipoles ($\hbar\omega_{\vec{k}} \ll kT$) is taken into consideration. Equation (133) establishes direct relationship between the basic characteristic of the medium $\phi_{\vec{k}}$ (or $\varphi_{\vec{k}}$) and the correlation of the dipole moments observed experimentally $S_{\vec{k}}$. In particular, it follows from eqn (133) that Pekar's approximation ($\phi_{\vec{k}} = \phi_0 = 4\pi/c$) points to the absence of space correlation between dipole moments:

$$S_{\vec{k}}^{(0)} \simeq \frac{kT}{\phi_0} = \frac{c}{4\pi} kT \equiv S_0 \tag{135}$$

or passing from the Fourier components to the functions of the coordinates

$$S^{(0)}(\vec{r} - \vec{r}') = S_0 \delta(\vec{r} - \vec{r}') = \frac{c}{4\pi} kT\delta(\vec{r} - \vec{r}') \tag{136}$$

It also follows from eqn (133) that if the correlation function is known, it is possible to determine the frequency dispersion law

$$\omega_{\vec{k}}^2 = \frac{\langle \varepsilon_{\vec{k}} \rangle_{av}}{F_0 S_{\vec{k}}} \simeq \omega_0^2 \cdot \frac{c}{4\pi} \cdot \frac{kT}{S_{\vec{k}}} = \omega_0^2 \cdot \frac{S_0}{S_{\vec{k}}} \tag{137}$$

For practical purposes, the correlation function can be conveniently approximated by the Gaussian distribution:

$$S(\vec{r}) = S_0 \sqrt{\left(\frac{2}{\pi r_0^2}\right)} \frac{1}{4\pi r^2} \cdot \exp\left(-\frac{r^2}{2r_0^2}\right) \tag{138}$$

where the parameter r_0 figures as the correlation radius of the solvent dipole moments. Calculation shows that the choice of other tentative correlation functions does not affect the qualitative results given below. The Fourier components for function eqn (138) are of the form:

$$S_{\vec{k}} = S_0 \sqrt{\left(\frac{\pi}{2}\right)} \frac{1}{kr_0} Erf\left(\frac{kr_0}{\sqrt{(2)}}\right) \tag{139}$$

where Erf is the probability function. According to eqn (137) the correlation function of the form eqn (138) corresponds to the following frequency dispersion law

$$\omega_{\vec{k}}^2 = \omega_0^2 \sqrt{\left(\frac{2}{\pi}\right)} \frac{kr_0}{Erf(kr_0/\sqrt{(2)})} \simeq \omega_0^2 \left(1 + \frac{k^2 r_c^2}{6}\right) \tag{140}$$

The dependence of normal vibration frequencies on the wave number k, determined by eqn (140) is shown in Fig. 12 by a solid curve. Equation (140) obtained by us for the dispersion law is valid only for small k (large wavelengths). In order to obtain a correct dependence of the frequencies on the wave vector at large k (small wavelengths) it is necessary to take into consideration also the correlation of polarization currents, which is important at short distances. If the correlation can be approximated by a Gaussian distribution in this case as well, the dispersion law valid for any wavelengths, can be written as

$$\omega_{\vec{k}}^2 = \omega_0^2 \cdot \frac{r_0}{l_0} \cdot \frac{Erf(kl_0/\sqrt{(2)})}{Erf(kr_0/\sqrt{(2)})}; \quad \omega_\infty^2 = \omega_0^2 \cdot \frac{r_0}{l_0} \tag{141}$$

7

where l_0 is the correlation radius of polarization currents (derivatives of dipole moments with respect to time). The dispersion law, corresponding to eqn (141) is shown on the plot (Fig. 12) by a dashed line. Later we shall show that only long polarization waves (small k), are of essential importance in the transfer process, the minimum wavelength coinciding in order of magnitude with the distance R_0, to which the charge is transferred (in the case of homogeneous reactions), or with the discharging ion radius (in heterogeneous

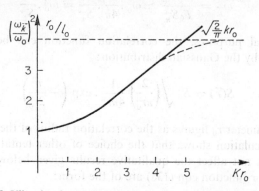

Fig. 12. Vibration frequencies dispersion law of polarization waves.

reactions). Pekar's approximation, which was actually used also by Marcus, is valid only in the case when the correlation radius of the dipole moments r_0 is less than the distance R_0. This is just the case, when according to eqn (140) we can neglect the frequency dispersion and use the Hamiltonian eqn (127). This approximation is the more accurate the larger is the ion radius. In the literature the charge together with the fluctuating polarization 'cloud' surrounding it, is usually called a polaron. If the condition

$$r_0 k_{max} \simeq \frac{r_0}{R_0} < 1 \tag{142}$$

is valid, we have a large radius polaron. In the case when the correlation radius of the dipole moments r_0 is less or of the order of magnitude of the distance to which charge is transferred, i.e. when the condition

$$r_0 k_{max} \simeq \frac{r_0}{R_0} \gtrsim 1 \tag{143}$$

is valid, we say that we have a small radius polaron. At present the theory of the small radius polaron is fairly thoroughly elaborated (see, e.g. References 38–41), frequency dispersion having been shown to be of importance in quantitative calculations of kinetic parameters (such as activation energy). However, in the case of a small radius polaron as well, the activation energy can be estimated in order of magnitude using Pekar's theory, assuming $\omega_{\vec{k}} \simeq \omega_0$. In this case, the activation energy thus calculated will be only several times higher, as we shall show later. This qualitative coincidence can be physically accounted for by a rather large number of the medium dipoles participating in the establishment of the activated state of the solvent. In fact, since the contribution of each dipole to the activation energy in order of magnitude must be equal to $\hbar\omega_0$ (vibrational energy quantum), the number of dipoles participating in the activation process can be estimated from the formula

$$N_a \simeq \frac{E_a}{\hbar\omega_0} \tag{144}$$

where E_a is the activation energy due to solvent fluctuation. For example, if we assume that $E_a \simeq 0.5$ eV, about 1.3×10^3 dipoles of the medium will participate in activation. By means of eqn (144) we can estimate the maximum extent of the solvent region, which is of importance in the charge transfer process:

$$R_{max} \simeq \left(\frac{3m_0 N_a}{4\pi\gamma}\right)^{1/3} \tag{145}$$

where m_0 is the dipole mass and γ is the solvent density. In particular, for water at $E_a \simeq 0.5$ eV we have $R_{max} \simeq 20$ Å.

It follows from the above that the wave numbers of most importance lie within the range

$$k_{min} \simeq \frac{1}{R_{max}} < k < k_{max} \simeq \frac{1}{R_0} \tag{146}$$

In other words, the charge transfer process is significantly affected by the portion of solvent located outside the ion in the region

$$r_0 \simeq R_0 < r < R_{max} \tag{147}$$

As regards theoretical calculations, the treatment of the first solvation shell of the ion (small r, i.e. large k) is the most difficult, since in this region the notion of specific polarization itself has only a qualitative significance. Strictly speaking, in this case one should use a discrete description of the solvent particles and take into consideration both vibrational motion and diffusion jumps of dipoles (jumpwise change of the coordinates of the centre of gravity of dipoles or of the dipole axis directions). The mean interval between jumps ($\tau_d \simeq 10^{-9}$ sec) being much larger, however, than the period of libration motions of a dipole ($\tau_0 = 2\pi/\omega_0 \simeq 2\pi \cdot 10^{-11}$ sec) we can ignore the effect of diffusional motion of solvent particles on the dynamic behaviour of the medium. Considering the libration motion of dipoles as a harmonic motion, we can approximately write for the energy of the particles in the first solvation shell

$$E = \tfrac{1}{2}\sum_i \hbar\omega_i(q_i^2 + g_i^2) \tag{148}$$

where q_i and g_i are the normal coordinates and momenta of dipoles, the normal frequencies ω_i coincide in order of magnitude with normal frequencies of libration vibrations ω_0, their dispersion being due to the interaction of dipoles.

The classical Hamiltonian eqn (124) in the continuum model can be also written in the form eqn (148) if we pass from $P_{\vec{k}_i}$ and $\dot{P}_{\vec{k}}$ to real canonically conjugate variables

$$
\begin{aligned}
g_{\vec{k}} &= \left(\frac{F_0}{2\hbar\omega_{\vec{k}}}\right)^{1/2} \cdot \begin{cases} i(\dot{P}_{\vec{k}} - \dot{P}_{\vec{k}}^{*}), \, k_x > 0 \\ \dot{P}_{\vec{k}} + \dot{P}_{\vec{k}}^{*}, \, k_x \leqslant 0 \end{cases} \\
q_{\vec{k}} &= \left(\frac{F_0\omega_{\vec{k}}}{2\hbar}\right)^{1/2} \cdot \begin{cases} i(P_{\vec{k}} - P_{\vec{k}}^{*}), \, k_x > 0 \\ P_{\vec{k}} + P_{\vec{k}}^{*}, \, k_x \leqslant 0 \end{cases}
\end{aligned} \tag{149}
$$

In these variables, eqn (124) can be written as

$$E = \tfrac{1}{2}\sum_{\vec{k}} \hbar\omega_{\vec{k}}(q_{\vec{k}}^2 + g_{\vec{k}}^2) \tag{150}$$

From the similarity of eqns (148) and (150) it is to be hoped that the use of a continuum model will enable us to take into account qualitatively also the solvent particles located near the discharging ion.[48,49]

It is known that we can pass from the classical Hamiltonian eqn (150) to the quantum Hamiltonian by formal substitution of the generalized momentum $g_{\vec{k}} \rightarrow -i(\partial/\partial q_{\vec{k}})$:

$$H_S(q) = \tfrac{1}{2}\sum_{\vec{k}} \hbar\omega_{\vec{k}}\left(q_{\vec{k}}^2 - \frac{\partial^2}{\partial q_{\vec{k}}^2}\right) \tag{151}$$

Finally it should be pointed out that below in statistical averaging of the solvent state, we shall substitute in the Gibbs distribution function the Hamiltonian eqn (151), which depends itself on the temperature in virtue of our model approximations. For instance, in Pekar's Hamiltonian eqn (127) the temperature dependence of the parameter c is due to the dielectric constant $\varepsilon_S(T)$ and in the presence of dispersion, the parameters r_0 and l_0, determining the correlation radii of dipole moments and their derivatives with respect to time, can be also temperature dependent (see eqn (141)). As an indirect criterion of the validity of our model, we can use the requirement that the Hamiltonian should depend slightly on temperature. As follows from estimates for the polar media of interest to us, this condition is always fulfilled quite well in practice. For instance, for water the relative variation of the energy value in eqn (127) does not exceed 1 per cent practically at any reasonable temperature values (see Reference 42).

2. Calculation of the Transition Probability

A detailed picture of an elementary act of ion discharge or ionization on the electrode is extremely complicated. In the general case, ion and electrode electrons, solvent and heavy particles (nuclei) forming part of a discharging ion participate in the process. More or less, all these components undergo changes in the discharge process. As we consider only the reactions in which electrochemical kinetics is determined by the charge transfer process, it is always necessary to take into account the change in the solvent state. Moreover, all processes of interest to us necessarily involve a change in the electron state in the electrode. As regards the remaining components of the system (inner electrons and heavy particles forming part of discharging ion), their role in the process can be determined only by considering a particular case of a discharging ion. For instance, for redox processes occurring without breaking or formation of chemical bonds, it is necessary to take into consideration only the electron participating in the process and the solvent, whereas in the hydrogen ions discharge, we should take into account not only the change in the proton state, but also the change in the state of the electrons holding the proton in the hydroxonium ion.

According to the quantum-mechanical perturbation theory the probability of one elementary electrochemical act occurring per unit time can be written as

$$W_{if} = \frac{2\pi}{\hbar} Av\sum |\langle \Psi_f(x, q, R)|V_{if}|\Psi_i(x, q, R)\rangle|^2 \delta(E_i - E_f) \qquad (152)$$

where Ψ_i and Ψ_f are the wave functions of initial and final states; (x, q, R) is the coordinates of the electrons, solvent and heavy particles, respectively; V_{if} is the perturbation responsible for the transitions. Summation is performed over all possible final states of the system, taking into consideration the conservation of energy law between the initial and final states, which is accomplished by introducing the δ-function into eqn (152); the symbol Av stands for statistical averaging over initial system states. The possibility of applying perturbation theory to our problem follows directly from the condition eqn (1) assumed by us earlier. In fact, eqn (152) is valid only for a small time interval $t < \tau_e$, where $\tau_e = W_{if}^{-1}$ is the mean time interval between transitions. However, since the ion resides in the reaction zone on the average during the time $\tau_d < \tau_e$, during the time of the 'impingement' of the ion on the electrode τ_d we can use eqn (152).

Actually, the calculation of W_{if} we shall perform within the framework of the adiabatic perturbation theory. The adiabatic theory is known to be used in the treatment of systems containing particles differing sharply in the velocities of their motion. Sometimes, adiabatic treatment is associated with the difference in the particle masses (as in the theory of molecules). However, this is true only in the case when mean kinetic energies of these particles are approximately equal:

$$\frac{mv^2}{2} \simeq \frac{MV^2}{2}; \quad \frac{V}{v} \simeq \left(\frac{m}{M}\right)^{1/2} \ll 1 \tag{153}$$

where m and M are the masses and v and V the velocities of light and heavy particles, respectively. It should be stressed that in the general case, we have to deal with the difference between mean particle velocities and therefore adiabatic approximation can be used even for a system containing slow and fast electrons. This is the case in solid state theory, when slow conduction electrons become separated from fast valence electrons. In the problem under consideration we can say beforehand that the electrons in ion and electrode constitute a fast subsystem with respect to the slow moving dipoles of the medium. An analysis of intramolecular vibrational motion of heavy particles forming part of the discharging ion is more complicated. If we assume the states of electrons and solvent to be fixed and calculate the energy levels of these heavy particles, from the appearance of the spectrum we can say what place is occupied by these particles in the adiabatic theory. If the characteristic interval between levels (e.g. the excitation energy) ΔE is greater than that between the solvent levels $\hbar\omega_0$, these particles, just as electrons, can be included in the fast subsystem. This is the case, for instance, in the consideration of the proton contained in the hydroxonium ion, or adsorbed on the electrode. In a harmonic approximation, the vibration frequency of the proton in hydroxonium ion is $\omega_i \simeq 5 \times 10^{14} \text{ sec}^{-1}$ and $\omega_a \simeq 2 \times 10^{14} \text{ sec}^{-1}$ for mercury electrode, which corresponds to the excitation energies

$$\Delta E_i \simeq \hbar\omega_i \simeq 0.3 \text{ eV} \quad \text{and} \quad \Delta E_a \simeq \hbar\omega_a \simeq 0.1 \text{ eV}$$

Thus, in both cases

$$\Delta E_{i,a} > \hbar\omega_0 \simeq 2.5 \times 10^{-3} \text{ eV}$$

If the excitation energy of internal degrees of freedom of ion or adsorbed particle is less than or of the order of $\hbar\omega_0$ only electrons should be included in the fast subsystem, whereas the solvent, together with heavy particles, forms a slow subsystem.

First we shall consider a theoretically more simple case, when $\Delta E > \hbar\omega_0$, i.e. when the slow subsystem consists only of the solvent. The Hamiltonian of the system in the initial state before discharge can be schematically written as:

$$\mathcal{H}^0_i(x, q, R) = H^0_e(x, R) + H_S(q) + V_{es}(x, R; q) \tag{154}$$

where the first term in the right hand side corresponds to the energy of the fast subsystem (electrons and ionic internal degrees of freedom); the second term is given by eqn (151) and the last term describes the interaction of the solvent with the fast subsystem and can be written as

$$V_{es}(x, R; q) = \sum_{\vec{k}} v_{\vec{k}}(x, R) q_{\vec{k}}$$

The form $v_{\vec{k}}$ depends on the concrete choice of an electrochemical system. If, for instance, the fast subsystem contains only one electron, we obtain for $v_{\vec{k}}(x)$ the following expression

$$v_{\vec{k}}(x) = e \left(\frac{8\pi c\hbar\omega_0^2}{k^2 \omega_{\vec{k}}^0 V_s}\right)^{1/2} \begin{Bmatrix} -\cos \vec{k}\,\vec{x}, & k_x > 0 \\ \sin \vec{k}\,\vec{x}, & k_x \leqslant 0 \end{Bmatrix} \tag{155}$$

Using this formula, we can find the interaction energy for each concrete case. Within the framework of the adiabatic theory, owing to the slow change of the solvent polarization (i.e. of coordinates q) with time, the state of the fast subsystem can be approximately determined with fixed values of the solvent coordinates (i.e. with fixed polarization of the medium):

$$\{H^0_e(x, R) + V_{es}(x, R; q)\}\psi_\alpha(x, R; q) = \varepsilon_\alpha(q)\psi_\alpha(x, R; q) \tag{156}$$

The physical sense of this equation is that the fast subsystem can instantaneously adjust itself in an 'equilibrium' manner to the slowly changing polarization of the medium. The slow fluctuation change of polarization involving a change in the interaction energy of the fast subsystem (e.g. electron) with the solvent, i.e. in the potential energy of the fast subsystem, the total energy of the fast subsystem $\varepsilon_\alpha(q)$ is a function of polarization of the medium (of coordinates q). In other words, the energy of the fast subsystem $\varepsilon_\alpha(q)$ will fluctuate with polarization. Figure 13 shows schematically the 'fluctuating' curves of potential energy with different solvent polarizations. In particular, curve (a) corresponds to the absence of polarization $q = 0$), i.e. to the ion in the gas phase. Curves (b) and (c) show various differences from zero polarizations of the medium. The subscript α for the energy of the

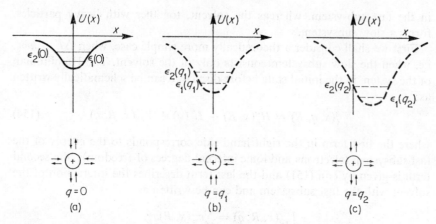

Fig. 13. Qualitative picture of the potential energy of fast subsystem for various solvent polarization values. Arrows show schematically the orientations of the solvent medium dipoles around the ion.

fast subsystem indicates that several quantum levels can exist in a given potential well.

With a given quantum state α of the fast subsystem, the state of the solvent (the slow subsystem) according to the adiabatic theory, is determined from the equation

$$\{H_S(q) + \varepsilon_\alpha(q)\}\chi_{\alpha n}(q) = E_{\alpha n}\chi_{\alpha n}(q) \tag{157}$$

where n is the number of the quantum state of the slow subsystem and $E_{\alpha n}$ is the total energy of the whole system in initial state. For physical interpretation of eqn (157), we shall write the Hamiltonian appearing in this equation and determining the total energy of the system as kinetic and potential energies:

$$H_S(q) + \varepsilon_\alpha(q) = -\tfrac{1}{2}\sum_{\vec{k}}\hbar\omega_{\vec{k}}\frac{\partial^2}{\partial q_{\vec{k}}} + U_\alpha(q) \tag{158}$$

where the potential energy is of the form

$$U_\alpha(q) = \tfrac{1}{2}\sum_{\vec{k}}\hbar\omega_{\vec{k}}q_{\vec{k}}^2 + \varepsilon_\alpha(q) \tag{159}$$

Thus, the total energy of the whole system can be found from the solution of the Schrodinger equation only for the slow subsystem if we include formally $\varepsilon_\alpha(q)$ in the potential energy of the solvent. Physically it means that along with the potential energy of dipoles (the first term in eqn (159)), we take into account the interaction of the solvent with the 'diffuse' cloud of the fast subsystem (the second term in eqn (159)). For this reason, in the adiabatic theory the function $U_\alpha(q)$ is usually called the potential energy surface. The interpretation for $U_\alpha(q)$ will prove to be somewhat different if the kinetic energy of the slow subsystem is assumed to be negligible. In this case $U_\alpha(q)$, will correspond to the total energy of the whole system with a given solvent polarization. Hence, another name for $U_\alpha(q)$ is the term of the system. If the fast subsystem contains only electrons, $U_\alpha(q)$ is called the electronic term. By analogy, if proton as well is included in the fast subsystem $U_\alpha(q)$ will be called the electron-protonic term, or simply, the term of the system.

Given the law of interaction of the slow and fast subsystems in the form eqn (155) we can write the expression for the term $U_\alpha(q)$ in a more convenient form:

$$U_\alpha(q) = \tfrac{1}{2}\sum_{\vec{k}}\hbar\omega_{\vec{k}}q_{\vec{k}}^2 + \varepsilon_\alpha(q_\alpha) + \sum_{\vec{k}}v_{\vec{k}\alpha}(\vec{q_k} - \vec{q_{k\alpha}})$$

$$= \varepsilon_\alpha(q_\alpha) + \tfrac{1}{2}\sum_{\vec{k}}\hbar\omega_{\vec{k}}q_{\vec{k}\alpha}^2 + \tfrac{1}{2}\sum_{\vec{k}}\hbar\omega_{\vec{k}}(\vec{q_k} - \vec{q_{k\alpha}})^2 \qquad (160)$$

where

$$\vec{q_{k\alpha}} = -\frac{v_{\vec{k}\alpha}}{\hbar\omega_{\vec{k}}} = -\frac{1}{\hbar\omega_{\vec{k}}}\int v_{\vec{k}}(x, R)|\psi_\alpha(x, R; q_\alpha)|^2 \, \mathrm{d}\vec{x} \, \mathrm{d}\vec{R} \qquad (161)$$

To explain the physical significance of the result obtained let us consider Fig. 14, where the dashed curve shows schematically the term of the pure

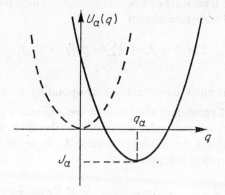

Fig. 14. Qualitative picture of the terms for the solvent, in the absence of ions (dashed curve) and in the presence of an ion (solid curve).

solvent in the absence of ion, when polarization perform harmonic fluctuation vibrations near the nonpolarized state $\vec{P}(\vec{r}) = 0$ (i.e. $\vec{q_k} = 0$). After the introduction of ion into the solvent, the interaction of the charge with the medium gives rise to a different from zero mean polarization $\vec{P_\alpha}(\vec{r})$, near which harmonic fluctuations will now occur. Normal coordinates, corresponding to mean polarization, we shall denote by $q_{\vec{k}\alpha}$ (see eqn (161); in Fig. 14 q_α). The third term in the right hand side of eqn (160) corresponds to purely vibrational potential energy. The second term corresponds to the work expended in producing mean polarization $\vec{P}_\alpha(\vec{r})$ and, finally, the first term corresponds to the energy of the fast subsystem, calculated with equilibrium polarization $\vec{P}_\alpha(\vec{r})$. Using the symbols of the preceding section, we can rewrite eqn (160) in the form:

$$U_\alpha(q) = J_\alpha + \tfrac{1}{2}\sum_{\vec{k}} \hbar\omega_{\vec{k}}(q_{\vec{k}} - q_{\vec{k}\alpha})^2 \tag{162}$$

where the minimum potential energy J_α for the initial state in the model under consideration can be written as:

$$J_\alpha = \varepsilon_f - e\varphi_m + \varepsilon_\alpha^0 + Z_\alpha e\psi_1 - \frac{\varepsilon_s - 1}{8\pi\varepsilon_s} \int D_\alpha^2 \, d\vec{r} \tag{163}$$

Here ε_α^0 is the energy of ion in the gas phase, Z_α is the ion charge and the last term is the ion solvation energy ($\vec{D}_\alpha$ is electrostatic induction set up by the ion).

The term of the final state can be considered in much the same way as the term of the initial state (see above)

$$U_{\alpha'}(q) = J_{\alpha'} + \tfrac{1}{2}\sum_{\vec{k}} \hbar\omega_{\vec{k}}(q_{\vec{k}} - q_{\vec{k}\alpha'})^2 \tag{164}$$

where $q_{\vec{k}\alpha'}$ are normal coordinates corresponding to mean polarization $\vec{P}_{\alpha'}(\vec{r})$ in the final state, and the form of the minimum potential energy is readily found for each particular electrochemical system. For instance, for a reaction occurring via an adsorption step, $J_{\alpha'}$ is equal to the adsorption energy, and for a redox reaction

$$J_{\alpha'} = \varepsilon_{\alpha'}^0 + Z_{\alpha'} e\psi_1 - \frac{\varepsilon_s - 1}{8\pi\varepsilon_s} \int D_{\alpha'}^2 \, d\vec{r} \tag{165}$$

Like the initial state term, in the final state the term is in the form of a parabola in $(N + 1)$-dimensional space, where N is the number of excited libration vibrational degrees of freedom of the solvent, i.e. the number of oscillators $(N \to \infty)$. Figure 15 shows schematically in the form of one-dimensional parabolas the terms for initial and final states.

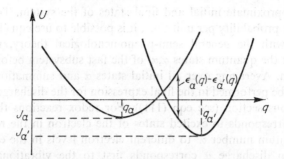

Fig. 15. Qualitative picture of the terms of initial and final states.

The unperturbed Hamiltonians of the solvent in initial and final states reducing to the sum of oscillator Hamiltonians, the corresponding energies are of the form:

$$E_{\alpha n} = J_\alpha + \sum_{\vec{k}} \hbar \omega_{\vec{k}} (n_{\vec{k}} + \tfrac{1}{2}); \quad E_{\alpha' n} = J_{\alpha'} + \sum_{\vec{k}} \hbar \omega_{\vec{k}} (n'_{\vec{k}} + \tfrac{1}{2});$$

$$n_{\vec{k}}, n'_{\vec{k}} = 0, 1, 2, \dots \quad (166)$$

where $n_{\vec{k}}$ and $n'_{\vec{k}}$ are the occupation numbers of oscillators with the wave vector $\vec{k}$ in initial and final states. Similarly, in an adiabatic approximation the unperturbed wave function reduce to the product of the wave functions of the fast and slow subsystems:

$$\Psi_{\alpha,n} = \psi_\alpha \Pi_{\vec{k}} \chi_{n_{\vec{k}}} (q_{\vec{k}} - q_{\vec{k}\alpha}); \quad \Psi_{\alpha',n'} = \psi_{\alpha'} \Pi_{\vec{k}} \chi_{n'_{\vec{k}}} (q_{\vec{k}} - q_{\vec{k}\alpha'}) \quad (167)$$

where χ are the ordinary oscillator wave functions, and ψ are determined as solutions of eqn (156). Naturally, the expressions given above for energies eqn (166) and wave functions eqn (167) are not accurate solutions of the total Hamiltonian of the system. In seeking for $E_{\alpha n}$ and $\Psi_{\alpha n}$, we ignored the interaction of the discharging particle with the electrode. For instance, in order to determine the unperturbed (approximate) initial state in a redox reactions we have to ignore the interaction with the electrode of the electron, localized at first in the reducing ion. This interaction causes the electron transfer from the ion to the electrode. Likewise, in considering the unperturbed final state, we do not take into account the energy of the interaction of the electron localized in the electrode with the oxidizing ion. Taking account of this interaction leads only to the electron transfer from the electrode to the ion. When considering hydrogen ion discharge, we ignore the interaction of proton first with the electrode, then with the water molecule, in order to determine approximate initial and final states of the system. To determine the discharge probability per unit time, it is possible to use eqn (152). But, in compliance with the general semi-phenomenological theory, we have to establish first the quantum states α,α' of the fast subsystem before and after the transition. Averaging over all initial states α and summation over final states α' can be performed in the final expression for the discharge current, as in the previous section (see eqn (111)). For redox reactions the quantum number α corresponds to excited states of the electron in the reducing ion and the quantum number α' to different electron levels in the electrode. In hydrogen ion discharge α corresponds first to the vibrational quantum number of the proton in hydroxonium ion and, second, to the number of the energy level ε_f, on which the electron of the metal was localized before the discharge. Similarly, α' contains the vibrational quantum number of the adsorbed hydrogen atom and the quantum number of the electron in the adsorbed particle. In the approximation of adiabatic perturbation theory the transition probability per unit time is determined by the formula

$$W_{\alpha\alpha'} = \frac{2\pi}{\hbar} |\langle\psi_{\alpha'}|V|\psi_\alpha\rangle|^2 Av_n \sum_{n'} |\langle\chi_{n'}|\chi_n\rangle|^2 \delta(J_{\alpha'} - J_\alpha + \varepsilon_n - \varepsilon_{n'}) \quad (168)$$

where ε_n and $\varepsilon_{n'}$ are excitation energies of the slow subsystem (solvent). If we use the wave functions eqn (167) and energies eqn (166) for oscillators, the probability $W_{\alpha\alpha'}$ can be calculated quite rigorously both in the absence of frequency dispersion[27,43,44] and for an arbitrary frequency dispersion.[51] In the general case, $W_{\alpha\alpha'}$ has a rather complex form and is expressed in terms of the Bessel function of imaginary argument (or the product of the Bessel functions of imaginary argument in the presence of frequency dispersion), the formula being much simpler, however, in two limiting cases, which we shall consider below.

(a) Classical Oscillators:

$$\hbar\omega_{\vec{k}} < kT \quad (169)$$

If the slow subsystem is classical, i.e. the condition eqn (169) is fulfilled, the transition probability assumes the form:

$$W_{\alpha\alpha'} = |V_{\alpha'\alpha}|^2 \left(\frac{\pi}{\hbar^2 kTE_r}\right)^{1/2} e^{-E_a/kT} \quad (170)$$

where the quantity

$$E_a = \frac{1}{4E_r}(J_{\alpha'} - J_\alpha + E_r)^2 \quad (171)$$

figures as the activation energy.

E_r is the reorganization energy of the solvent:

$$E_r = \tfrac{1}{2}\sum_{\vec{k}}\hbar\omega_{\vec{k}}(q_{\vec{k}\alpha'} - q_{\vec{k}\alpha})^2 \tag{172}$$

Thus the assumption of the classical nature of motion of the medium dipoles leads to the Arrhenius equation for the transition probability. Moreover, as a simple mathematical analysis shows, the quantity E_a determined by eqn (171) coincides geometrically with the difference of energies at the saddle point on the intersection surface of the terms and at the minimum point of the initial term (see Fig. 16(a)). This fact allows us to make use of the analogy

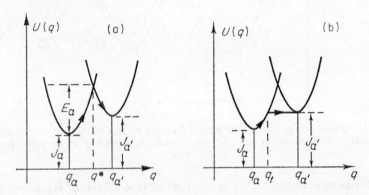

Fig. 16. Interpretation of the classical (a) and quantum (b) transitions of the system from initial to final state.

with the particle transition from one potential well to another in classical physics. The fluctuation motion of the medium dipoles produces an activated state q^*, i.e. the system rises to the top of the potential barrier. At the moment when the activated state q^* is established, the fast subsystem is able to accomplish a quantum ('tunnel') transition from initial to final terms, whereupon the classical subsystem 'falls' into the final well. In Fig. 16(a) this reaction path is shown by a line with arrows. At this point, we have to make a very important reservation. Although by tradition we have used the words 'tunnel' transition, it does not mean that in calculating the probability of a jump of the fast subsystem from term to term in the activated state we can use Gamov's equation[45]

$$W = \exp\left(-\frac{2}{\hbar}\int_a^l |p|\,dx\right) \tag{173}$$

as some authors do (see e.g. Reference 46). This becomes evident if we consider the form of potential energy of the fast subsystem, since the 'tunnel' transition must occur just in this field. As we have already said in discussing Fig. 13, the potential energy of the fast subsystem fluctuates (i.e. varies in time) with the motion of dipoles of the medium. Figure 17 shows qualitatively

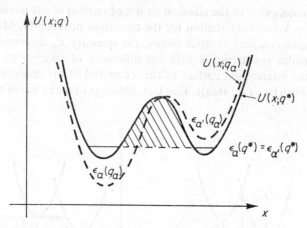

Fig. 17. Qualitative picture of the potential energy of fast subsystem with equilibrium polarization (dashed curve) and with polarization corresponding to the activation state (solid curve).

the forms of potential energy at two polarization values: q^* is corresponding to the activated state of the solvent polarization (solid curve) and q_α is corresponding to the initial equilibrium polarization (dashed curve). In the case of equilibrium polarization q_α the energy levels of the fast subsystem do not coincide (see also Fig. 15) and therefore the quantum transition is not favoured. In the case of activated polarization, the levels of the fast subsystem become equal ($\varepsilon_\alpha(q^*) = \varepsilon_{\alpha'}(q^*)$) and the transition probability is maximal. Here the probability is determined not only by the shape of the potential hump (shaded area in Fig. 17), but mainly by the time interval during which the levels $\varepsilon_\alpha(q^*)$ and $\varepsilon_{\alpha'}(q^*)$ are close enough to each other, i.e. by the rate of change of polarization with time. For instance, if the polarization changes very quickly ($\vec{P} \to \infty$), practically no transition occurs and vice versa, with a very slow change of polarization ($\vec{P} \to 0$) the transition will occur with a large probability. The problem of particle transition with the potential varying with time is known as the Landau-Zener problem.[45] As regards eqn (173) it is used only for the particle motion with the potential constant in time (e.g. the problem of nuclear α-decay). To avoid misunderstanding, in what follows we shall use the term quantum transition instead of 'tunnel' transition.

It is clear from eqns (170) and (171) that the basic parameter of the theory is the reorganization energy E_r, for an accurate calculation of which it is necessary to know the wave functions of the subsystem and the law of the interaction of the charge with the solvent (see eqn (161)). For instance, for redox reactions we can use eqn (155) and write for E_r:

$$E_r = \frac{2\pi c e^2}{V_S} \sum_{\vec{k}} |\int \{|\psi_\alpha|^2 - |\psi_{\alpha'}|^2\} e^{i\vec{k}\cdot\vec{r}} \, d\vec{r}|^2 \cdot \frac{\omega_0^2}{k^2 \omega_{\vec{k}}^2} \tag{174}$$

In the case of homogeneous reactions $|\psi_\alpha|^2$ and $|\psi_{\alpha'}|^2$ are localized in the ions of the reducing and oxidizing agents separated by the distance R_0. By means of the Fourier theory, it can be shown that the integral contained in eqn (174) has its largest value at $k \lesssim R_0^{-1}$ (see eqn (142)). In the case of heterogeneous reactions, the electron in the electrode is delocalized and $|\psi_{\alpha'}|^2 \sim 1/V_e \rightarrow 0$, therefore

$$E_r = \frac{2\pi c e^2}{V_S} \sum_{\vec{k}} |\int |\psi_\alpha|^2 e^{i\vec{k}\cdot\vec{r}} \, d\vec{r}|^2 \cdot \frac{\omega_0^2}{k^2 \omega_{\vec{k}}^2} \tag{175}$$

Since $|\psi_\alpha|^2$ varies significantly over the dimensions of the discharging ion R_0, the maximum value of the integral in eqn (175) will be reached at $k \lesssim R_0^{-1}$. By a vector analysis it is possible to transform eqn (174) identically to the form

$$E_r = \frac{c}{8\pi} \cdot \sum_{\vec{k}} \frac{\omega_0^2}{\omega_{\vec{k}}^2} |\vec{D}_{\vec{k}\alpha} - \vec{D}_{\vec{k}\alpha'}|^2 = \frac{2\pi}{c} \cdot \sum_{\vec{k}} \frac{\omega_0^2}{\omega_{\vec{k}}^2} |\vec{P}_{\vec{k}\alpha} - \vec{P}_{\vec{k}\alpha'}|^2 \tag{176}$$

where $\vec{P}_{\vec{k}\alpha}$ and $\vec{P}_{\vec{k}\alpha'}$ are the Fourier components of equilibrium polarizations, and $\vec{D}_{\vec{k}\alpha}$ and $\vec{D}_{\vec{k}\alpha'}$ the inductions related to $\vec{P}_{\vec{k}\alpha}$ and $\vec{P}_{\vec{k}\alpha'}$ by the formula

$$\vec{D}_{\vec{k}\alpha(\alpha')} = \frac{4\pi}{c} \vec{P}_{\vec{k}\alpha(\alpha')} \tag{177}$$

In particular, it is evident from eqn (176) that in the absence of frequency dispersion, the reorganization energy coincides with the expression obtained by Marcus[11] and by the author of this article[24] earlier:

$$E_r^{loc} = \frac{2\pi}{c} \int \{\vec{P}_\alpha(\vec{r}) - \vec{P}_{\alpha'}(\vec{r})\}^2 \, d\vec{r} \tag{178}$$

As shown above (see eqns (140) and (141)), in the presence of dispersion ω_0 corresponds to minimum frequency in the oscillator spectrum ($\omega_{\vec{k}} \geqslant \omega_0$) and hence eqn (178) gives the upper limit of the reorganization energy. Using eqn (141) it is possible to find the lower limit of E_r:

$$\frac{l_0}{r_0} E_r^{loc} \leqslant E_r \leqslant E_r^{loc} \tag{179}$$

Since the correlation radius of dipole moments r_0 can exceed the correlation radius of the rates of change of dipole moments l_0 only several times, the substitution of E_r^{loc} for E_r might be approximately justified. If, however, we know experimentally the correlation radii of dipole moments of different solvents, using eqns (176) and (140) it is possible to calculate the change in the reorganization energy in various media. It can be pointed out qualitatively that in media with a strong correlation of dipole moments E_r must decrease.[50]

(b) *Quantum Oscillators*:

$$\hbar \omega_{\vec{k}} > kT \tag{180}$$

When the slow subsystem is a quantum system we have for $W_{\alpha\alpha'}$

$$W_{\alpha\alpha'} = \text{constant} \begin{cases} \exp\left(J_\alpha - J_{\alpha'}/kT\right), & J_{\alpha'} > J_\alpha \\ 1, & J_{\alpha'} < J_\alpha \end{cases} \tag{181}$$

where we have not written an expression for the constant (see e.g. Reference 27). This formula also has a simple physical sense. If $J_{\alpha'} > J_\alpha$ the slow subsystem first fluctuates from the initial equilibrium state q_α to the point q_t (see Fig. 16(b)), which makes it possible for the quantum subbarrier transition of the slow subsystem to occur. Thus, both the slow and the fast subsystems perform a quantum transition. In Fig. 16(b) the reaction path described above is shown by a heavy line with arrows. Evidently, when $J_{\alpha'} < J_\alpha$ a subbarrier ('tunnel') transition is possible directly from the initial equilibrium state q_α. Although in the description of the solvent the criterion of the classical behaviour eqn (169) is always fulfilled, an analysis of case (b) is of fundamental importance to us. Since for oscillators $\hbar \omega_{\vec{k}}$ is the distance between the levels, it can be assumed from the above consideration that the criterion of the classical behaviour of one or the other of the subsystems is the condition (see eqn (14)):

$$\Delta E < kT \tag{182}$$

where ΔE is the characteristic distance between the levels of the respective subsystem. To verify this assumption we considered first the linear terms shown in Fig. 10, which permit of a quantum calculation. It proved that the criterion eqn (182) is valid in this case as well, if ΔE is understood to mean the distance from the first excited level to the ground level. Finally, by means of the quasi-classical wave functions we calculated the transition probability for one-dimensional terms of a rather general form. It was shown in this case as well that when condition eqn (182) is valid, a transition of the system over the barrier occurs, whereas with the reverse condition a subbarrier ('tunnel') transition of the system takes place. A very important conclusion can be drawn from what has been just said, which we shall give as a prescription. In considering the internal degrees of freedom of the particles participating in the reaction (before and after reaction), they should be divided into classical and quantum according to criterion eqn (182). Then, in an adiabatic treatment, all the quantum degrees of freedom should be included in the fast subsystem and all the classical degrees of freedom in the slow subsystem:

$$\Delta E_{\text{quant}} > kT > \Delta E_{\text{cl}} \tag{183}$$

Thus, condition eqn (183) allows us to divide the whole system into only two subsystems: a slow and a fast subsystem. This is due to the fact that the condition $\Delta E_{\text{quant}} > \Delta E_{\text{cl}}$ in terms of an adiabatic approximation means at the same time that the quantum subsystem is faster than the classical subsystem. With such an approach, the potential energy surfaces should be automatically introduced only in the case of a classical subsystem, whereas the quantum subsystem will contribute to the minimum potential energy $J_{\alpha,\alpha'}$. Here we have to make a reservation. Above we assumed for simplicity that a particle which was 'classical' before the reaction remains so after it as well. The case when a 'classical' particle becomes a 'quantum' one as the result of the reaction (or vice versa) permits of a theoretical calculation, but here we shall not give the results of such a treatment. It should be noted, however, that in spite of the appearance of some interesting effects, the general picture described above remains valid.[51,52]

In order to calculate the transition probability it is possible to use eqn (168), where the exchange integral (the first factor) is completely determined by the quantum subsystem, and the rest of the formula gives the Arrhenius activation factor. In this case $\chi_{n,n'}$ and $\varepsilon_{n,n'}$ should be understood to mean respectively the wave functions and excitation energies of the whole classical subsystem. It is clearly evident from eqn (168) that the transition probability $W_{\alpha\alpha'}$ depends on ΔJ (the last factor in eqn (168)) and on the parameters determining the shape of the term (through $\varepsilon_{n,n'}$ and $\chi_{n,n'}$) as was assumed in the previous section.

The transition of the classical subsystem occurring above the potential barrier, the value of $W_{\alpha\alpha'}$ can be calculated by a method which is a generalized version of the Landau-Zener method developed for one-dimensional terms. Below we shall give only the basic steps of the derivation of this formula for

$W_{\alpha\alpha'}$. First of all, we shall write in a very general form the expressions for the energies corresponding to initial and final states of the classical subsystems:

$$E_{\alpha,\alpha'} = \sum_{i=1}^{N} \frac{M_i V_i^2}{2} + J_{\alpha,\alpha'} + U_{\alpha,\alpha'}(X_i) \tag{184}$$

where M_i, V_i, X_i are the mass, velocity and coordinate of the ith particle of the classical subsystem, respectively. Now let us pass to dimensionless coordinates and momenta

$$q_i = X_i \left(\frac{M_i kT}{h^2}\right)^{1/2}; \quad P_i = V_i \left(\frac{M_i}{kT}\right)^{1/2} \tag{185}$$

In these canonically conjugate variables, the energies assume the form

$$E_{\alpha,\alpha'}(P, q) = kT \sum_{i=1}^{N} \frac{P_i^2}{2} + J_{\alpha,\alpha'} + U_{\alpha,\alpha'}(q_i) \tag{186}$$

and the quantities

$$\dot{q}_i = \frac{kT}{h} P \tag{187}$$

appear as the particle velocities.

In the N-dimensional space $\{q_i\}$ the activated state corresponds to the points $\{q_i^*\}$, lying on the surface S^*; the equation for this surface is found from the condition of the intersection of the terms:

$$U_\alpha(q_i^*) = U_{\alpha'}(q_i^*) + J_{\alpha'} - J_\alpha; \quad \phi(q^*) = 0 \tag{188}$$

For clarity, the surface S^* is shown in Fig. 18 as a curve $\phi(q_1^*, q_2^*) = 0$ for the case when $N = 2$. Now we have to find the probability of the system located at the term $U_\alpha(q_i)$ passing in the time dt to the term of the final state $U_{\alpha'}(q_i)$:

$$dW_{\alpha\alpha'} = W_{\alpha\alpha'}\, dt \tag{189}$$

where $W_{\alpha\alpha'}$ is the transition probability in unit time sought for. With arbitrary (but fixed) velocities of the motion of particles $\dot{q}_i = (kT/h)P_i$ the system can pass through the intersection surface of the terms S^* in the time dt, if the particles of the classical subsystem are contained in the volume element

$$dV_q^* = \prod_{i=1}^{N} dq_i^* = dS^* \vec{q}_{n*}\, dt = \frac{kT}{h} dS^* \vec{P}_{n*}\, dt \tag{190}$$

where $\vec{P}$ is the N-dimensional momentum vector of the system with the components $(P_1, \ldots, P_N)$, the subscript n^* indicates that we have a projection of vector $\vec{P}$ onto a direction normal to the surface S^* at the point $(q_1^*, \ldots, q_N^*)$ dS^* is an element of the activated surface S^* near the point $(q_1^*, \ldots, q_N^*)$. Figure 18 shows for a two-dimensional case the vector $\vec{P}$ and its normal and tangential components, the region dV_q^* eqn (190) is shaded. The probability of the system initially located at the term $U_\alpha(q)$ being localized in the space

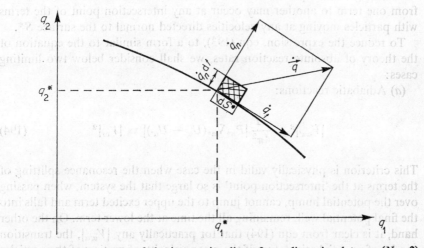

Fig. 18. Schematic picture of the intersection line of two-dimensional terms $(N = 2)$. $\dot{q}_n$ and $\dot{q}_t$ are normal and tangential components of the two-dimensional vector $\vec{q}$, $dV_q^* = \dot{q}_n$ $dt\, dS^*$ (shaded area).

dV_q^* (see eqn (190)) and of the particles having given momenta $(P_1 \ldots P_N)$ can be found from the Gibbs distribution:

$$W(P, q^*)\, dV_q^*\, d\vec{P} = e^{-E_\alpha(p, q^*)/kT} \cdot \frac{kT}{h} dS^* P_{n*}\, dt\, d\vec{P} \left[\int e^{-E_\alpha(p, q)/kT} d\vec{q}\, d\vec{p} \right]^{-1}$$

(191)

The probability that upon transition through an activated state at the point $(q_1^*, \ldots, q_N^*)$ at the velocity $(P_1, \ldots, P_N)$ the system will pass to the term of the final state can be found by a method similar to that of Landau-Zener

$$W_{LZ}(q^*, P) = 1 - \exp\left\{ -\frac{4\pi^2 |V_{\alpha'\alpha}|^2}{kT |P_{n*} \nabla_{n*}(U_\alpha - U_{\alpha'})|} \right\}$$

(192)

where $V_{\alpha'\alpha}$ is the exchange integral of the quantum subsystem (see eqns (168) and (170)) and the gradient ∇_n^* is taken at the point $(q_1^*, \ldots, q_N^*)$ in the direction normal to the surface S^*. Now the expression for $W_{\alpha\alpha'}$ can be written as

$$W_{\alpha\alpha'}\, dt = \int_{(N)} \ldots \int d\vec{P} \int_{(N-1)} \ldots \int dS^* \frac{kT}{h} P_{n*}\, dt W(P, q^*) W_{LZ}(q^*, P) \quad (193)$$

where integration is performed over momenta and S^* because the transition from one term to another may occur at any intersection point of the terms with particles moving at any velocities directed normal to the surface S^*.

To reduce the expression, eqn (193), to a form similar to the equation of the theory of absolute reaction rates, we shall consider below two limiting cases:

(a) Adiabatic reactions:

$$|V_{\alpha\alpha'}|^2 \gg \frac{kT}{4\pi^2} |P_{n*}\nabla_{n*}(U_\alpha - U_{\alpha'})| \equiv |V_{cr}|^2 \quad (194)$$

This criterion is physically valid in the case when the resonance splitting of the terms at the 'intersection point' is so large that the system, when passing over the potential hump, cannot jump to the upper excited term and falls into the final potential well, remaining all the time at the lower term. On the other hand, it is clear from eqn (194) that for practically any $|V_{\alpha'\alpha}|$, the transition can be of an adiabatic nature if the mean velocity of motion of the particles of the classical subsystem is small enough ($P_{n*} \to 0$). If eqn (194) is fulfilled W_{LZ} in eqn (192) can be substituted by 1. As has been said earlier, the activation energy for a classical subsystem (in terms of the Arrhenius dependence of the transition probability on temperature) appears as the quantity

$$E_a = U_\alpha(q_S^*) \quad (195)$$

where q_S^* is the saddle point on the activation surface S^*. If the potential energy in the activated state is reckoned from the minimum value

$$U_a(q^*) = U_\alpha(q^*) - U_\alpha(q_S^*) \quad (196)$$

eqn (193) can be rewritten as

$$W_{\alpha\alpha'} = \frac{kT}{h} \frac{e^{-Ea/kT} \int d\vec{P} dS^* P_{n*}\, e^{-\frac{1}{2}\Sigma P_i^2 - (1/kT)U_a(q^*)}}{\int d\vec{P}\, d\vec{q}\, e^{-\frac{1}{2}\Sigma P^2 - (1/kT)U_\alpha(q)}} \quad (197)$$

For this expression to be written in the form of the equation of the absolute reaction rates theory, the last factor in it should be identified formally with the entropy factor:[53]

$$e^{Sa/k} = \frac{\displaystyle\int_{(P_{n*} > 0)} d\vec{P}\, dS^* P_{n*}\, e^{-\frac{1}{2}\vec{P}^2 - (1/kT)U_a(q^*)}}{\int d\vec{P}\, d\vec{q}\, e^{-\frac{1}{2}\vec{P}^2 - (1/kT)U_\alpha(q)}} \tag{198}$$

It should be stressed that in eqn (197) only the Arrhenius factor containing the activation energy is of real physical significance, whereas in the general case the origin of the entropy factor is only of a formal mathematical nature. This is accounted for by the fact that in eqn (198) the momenta and coordinates of particles in the activated state are not separated due to the presence of the factor P_n^* and hence it is impossible to determine the state of the system for which the entropy could be calculated by means of general thermodynamic formulae. In a sense, the distortion of the activation surface S^* gives rise to noninertial (centrifugal) forces operating in the activated state. The problem is greatly simplified if the intersection surface of the terms is plane (in Fig. 18 the curve $\phi = 0$ in this case is a straight line). It can be readily seen that in the general case, this can be true only of parabolic terms, when the terms of initial and final states have an identical shape and are shifted with respect to one another (without turning). The approximation of S^* by a plane can be used, however, also in the case of a slight distortion, or if only a small region near the saddle point contributes to the integral, eqn (198). Of a particularly simple form is the entropy of activation for parabolic terms in the absence of frequency dispersion $\omega_i = \omega_0$:

$$S_a = k \ln \frac{\hbar \omega_0}{kT} \tag{199}$$

This result, obtained by us from eqn (198) can be also derived by a strictly thermodynamic method, if we take into consideration that the activated state differs from the initial state by the absence of one vibrational degree of freedom:

$$S_a = -k \ln \int_{-\infty}^{+\infty} e^{-\frac{1}{2}P^2} dP \int_{-\infty}^{+\infty} e^{-(\hbar^2 \omega_0^2 q^2 / 2k^2 T^2)} dq = k \ln \frac{\hbar \omega_0}{kT} \tag{200}$$

In the presence of an arbitrary frequency dispersion, the activation entropy is of the form similar to eqn (199), where ω_0 is substituted by some effective frequency:[27]

$$S_a = k \ln \frac{\hbar \omega_{\text{eff}}}{kT}; \quad \omega_{\text{eff}}^2 = \sum_i \omega_i^2 \frac{E_{ri}}{E_r} \tag{201}$$

where E_{ri} is the reorganization energy of the ith oscillator, and E_r is the total reorganization energy.

(b) Nonadiabatic reactions:

$$|V_{\alpha'\alpha}|^2 \ll |V_{cr}|^2 \tag{202}$$

In the case of the resonance splitting of the terms being slight there is a large probability that when the system passes the 'intersection point' it can remain at the initial term. Physically, this effect is due either to a small overlapping the wave functions of the quantum subsystem $(V_{\alpha'\alpha} \to 0)$, or to a large mean velocity of motion of classical particles $(P \to \infty)$. Thus, the transition from term to term occurs only after a repeated passing through the 'intersection point'. For a nonadiabatic transition eqn (193) can be written as

$$W_{\alpha\alpha'} = \frac{kT}{h} \kappa_{\alpha\alpha'} \, e^{-(E_a/kT)} \, e^{S_a/k} \tag{203}$$

where E_a is the activation energy (as understood above). S_a as before is determined by eqn (198) and the transmission coefficient is determined from the relation:

$$\kappa_{\alpha\alpha'} = \frac{\sqrt{(2\pi)}4\pi^2|V_{\alpha'\alpha}|^2}{kT} \int \frac{\exp\left[-(1/kT)U_a(q^*)\right]}{|\nabla_{n*}(U_\alpha - U_{\alpha'})|} \, dS^* \Big/ \int e^{-(1/kT)U_a(q^*)} \, dS^*$$

$$= \frac{\sqrt{(2\pi)}4\pi^2|V_{\alpha'\alpha}|^2}{kT} \left\langle \frac{1}{|\nabla_{n*}(U_\alpha - U_{\alpha'})|} \right\rangle_{av} \tag{204}$$

In the general case, in order to calculate $\kappa_{\alpha\alpha'}$ it is necessary to know the concrete form of the terms. But if the intersection surface of the terms is approximated by a plane, the transmission coefficient can be readily estimated. In particular, we obtain from eqn (204) the following expression:

$$\kappa_{\alpha\alpha'} = |V_{\alpha'\alpha}|^2 \left(\frac{4\pi^3}{\hbar^2\omega_{\text{eff}}^2 kTE_r}\right)^{1/2} \tag{205}$$

where ω_{eff} is determined in eqn (201). Comparing eqn (204) with the determination of the critical value of the exchange integral V_{cr} (see eqn (194)), we see that if P_{n*} is replaced by the mean momentum $\langle|P|\rangle_{av} = \sqrt{(2/\pi)}$, we obtain for $\kappa_{\alpha\alpha'}$ for nonadiabatic transitions the condition:

$$\kappa_{\alpha\alpha'} = \frac{2|V_{\alpha'\alpha}|^2}{\langle|V_{cr}|^2\rangle_{av}} \ll 1 \tag{206}$$

Thus, we can consider the transition probability to be always determined by eqn (203) but for adiabatic reactions $\kappa_{\alpha\alpha'}$ should be replaced by 1.

It can be readily seen that upon substitution of eqns (201) and (205) into eqn (203) we obtain exactly the same expression as we derived from a rigorous quantum-mechanical treatment, eqn (170). Thus, the application of the quantum-mechanical perturbation theory gives an expression for the transition probability only in nonadiabatic processes. The advantage of the semiclassical treatment developed above is that it permits us to advance further than the perturbation theory and, on the whole, to obtain an expression for the probability of adiabatic transitions. Unfortunately, at present this fact can be made use of only in the calculation of homogeneous reactions. In the investigation of electrochemical processes we used essentially the one-electron approximation. In this case, due to the delocalization of the electron in the electrode, the one-electron exchange integral is in inverse proportion to the electrode volume (see eqn (17)) and hence the condition of a nonadiabatic transition, eqn (202) is valid. We obtain a final expression for the current, since the electron level density is directly proportional to the electrode volume (see eqn (4)). In the treatment of adiabatic electrochemical reactions, the reorganization of the many-electron state, is of great importance even if the metal electrons can be assumed to be noninteracting. This problem requires a serious theoretical investigation.[54] At present we can readily formulate only a prescription for order-of-magnitude estimation of the current: in adiabatic processes unity should be substituted for the product of the one-electron transmission coefficient κ_F the level density ρ_F (see eqn (4)) and the range of levels contributing to the current:

$$\kappa_F \rho_F kT \to 1 \tag{207}$$

IV. QUANTUM THEORY OF ELECTROCHEMICAL PROCESSES

In this section we shall consider some electrochemical systems for which the current will be calculated on the basis of the general semi-phenomenological theory presented in Section III. We shall use the continuum model for the solvent and eqn (170) for the transition probability.

1. Redox Reactions at the Metal Electrode

Redox reactions occurring without breaking or formation of any chemical bonds of reactants are the simplest case as regards theoretical calculations. Calculation of the current for these systems is described at length by many authors (see, e.g. References 42–44, 47–49). Therefore, we shall dwell here only on some fundamental points.

According to the theory presented in the previous section, redox reactions involve electron (quantum subsystem) and solvent (classical subsystem). Therefore, in calculating the electron transfer probability, we should proceed from consideration of the electronic terms of the solvent, eqns (162) and (164). For the process of electron transfer from the electrode to the oxidizing ion, the difference of minimum potential energies at the beginning and at the end of the reaction can be written as (see eqns (163) and (165))

$$\Delta J(\varepsilon_f) = (\varepsilon_f - e\varphi_m - E_{solv}^0) - (-\varepsilon_i - E_{solv}^R)$$

$$= (\varepsilon_f - \varepsilon_F) - e\eta + \Delta J_{0F}$$

$$= (\varepsilon_f - \varepsilon_F) - e\eta + kT \ln \frac{C_R}{C_0} \tag{208}$$

where ε_i is the electron binding energy in the reducing ion; $E_{solv}^{0,R}$ is the solvation energy of the oxidizing and reducing ions and C_0, C_R are the surface concentrations of ions (for simplicity we assume the ψ_1-effect to be absent). Due to the symmetry of parabolic terms of the solvent, the reorganization energies for forward and reverse processes coincide and are determined from eqn (172).

For all redox systems at reasonable values of ion concentrations, the quantity ΔJ_{0F} is much less than E_r. Therefore, over a very wide overvoltage range the eqn (82) is valid. This means that electrochemical kinetics of all redox reactions is confined to the normal region. According to eqns (171) and (208) the activation energy for redox systems can be written as

$$E_a = \frac{(\varepsilon_F - \varepsilon_f + e\eta - kT \ln (C_R/C_0) + E_r)^2}{4E_r} \tag{209}$$

The level ε^* making the chief contribution to the current, according to eqns (53), (84), and (209) is determined from the equation

$$1 - n(\varepsilon_f) = \frac{1}{2} + \frac{e\eta}{2E_r} - \frac{kT}{2E_r} \ln \frac{C_R}{C_0} - \frac{\varepsilon_f - \varepsilon_F}{2E_r} \tag{210}$$

As has been already pointed out above, the third term in the right hand side of this equation for redox systems is very small and can be ignored. Then from eqn (210) we obtain that $\varepsilon^* = \varepsilon_F$ at the equilibrium potential ($\eta = 0$), i.e. $\eta^F = 0$. According to eqn (93) the transfer coefficient for redox reactions can be written as

$$\alpha^* = \frac{1}{2} + \frac{e\eta}{2E_r}, \quad |e\eta| < E_r \tag{211}$$

In order to calculate the current by means of eqn (92) it is necessary only to find the free energy for the Fermi electron at the equilibrium potential ($\eta^F = 0$). According to eqns (201) and (209) we have

$$F_a(\varepsilon_F - e\eta^F) = \frac{1}{4E_r}\left(E_r - kT\ln\frac{C_R}{C_0}\right)^2 - kT\ln\frac{\hbar\omega_{\text{eff}}}{kT}$$

$$\simeq \frac{E_r}{4} - \frac{kT}{2}\ln\frac{C_R}{C_0} - kT\ln\frac{\hbar\omega_{\text{eff}}}{kT} \tag{212}$$

The final expression for the current density is obtained if we substitute into eqn (92) the transmission coefficient κ, eqn (205), the width of the levels contributing to the current $\Delta\varepsilon^* \simeq \sqrt{(8\pi)}kT$ eqn (86) and the free activation energy eqn (212):

$$\overrightarrow{i} = \frac{\pi e\rho_F}{\hbar}\sqrt{(C_R C_0)}\left(\frac{2kT}{E_r}\right)^{1/2}|V_{iF}|^2\, e^{-[(E_r + e\eta)^2/4E_r kT]}$$

$$= i_0\, e^{-(e\eta/2kT) - [(e\eta)^2/4E_r kT]} \tag{213}$$

where i_0 is the exchange current and V_{iF} the electron exchange integral. The total current density with an allowance made for the back reaction can be found by means of eqn (48)

$$i = \overleftarrow{i} - \overrightarrow{i} = i_0(e^{e\eta/2kT} - e^{-(e\eta/2kT)})\, e^{-[(e\eta)^2/4E_r kT]} \tag{214}$$

According to eqn (207) the exchange current for adiabatic reactions can be calculated in order of magnitude from the relation:

$$i_0^{ad} \simeq \frac{e\omega_{\text{eff}}}{2\pi}\sqrt{(C_R C_0)}\, e^{-(E_r/kT)} \tag{215}$$

The deviation of the transfer coefficient from the constant value 0·5 is of fundamental importance, since in an experimental verification of eqn (211) we can say with great certainty that in electrochemical kinetics the solvent plays the main dynamic part. Moreover, eqn (211) shows that in redox reactions the electron transfers from the Fermi level are of essential importance only with a strictly equilibrium potential ($\eta = 0$), whereas when overvoltage is applied the chief contribution to the current is made by the levels located near ε_F. To verify eqns (211) and (213) we have analysed the data obtained by Frumkin and collaborators,[50] who investigated the electroreduction of $Fe(CN)_6^{3-}$ anions on a mercury electrode. The comparison of the theoretical and experimental curves (see Fig. 19) shows qualitative agreement of theory with experiment. The greatest discrepancy is observed in the range of the potentials at which other effects can come into play[50] (such as the formation

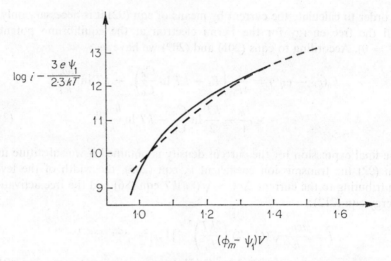

$\log i - \dfrac{3e\,\psi_1}{2\cdot3\,kT}$

$(\phi_m^- - \psi_1)V$

Fig. 19. Theoretical (dashed curve) and experimental (solid curve) polarization curves for the electroreduction reaction of $Fe(CN)_6^{3-}$ anions on mercury electrode.

of ion pairs). A specific feature of this reaction is that the $Fe(CN)_6^{3-}$ ion is rather large and the quantity E_r relatively small. Therefore, according to the predictions of the theory, already with not very large overvoltages a deviation from the linear Tafel equation should be observed, which is actually the case in experiment. Later Parsons and Passeron[51] carried out measurements on the Cr^{3+}/Cr^{2+} system at a mercury electrode to test relations, eqn (211), which showed theory to agree with experiment within experimental error.

We have considered above redox systems in which an elementary electrochemical act involves transfer of only one electron. In more complex redox systems, along with the one electron process, reactions involving the transfer of two or more electrons can occur. Below we shall give for illustration a calculation of the current due to the transfer of two electrons from the electrode to the oxidizing ion. In this case, as before, we shall assume that in an elementary act the chemical bonds of the discharging ion are unchanged. Then, similarly to eqn (49) the direct current can be written as

$$\vec{i}_2 = 2eC_0 \frac{kT}{h}\,\kappa\,e^{S_a/k}\rho_F^2 \int d\varepsilon_1\,d\varepsilon_2 n(\varepsilon_1)n(\varepsilon_2)\,e^{-[E_a(\varepsilon_1,\varepsilon_2)/kT]} \qquad (216)$$

where the activation energy E_a depends on the difference of minimum potential energies at the beginning and end of the reaction

$$\Delta J(\varepsilon_1, \varepsilon_2) = (\varepsilon_1 - \varepsilon_F) + (\varepsilon_2 - \varepsilon_F) - 2e\eta + \Delta J_{0F} \qquad (217)$$

In the treatment of two electron processes it is possible to use the general ideas of the semi-phenomenological theory presented in Section II. In particular, the microscopic transfer coefficient should be determined according to the following relation (see eqn (55)):

$$\alpha(\varepsilon_1, \varepsilon_2) = \frac{\partial E_a}{\partial 2e\eta} = -\frac{\partial E_a}{\partial \varepsilon_{1,2}} = -\frac{\partial E_a}{\partial \varepsilon} \tag{218}$$

where $\varepsilon = \varepsilon_1 + \varepsilon_2$ is the energy of two electrons in the electrode. Similarly to eqn (56) the following quantity will appear as the experimental transfer coefficient

$$\alpha_{exp} = -kT\frac{\partial \ln \vec{i}_2}{\partial 2e\eta} \tag{219}$$

for which as can be readily shown in the general form, the following relation is valid (see derivation of eqns (57) and (60)):

$$\alpha_{exp} = \langle\alpha(\varepsilon_1, \varepsilon_2)\rangle_{av} = 1 - \langle n(\varepsilon_1)\rangle_{av} = 1 - \langle n(\varepsilon_2)\rangle_{av} \tag{220}$$

Here averaging is understood to mean that the corresponding quantity is taken with the weight factor proportional to the elementary current associated with the electron transfer from the levels ε_1 and ε_2. For instance,

$$\langle\alpha(\varepsilon_1, \varepsilon_2)\rangle_{av} = \iint \alpha(\varepsilon_1, \varepsilon_2)\frac{\vec{di}_2(\varepsilon_1, \varepsilon_2)}{\vec{i}_2} \tag{221}$$

It follows from eqn (221) that the transfer coefficient determined by means of eqn (219) always lies within the range from zero to 1.

In calculating the current $\vec{i}_2$ it is convenient in eqn (216) to pass to new variables

$$\varepsilon = \varepsilon_1 + \varepsilon_2; \quad \varepsilon' = \varepsilon_1 - \varepsilon_2 \tag{222}$$

and to perform accurate integration with respect to ε'. As a result, we obtain the formula

$$\vec{i}_2 = 2eC_0\frac{kT}{h}\kappa_F e^{Sa/k}\rho_F^2 kT\int d\varepsilon\mu(\varepsilon) e^{-[E_a(\varepsilon)/kT]} \tag{223}$$

where

$$\mu(\varepsilon) = \frac{\varepsilon - 2\varepsilon_F}{kT}\left[\exp\left(\frac{\varepsilon - 2\varepsilon_F}{kT}\right) - 1\right]^{-1} \tag{224}$$

Now the integral contained in eqn (223) can be calculated by the Laplace method:

$$\vec{i_2} = 2eC_0 \frac{kT}{h} \kappa_F e^{S_a/k} \rho_F^2 kT \mu(\varepsilon^*) e^{-[E_a(\varepsilon^*)/kT]} \Delta\varepsilon^* \tag{225}$$

where ε^* corresponds to the optimum energy of two electrons making the maximum contribution to the current and is determined as the solution of the equation

$$\alpha(\varepsilon) = 1 - \nu(\varepsilon) \tag{226}$$

where

$$\nu(\varepsilon) = \frac{kT}{\varepsilon - 2\varepsilon_F} [1 - \mu(\varepsilon)] \tag{227}$$

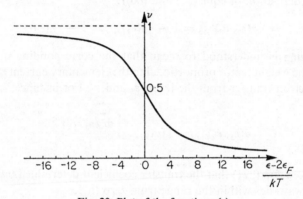

Fig. 20. Plot of the function $\nu(\varepsilon)$.

The function $\nu(\varepsilon)$ shown graphically in Fig. 20 is formally very similar to the Fermi distribution function $n(\varepsilon_f)$ (see Fig. 8) and differs from it only by a more smooth behaviour. Therefore, from the comparison of eqns (84) and (226), we can speak of there being an analogy between one-electron and two-electron processes, if we understand ε to mean the energy of both electrons and take into consideration that the chemical potential of such system is equal to $2\varepsilon_F$. Strictly speaking, the above analogy was contained in eqns (223) (cf. eqn (49)) and (225) (cf. eqn (85)). In particular in two electron processes the quantity (cf. eqn (90))

$$E_a^{\text{exp}} = -kT \ln \mu(\varepsilon^*) + E_a(\varepsilon^*) \tag{228}$$

appears as the experimental activation energy and the transfer coefficient α_{exp} is given by the relation (see derivation of eqn (91)):

$$\alpha_{exp} = \frac{\partial E_a^{exp}}{\partial 2e\eta} = \alpha(\varepsilon^*) = 1 - \nu(\varepsilon^*) \tag{229}$$

The treatment given above was based on the general semi-phenomenological theory and did not contain any model assumptions. If we use now the model of the solvent described in Section III and the expression for the quantum-mechanical transition probability eqn (170), eqn (226) can be rewritten as (cf. eqn (210)):

$$1 - \nu(\varepsilon) = \tfrac{1}{2} + \frac{2e\eta}{2E_r} - \frac{kT}{2E_r} \ln \frac{C_R}{C_0} - \frac{\varepsilon - 2\varepsilon_F}{2E_r} \tag{230}$$

Assuming as before the one before the last term in the right hand side of this equation to be very small and that, in addition ε^*, differs little from $2\varepsilon_F$ (this is valid at low overvoltages $|2e\eta| < E_r$), we can solve eqn (230)

$$\alpha_{exp} = \alpha(\varepsilon^*) \simeq \tfrac{1}{2} + \frac{2e\eta}{2E_r}; \quad |2e\eta| < E_r \tag{231}$$

Knowing α_{exp} we can find by the method described in Section II (see eqns (101)–(103)) the experimental activation energy

$$E_a^{exp} = \frac{(E_r + 2e\eta)^2}{4E_r} - \frac{kT}{2} \ln \frac{C_R}{C_0} \tag{232}$$

In this case we made use of the fact that, just as in the one-electron processes, the kinetics of the two-electron discharge is confined to the normal region, since according to eqn (230) at the equilibrium potential ($\eta = 0$) the transitions occur mainly from the level $\varepsilon_F(\varepsilon^* = 2\varepsilon_F)$. We obtain a final expression for the current density, if we substitute into eqn (225) the activation entropy (see eqn (201)) and

$$\mu(\varepsilon^*) \simeq 1; \quad \Delta\varepsilon^* \simeq 2\sqrt{(6\pi)}kT \tag{233}$$

As the result we have

$$i_2 = \overleftarrow{i_2} - \overrightarrow{i_2} = i_0(e^{2e\eta/2kT} - e^{-(2e\eta/2kT)}) \, e^{-[(2e\eta)^2/4E_rkT]} \tag{234}$$

where the exchange current is determined by the formula:

$$i_0 = \frac{2\pi}{\hbar} \cdot 2e\rho_F^2 \sqrt{(3)}kT\sqrt{(C_RC_0)} \left(\frac{2kT}{E_r}\right)^{1/2} |V_{iF}|^2 \, e^{-(E_r/4kT)} \tag{235}$$

To estimate the exchange current in the case of an adiabatic process we have to make in eqn (235) the following substitution

$$\frac{1}{4} \kappa_F \rho_F^2 \left[\sqrt[4]{\left(\frac{3\pi}{2} \right)} kT \right]^2 \to 1 \tag{236}$$

whereupon we obtain

$$i_0^{ad} \simeq \frac{2e\omega_{eff}}{2\pi} \sqrt{(C_R C_0)} \, e^{-(E_r/4kT)} \tag{237}$$

A formal similarity in the expressions for the current in one electron and two electron processes (upon substitution of 2e for e) is accounted for by the fact that in the scheme calculated above the transitions of two electrons were considered as being simultaneous and independent of each other. Nevertheless, the values of the current in one electron and two electron transitions can differ essentially since the reorganization energies of the solvent in these processes are also essentially different. For instance, if we assume the quantity E_r to be proportional to the charge being transferred squared, in two electron processes E_r will be four times as large as in one electron transitions. If redox reactions occur by steps, the total current can be found from the formulae obtained above by the methods of formal kinetics only in the case when the one electron and two electron transitions occur independently. The case is much more complicated in the presence of a correlation between the transitions of the first and second electrons from one ion, i.e. when these transitions are separated by a time interval $\Delta t < \tau_d$. In such case, after the transition of the first electron there is not enough time for the ion to diffuse into the solution and thus the transition of the second electron may occur under non-equilibrium conditions. Further theoretical consideration is required for the investigation of such processes.

2. Redox Reactions on a Semi-conductor Electrode

The quantum-mechanical theory of redox reactions on a semi-conductor electrode is presented in detail in References 52–58 where electrochemical kinetics was considered both on intrinsic[52,55] and impurity semiconductors.[57] Some steady state processes occurring on a semiconductor electrode were studied[58] (e.g. the steady state photoeffect). In addition, in Reference 53, the kinetics of redox reactions was studied on a metal electrode coated with a thin semiconductor film. In the present article we shall not give in detail the results of the above mentioned studies (see also reviews[42–44]) but will only consider briefly the general physical picture and the statement of the problem within the framework of the general semi-phenomenological theory developed in Section II.

The physical difference in the mechanism of the passage of electrochemical current at the metal and semiconductor electrodes is due to the fact that in the latter case the Fermi level lies in the forbidden band and the occupation of the energy levels by electrons depends strongly on the temperature and potential distribution in the system. Figure 21 shows schematically the

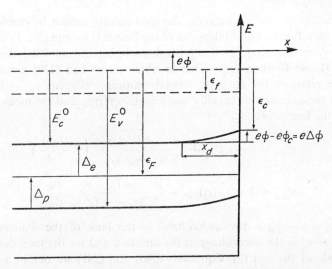

Fig. 21. Schematic picture of the electron energy spectrum in semiconductor in the presence of a contact with electrolyte solution. ψ and ψ_c are the potential in the bulk of semiconductor and at the contact, $\Delta_{e,p}$—the distance from the bottom of the conduction band and the top of the valence band to the Fermi level ε_F. X_d—the Debye screening distance.

structure of the electron energy spectrum in the case of anodic polarization. Below for definiteness we shall consider practically the most important case of the non-degenerate semiconductor surface, i.e. the case when the Fermi level does not intersect with the bent bands

$$|e\Delta\varphi| < \Delta_{e,p} \qquad (238)$$

(for designations see Fig. 21). In this case the electron concentration in the conduction band and the hole concentration in the valence band are very small, so that the main potential drop between solution and semiconductor will occur on the electrode at a sufficiently large distance (e.g. the Debye screening distance X_d for Ge is of the order of 10^{-4}–10^{-5} cm). Because of this fact, not all electrons and holes can participate in electrochemical kinetics. For example, in the case of anodic polarization from the conduction band only the electrons with energies greater than $\varepsilon_c = E_c^0 + e\Delta\varphi$ participate in the kinetics, since the other electrons would have to travel 'tunnelwise' to a large distance of the order of X_d. Thus, the electron component of the total discharge current, which is due only to transitions from the conduction band, can be written as

$$\vec{i}^e = eC_0 \frac{kT}{h} \kappa_e e^{S_a/k} \int_{\varepsilon_c}^{\infty} d\varepsilon_f \rho_e (\varepsilon_f - e\Delta\varphi) n(\varepsilon_f) e^{-[E_a(\Delta J(\varepsilon_f))/kT]} \tag{239}$$

where, unlike the metal electrode, the level density cannot be considered as being a slow function and taken out of the integral (see eqn (9)). In much the same way as was done in the general semi-phenomenological theory (see eqns (57) and (60)), it is possible to obtain from eqn (239) a number of accurate relations for an experimental transfer coefficient α^e_{exp}, the mean value of the microscopic transfer coefficient $\langle \alpha^e(\varepsilon_f) \rangle_{av}$ and the mean occupation of the levels $\langle n(\varepsilon_f) \rangle_{av}$

$$\alpha^e_{exp} = \langle \alpha^e(\varepsilon_f) \rangle_{av} + \left\langle \frac{kT}{2(\varepsilon_f - \varepsilon_c)} \right\rangle_{av} \left(1 - \frac{d\eta_c}{d\eta} \right) \tag{240}$$

$$\alpha^e_{exp} = 1 - \langle n(\varepsilon_f) \rangle_{av} - \left\langle \frac{kT}{2(\varepsilon_f - \varepsilon_c)} \right\rangle_{av} \cdot \frac{d\eta_c}{d\eta} \tag{241}$$

where $\eta = \varphi - \varphi_0$ is the overvoltage in the bulk of the semiconductor, $\eta_c = \varphi_c - \varphi_c^0$ is the overvoltage at the interface and for the level density use was made of the eqn (9). Equations (240) and (241) we obtain a relation similar to eqn (60)

$$\langle \alpha^e(\varepsilon_f) \rangle_{av} = 1 - \langle n(\varepsilon_f) \rangle_{av} - \left\langle \frac{kT}{2(\varepsilon_f - \varepsilon_c)} \right\rangle_{av} \tag{242}$$

Comparing eqns (61) and (242) we see that the difference in the metal and semiconductor spectra leads to the appearance of one more term in eqn (242). To understand this result it should be recalled that eqn (61) reflects physically the competition of two tendencies in the case of a reaction at a metal electrode. On one hand, the transitions from high electron energy levels for which the activation energy is small are favoured. On the other hand, the occupation of these levels is very small. In the case of semiconductor electrodes, a third factor enters into competition, i.e. the level density ρ_e, which is zero in the forbidden band and increases rather sharply near the bottom of the conduction band. To analyse this question it is convenient to use the expression for the electron discharge current obtained in the calculation of the integral in eqn (239) by the Laplace method:

$$\vec{i}^e = eC_0 \frac{kT}{h} \kappa_e e^{S_a/k} \rho_e(\varepsilon^*) n(\varepsilon^*) e^{-[E_a(\varepsilon^*)/kT]} \Delta\varepsilon^* \tag{243}$$

where ε^* is the level making the main contribution to the current and $\Delta\varepsilon^*$ is the width of the energy levels near ε^* participating in the reaction. It can be readily shown by a direct calculation that the accurate relations obtained above, eqns (240)–(242) are now valid for the level ε^*. For instance from eqn (242) we obtain an equation for the determination of ε^*:

$$\alpha^e(\varepsilon^*) = 1 - n(\varepsilon^*) - \frac{kT}{2(\varepsilon^* - \varepsilon_c)} \qquad (244)$$

It is physically evident that the main contribution to the current will be made by the levels located in the immediate vicinity of ε_c. Therefore, if we assume the potential at the interface to depend very slightly on overvoltage (this is valid at not too large surface state densities of semiconductors) we shall obtain from eqn (241)

$$\alpha^e_{\exp} \simeq 1 \qquad (245)$$

This result is due to the fact that at $\varepsilon^* \simeq \varepsilon_c$ and $d\eta_c/d\eta \rightarrow 0$ the activation energy and hence $\alpha^e(\varepsilon^*)$ practically do not depend on overvoltage

$$E_a(\Delta J(\varepsilon^*)) \simeq E_a(\Delta J(\varepsilon_c)) = E_a(\varepsilon_c - \varepsilon_F - e\eta + \Delta J_{0F})$$
$$= E_a(\Delta_e + e\Delta\varphi_0 - e\eta_c + \Delta J_{0F}) \simeq \text{constant} \qquad (246)$$

On the other hand, according to eqn (244) the level density will not depend on overvoltage either

$$\rho_e(\varepsilon^*) \sim \sqrt{(\varepsilon^* - \varepsilon_c)} = \left[\frac{kT}{2(1 - \alpha^e(\varepsilon^*))}\right]^{1/2} \simeq \text{constant} \qquad (247)$$

Thus, the total dependence of the current on overvoltage will be determined by the electron concentration at the contact:

$$n(\varepsilon^*) \simeq e^{-(\varepsilon^* - \varepsilon_F/kT)} \simeq e^{-(\varepsilon_c - \varepsilon_F/kT)} \simeq e^{-(e\eta + \Delta_e + e\Delta\varphi_0/kT)} \qquad (248)$$

which gives directly relation eqn (245).

It is clear from the above consideration that unlike the metal electrode, the experimental transfer coefficient for semiconductors can differ essentially from the mean microscopic transfer coefficient in the Brønsted relation. In order to calculate $\alpha^e(\varepsilon^*)$ it is possible to use the model described in the previous section. As shown by calculation, in this case

$$\alpha^e(\varepsilon^*) = \frac{\frac{1}{2} - (\Delta_e + e\Delta\varphi_0 - e\eta_c + kT \ln C_R/C_0)}{2E_r} \qquad (249)$$

8

and for the redox systems in which the reorganization energy is larger than the halfwidth of the forbidden band Δ_e and the equilibrium bending of the band $e\Delta\varphi_0$, the discharge will occur in the normal region ($\alpha^e(\varepsilon^*) \simeq 0, 5$). An interesting situation may arise in the opposite case of small E_r, when $\alpha^e(\varepsilon^*) < 0$. If the surface state density is large enough, the overvoltage at the interface will vary considerably[42,59] and according to eqn (240)

$$\alpha^e_{exp} \simeq \alpha^e(\varepsilon^*) < 0 \qquad (250)$$

i.e. the polarization curve for the electron current may have a descending branch (for the discussion of this problem see also Gerischer's article[60]).

Up to the present we have considered only one component of the total current. The reverse current can be found by means of eqn (48), which is of a general nature. The hole current $\overrightarrow{i^p}$ associated with the electron transfers from the valence band to the solution can be calculated using simple physical considerations, which are confirmed by direct calculations. In fact, the quantity $\overrightarrow{i^p}$ is equivalent to the reverse current of positive holes when the oxidizing ion is replaced by the reducing ion. Therefore, $\overrightarrow{i^p}$ can be obtained from $\overrightarrow{i^e}$ by multiplying it by $e^{e\eta/kT}$ (see eqn (48)), taking into consideration the plus sign of the hole) and with the substitutions:

$$C_R \rightleftarrows C_0; \quad \Delta_e \rightleftarrows \Delta_p; \quad \kappa_e \rightleftarrows \kappa_p; \quad \eta \rightleftarrows -\eta \qquad (251)$$

3. Hydrogen Overvoltage

The quantum-mechanical theory of electrode reactions developed in the previous section permits us to calculate practically any process of proton transfer in polar liquids. As an illustration we shall consider below, hydrogen ion discharge on electrodes with a high overvoltage value. Let us assume that the slowest process is proton transition from the hydroxonium ion to the electrode

$$H_3O^+ + e_f \rightarrow H_{ads} + H_2O \qquad (252)$$

In addition, let us assume hydrogen atom desorption to follow the electrochemical mechanism:

$$H_{ads} + H_3O^+ + e_f \rightarrow H_2 + H_2O \qquad (253)$$

(Other possible mechanisms of hydrogen discharge and desorption are given in the book[61].) The most representative case satisfying the above conditions seems to be hydrogen ion discharge on the mercury electrode.

Both in initial and final states, the proton forms a sufficiently strong chemical bond with the water molecule and electrode respectively. As shown by the analysis of experimental data, the proton vibration frequencies in these states are

$$\omega_i \simeq 5 \times 10^{14} \sec^{-1} > \frac{kT}{\hbar}; \quad \omega_a \simeq 2 \times 10^{14} \sec^{-1} > \frac{kT}{\hbar} \qquad (254)$$

where the quantity ω_a is taken for the mercury electrode.[36] The frequencies in eqn (254) correspond to proton motion along an axis normal to the electrode. There exist also other forms of vibrations in a real system (such as bending of bonds). Taking account of these vibrations does not present any fundamental difficulties, but, as shown by estimates, they do not affect to any considerable degree the transition probability. The physical reason for this is that as a result of transfer of the proton its equilibrium coordinate changes significantly only in the direction normal to the electrode. Therefore, overlap of the proton wave functions characterizing its transverse motion is very large and the corresponding overlap integral is close to unity. Owing to this peculiarity of the behaviour of the wave functions, it is possible in the case of H_3O^+ ion discharge to take into account only the proton nearest to the electrode. A somewhat different approach will give the same result, if we take into consideration eqn (1), according to which the H_3O^+ ion has an opportunity to 'impinge' upon the electrode several times before the discharge occurs. Thus, the hydroxonium ion will frequently happen to be near the electrode in a position 'convenient' for the reaction. To make an allowance for this situation, it is sufficient to introduce into the expression for the current a steric factor of the order of 1. However, if the following condition is fulfilled

$$\tau_e \gg \tau_d \gg \tau_r \qquad (255)$$

where τ_r is the mean time of turning of H_3O^+ ions, the steric factor will be exactly equal to 1.

According to the preceding section (see eqn (183)), eqn (254) means that proton should be included in the quantum subsystem along with electron. A physical interpretation of the reaction path for the case of the proton discharge is given in Section III (see Fig. 16(a)) if $U(q)$ is understood to mean the electron-protonic term and q—the solvent coordinates. However, since the problem of proton transfers has been discussed more than once in the literature[62] in terms of the electron term, in Fig. 22(a) we show the 'reaction path' corresponding to the theory developed in the previous section. Here, unlike earlier theories (see e.g. Reference 62) which did not take into account the dynamic role of the solvent, the electronic terms in Fig. 22(a) are plotted against the coordinates of proton R and solvent q. In addition, for illustration a case is considered when the proton in initial and final states performs harmonic motion with the same frequency. This approximation is particularly helpful in that it permits of a rigorous analytical calculation supporting the interpretation given below.[63-64] In virtue of the quantum nature of the proton behaviour, only the medium dipoles can perform classical fluctuations. Therefore, we can assume the proton coordinate in the initial state to be rigidly fixed ($R = R_{i0}$) and the motion along the electron term to occur only towards q. When due to fluctuation the solvent coordinate assumes the value q^*, the system will have the energy $J_i + \frac{1}{2}\hbar\omega_0(q^* - q_{i0})^2$ (Section 01 in Fig. 22(a)); if the proton in initial state was at the excited state with the

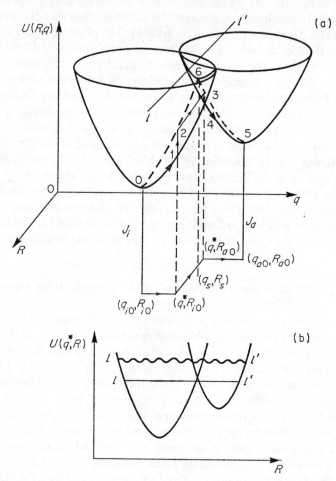

Fig. 22. Schematic diagram of the 'reaction path' for a proton transition: (1) according to the theory of absolute reaction rates (dashed curve); (2) the true reaction path (solid curve).

energy $\varepsilon_e^{(i)} = \hbar\omega_p(l + \frac{1}{2})$ (Section 12 in Fig. 22(a)) the total energy of the system in the activated state q^* will be

$$U_i(q^*, R_{i0}) + \varepsilon_e^{(i)} = J_i + \tfrac{1}{2}\hbar\omega_0(q^* - q_{i0})^2 + \tfrac{1}{2}\hbar\omega_p(2l + 1) \quad (256)$$

The solvent being a slow subsystem with respect both to electron and proton, the transition from term to term occurs with a fixed coordinate q^* (the Franck-Condon principle). The transition from the initial to final terms involves, in addition to the change in the electronic state, a change in the proton coordinate (Section 23 in Fig. 22(a)). To illustrate this stage Fig. 22(b) shows the intersection of the electron terms with the plane ll' parallel to the plane UOR and passing the point (q^*, R_{i0}). Since $U(q^*; R)$ is the proton potential energy the nature of the proton transition should be explained on this plot (Fig. 22(b)). The proton transition can be both subbarrier (solid line in Fig. 22(b)) above the barrier (wiggly line in Fig. 22(b)). It should be stressed that in both cases we have to deal with a purely quantum proton transfer. Moreover, as has been already stated in the previous section, even in the case of a subbarrier transition, the probability is not determined by the Gamow factor, but is calculated from the Landau-Zener equation. At the final term, the proton has the excitation energy $\varepsilon_e^{(a)} = \hbar\omega_p(l' + \frac{1}{2})$ (Section 34 in Fig. 22(a)) and the total energy of the system is determined by the formula:

$$U_a(q^*; R_{a0}) + \varepsilon_e^{(a)} = J_a + \tfrac{1}{2}\hbar\omega_0(q^* - q_{a0})^2 + \tfrac{1}{2}\hbar\omega_p(2l' + 1) \quad (257)$$

After the transition from term to term, the system relaxes classically from the point (q^*, R_{a0}) to the equilibrium point of final state (q_{a0}, R_{a0}). By conservation of energy, the quantum transition in Section 23 of Fig. 22(a) can occur only if the energies of the system before and after the reaction coincide (see also Fig. 22(b))

$$J_i + \tfrac{1}{2}\hbar\omega_0(q^* - q_{i0})^2 + \varepsilon_e^{(i)} = J_a + \tfrac{1}{2}\hbar\omega_0(q^* - q_{a0})^2 + \varepsilon_e^{(a)} \quad (258)$$

Actually, this relation is the equation for the determination of the activation configuration of the solvent q^*. After solution of eqn (258) we obtain for the activation energy the expression:

$$E_a^{ll'} = \frac{(E_S - \Delta J^{ll'})^2}{4E_S} \quad (259)$$

where E_S is the reorganization energy of the solvent

$$E_S = \tfrac{1}{2}\hbar\omega_0(q_{i0} - q_{a0})^2 \to \tfrac{1}{2}\sum_{\vec{k}}\hbar\omega_{\vec{k}}(\vec{q}_{\vec{k}\,i} - \vec{q}_{\vec{k}\,a})^2 \quad (260)$$

and

$$\Delta J^{ll'} = (J_i + \varepsilon_e^{(i)}) - (J_a + \varepsilon_{e'}^{(a)})$$
$$= (\varepsilon_f - \varepsilon_F) + e\eta_c + \Delta J_{0F} + (\varepsilon_e^{(i)} - \varepsilon_{e'}^{(a)}) \tag{261}$$

(η_c is the cathodic overvoltage eqn (43)).

In Fig. 22(a) for comparison is shown, by a dashed line, the 'transition path' according to the absolute reaction rate theory, when the reaction coordinate passes through the saddle point (q_S, R_S). Physically, this path corresponds to the classical fluctuation motion both of solvent and proton, when during the reaction the proton chemical bond stretches. In this case the activation energy differs from eqn (259) and is of the form

$$E_a^* = \frac{(E_S + E_p - J_i + J_a)^2}{4(E_S + E_p)} \tag{262}$$

where E_p is the proton reorganization energy:

$$E_p = \frac{M_p \omega_p^2}{2} (R_{i0} - R_{a0})^2 \tag{263}$$

(M_p is the proton mass.) An accurate quantum-mechanical calculation shows this result to be valid only in the case when $\hbar\omega_p < kT$. Since in practice the reverse condition is fulfilled (see eqn (254)), the application of the 'standard' absolute reaction rate theory to the proton transfer processes should be considered as incorrect.

In calculating the discharge current, we shall proceed from the results of the semi-phenomenological theory, into which we shall substitute an expression for the transition probability in eqn (170). Moreover, in accordance with the general theory, we shall first find the discharge current assuming the proton to be at the beginning and at the end of the reaction at the unexcited levels: $\varepsilon_e^{(i)} = \varepsilon_{e'}^{(a)} = 0$. Since in the mechanism under consideration there is for each reaction act, eqn (252), one electrochemical desorption act eqn (253) (condition of steadiness) the total discharge current will be twice as large as the current due only to the process, eqn (252):

$$\vec{i}_c^0 = 2eC_{H^+}^0 \frac{\omega_{eff}}{2\pi} \kappa_{00}\rho_F \int n(\varepsilon_f) \exp\left[-\frac{(E_S - \Delta J)^2}{4E_S kT}\right] d\varepsilon_f \tag{264}$$

where $C_{H^+}^0$ is the surface density of the H_3O^+ ions on the outer side of the Helmholtz layer (for simplicity, the ψ_1-effect is assumed to be absent) and κ_{00} is the transmission coefficient calculated using the wave function of the proton in unexcited state. Even for a continuum model of the solvent, the integral in eqn (264) is not calculated accurately, but assuming the parameter

E_S to have a definite value, we can calculate numerically the current as a function of the overvoltage. Such a calculation of the current at $E_S = 2\,\text{eV}$ was carried out by means of a computer, the result obtained having been shown to be practically the same as in the case of an approximate analytical treatment. Therefore, we shall use the Laplace method for the calculation of the integral in eqn (264). According to eqn (84) the level ε^* making the main contribution to the current is found from the equation

$$\alpha(\varepsilon^*) = 1 - n(\varepsilon^*) = \tfrac{1}{2} - \frac{\Delta J_{0F}}{2E_S} - \frac{e\eta_c}{2E_S} - \frac{\varepsilon_f - \varepsilon_F}{2E_S} \tag{265}$$

With respect to ε^* this transcendental equation cannot be solved analytically, but the dependence of the transfer coefficient $\alpha^* = \alpha(\varepsilon^*) = \alpha_{\text{exp}}$ on overvoltage can be readily found from it as the inverse function:

$$e\eta_c = E_S - \Delta J_{0F} - 2E_S\alpha^* + kT\ln\frac{1 - \alpha^*}{\alpha^*} \tag{266}$$

The dependence of the transfer coefficient on overvoltage corresponding to eqn (266) is shown in Fig. 23, where the parameter E_S was assumed to be 2 eV. It is evident from the plot that α^* in the range $|e\eta_c + \Delta J_{0F}| < E_S$ is approximated quite well by the linear function of overvoltage:

$$\alpha^* \simeq \tfrac{1}{2} - \frac{e\eta_c + \Delta J_{0F}}{2E_S}; \quad |e\eta_c + \Delta J_{0F}| < E_S \tag{267}$$

The overvoltage range in which the transfer coefficient has an approximately constant value depends essentially on the reorganization energy of the solvent E_S. For illustration, we can consider two cases: (a) $E_S = 2\,\text{eV}$. Then in the overvoltage range $|e\eta_c + \Delta J_{0F}| < 0.4\,\text{eV}$ we have $\alpha^* = 0.5 \pm 0.1$; (b) $E_S = 4\,\text{eV}$. Here we have $\alpha^* = 0.5 \pm 0.1$ already in the overvoltage range 1·6 V ($|e\eta_c + \Delta J_{0F}| < 0.8\,\text{eV}$).

From eqn (266) we can readily find the experimental activation energy as the function of the transfer coefficient:

$$E_a^{\text{exp}} = -\int\alpha^*(e\eta_c)\,\mathrm{d}e\eta_c = \int_0^{\alpha^*} \alpha^*\left(2E_S + \frac{kT}{\alpha^*(1 - \alpha^*)}\right)\mathrm{d}\alpha^*$$

$$= \alpha^{*2}E_S - kT\ln(1 - \alpha^*) \tag{268}$$

As the result, the final expression for the discharge current without an allowance for proton excited states assumes the form:

$$\vec{i}_c^0 = 2eC_{H^+}^0\,\frac{\omega_{\text{eff}}}{2\pi}\,\kappa_{00}\rho_F(1 - \alpha^*)\,e^{-\alpha^{*2}(E_S/kT)}\,\Delta\varepsilon^* \tag{269}$$

where the width of the levels $\Delta \varepsilon^*$ making the main contribution to the current is determined by the eqn (86):

$$\Delta \varepsilon^* = \sqrt{(2\pi)}kT \left[\frac{kT}{2E_S} + \alpha^*(1 - \alpha^*) \right]^{-1/2} \quad (270)$$

Equation (269) together with eqn (266) gives the dependence of the discharge current on overvoltage over the whole potential range. Figure 24 shows the

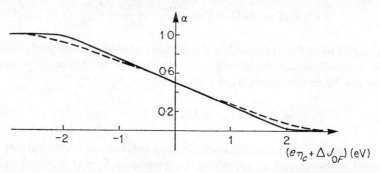

Fig. 23. Theoretical dependence of the transfer coefficient on overvoltage without an allowance (solid curve) and with an allowance (dashed curve) for the proton excited states. $E_r = 2$ eV, $R_{i0} - R_{a0} = 10^{-8}$ cm, $kT = 0.025$ eV.

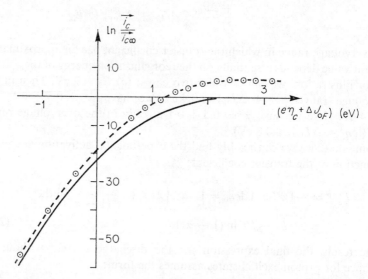

Fig. 24. Theoretical polarization curves for hydrogen ion discharge without allowance (solid curve) and with allowance (dashed curve) for the proton excited states. $E_r = 2$ eV, $R_{i0} - R_{a0} = 10^{-8}$ cm, $kT = 0.025$ eV.

polarization curve in a semi-logarithmic scale calculated from eqns (266) and (269) at $E_S = 2$ eV, the logarithm of the ratio of the discharge current to the limiting activationless current being plotted along the ordinate axis

$$\vec{i}^0_\infty = 2eC^0_{H^+} \frac{\omega_{eff}}{2\pi} \kappa_{00} \rho_F \sqrt{(4\pi k T E_S)} \tag{271}$$

In compliance with the general semi-phenomenological theory, the whole discharge current-overvoltage curve can be divided into three characteristic regions: barrierless ($\alpha^* \simeq 1$), normal ($\alpha^* \simeq 0.5$) and activationless ($\alpha^* \simeq 0$). It should be noted that in plotting the theoretical polarization curve, it is possible to calculate the current only for the overvoltage range $e\eta_c + \Delta J_{OF} > 0$, since in the range $e\eta_c + \Delta J_{OF} < 0$, the curve is completed automatically by means of the identity

$$\ln \vec{i}^0_c(-e\eta_c - \Delta J_{OF}) = \ln \vec{i}^0_c(e\eta_c + \Delta J_{OF}) - \frac{e\eta_c + \Delta J_{OF}}{kT} \tag{272}$$

which follows directly from eqns (266) and (269)–(271).

In conclusion, we shall consider the role of the excited states in the kinetics of the proton discharge. According to eqn (111) the total discharge current can be written as:

$$\vec{i}_c = \sum_{ll'} \frac{\kappa_{ll'}}{\kappa_{00}} \cdot \vec{i}^0_c(e\eta_c + \varepsilon_e^{(i)} - \varepsilon_{e'}^{(a)}) \, e^{-(\varepsilon_l^{(i)}/kT)} \tag{273}$$

First of all, it can be readily seen from eqn (273) that the excited states should be of most importance in the activationless and barrierless regions. In fact, the ratio of the current $\vec{i}^{ll'}_c$ due to the proton transfer from the excited level $\varepsilon_e^{(i)}$ to the excited level $\varepsilon_{e'}^{(a)}$ can be written as:

$$\frac{\vec{i}^{ll'}_c}{\vec{i}^0_c} \simeq \frac{\kappa_{ll'}}{\kappa_{00}} \begin{cases} \exp\left(-\dfrac{\varepsilon_{e'}^{(a)}}{kT}\right) & \text{(barrierless region)} \\[2mm] \exp\left(-\dfrac{\varepsilon_e^{(i)} + \varepsilon_{e'}^{(a)}}{2kT}\right) & \text{(normal region)} \\[2mm] \exp\left(-\dfrac{\varepsilon_e^{(i)}}{kT}\right) & \text{(activationless region)} \end{cases} \tag{274}$$

For analysis of this formula let us first consider the case when $\kappa_{00} \simeq 1$. Since the overlapping of the proton wave function is larger in excited states than in the ground state, evidently $\kappa_{ll'} \simeq 1$. According to eqn (272) in the barrierless region the contribution of the excited proton levels in the adsorbed state is exponentially small and the transitions can be assumed to occur to the ground level of the final state ($\varepsilon_e^{(q)} = 0$). However, the excited states of the proton in the H_3O^+ ion will make the same contribution to the current as the ground state. In the normal region the contribution of all excited levels (both of initial and final states) is exponentially small. Therefore, it is possible to take into consideration only the transitions from one ground level to another ($\varepsilon_e^{(i)} = \varepsilon_e^{(q)} = 0$). Finally, in the activationless region the contribution of the excited proton levels of the proton in the H_3O^+ ion is exponentially small and of essential importance are the transitions to the excited levels of the final state.

If the proton transitions for lower levels are of a nonadiabatic nature ($\kappa_{00} < 1$), the relative role of excited states in different kinetic regions is the same as in adiabatic processes considered above. However, for a qualitative analysis, the change of the transmission coefficient for different excited levels becomes of importance now.

In order to evaluate quantitatively the role of the proton excited states, we have calculated the discharge current with the use of a computer taking into account the ground and two excited levels. The Morse potential parameters used are as follows:

$$\left.\begin{array}{ll} \omega_a = 10^{14} \text{ sec}^{-1}; & D_a = 1.3 \text{ eV} \\ \omega_i = 5 \cdot 10^{14} \text{ sec}^{-1}; & D_i = 7.8 \text{ eV} \end{array}\right\} \tag{275}$$

The distance of proton transfer was assumed to be 1 Å. The remaining parameters are: $E_S = 2$ eV, $kT = 0.025$ eV. The dotted curve on Fig. 24 corresponds to the case when the proton excited states are taken into account, and the solid one—to the case of the proton transfer between the ground vibrational states. The dependence of the transfer coefficient on overvoltage for these two cases is shown on Fig. 23.

The discussion of the results of the present investigation with Professor Levich and Dr Kuznetsov, who have read my manuscript and made a number of valuable comments, has been exceedingly helpful to me. I wish also to thank my colleagues Dr Harkats, Dr German, Vorotyntsev, Kats, Shvets, and Konkina, who helped me in writing this article.

4. Further Developments of the Theory

In references 65 and 66 there are further developments of semi-phenomenological theory. The most important new results are described in refs. 67–69, where the dielectric formalism for description of polar solvents is developed. All the parameters of chemical and electrochemical reaction rates connected with the solvent reorganization are expressed in terms of the complex dielectric permittivity $\varepsilon(\kappa,\omega)$, i.e. the time dispersion of ε is taken into account along with the space dispersion of ε. For example, the reorganization energy can be written as [cf. eqn. (176) above]:

$$E_r = \frac{1}{8\pi} \sum_k |\vec{D}_{k\alpha} - \vec{D}_{k\alpha'}|^2 \cdot \frac{2}{\pi} \int_0^{kT/\hbar\alpha(1-\alpha)} d\omega \, \frac{\mathrm{Im}\varepsilon(\kappa,\omega)}{\omega \, |\varepsilon(\kappa,\omega)|^2}$$

The general expressions for the transition probability in the harmonic approximation are given in ref. 70 for the case of arbitrary change of frequencies and normal coordinates; this includes consideration of the mutual transformations of classical and quantum degrees of freedom in the course of reaction. The effect of deviation of the potential energy from harmonicity on the transition probability is discussed in ref. 71. Further development of the semiclassical method of calculation of transition probabilities is given in ref. 72, where in particular a closed expression for the entropy of activation is obtained, viz.:

$$e^{S_a/k} = \frac{\hbar\omega_{\mathrm{eff}}}{kT} \cdot Z_i^{-1} \left\{ \det \| G_{kl} \| \right\}^{-1/2}$$

$$\| G_{kl} \| = \| (1-\alpha)\Omega_k^{*2} \delta_{kl} + \alpha \sum_n \Omega_m'^{*2} \tau_{kn} \tau_{ln} \|$$

where Z_i is the partition function for intermolecular degrees of freedom, and α is the transfer coefficient. For the calculation of Ω_k^* and $\Omega_k'^*$ one must approximate the initial and final potential energy surface near the saddle point by quadratic forms, τ_{kn} is the orthogonal matrix connecting the initial and final normal coordinates. Ω_k^* and $\Omega_k'^*$ are the 'frequencies' in these quadratic forms.

The theory of adiabatic electrochemical redox reactions is given in ref. 73. The theory of electrochemical isotope effects is developed in ref. 74. This theory has been confirmed by experimental results obtained by Krishtalik and co-workers,[75] which have shown that proton transfer is of a quantum nature.

REFERENCES*

1. LIBBY, W. F.: *J. Phys. Chem.*, **56**, 863 (1952).
2. PLATZMAN, R., and FRANCK, J.: *Z. Physik.*, **138**, 411 (1954).
3. LAX, M.: *J. Chem. Phys.*, **20**, 1752 (1952).
4. PEKAR, S. I.: *Investigations of Electronic Theories of Crystals* (Russian edition) (Fizmatgiz, Moscow, 1951).
5. HUANG, K., and RHYS, A.: *Proc. Roy. Soc.*, **A204**, 406 (1950).
6. KRIVOGLAZ, M. A.: *Zh. Eksp. Teor. Fiz.* (Soviet Physics—JETP) **25**, 191 (1953).
7. FRÖHLICH, H.: *Advances in Physics*, **3**, 325 (1954).

* Page number in references to Russian journals are to those in the Russian editions.

8. KUBO, R., and TOYOZAWA, Y.: *Progr. Theor. Phys.*, **13**, 160 (1955).
9. DAVYDOV, A. S.: *Zh. Eksp. Teor. Fiz.* (Soviet Physics—JETP), **18**, 913 (1948).
10. DEIGEN, M. F.: *Trudy Inst. Fiz. Akad. Nauk Ukr. SSR*, pt. 5, 119 (1954).
11. MARCUS, R. A.: *J. Chem. Phys.*, **24**, 966, 979 (1956).
12. MARCUS, R. A.: *J. Chem. Phys.*, **26**, 867 (1957).
13. MARCUS, R. A.: *Can. J. Chem.*, **37**, 155 (1959).
14. MARCUS, R. A.: *Trans. Symp. Electrode Processes*, Edited by E. Yeager, 239 (1961).
15. MARCUS, R. A.: *Disc. Farad. Soc.*, **29**, 21 (1960).
16. MARCUS, R. A.: *J. Phys. Chem.*, **67**, 853, 2889 (1963).
17. MARCUS, R. A.: *J. Chem. Phys.*, **38**, 1335, 1858 (1963); **39**, 1734 (1963).
18. MARCUS, R. A.: *J. Chem. Phys.*, **41**, 603, 2624 (1964).
19. MARCUS, R. A.: *J. Chem. Phys.*, **43**, 679 (1965).
20. HUGENHOLTZ, N. M., and HOVE, L. VAN: *Physica*, **24**, 363 (1958).
21. BONCH-BRUEVICH, V. L., and GLASKO, V. B.: 'Vestnik', Moscow State University, **5**, 91 (1968).
22. BONCH-BRUEVICH, V. L., and GLASKO, V. B.: *Dokl. Akad. Nauk*, **124**, 1015 (1959).
23. BONCH-BRUEVICH, V. L., and TYABLIKOV, S. B.: *Green's Function Methods in Statistical Mechanics* (Russian edition) (G.I.F.M.L., Moscow, 1961).
24. LEVICH, V. G., and DOGONADZE, R. R.: *Dokl. Akad. Nauk*, **124**, 123 (1959).
25. LEVICH, V. G., and DOGONADZE, R. R.: *Dokl. Akad. Nauk*, **133**, 158 (1960).
26. DOGONADZE, R. R.: *Dokl. Akad. Nauk*, **133**, 1368 (1960).
27. LEVICH, V. G., and DOGONADZE, R. R.: *Coll. Szech. Chem. Comm.*, **29**, 193 (1961).
28. DOGONADZE, R. R.: *Dokl. Akad. Nauk*, **142**, 1108 (1962).
29. DOGONADZE, R. R., KUZNETSOV, A. M., and CHERNENKO, A. A.: *Elektrokhimiya*, **1**, 1434 (1965).
30. VOROTYNTSEV, M., KATZ, V. M., and KUZNETSOV, A. M.: *Elektrokhimiya*, **7**, No. 1 (1971).
31. LEVICH, V. G.: *Course of Theoretical Physics*, Vol. 1 (Russian edition) (Fizmatgiz, Moscow, 1962).
32. KRISHTALIK, L. I.: *Zh. Fiz. Khim.*, **33**, 1715 (1959).
33. KRISHTALIK, L. I.: *Zh. Fiz. Khim.*, **34**, 117 (1960).
34. KRISHTALIK, L. I.: *Zh. Fiz. Khim.*, **39**, 642 (1965).
35. KRISHTALIK, L. I.: *Elektrokhimiya*, **2**, No. 10 (1966).
36. KRISHTALIK, L. I.: *Elektrokhimiya*, **2**, 1176 (1966).
37. *Modern Aspects of Electrochemistry*, Edited by J. O. M. Bockris (Butterworth's Scientific Publications, London, 1954).
38. HOLSTEIN, T.: *Ann. Phys.*, **8**, 325 (1959).
39. DOGONADZE, R. R., and CHIZMADZMEV, YU. A.: *Fiz. Tverd. Tela.* (Solid State Physics), **3**, 3712 (1961).
40. DOGONADZE, R. R., CHIZMADZNEV, YU. A., and CHERNENKO, A. A.: *Fiz. Tverd. Tela.* (Solid State Physics), **3**, 3720 (1961).
41. LANG, I. G., and FIRSOV, YU. A.: *Zh. Eksp. Teor. Fiz.* (Soviet Physics—JETP), **43**, 1843 (1962).
42. DOGONADZE, R. R., KUZNETSOV, A. M., and CHERNENKO, A. A.: *Usp. Khimii.*, **34**, 1779 (1965).
43. LEVICH, V. G.: *Advances in Electrochemistry and Electrochemical Engineering*, **4**, 249 (1966).
44. LEVICH, V. G.: *Electrochemistry: the Universal Science*, 1965 (Russian edition) (VINITI, Moscow, 1967).
45. LANDAU, L. D., and LIFSCHITZ, E. M.: *Quantum Mechanics* (Fizmatgiz, Moscow, 1963).
46. MARCUS, R. J., ZWOLINSKY, B. J., and EYRING, H.: *J. Phys. Chem.*, **58**, 432 (1954).
47. DOGONADZE, R. R., and CHIZMADZHEV, YU. A.: *Dokl. Akad. Nauk*, **144**, 1077 (1962).
48. DOGONADZE, R. R., and CHIZMADZHEV, YU. A.: *Dokl. Akad. Nauk*, **145**, 849 (1962).
49. LEVICH, V. G., and DOGONADZE, R. R.: in *Basic Problems of Contemporary Theoretical Electrochemistry* (Russian Edn.) (Mir, Moscow, 1965).

50. FRUMKIN, A. N., PETRY, O. A., and NIKOLAEVA-FEDOROVICH, N. N.: Electrochim. Acta, 8, 177 (1963).
51. PARSONS, R., and PASSERON, E.: J. Electroanalyt. Chem., 12, 524 (1966).
52. DOGONADZE, R. R., and CHIZMADZHEV, YU. A.: Dokl. Akad. Nauk, 150, 333 (1963).
53. DOGONADZE, R. R., and KUZNETSOV, A. M.: Izvest. Akad. Nauk SSSR, Ser. Khim., 2140 (1964).
54. DOGONADZE, R. R., and KUZNETSOV, A. M.: Izvest. Akad. Nauk SSSR, Ser. Khim., 1885 (1964).
55. DOGONADZE, R. R., KUZNETSOV, A. M., and CHIZMADZHEV, YU. A.: Zh. Fiz. Khim., 38, 1195 (1964).
56. DOGONADZE, R. R., and KUZNETSOV, A. M.: Elektrokhimiya, 1, 742 (1965).
57. DOGONADZE, R. R., and KUZNETSOV, A. M.: Elektrokhimiya, 1, 1008 (1965).
58. DOGONADZE, R. R., and KUZNETSOV, A. M.: Elektrokhimiya, 3, 280 (1967).
59. GREEN, M.: in Modern Aspects of Electrochemistry, Vol. 2, ed. J. O'M. Bockris (Butterworths, London, 1954).
60. GERISCHER, H.: in Advances in Electrochemistry and Electrochemical Engineering, Vol. 1, ed. P. Delahay (Interscience, London and New York, 1961).
61. BOCKRIS, J. O'M.: in Modern Aspects of Electrochemistry, Vol. 1, ed. J. O'M. Bockris (Butterworths, London, 1954).
62. HORIUTI, J., and POLANYI, M.: Acta Physiochim SSSR, 2, 505 (1935).
63. DOGONADZE, R. R., and KUZNETSOV, A. M.: Elektrokhimiya, 2, 1324 (1967).
64. DOGONADZE, R. R., KUZNETSOV, A. M., and LEVICH, V. G.: Electrochim. Acta, 13, 1025 (1968).
65. DOGONADZE, R. R., and KUZNETSOV, A. M.: Elektrokhimiya, 7, No. 2 (1971).
66. VOROTYNTSEV, M. A., DOGONADZE, R. R., and KUZNETSOV, A. M.: Elektrokhimiya, 7, No. 3 (1971).
67. DOGONADZE, R. R., and KUZNETSOV, A. M.: Elektrokhimiya, 7, No. 3 (1971).
68. DOGONADZE, R. R., KUZNETSOV, A. M., and LEVICH, V. G.: Dokl. Akad. Nauk, 188, 383 (1969).
69. VOROTYNTSEV, M. A., DOGONADZE, R. R., and KUZNETSOV, A. M.: Dokl. Akad. Nauk, 195, No. 5 (1970).
70. VOROTYNTSEV, M. A., and KUZNETSOV, A. M.: Vestnik Moscow Univ., Ser. Phys., 2, 146 (1970).
71. DOGONADZE, R. R., and KUZNETSOV, A. M.: Dokl. Akad. Nauk, 194, 116 (1970).
72. DOGONADZE, R. R., and URUSHADZE, Z. D.: J. Electroanalyt. Chem., 1971 (in press).
73. DOGONADZE, R. R., KUZNETSOV, A. M., and VOROTYNTSEV, M. A.: J. Electroanalyt. Chem., 25, app. 17 (1970).
74. GERMAN, E. D., DOGONADZE, R. R., KUZNETSOV, A. M., LEVICH, V. G., and KHARKATZ, YU. I.: Elektrokhimiya, 6, 350 (1970).
75. KRISHTALIK, L. I., and TSIONSKY, V. M.: Elektrokhimiya, 5, 1019, 1184, 1501 (1969).

LIST OF SYMBOLS

av	Symbol of statistical averaging
a_0	Radius of localization of electron in ion or adsorbed particle
c	Basic parameter of the continuum model of dielectric in the absence of frequency dispersion, eqn (128)
$C_{H^+}^0$	Surface concentration of H_3O^+ ions in the Helmholtz layer
C_R; C_0	Surface concentration of reducing and oxidizing ions in the Helmholtz layer
$C_{0S}^{1,2}$	Surface concentration of ions in the Helmholtz layer at equilibrium potential before and after discharge

C_S	Concentration of discharging ions in the reaction zone
$\vec{D}$	Vector of electric induction
E	Classical Hamiltonian (energy) of polar solvent (eqn (120) and subsequently)
$E; E'$	Variables corresponding to total energies of the system during chemical reactions (Fig. 3–4)
E_a	Activation energy
E_a^*	Macroscopic value of activation energy
E_a^{exp}	Experimental activation energy
E_c^0	Energy corresponding to the bottom of the conduction band in semiconductor
ε_i	Ionization energy of ion
E_k	Energy levels of classical subsystem reckoned from the ground state
E_0	Energy of the ground state of classical subsystem
E_p	Reorganization energy of proton
E_r	Reorganization energy of classical subsystem
$E_{r1}; E_{r2}$	Reorganization energies for forward and back reactions
E_{ri}	Reorganization energy of the ith oscillator
E_r^{loc}	Reorganization energy of solvent in the absence of frequency dispersion
E_s	Reorganization energy of solvent
E_{solv}	Solvation energy of ion
E_v^0	Top of the valence band in semiconductor
$E_{\alpha n}$	Energy of the total system in the adiabatic perturbation theory
$E_\alpha(p,q)$	Hamiltonian for classical subsystem
ΔE	Difference between energies of the first excited level and the ground state
e_f	Electron localized at the level ε_f in electrode
$e\Delta\varphi$	Bending of semiconductor bands at interface
$e\Delta\varphi_0$	Equilibrium bending of semiconductor bands at interface
$F(\vec{r} - \vec{r}')$	Function characterizing the fluctuation energy of solvent
$F(\varepsilon_f)$	Free energy with fixed state of electron in electrode
F_a	Free energy of activation
$\Delta F(\varepsilon_f)$	Free energy of electrode reaction with fixed electron state
ΔF_{0F}	Free energy of reaction at the equilibrium potential of electrode
ΔF^*	Macroscopic value of free energy of electrode reaction
$(f\vec{r} - \vec{r}')$	Function characterizing the fluctuation energy of solvent
$\mathscr{F}$	Free energy calculated from the spectrum of classical subsystem reckoned from the ground state
$g(\varepsilon)$	Statistical weight for level
g_i	Generalized dipole momentum in ion solvation shell
$g_{\vec{k}}$	Generalized polarization wave momentum in solvent
$\mathscr{H}^0$	Quantum-mechanical unperturbed Hamiltonian of the system
H_e^0	Quantum-mechanical Hamiltonian of fast subsystem
$H_S(q)$	Quantum-mechanical Hamiltonian of polar solvent
$h(\hbar)$	Planck's constant (divided by 2π)
$\vec{i}^{e,p}$	Electron and hole discharge currents on semiconductor
i_c	Total cathodic current
$\vec{i}_c; \overleftarrow{i}_c$	Cathodic currents for forward and back processes

$i_0(i_0^{ad})$	Exchange current (for adiabatic process)
$\vec{i_2}$	Discharge current with transfer of 2 electrons
J	Minimum potential energy of classical subsystem
$\Delta J(\varepsilon_f)$	Change of minimum potential energy of classical subsystem during reaction
$\Delta J^{rr'}(\varepsilon_f)$	Change of minimum potential energy of Classical subsystem with fixed states of quantum subsystem before (r) and after (r') reaction
K	Reaction rate constant in the gas phase
K_0	Equilibrium constant of electrode reaction
$K_{0i,f}$	Kinetic energy of particles in the gas phase
k	Boltzmann constant
k_f	Wave number of electron in electrode
$\vec{k}$	Wave vector of polarization wave
$l; l'$	Quantum numbers of excited states of proton in H_3O^+ ion and H_{ads}.
l_0	Correlation radius of polarization currents in solvent
$M; m$	Masses of heavy and light particles in the adiabatic perturbation theory
M_i	Mass of the ith particle of classical subsystem
M_p	Mass of proton
m	Mass of electron
m_0	Mass of dipoles of solvent
m_e^*, m_p^*	Effective masses of electrons and holes in semiconductor
N_a	Number of dipoles contributing to activation energy
$n(\varepsilon_f)$	Fermi–Dirac distribution function
$\bar{n}_{\vec{k}}$	Mean occupation numbers of oscillator levels
Ox	Oxidizing ion
$\vec{P}; \dot{\vec{P}}$	Specific polarization and polarization current
$\vec{P}_{en}$	Electronic polarization due to atomic nuclei of solvent
$\vec{P}_{ir}$	Infrared (inertial) polarization
$\vec{P}_{\vec{k}}; \dot{\vec{P}}_{\vec{k}}$	Fourier amplitudes of polarization waves and polarization current
$\vec{P}_n$	Polarization due to dipole moments of solvent nuclei
$\vec{P}_{op}$	Optical (inertialess) polarization
$\vec{P}_{11}$	Longitudinal polarization
P_i	Generalized momentum of the ith particle of classical subsystem
Q	Statistical sum
q_i	Normal coordinate of classical subsystem (also in eqn (148) normal coordinate of dipole in solvation shell)
$q_{\vec{k}}$	Normal coordinate of polarization wave
$q_{k0}(\varphi_m)$	Equilibrium normal coordinate of classical subsystem versus electrode potential
$q_{\vec{k}\alpha}$	Equilibrium normal coordinate of polarization wave
q_s^*	Saddle point activated state
q_t	Normal coordinate of solvent at which the 'tunnel' transition of quantum subsystem is possible
q^*	Normal coordinate of solvent corresponding to activated state
$R_{i0}; R_{a0}$	Equilibrium coordinate of proton in H_3O^+ ion and in H_{ads}

Red	Reducing ion
r_0	Correlation radius of dipole moments of solvent
$S(\vec{r})$	Correlation function of dipole moments of solvent
S_a	Entropy of Activation
S^*	Intersection surface of terms
T	Temperature (degrees Kelvin)
$U(q_i\,\varphi_m)$	Term of classical subsystem versus electrode potential
$U_a(q^*)$	Potential energy of classical subsystem in activated state
U_0	Depth of potential well for electrons in metal
$U_{\alpha,\alpha'}(q)$	Terms of classical subsystem corresponding to the states α and α' of quantum subsystem
$V;v$	Velocities of heavy and light particles in the adiabatic perturbation theory
$V(q;\varphi_m)$	Energy of interaction of classical subsystem with electric field of electrode
V_e	Electrode volume
V_{es}	Energy of interaction of electron with solvent
V_{cr}	Critical value of exchange integral separating adiabatic and non-adiabatic reactions
V_S	Solvent volume
$V_{\alpha'\alpha}$	Exchange integral
$\vec{v_k}$	Function determining the interaction of solvent with quantum subsystem
v_i	Velocity of the ith particle of classical subsystem
W	Transition probability per unit time
W_{LZ}	Transition probability from term to term in the Landau–Zener theory
X_d	Debye screening distance for semiconductors
X_i	Cartesian coordinate of the ith classical subsystem
Z_α	Ionic charge
$\alpha(\varepsilon_f)$	Microscopic transfer coefficient in the Brønsted equation
$\alpha_{\exp}$	Experimental transfer coefficient
$\alpha^e; \alpha^e_{\exp}$	Transfer coefficients for electronic current on semiconductor
$\beta_{\exp}$	Experimental transfer coefficient for reverse current
γ	Parameters determining the geometric form and relative position of terms
γ_0	Solvent density
$\Delta_{e,p}$	Distance from the bottom of conduction and the top of valence bands to Fermi level
$\delta(x)$	Dirac delta function
ε_a	Electron energy in adsorbed particle
$\varepsilon_{al}; \varepsilon_{bl}; \varepsilon_n$	Electron levels in metals for which the discharge is activationless, barrierless and normal
ε_c	Minimum energy of conduction electron at interface
ε_f	Arbitrary electron level in electrode
ε_F	Fermi level of electron in electrode
ε_i	Quantum levels of electron in ion
$\varepsilon_e^{(i)}; \varepsilon_{e'}^{(a)}$	Excitation energies of proton in H_3O^+ ion and H_{ads}
ε_n	Excitation energies of slow subsystem
ε_0	Optical dielectric constant of solvent

$\varepsilon_r^{(1)}; \varepsilon_r^{(2)}$	Excited levels of quantum subsystem before (r) and after reaction (r')
ε_s	Static dielectric constant of solvent
$\varepsilon_\alpha(q)$	Quantum energy of fast subsystem
ε^*	Electron level in electrode making maximum contribution to current
$\Delta\varepsilon^*$	Range of levels near ε^* contributing to current
η	Overvoltage
η_c	Cathodic overvoltage (also overvoltage at interface for redox reactions on semiconductor)
η^F	Overvoltage at which main contribution to current is made by electron at Fermi level
η_s	Overvoltage at the point where discharging ion is localized
κ	Transmission coefficient
$\mu(\varepsilon)$	Two-electron distribution function for metals, eqn (224)
$\nu(\varepsilon)$	Analogue of occupation number of electrons in metal for 2-electron processes, eqn (227)
ρ	Density of bound charges of polar medium
$\rho(\varepsilon_f)$	Density of energy levels of electron in electrode
σ	Cross-section for gas reactions
τ_d	Mean time interval between diffusion jumps of ion in solution
τ_e	Mean time necessary for discharge of one ion
τ_r	Mean time of ion turning in solution
$\phi(\vec{r}); \varphi(\vec{r})$	Functions characterizing fluctuation energy of solvent
$\varphi_c(\varphi_c^0)$	(Equilibrium) potential at semiconductor interface
φ_m	Galvani potential of metal electrode
φ_0	Equilibrium Galvani potential of electrode
φ_s	Potential at the point where discharging (discharged) ion is localized
$\chi_{\alpha n}(q)$	Wave function of slow subsystem in the adiabatic perturbation theory
$\Psi_f(x)$	Wave function of electron in electrode
Ψ_1'	Psi-prime potential
$\Omega_k(\varphi_m)$	Normal frequencies of classical subsystem depending on electrode potential
$\omega_a; \omega_i$	Vibration frequencies of proton in H_{ads} and H_3O^+ ion
ω_{eff}	Effective frequency of polarization waves in solvent
$\omega_{\vec{k}}$	Normal frequencies of polarization waves in solvent
ω_0	Limiting infrared frequency of polarization waves in solvent
ω_p	Proton vibration frequency
ω_∞	Maximum frequency of polarization waves in solvent

THE RATES OF REACTIONS INVOLVING ONLY ELECTRON TRANSFER, AT METAL ELECTRODES

J. M. Hale

I. INTRODUCTION

This chapter is concerned with the rates of the simplest class of electrode reactions which comprises those involving only the addition of an electron e to an oxidized species O, or the removal of an electron from a reduced species R:

$$O + e \rightleftarrows R$$

All reactions complicated by the rupture or formation of bonds, or by the adsorption or crystallization of one of the participating species are thereby excluded from consideration. This restriction is made because it is for this class only that a quantitative theory of the electrode reaction rate has been formulated,[1-10] and a primary purpose of this work is to describe the agreement or conflict which exists between the experimental data which has been gathered and the theory as it stands at present.

Examples of redox couples which can be included in this class are frequently encountered when experimental conditions are suitably chosen. Thus it has

229

been discovered[11,12] that the reductions of organic molecules at an electrode in contact with an aprotic medium usually involve only electron addition, and therefore are eligible for consideration here, whilst the oxidation of the same molecule needs to be excluded because the radical cation produced is usually unstable and reacts rapidly with the solvent.[12,13] Similarly a large number of coordination compounds of the transition metal ions can be included provided that an excess of the ligand is present in the electrolyte.[14]

The rates of electron exchange between an electrode and such a redox couple is commonly found to be rapid,[14,15] indeed in many cases it has been described to be 'reversible' meaning that its velocity exceeded the limit imposed by the experimental technique chosen for its study. It has been demonstrated during recent years, however, that modern methods of investigation of rapid electrode reactions, such as that of measurement of faradaic impedance, are in fact capable of determining the absolute rates of these processes. These experimental developments are linked with the names of Randles, Hoijtink, Peover, and others.

The close similarity in all respects except the number of bound electrons, of the oxidized and reduced states of the redox system, coupled with the fact that a very weak electronic interaction of them with the electrode is required to permit electron transfer, makes possible the *a-priori* theoretical calculation of reaction rate. Conversion of the oxidized species into the reduced species requires the reorganization of the solvent in the immediate neighbourhood of the reacting species, together with some changes of bond length within the reactant. Extensive theoretical work, described elsewhere in this volume, has produced a means of calculation of the activation free energy of this process which is the starting point for the comparisons made herein. These theoretical developments were first associated with the names of Hush[1-3] and Marcus,[4-7] with later valuable contributions made by Levich, Dogonadze and their coworkers.[8-10]

In spite of the fact that the theory has been extended to include reactions at semiconductor electrodes,[16-21] the comparisons of this chapter have been made only for reactions taking place at metals.

II. THE THEORY OF ELECTRODE REACTION RATES, AND THE FREE ENERGY OF ACTIVATION

According to the theory of electron transfer reactions at electrodes,[7,22] the heterogeneous rate constant k_f (cm sec^{-1}) may be expressed in the following manner:

$$k_f = \kappa Z \exp \left\{ - \frac{\Delta G_f^*}{kT} \right\} \tag{1}$$

where

$$\Delta G_f^* = w^* + \frac{[\lambda + w - w^* + e(E - E_c^0)]^2}{(4\lambda)} \qquad (2)$$

The symbols appearing in these equations have the following meaning: Z the thermal velocity of reacting particles $\approx (kT/2\pi m)^{1/2}$, where k is Boltzmann's constant, T the temperature, and m the reduced mass of the reactant. $Z \sim 10^4$ cm sec^{-1}. ΔG_f^* the free energy of activation of the forward reaction. w^* the work required to transport the reactant from the bulk of the electrolyte to the plane at which electron transfer occurs (the pre-electron transfer state). w is similarly defined for the product. E, the electrode potential, and E_c^0 the standard electrode potential of the redox couple, both referred to the same reference electrode. λ, the work required to reorganize the environment about the reactant, such that all atoms are rearranged to adopt the positions taken about the product species at equilibrium. e the electronic charge. κ the transmission coefficient introduced in order to account for possible non-adiabaticity of the electrode reaction. Non-adiabatic reactions are those of which the rate is partially controlled by the tunnelling velocity of the electron through the potential energy barrier at the interface. For such reactions[8,9,22,23] the transmission coefficient is defined by an equation of the form:

$$\kappa \approx \sqrt{\left(\frac{2\pi m}{kT}\right)} \frac{4\pi^2 \rho x_p}{h} \Delta^2 \qquad (3)$$

where x_p is the separation between the reactant and the electrode at the instant of electron transfer, ρ is the number of energy levels in the conduction band of the electrode per unit range of energy, h is Planck's constant, and Δ is the energy difference between the upper and lower states derived from the interaction of the degenerate reactant and product systems in the transition state.[1,22] When Δ is very small, $\kappa \ll 1$, and the reaction is non-adiabatic in the sense described above. If, however, Δ turns out to be so large that κ is predicted by eqn (3) to be of the order of or greater than unity, then the reaction is said to be adiabatic and $\kappa = 1$ is introduced into eqn (1). The critical value of Δ which separates the adiabatic reactions from the non-adiabatic ones, would appear to be:

$$\Delta^{cr} = \left(\frac{kT}{2\pi m}\right)^{1/4} \left(\frac{h}{4\pi^2 \rho x_p}\right)^{1/2}$$

This has a magnitude of about 10^{-13} eV if all of those electrons in the conduction band of the metal having the correct energy can take part in the electrode reaction.

It is to be expected that Δ varies strongly with the distance x_p separating the reactant from the electrode, so that $\Delta < \Delta^{cr}$ for $x_p > x_p^{cr}$ and $\Delta > \Delta^{cr}$ for $x_p < x_p^{cr}$. Some indication about the manner of this dependence and the magnitude of x_p^{cr} would be particularly valuable for the understanding of the electrode reactions of those species repelled from the neighbourhood of the electrode, for if the critical distance x_p^{cr} can be greater than the range (about 8 Å in 1 M electrolytes) of these forces, then such repulsion might be irrelevant.

Unfortunately it is difficult to make any general comments about the dependence of Δ upon x. The interaction energy is given by

$$2\Delta = \int \phi_a H_{\text{int}} \phi_b \, dV$$

where ϕ_a and ϕ_b represent orthogonal orbitals, centred on the metal surface and upon the solution-phase species respectively, between which the electron is exchanged. H_{int} is the one electron interaction operator. The volume integration has its largest contribution from a region where the amplitudes of each of the wave functions is very small. Theoretical estimates of Δ are expected to be very poor in such a situation, since it is known that the methods of calculation of approximate wavefunctions, even of atoms, yield inaccurate results at points far from the nucleus and the present problem is complicated by the presence of the solvent, ligands, and other ions in solution. Illustrative calculations have been made, however, employing Slater functions to represent ϕ_a and ϕ_b.[24] Dogonadze made the calculation for exchange between two $3d_{z^2}$ orbitals on Fe^{2+} and Fe^{3+} ions, de Hemptinne for Cu(4s) and C(2s) orbitals, and Hush for the molecular orbitals of $Cr(H_2O)_6^{3+}$ and $Cr(H_2O)_6^{2+}$ ions. These calculations suggest that Δ becomes smaller by about one order of magnitude for each additional 0·7–1 Å of intervening separation and that $\Delta \sim 10^{-13}$ eV at a separation of about $x_p^{cr} \approx 15$ Å.

Hence we may conclude that electrode reactions involving only electron transfer can be adiabatic even though the reaction site of the electroactive species is situated outside the diffuse layer (see Section III(1) for nomenclature). The assumption usually made, although perhaps not explicitly, that at most one or two sheaths of solvent molecules can separate the reactant from the electrode at the instant of electron transfer therefore seems not to be justified. Furthermore, the calculations of activation energies of simple electrode reactions presented in Section IV, tend to support the point of view that reaction sites are often, in practice, outside the influence of coulombic interaction with the electrode. Of course it may be that the influence of the diffuse layer is to favour a reaction plane closer to the metal, through a decrease in the magnitude of ΔG^*; one of the conclusions drawn in Section III of this chapter, however, is that no really reliable method is known with which such a situation might be investigated theoretically.

The comparison of experiment and theory presented in Section IV is made at the standard potential E_c^0 where $k_f = k_s$ the standard rate constant, and is based directly upon eqns (1) and (2). The numerical magnitudes are compared of the quantities ΔG^*_{exp} and ΔG^*_{th}, representing experimental and theoretical free energies of activation, defined as follows:

$$\Delta G^*_{exp} = kT \ln \left(\frac{k_s}{10^4} \right) \tag{4}$$

$$= 0{\cdot}05915 \ (4 - \log_{10}k_s) \ \text{eV at } 25°$$

$$= 0{\cdot}06013 \ (4 - \log_{10}k_s) \ \text{eV at } 30°$$

and

$$\Delta G^*_{th} = w^* + \frac{[\lambda + w - w^*]^2}{4\lambda} \tag{5}$$

One electron volt (eV) per molecule is $23{\cdot}062$ kcal mole^{-1}. The crucial step in this comparison, naturally, is the calculation of the quantities w, w^*, and λ.

In some publications authors have described the temperature dependence of a rate constant and have interpreted it in terms of the 'Arrhenius' equation:

$$k_s = A \exp \left\{ - \frac{E_a}{kT} \right\}$$

where E_a, the activation energy, and A the pre-exponential factor are independent of temperature over the experimental range. The pairs of quantities A, Z, and E_a, ΔG^* are not expected to be identical because of the presence of temperature dependent terms in ΔG^*, but the relationship between them may be deduced.[25] We define an entropy of activation as follows:

$$\Delta S^* = - \frac{\partial}{\partial T}(\Delta G^*)$$

then, from the definition of E_a, we find

$$E_a = - \frac{\partial(\ln k_s)}{\partial(1/T)} = \Delta G^* + T\Delta S^*$$

and

$$A = \kappa Z \exp \left(\frac{\Delta S^*}{k} \right)$$

III. THE CALCULATION OF THE ACTIVATION FREE ENERGY ΔG_{th}^*

1. Models of the System

In order to calculate the work terms w^* and w, and the reorganization energy λ appearing in eqn (5) it is necessary to have a reliable model of the system of ion, solvent, electrolyte, and electrode. The energy of one of the constituent ions of an electrolyte at infinite dilution, is usually calculated from an electrostatic formula valid for a uniformly charged sphere in a uniform dielectric:[26]

$$\text{Energy} = \frac{(ze)^2}{2a\varepsilon_s} = \frac{7 \cdot 2z^2}{a\varepsilon_s} \text{ eV} \tag{6}$$

where ze is the total charge on the ion, a is the radius of the ion, ε_s is the dielectric constant of the solvent. When the second version of eqn (6) is used, the radius must be introduced in Å. The Born equation for the free energy of solvation of an ion at infinite dilution, ΔG_s, is derived from eqn (6)

$$\Delta G_s = \frac{(ze)^2}{2a} \left(\frac{1}{\varepsilon_s} - 1 \right) \tag{7}$$

Peover[27] has submitted this equation to an experimental test, by comparing its estimate of the solvation energy with that measured from polarographic half-wave potentials for a series of radical ions of aromatic compounds. He followed Lyons' suggestion[28] that the molar volume be used for the calculation of a, assuming each molecule to be a sphere:

$$\tfrac{4}{3}\kappa a^3 = \frac{1}{N} \times \frac{M}{\rho}$$

N is Avogadro's number, M is the molecular weight and ρ the density of the pure aromatic compound. Peover's results[27] are reproduced in Table 1.

In view of the coarseness of other approximations made in the derivation of the equations of electron transfer theory, and in the treatment of the remaining constituents of the electrode-electrolyte system, the agreement between theory and experiment revealed in Table 1 is considered adequate.

In two situations, at least, such a simple minded approach to the solvation energy must fail. A very small ion causes appreciable saturation of the dielectric in its neighbourhood, and Born's formula overestimates ΔG_s. The simplest procedure to follow when this condition holds, is to divide the environment of the ion into two regions, an inner spherical shell within which a number of solvent molecules are assumed to be fully orientated by the ionic

Table 1. *Comparison of experimental and theoretical free energies of solvation in acetonitrile of organic compounds*[27]

Compound	ΔG_s (experimental) (eV)		ΔG_s (theory) (eV)
	Anion	Cation	
Naphthalene	1·91		1·98
Triphenylene	1·69	1·84	1·72
Phenanthrene	1·66	1·73	1·79
Chrysene	1·72	1·87	1·69
Anthracene	1·93	1·82	1·8?
Pyrene	1·78	1·75	1·76
Acetophenone	2·23		1·95
Benzaldehyde	2·34		2·05
1:4 Benzoquinone	2·65		2·20
Pyrrole		2·5	2·31
Pyridine		2·5	2·21
Benzene		2·3	2·14
Aniline		2·1	2·12
p. Toluidine		2·5	2·04
N:N.Dimethyl Aniline		2·0	1·90
Diphenylanine		2·0	1·81

field, and the volume outside this in which saturation is ignored.[29] Then the inner region is treated microscopically in some appropriate manner, and the 'macroscopic' approach used above is reserved for the volume outside the saturated sphere.

The other situation in which application of Born's formula for ΔG_s yields a poor result, arises when the molecular ion has a very non-uniform charge density over its surface.[30-34] Consider the anion of an aromatic molecule which contains a heteroatom; a nitrocompound, or a quinone provide suitable examples. Rarely could these be adequately considered as point charges or uniformly charged spheres since accumulation of charge occurs at the heteroatom.

The normally random distribution of ions of the indifferent electrolyte is perturbed by the presence of a charged body, such as the electrode or an ion, so that a 'space charge' layer containing an equal and opposite charge to that on the body forms in the neighbourhood of the body. As is usual, we refer to this space charge as a diffuse double layer in the vicinity of the electrode,[35,36] and as the 'self atmosphere',[37,38] in the case of the ion. It is possible to describe these space charges in several ways, but the method adopted most frequently in electrochemistry is that originally suggested by Gouy.[35] The concentration of ions of type i, n_i, at any point is expressed by means of the Boltzmann law, as

$$n_i = n_i^s \exp\left(-\frac{w_i}{kT}\right) \tag{8}$$

where w_i is the work required to transport the ion from the bulk of the solution, where the concentration of ions is n_i^s, to the point in question. Then the space charge density is:

$$\rho = \sum_i z_i e n_i \tag{9}$$

where z_i is the valency of ion i. This method, together with some additional assumptions about the calculation of w_i, has been used to compute the activity coefficients of ions,[37] and the differential capacitance of the diffuse layer.[35,36] A great deal has been written about its advantages and limitations,[38,39] but it provides a reasonable representation of the indifferent electrolyte in not too concentrated solutions.[38]

A metal electrode is invariably modelled by a conducting plane in electrochemical calculations. Its surface will be referred to by the abbreviation ESP signifying the electrode surface plane. The plane of closest approach of ions to the electrode[40,41] is called the outer Helmholtz plane (OHP). This plane is supposed to pass through the centres of solvated ions which are in contact with the electrode or with a layer of solvent molecules specifically adsorbed on the metal surface, thus it is situated some 3–6 Å away from the ESP. Between the ESP and the OHP is the inner Helmholtz layer, which in the absence of specifically adsorbed ions contains only oriented solvent molecules.[42] Thus it has been modelled as a charge free layer having a dielectric constant much lower than that of the bulk solvent. In the remainder of this chapter its thickness will be symbolized by δ and its dielectric constant by ε_1; also ε_2 is used in place of ε_s as the dielectric constant of the bulk solvent. Some idea about the magnitude of ε_1/δ may be gained from the differential capacity versus electrode potential curve, for a dropping mercury electrode.[44,45] A potential independent capacitance $\varepsilon_1/4\pi\delta$ is observed when the electrode bears a large negative charge provided that cations of the electrolyte cannot become specifically adsorbed. Furthermore, δ may be guessed quite reasonably with the aid of molecular models so yielding an estimate of ε_1. Some results are recorded in Table 2. ε_0, the square of the refractive index of the solvent, is known as the optical dielectric constant and is the appropriate one

Table 2. *Some properties of solvents:* DMSO ≡ *dimethyl sulphoxide;*
D.M.F. ≡ *NN dimethyl formamide;* MeCN ≡ *acetonitrile*

Solvent	ε_0	ε_1	ε_2	δ (Å)
Water	1·79	6	78·3	5·0
DMSO	2·17	5·5	46·7	6·8
DMF	2·04	4·1	36·7	7·0
MeCN	1·81	5·5	36·7	6·6

to use when electric fields vary so rapidly that only the electronic polarization of the medium can 'follow' the change.

2. Calculation of the Work Terms

The work of transfer of an ion, charge ze, from the interior of an electrolytic solution to a point x_p in the neighbourhood of a charged metallic electrode is usually approximated by $w_z = ze\phi(x_p)$ where $\phi(x_p)$ is calculated from the Gouy-Chapman version of diffuse double layer theory. Thus the terms w and w^* appearing in eqn (5) can be calculated according to this approximation after evaluation of the Gouy-Chapman potential at the site of the ion from:[46,47]

$$\phi(x_p) = 4 \frac{kT}{z_b e} \tanh^{-1} \left\{ \exp\left[-\kappa_D(x_p - \delta)\right] \tanh\left(\frac{z_b e \phi_2}{4kT}\right) \right\} \tag{10}$$

Here, ϕ_2 is the potential of the OHP with respect to the bulk of the solution, and κ_D, the Debye reciprocal length is defined

$$\kappa_D = \sqrt{\left(\frac{8\pi n^s z_b^2 e^2}{\varepsilon_2 kT}\right)} = 50\cdot29 \times 10^8 (\varepsilon_2 T)^{-1/2} \sqrt{(C)} \tag{11}$$

where C is the concentration of electrolyte in moles/l.

These formulae are written for a $z_b - z_b$ indifferent electrolyte, present at concentration n^s/N moles/l. N is Avogadro's number. If z differs from z_b then the concentration of the reactant must be much smaller than that of the indifferent electrolyte. The dependence of ϕ upon x_p is illustrated in Fig. 1 for $ze\phi_2/(4kT) = 0\cdot25, 0\cdot5, 0\cdot75, 1\cdot0, 1\cdot25,$ and $1\cdot5$. ϕ_2 is given in terms of the charge density q_m on the ESP by[48]

$$\phi_2 = \frac{2kT}{z_b e} \sinh^{-1} \left\{\frac{2\pi z_b e q_m}{\kappa_D \varepsilon_2 kT}\right\}$$

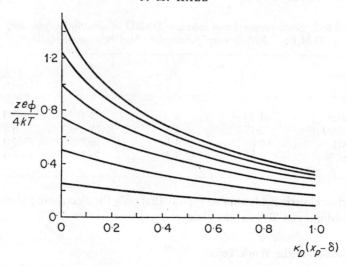

Fig. 1. The distribution of average potential in the diffuse layer for various ϕ_2 potentials.

Finally, q_m can be computed at any electrode potential E by integration of the experimental differential capacity curve $C(E)$ of the base electrolyte

$$q_m = \int_{E_{pzc}}^{E} C(E)\, dE$$

Here E_{pzc} is the potential of zero charge of the metal in the electrolyte.

Whilst the above procedure certainly estimates the major part of the electrostatic term to be included in w, it does have some failings which become most important in dilute electrolyte solutions. Williams[49] achieved a partial improvement upon taking account of the modification of the potential distribution in the neighbourhood of the ion itself resulting from its proximity to the electrode. In his model the electrolyte was considered uniform up to the electrode surface. He evaluated the incremental work of transfer of the ion, given by:[50]

$$w' = \int_0^{ze} \left\{ \lim_{r \to 0} \left(\psi' - \frac{ze}{\varepsilon_2 r} \right) - \lim_{r \to 0} \left(\psi' - \frac{ze}{\varepsilon_2 r} \right)_{x_p \to \infty} \right\} d(ze) \tag{12}$$

where ψ' is the perturbation potential at any point arising from the presence of the ion positioned at x_p, and $ze/(\varepsilon_2 r)$ is the potential which would exist at this point in the absence of other ions of the electrolyte. The limit $r \to 0$ was taken because the ions were modelled as point charges, the self energy, which then becomes infinite, cancels out of the integrand. It is noteworthy that in Debye's approximation

$$\lim_{r \to 0} \left(\psi' - \frac{ze}{\varepsilon_2 r} \right)_{x_p \to \infty} = \kappa_D \left(\frac{ze}{\varepsilon_2} \right)$$

After a lengthy calculation the following expression for w' was obtained:*

$$w' = - \frac{(ze)^2}{4\varepsilon_2 x_p} \exp(-2\kappa_D x_p)$$

$$+ \frac{\kappa_D(ze)^2}{\varepsilon_2(1 - y_p)^2} \left\{ 2y_0 \left(1 - \frac{y_p}{y_0} \right)^2 \alpha_0 \exp(-4\kappa_D x_p \alpha_0) E_1[4\kappa_D x_p(\alpha_0 + \tfrac{1}{2})] \right.$$

$$\left. - y_p[E_1(4\kappa_D x_p) + \gamma + \ln(4\kappa_D x_p)] \right\} \quad (13)$$

The new symbols appearing in this equation are as follows:

$$E_1(x) = \int_x^\infty \frac{\exp(-t)}{t} \, dt$$

is the exponential integral function;[51]

$$\gamma = 0.5772 \ldots$$

is Euler's constant;

$$\alpha_0 = \frac{1 + y_0}{2(1 - y_0)} = \cosh \left(\frac{(z_b e\phi_m/2kT)}{2} \right)$$

$$y_p = y_0 \exp(-2\kappa_D x_p)$$

and

$$y_0 = \tanh^2 \frac{z_b e\phi_m}{4kT}$$

ϕ_m is the potential of the ESP with respect to the bulk of the solution.

The first term in eqn (13) represents an interaction with the image of the ion in the electrode,† and the second term, a modification to the interaction with the self atmosphere due to the change in the ionic strength of the medium within the diffuse layer.

w' appears to be negative under conditions normally encountered. The dependence of w' at $2\kappa_D x_p = 0.1$ upon the charge on the metal surface,

---

* Apart from differences of notation, eqn (13) differs from eqn (11) of Williams' paper[49] by having $4\kappa_D \alpha_0 x_p$ as the argument of the exponential function in the second term, in place of $2\kappa_D \alpha_0 x_p$ as written incorrectly by Williams.

† Marcus,[5] in an unpublished investigation, reached a similar expression for the image force in the presence of an indifferent electrolyte.

represented by ϕ_m, is illustrated in Fig. 2; the sign of ϕ_m has no influence upon w'. Since $e^2\kappa_D/2\varepsilon_2$ has a value of about 10^{-2} eV in 0·1 M aqueous electrolytes, we conclude that w' has a magnitude of $-0·1$ eV when $\phi_m \sim 0·1$ V at this point some 0·5 Å away from the metal surface. At $2\kappa_D x_p = 1·0$, which corresponds roughly to the position of the OHP, $w' \sim 0·005$ eV, a very small term.

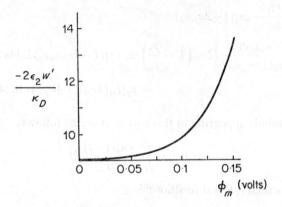

Fig. 2. Dependence of w' upon ϕ_m at $2Kx_p = 0·1$.

A further improvement of the estimate of w may be gained upon taking account of the structure of the interphase between the bulk of the metal and the bulk of the electrolytic solution. An electrostatic calculation of the potential distribution in such a system requires the solution of Laplace's equation 'in the inner layer', and Poisson's equation 'in the bulk of the electrolyte', with appropriate boundary conditions on the OHP to join their solutions. Evaluation of w at the level of approximation described by Williams yields the following formula:

$$w = ze\phi + w' + w''$$

where

$$w'' = \frac{\kappa_D(ze)^2}{2\varepsilon_2} \int_0^\infty \frac{f(\alpha) \exp\left[-4\kappa_D(x_p - \delta)\alpha\right]}{\alpha - \alpha_0} \, d\alpha \tag{14}$$

Here:

$$f(\alpha) = \frac{4\alpha(\alpha^2 - \tfrac{1}{4})}{(\alpha + \alpha_0)[1 + \zeta(\alpha)] + y_0/(1 - y_0)^2}$$

$$\left[1 + \frac{2\alpha}{(\alpha + \alpha_0)} \frac{[(2\alpha_0 - 1)/(2\alpha - 1) - (2\alpha_0/\alpha)y_p + (2\alpha_0 + 1)/(2\alpha + 1)\,y_p^2]}{1 - 2y_p + y_p^2}\right]$$

and

$$\zeta(\alpha) = \frac{\varepsilon_1}{\varepsilon_2} \sqrt{(\alpha^2 - \tfrac{1}{4})} \coth \left[2\kappa_D \delta \sqrt{(\alpha^2 - \tfrac{1}{4})} \right]$$

In the definitions of α_0 and y_0 the potential of the OHP, ϕ_2, replaces ϕ_m. Also, $x_p - \delta$ replaces x_p. At $\phi_2 = 0$ y_0 and y_p vanish and $\alpha_0 = \tfrac{1}{2}$. Although the integral for w'' cannot be performed analytically, it may be evaluated by numerical quadrature after removal of the singularity of the integrand at $\alpha = \alpha_0$.[52] The plot of $(w' + w'')/kT$ versus $(x_p - \delta)/\delta$ illustrated in Fig. 3 was computed after rearrangement of the integrand into the form:

$$w'' = \frac{\kappa_D (ze)^2}{2\varepsilon_2} \left\{ -f(\alpha_0) \exp\left[-4\kappa_D \alpha_0 (x_p - \delta) \right] Ei[4\kappa_D x_p (\alpha_0 - \tfrac{1}{2})] \right.$$

$$\left. + \exp(-2\kappa_D x_p) \int_0^\infty \frac{\{f[(t/4\kappa_D x_p) + \tfrac{1}{2}] - f(\alpha_0)\}}{t - 4\kappa_D x_p (\alpha_0 - \tfrac{1}{2})} \exp(-t)\, dt \right.$$

$Ei(x)$ is the real part of the extended exponential integral function

$$Ei(x) = -\int_{-x}^\infty \frac{\exp(-t)}{t} dt$$

The 10 point Gaussian-Laguerre quadrature formula[53] was used with completely satisfactory results.

Inspection of Fig. 3 reveals that w'' is much more positive than w' is negative in the region of the bulk electrolyte close to the OHP, in other words, the attraction of an ion to its image in the electrode is more than compensated by

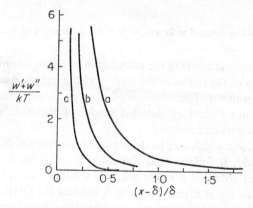

Fig. 3. Contribution to energy of an ion from self atmosphere as a function of distance from the outer Helmholtz plane, for three values of $2\kappa_D\delta$: (a) 0·329, (b) 1·04, (c) 3·29.

some repulsion from the inner layer. This positive contribution to the work of transfer of an ion into the diffuse layer may be ascribed to a loss of solvation energy due to penetration of the electric field of the ion into the saturated dielectric. Its net effect, presumably, is to make the inner layer thicker than the sum of a crystallographic radius and the diameter of a solvent molecule, because this electrostatic force reinforces the repulsion due to the contact of electron shells.

Increase of the concentration of the indifferent electrolyte tends to shield the ion from the influence of the inner layer, so that the repulsive force is not experienced until the ion reaches a point closer to the plane of closest approach. This conclusion may also be drawn by inspection of Fig. 3. Finally, the manner of variation of the work of transfer of an ion to the plane at $2\kappa_D(x - \delta) = 0.1$ with ϕ_2 is illustrated in Fig. 4.

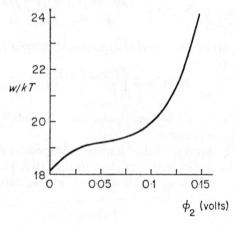

Fig. 4. Energy of ion situated at $2\kappa_D(x_p - \delta) = 0.1$ as function of ϕ_2. $2\kappa_D\delta = 1.04$.

Additional improvements in the calculation of w can be envisaged, such as the introduction of the radius of the ion[54] and the inclusion of van der Waals interactions. A number of interactions of this type are considered by Anderson and Bockris[55] but their very detailed model of the interface requires the choice of many parameters.

Two conclusions which can be drawn from the foregoing analysis deserve emphasis. Firstly, it follows that the calculation of work terms from the Gouy-Chapman potential is really not very reliable, unless there is evidence that the reaction site is situated several Å beyond the OHP. Secondly it has been shown that image forces upon ions within the diffuse layer are completely negligible.

3. The Reorganization Energy λ

λ measures the work expended in the reorganization of the atoms and molecules in the environment of the reactant from their positions at equilibrium to the positions occupied about the product species when it is at equilibrium.

In their derivations of formulae for λ, Hush and Marcus treated separately the behaviour of the dielectric surrounding the reactant, and that of the molecule or coordinated ion within the 'charged sphere'. As mentioned earlier (p. 235) this enables a microscopic treatment to be made of changes of bonds lengths and angles within the molecule, and a macroscopic electrostatic treatment of the unsaturated dielectric outside the sphere. These separate contributions to λ have been symbolized by λ_i and λ_0 respectively, representing 'inner' and 'outer' terms.

In his initial study of the problem, Marcus[4,8,56] expressed the electrostatic contribution to λ in the following fashion:

$$\lambda_0 = \frac{1}{8\pi} \int \left(\frac{1}{\varepsilon_0} - \frac{1}{\varepsilon_2} \right) (D^* - D)^2 \, dV \tag{15}$$

where the integral over volume extends to infinity, D is the dielectric displacement arising from the charge on the product ion, and D^* is that arising from the reactant ion.

Marcus evaluated the integral for λ_0 after choosing D as follows:

$$D \approx -Ze \left(\frac{1}{r_1^2} - \frac{1}{r_2^2} \right)$$

Z is here the charge on the ion, r_1 the distance from the field point to the ion and r_2 that to the image of the ion in the electrode. He then found

$$\lambda_0 = \frac{(ne)^2}{2} \left(\frac{1}{a} - \frac{1}{R} \right) \left(\frac{1}{\varepsilon_0} - \frac{1}{\varepsilon_2} \right) \tag{16}$$

where n is the number of electrons transferred and R twice the distance from the ion to the electrode.

It is likely that Marcus' estimate of λ_0 may exaggerate the influence of the image in the metal, since the charge induced on the metal by the presence of the ion can be equal and opposite to the charge on the ion only in the absence of a volume space charge in the solution. It follows from the electrostatic calculations described in the previous section that an ion is virtually shielded from the electrode, when more than 2 or 3 Å of 0·1 M electrolyte solution

9

intervenes between them. In such cases it is valid to neglect completely the polarizing influence of the image, and so to use:

$$D = - \frac{Ze}{r^2} \quad \text{and} \quad \lambda_0 = \frac{(ne)^2}{2a} \left(\frac{1}{\varepsilon_0} - \frac{1}{\varepsilon_2} \right) \tag{17}$$

which is the limit attained by Marcus' formula as $R \to \infty$. Since in practice, $R \sim 2a$, these estimates of λ_0 differ by a factor approximately equal to 2.

In a later paper, Marcus[7] made a more generally applicable derivation of the electrostatic contribution to the free energy of activation which allowed calculation of the effects of reorganization upon the charge in the self atmosphere of the ion. If the limit $R \to \infty$ is taken once more in his expression for the corrected value of λ_0, we find:

$$\lambda_0 = \frac{(ne)^2}{2a} \left(\frac{1}{\varepsilon_0} - \frac{1}{\varepsilon_2} \right) + \frac{(ne)^2}{2\varepsilon_2} \times \frac{\kappa_D}{1 + \kappa_D a} \tag{18}$$

Although it often happens that $\kappa_D \sim 1/a$ the second term here is always much smaller than the first because $\varepsilon_0 \ll \varepsilon_2$. In a typical case the correction amounts to about 2 or 3 per cent and can be neglected.

There has not yet been published any estimate of the effect of the proximity of the inner layer upon the reorganization energy. It is expected from qualitative reasoning that less reorganization energy should be required as the ion nears the OHP, since the saturated dielectric of the inner layer is less polarized by the ion than is the bulk dielectric.

The contribution to λ originating with the bond length and angle changes within the molecule or coordinated ion, designated λ_i, was calculated by Marcus[7,58] and Hush[1,3] from a microscopic model of the system, in contrast to their macroscopic model of the medium surrounding the sphere. Hush dealt specifically with an ion coordinated to dipolar ligands, whilst Marcus gave a more general treatment equally applicable to a neutral molecule and a complex ion. Sutin[59,60] has pointed out that these two approaches are complimentary, as far as complex ions are concerned, since the pointcharge—dipole model introduced by Hush may be used to calculate those force constants needed for Marcus' formula which may be unavailable from the literature.

Marcus'[61] expression for λ_i is:

$$\lambda_i = \sum_j \frac{f_j f_j^*}{f_j + f_j^*} (\Delta q_j)^2 \equiv 6 \cdot 25 \sum_j \frac{f_j f_j^*}{f_j + f_j^*} (\Delta q_j)^2 \tag{19}$$

where f_j and f_j^* are the force constants of the jth normal coordinate in the product and reactant respectively, and Δq_j is the change in coordinate accompanying reaction. In the second version of eqn (19), the f's must be expressed as mdyne/Å, and Δq in Å.

It commonly occurs that neither the force constants of bonds in the anion radicals derived from organic molecules, nor the differences of lengths of corresponding bonds Δq_j in such molecules and their anions, are known from experiment. Well tested relationships exist, however, between bond order and bond length[62,63] or force constant,[64] hence fairly reliable data may be collected for the majority of simple molecules. For carbon-carbon links one can use, for example, the following pair of relationships:

$$\Delta q_j = 0.18(p_j - p_j^*) \, (\text{Å}) \tag{20}$$

where

$$p_j = \sum_m n_m c_{r,m} c_{s,m}$$

and

$$f_j = 5.47(1 + p_j) - 2.06 \, \text{mdyne/Å} \tag{21}$$

Equation (20), in which p_j is the π bond order of the jth bond, was suggested by Coulson and Golebiewsky,[63] and eqn (21) was found to fit force constant—bond order data quoted by Gordy.[64] n_m is the number of electrons in the mth molecular orbital, and $c_{r,m}, c_{s,m}$ the Hückel coefficients of the carbon atom forming the C-C link, in the mth orbital.

IV. THE ABSOLUTE RATES OF ELECTRODE REACTIONS

1. Experimental Results

The experimental information on rates of electrode reactions involving only electron transfer is summarized in Table 3. The systems are identified by means of their oxidized components only.

With the exception of the results for $V(H_2O)_6^{3+}$, $Fe(H_2O)_6^{3+}$, $Fe(CN)_6^{3-}$, and WO_4^{2-} the recorded rate measurements of inorganic ions were made by analysis of steady-state current-voltage curves. These are notoriously sensitive to the state of the electrode surface, especially when solid metal electrodes are employed, and hence the precision of the experimental measurement is usually stated to be low by the authors. Even when the method of study is a transient one, so that the electrode may be exposed to the solution for less time, the reproducibility of measurements of the rate of the same reaction by different authors suggests that the error may amount to a factor of about two in the rate constant. Finally, attention is drawn to the point made by James[87] and Greef,[73] that because platinum becomes oxidized in aqueous solutions under applied anodic potentials in excess of about 0·75 V SCE, measurements of the reduction rates of species at potentials more anodic than this suffer interference from the parallel growth of oxide on the electrode.

Table 3. *Experimental information on rates of electrode reactions involving only electron transfer.* W = *Water;* TBAP $\equiv$ NBu$_4$ClO$_4$; TBAI $\equiv$ NBu$_4$I; TEAP $\equiv$ NEt$_4$ClO$_4$. α *is the transfer coefficient. Standard potentials E_c^0 are referred to a saturated calomel electrode, except for Azobenzene result which refers to a Ag/Ag$^+$ electrode in D.M.F.*[86]

System	Electrode	Medium	E_c^0(SCE)	k(cm sec^{-1})	α	ΔG_{exp}^*	Ref.
V(H$_2$O)$_6^{3+}$	Hg	W/1 M HClO$_4$/20°	-0.49	$4 \cdot 10^{-3}$	0·54	0·379	14, 65 66
Cr(H$_2$O)$_6^{3+}$	Hg	W/0·1 M HClO$_4$/25°	-0.72	$1 \cdot 4 . 10^{-5}$	0·45	0·541	67–69
Mn(H$_2$O)$_6^{3+}$	Pt	W/4 M HClO$_4$/22°	1·2	$6 \cdot 10^{-4}$	0·26	0·427	70
Fe(H$_2$O)$_6^{3+}$	Pt	W/1 M HClO$_4$/20°	0·5	$5 \cdot 10^{-3}$	0·42	0·373	14, 71
Co(H$_2$O)$_6^{3+}$	Pt	W/5·6 M HClO$_4$/2°	1·14	$1 \cdot 83 . 10^{-7}$	0·47	0·586	72
Ce(H$_2$O)$_6^{4+}$	Pt	W/1 M HClO$_4$/22°	1·5	$2 \cdot 10^{-5}$	0·25	0·515	73
Fe(CN)$_6^{3-}$	Pt	W/1 M KCl/20°	0·2	$9 \cdot 10^{-2}$	0·5	0·299	74–76
MnO$_4^-$	Graphite	?	?	$1 \cdot 10^{-2}$	—	0·355	77
Co(NH$_3$)$_6^{3+}$	Hg	W/0·1 M NaClO$_4$/25°	-0.3	$5 \cdot 10^{-6}$	0·61	0·551	78, 79
WO$_4^{2-}$	Pt	W/1 M H$_3$PO$_4$/25°	-0.02	$1 \cdot 10^{-2}$	—	0·355	80
ClO$_2$	Graphite	W/0·5 M Na$_2$SO$_4$/25°	0·7	$1 \cdot 7 . 10^{-2}$	—	0·341	81
O$_2$	Hg	DMSO/0·5 M TBAP/30°	-0.75	5	—	0·195	82, 83
O$_2$	Hg	MeCN/0·1 M TBAP/30°	—	0·8	—	0·242	82
Naphthalene	Hg	DMF/0·1 M TBAI/30°	-2.49	1·0	0·56	0·237	84
Anthracene	Hg	DMF/0·1 M TBAI/25°	-1.95	4	—	0·2	15
Tetracene	Hg	DMF/0·1 M TBAI/30°	-1.58	1·64	0·52	0·224	84
Perylene	Hg	DMF/0·1 M TBAI/25°	-1.67	4	—	0·2	15
[Anthracene]$^-$	Hg	,,	-2.55	$9 \cdot 1 . 10^{-3}$	—	0·357	15
[Perylene]$^-$	Hg	,,	-2.26	$8 \cdot 7 . 10^{-3}$	—	0·359	15
trans Stilbene	Hg	DMF/0·1 M TBAI/30°	-2.15	1·22	0·58	0·235	84
Cyclooctatetraene	Hg	DMF/0·1 M TPAP/25°	-1.62	$8 \cdot 7 . 10^{-3}$	—	0·359	85
Azobenzene	Hg	DMF/0·1 M TEAP/—	-1.81	0·5	0·37	0·255	86

All of the reactions of organic species tabulated happen to be reduction reactions, since the radical cations resulting from oxidation reactions are extremely reactive and rapidly oxidize most solvents.[12,88,89] Furthermore, mercury becomes oxidized under the anodic potential which must be applied to oxidize most neutral organic compounds, so that the advantages which accrue from the use of the dropping mercury electrode can be realized only in the study of reduction processes. Reductions of organic molecules are normally sufficiently fast to be polarographically reversible. This means that the plots of current density flowing through a dropping mercury electrode versus the potential applied are quite independent of the rate of reduction of the electroactive species. It has been found, however, that the measurement of the reactive and resistive components of the faradaic impedance of the electrode, by means of a bridge, is normally able to detect the influence of the electrode reaction rate. This was the method used at the source of the information on organics contained in Table 3. The method is particularly informative, since it reveals any interference from adsorption of the organic molecule at the electrode in a direct manner.[90-94]

In contrast to their behaviour in aqueous solution, it appears that aromatic molecules are not strongly adsorbed on to mercury from dimethyl formamide solution.[84,86] Furthermore, careful experimentation has failed to discover any blocking action arising from coverage of the negatively charged mercury surface by tetraalkyl ammonium ions used in the indifferent electrolyte.[66,84] These advantages stemming from the application of non-aqueous 'aprotic' solvents in organic electrochemistry, have only recently been realized, and have awakened widespread interest in the kinetics of electrode reactions of organic substances.[12]

2. Theoretical Results

In Table 4, the values have been recorded of λ_0, λ_i, and ΔG_{th}^*, assuming that the energies of the reacting species are not influenced by interaction with the electrode.

Table 4. *Theoretical results for electrode reactions involving only electron transfer*

System	a (Å)	λ_i (eV)	λ_0 (eV)	ΔG_{th} (eV)
$V(H_2O)_6^{3+}$	3·48	0·405	1·149	0·389
$Cr(H_2O)_6^{3+}$	3·49	0·583	1·146	0·432
$Mn(H_2O)_6^{3+}$	3·56	0·744	1·123	0·467
$Fe(H_2O)_6^{3+}$	3·51	0·461	1·140	0·400
$Co(H_2O)_6^{3+}$	3·48	0·412	1·149	0·390
$Ce(H_2O)_6^{4+}$	3·74	0·102	1·069	0·293
$Fe(CN)_6^{3-}$	4·65	0·379	0·860	0·310
MnO_4^-	3·09	(0·25)	1·294	0·386
$Co(NH_3)_6^{3+}$	3·35	0·212	1·194	0·352
WO_4^{2-}	3·19	(0·25)	1·255	0·376
ClO_2	2·35	0·31	1·70	0·503
O_2 (DMSO)	2·00	0·15	1·581	0·458
O_2 (MeCN)	2·00	0·15	1·691	0·460
Naphthalene	3·54	0·046	0·943	0·247
Anthracene	3·84	0·025	0·869	0·223
Tetracene	4·12	0·054	0·810	0·216
Perylene	4·22	0·05	0·790	0·210
Anthracene⁻	3·84	0·03	0·869	0·225
Perylene⁻	4·22	0·05	0·790	0·210
Stilbene	4·20	0·03	0·795	0·206
Cyclooctatetraene	3·56	0·622	0·936	0·375
Azobenzene	4·20	0·07	0·795	0·216

The data which was used for the calculation of λ_i of the inorganic species is summarized in Table 5. Only for a few of the systems are the bond lengths

Table 5. *Data required for calculation of λ_i of inorganic systems*

System	f_0 (mdyne/Å)	f_R (mdyne/Å)	Δq (Å)
$V(H_2O)_6^{3+}$	(1·31)	(0·78)	0·15[59]
$Cr(H_2O)_6^{3+}$	1·31[95]	(0·78)	0·18[59]
$Mn(H_2O)_6^{3+}$	(1·31)	0·8[95]	0·2[59]
$Fe(H_2O)_6^{3+}$	(1·31)	0·76[95]	0·16[59]
$Co(H_2O)_6^{3+}$	(1·31)	0·64[46]	0·16
$Fe(CN)_6^{3-}$	1·73[97,98]	2·43[97,98]	(0·1)
MnO_4^-	5·29[99]	4·72[99]	(0·05)
$Co(NH_3)^{3+}$	1·05[100]	0·33[100]	0·15[101]
WO_4^{2-}	(5·29)	(4·72)	(0·05)
ClO_2	7·4[102]	(7·4)	0·082[103]
O_2	11·76[104]	7·72[105]	0·073[104,106]

and the metal-ligand stretching force constants known for both of the valence states of the redox system. For the remainder, an estimate was made on the basis of an analogy with ions of similar structure. In some cases, as for example in the assessment of Δq for MnO_4^-, such estimates might be in error by a factor of 2; but in the majority of the systems it is expected that λ_i is predicted within about 10 per cent.

These calculations for the inorganic species are simplified because each of the bonds stretched are identical; it is natural to assume that a symmetrical vibration is then activated. Such symmetry can rarely be preserved during the reduction of organic species since the strengths of the various links vary with their position in the molecule. In general, therefore, more information is required about the oxidized and reduced forms of an organic redox system, in order that a reasonably accurate estimate of λ_i can be made.

For the polyacenes it was assumed that only bond stretching was of any importance; then eqns (20) and (21) were employed in conjunction with eqn (19). The Hückel coefficients used were those compiled in the 'Dictionary of π electron calculations'.[107] A totally analogous procedure was used for azobenzene, with the refinement that a semiempirical relationship between force constants f_j and bond lengths l_j of $N-N$ links suggested by Decius[108] was used:

$$f_j = \frac{38\cdot5}{l_j^6} \text{ mdyne/Å}$$

where l_j is in Å.

A more complicated case, that of cyclooctatetraene was treated in a totally different manner. Considerable internal reorganization is required in this molecule since it is known that the neutral molecule exists as a tub-shaped conformation, whilst it is strongly suspected that the anion is planar.[109,110] Hence λ_i measures the work required to 'flatten' the molecule. Whilst it may not be too grave an assumption to approximate the potential energy curve for the molecule as a parabolic function of the atomic displacements for configurations near equilibrium, it would be unrealistic to use such an approach for the large scale change of geometry needed in this reaction. Hence eqn (19) for λ_i was rejected in favour of an experimental estimate made by Allendoerfer and Rieger.[85] They assumed that the planar configuration was a transition state in bond isomerization, and therefore used the activation energy of this reaction measured by Anet,[111] as a measure of the energy required to convert the tub to the planar conformation.

Equation (17) was used for the calculation of λ_0, since all experimental measurements were obtained in an excess of indifferent electrolyte and it therefore was deemed necessary to ignore any effects of an image in the electrode. ε_0 and ε_2 data for all solvents needed were recorded in Table 2. The radii a of inorganic species were taken, wherever possible, to be the sums of bond lengths and van der Waals radii[112] of ligands. Those of organic molecules were calculated from molar volumes.

Finally, the theoretical activation energies were calculated from

$$\Delta G_{th}^* = \frac{\lambda_i + \lambda_0}{4}$$

which is the limiting form of eqn (5) reached by neglecting work terms w and w^*.

3. Comparison of Theory and Experiment

Out of the twenty-two systems which can be compared in Tables 3 and 4, thirteen have energies of activation ΔG_{exp}^* and ΔG_{th}^* differing by about 10 per cent or less. These include all the neutral organic molecules and most of the transition metal complex ions. Presumably, therefore, all of the assumptions contained in the electron transfer theory, and the means of calculation of the reorganization energies, are adequate for these reactions. It must be emphasized that very little effort is involved in the calculation of the activation energy at the level of approximation described, so that the excellent correlation observed is rewarding. Also we wish to draw particular attention to the indication that these reactions proceed with the solution phase reactant far enough away from the OHP that coulombic work terms can be neglected.

The overestimation of the activation energies for reduction of oxygen in non-aqueous solvents, and of chlorine dioxide in aqueous solution might be due to the neglect of saturation of the environment of the ions produced in these reactions. These molecules are the smallest included among the systems treated, and the method of calculation of λ_0 leads to large estimates of the work of reorganization of the medium. Some 'primary solvation' of ions with radii less than 2·5 Å is commonly observed[38,113] however, so that the assumption of a uniform dielectric outside spheres of this radius is prone to failure.

The rates of reduction of the hexaquo Ce^{4+} and Co^{3+} ions are badly overestimated; these ions are fairly large, leading one to expect that the part of the activation energy to be expended upon reorganization of the solvent should be small. The explanation in this case might be connected with the state of the electrode surface at which the measurements were made, for these ions are so powerfully oxidizing that even the noble metals are covered by a thin layer of oxide in the region of potential within which rates are measured. The authors[72,73] who have studied these reductions, and those of Tl^{3+},[87] $IrCl_6^{3-}$,[70] and Cl_2,[70,114] which are similarly complicated, have stressed the fact that electron transfer was occurring through a layer of oxide, and that the absolute rate of the electrode reaction depended critically upon the history of the electrode. It would be most unlikely that the electrode surface would remain uniformly accessible in this condition, for it is certain that the oxide film would influence the magnitude of the transmission coefficient κ.[115]

It is predicted by calculation that the anions of aromatic hydrocarbons should be reduced as rapidly as the parent molecules, yet the experimental measurements of Aten and Hoijtink[15] suggest that addition of the second electron is much slower than of the first. A contrary result was obtained by Allendoerfer and Rieger,[85] however, who showed that reduction of the cyclooctatraene anion is actually much faster than of the molecule. The latter authors advanced the tentative suggestion that the dianions of the aromatic hydrocarbons are sufficiently reactive to extract protons from the DMF molecule; they showed, furthermore, that this would lead to an apparently low value for the electrode reaction rate constant when the method of analysis of Aten and Hoijtink was used. This interpretation was confirmed recently by Hoijtink and Buthker in an unpublished investigation.

The reduction of the cobalt(III) hexammine complex proceeds anomalously slowly. It was proposed by Laitinen and Kivalo[79] that this reduction involves an electronically excited state of the Co(II) ions as an intermediate. This parallels suggestions made more recently by Vlcek[116] concerning the reductions of $Eu(H_2O)_6^{3+}$ and VO^{2+}.

In each of these examples the electronic structures of the oxidized and reduced states of the redox system are completely different and it is supposed that the electron addition step, involving some 'outer orbital' for which the transmission coefficient can be unity, is followed by a rapid internal electronic rearrangement. The foregoing theory may be used to calculate the rate of the electron transfer under these circumstances, since it is necessary only to introduce the excitation energy of the product as a contribution to w. One commonly finds that the activation energy is then predicted to be *very* large for such excitation energies are commonly of the order of one or two electron volts. Mechanisms for electrode reactions passing through electronically excited states are probably irrelevant, therefore, in most cases.

4. Reductions of Sterically Hindered Stilbenes

Evidence of the importance of steriochemical factors in the reactions of large organic molecules has been gathered by Dietz and Peover,[84] from a consideration of the rates of reduction of various substituted stilbenes. It is known that steric hindrance in the two-substituted stilbenes is relieved by a rotation θ about the 1-α single bond, and it follows from molecular orbital theory that the energy of the first antibonding orbital, which receives the excess electron in the anion, is raised by this rotation. Hence it is to be expected that reduction should be accompanied by a rotation such that θ becomes reduced; the energy required for this must be a major contribution to λ_i, and might be assumed to increase with θ.

The general trend of the results of AC impedance measurements at the dropping mercury electrode in DMF/0·1 M TBAI at 30°C, quoted by Dietz and Peover[84] and recorded in Table 6, are in accord with this qualitative

Table 6. *Reduction rates of hindered stilbenes*[84]

System	$E_{\frac{1}{2}}$ (volts vs SCE)	k (cm sec^{-1})	α	$\Delta G^*_{\exp}$ (eV)	θ
trans stilbene	−2·15	1·22	0·58	0·235	0
α-Me trans stilbene	−2·26	0·43	0·45	0·262	35°
2,4,6 Me$_3$ trans stilbene	−2·28	0·52	0·57	0·258	28°–38°
2,4,6,2^1,4^1, 6^1 Me$_6$ trans stilbene	−2·46	0·18	0·46	0·285	54°
cis stilbene	−2·18	0·42	0·44	0·263	28°

reasoning. As these authors have pointed out, a quantitative treatment requires information concerning the conformation of the anions, which is lacking at present.

5. Reductions of Nitrocompounds

The rates of electroreduction at the dropping mercury electrode of a series of nitrocompounds in DMF/0·1 M TBAI at 30°, were studied with the AC impedance method by Peover and his collaborators.[117] These results are presented in Table 7. Although the rate constants themselves are not given, because the diffusion coefficients of these compounds in DMF are not yet available, it is evident that the reduction rate has a tendency to decrease with increasing size of the molecule in contrast to the behaviour of the aromatic hydrocarbons, for example. If the diffusion coefficients were available, it is likely that this trend would appear even more pronounced.

Peover proposed that the non-uniform distribution of charge, in the anion radicals derived from the nitrocompounds might be responsible for this phenomenon. Consider a molecule which may be modelled by two spheres, radii a_1 and a_2 separated by a distance R, and suppose that in the anion radical the spheres bear charges q_1 and q_2 where $q_1 + q_2 = -e$. We evaluate the reorganization energy λ_0 associated with the charging of these spheres, making the approximation that

$$D^* = 0 \quad \text{and} \quad D = -\frac{q_1}{r_1^2} - \frac{q_2}{r_2^2}$$

where r_1 and r_2 are the distances of the field point from the centres of the spheres. Then

$$\lambda_0 = \tfrac{1}{2} \left(\frac{1}{\varepsilon_0} - \frac{1}{\varepsilon} \right) \left(\frac{q_1^2}{a_1} + \frac{q_2^2}{a_2} + \frac{2q_1 q_2}{R} \right) \tag{22}$$

Let us now apply this model to the nitro compounds, with the sphere radius a_1 representing the nitro group and that of radius a_2 representing the remainder of the molecule.

We choose $a_1 = 2·3$ Å and a_2 as recorded in Table 7. Since $a_1 < a_2$, λ_0 tends to increase with q_1; in other words, more extensive reorganization of the solvent is required about a molecule having charge localized upon the small nitrogroup, than about one in which the charge is delocalized over the remainder of the molecule.

Peover analysed the electron-spin resonance spectra of the nitroanions in DMF, and so found the nitrogen coupling constants α_N which are proportional to q_1 the charge localized on the nitro group; these are also recorded in

Table 7. *Reduction rates of nitro-compounds*

System	a (Å)	a_2 (Å)	$k/\sqrt{2}D$ sec$^{\frac{1}{2}}$	α_N	ΔG^*_{th} (eV)
p-nitrotoluene	3·485	3·48	415	10·06	0·233
o-nitrotoluene	3·60	3·48	415	10·8*	0·225
p-nitroaniline	3·376	3·31	280	11·5	0·231
o-tert-butyl-nitrobenzene	4·08	3·95	100	14·9*	0·239
nitromesitylene	3·90	3·81	62	16·0	0·248
nitrodurene	4·11	3·99	35	19·1	0·274
aninonitrodurene	4·16	4·11	41	20·6	0·291
tert-butyl nitrate	3·58	3·37	1·8	26·4*	0·362
2-nitropropane	3·295	3·07	2·5	25	0·347

* α_N measured in acetonitrile.

Table 7. He supposed that the α_N characteristic of the aliphatic nitro compounds, $\sim$26, represented the complete localization of the electron on the nitrogroup. Then $q_1 = -(\alpha_N/26)e$.

We further choose $R = 5$ Å a constant in the series, and ignore any internal reorganization energy, $\lambda_i = 0$; then $\Delta G^*_{th} = \lambda_0/4$ may be computed. The theoretical results are recorded in column 6 of Table 7, and a plot of $\log_{10}[k/\sqrt{(2D)}]$ versus ΔG^*_{th} illustrated in Fig. 5. The predicted values of ΔG^*_{th} are certainly of the correct order of magnitude, and increase in the direction required by the order of rate constants. There are some indications that the range of variation of ΔG^*_{th} is not as wide as ΔG^*_{exp}, perhaps because

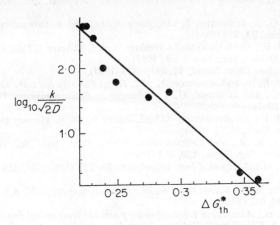

Fig. 5. Reduction of aromatic nitrocompounds: model of two charged spheres.

some internal activation energy is expended to rotate the plane of the nitro group towards the plane of the ring, but it is likely that a major contribution to the activation energy has been satisfactorily accounted for. Similar reasoning explains the lower reduction rate of azobenzene than that of stilbene, though a more complicated model of the anion would be needed in this example.

ACKNOWLEDGEMENT

The author is pleased to express his gratitude to Drs N. S. Hush and Roger Parsons, for helpful suggestions concerning images of ions in the diffuse layer, and to Dr M. Peover for making available some of his experimental results before publication.

REFERENCES

1. HUSH, N. S.: *J. Chem. Phys.*, **28**, 962 (1958).
2. HUSH, N. S.: *Z. Elektrochem.*, **61**, 734 (1957).
3. HUSH, N. S.: *Trans. Faraday Soc.*, **57**, 557 (1961).
4. MARCUS, R. A.: O.N.R. Technical Report No. 12, Project NR-051-331 (1957).
5. MARCUS, R. A.: *Can. J. Chem.*, **37**, 155 (1959).
6. MARCUS, R. A.: *Trans. Symp. Electrode Processes*, 239–45 (1959), Edited by E. Yeager (John Wiley and Sons, New York, N.Y., 1961).
7. MARCUS, R. A.: *J. Chem. Phys.*, **43**, 679 (1965).
8. DOGONADZE, R. R., and CHIZMADZHEV, Y. A.: *Proc. Russ. Acad. Sci., Phys. Chem. Sect.*, **145**, 563 (1962).
9. DOGONADZE, R. R., KUZNETSOV, A. M., and CHERNENKO, A. A.: *Russ. Chem. Reviews*, **34**, 759 (1965).
10. LEVICH, V. G.: *Advances in Electrochemistry and Electrochemical Engineering*, Vol. 4, Chapter 5, Edited by P. Delahay (Interscience Publishers, Inc., New York, N.Y., 1966).
11. HOIJTINK, G. J., SCHOOTEN, J. VAN, BOER, E. DE, and AALBERSBERG, W. Y.: *Rec. Trav. Chim.*, **73**, 355 (1954).
12. PEOVER, M. E.: *Electroanalytical Chemistry*, Vol. 2, Chapter 1, Edited by A. J. Bard (Marcel Dekker, Inc., New York, 1967).
13. LUND, H.: *Acta Chim. Scand.*, **11**, 491, 1323 (1957).
14. RANDLES, J. E. B., and SOMERTON, K. W.: *Trans. Faraday Soc.*, **48**, 937 (1952).
15. ATEN, A. C., and HOIJTINK, G. J.: *Advances in Polarography*, p. 777 (Pergamon, Oxford, 1961).
16. DEWALD, J. F.: *Semiconductors*, 727–52, Edited by N. B. Hannay (Reinhold Publ. Co., New York, N.Y., 1959).
17. DOGONADZE, R. R., and CHIZMADZHEV, Y. A.: *Proc. Acad. Sci. U.S.S.R., Phys. Chem. Sect., Eng. Trans.*, **150**, 402 (1963).
18. GERISCHER, H.: *Z. Physik Chem. (Frankfurt)*, **26**, 223 (1960); **26**, 325 (1960); **27**, 48 (1961).
19. DOGONADZE, R. R., KUZNETSOV, A. M., and CHIZMADZHEV, Y. A.: *Russ. J. Phys Chem.*, **38**, 652 (1964).
20. GERISCHER, H.: *Advances in Electrochemistry and Electrochemical Engineering*, Vol. 1, Chapter 4, Edited by P. Delahay (Interscience Publishers Inc., New York, N.Y., 1960).

21. MEHL, W., and HALE, J. M.: *Advances in Electrochemistry and Electrochemical Engineering*, Vol. 6, Chapter 5, Edited by P. Delahay (Interscience Publishers Inc., New York, N.Y., 1967).
22. MARCUS, R. A.: *Ann. Rev. Phys. Chem.*, 15, 155 (1964).
23. HALE, J. M.: *J. Electroanal. Chem.*, 19, 315 (1968).
24. DOGONADZE, R. R.: *Proc. Russ. Acad. Sci.*, *Phys. Chem. Sect.*, 133, 765 (1960); X. de Hemptinne, *Bull. Soc. Chim. France*, 2328 (1964); N. S. Hush, *Electrochim. Acta*, 13, 1016 (1968).
25. MARCUS, R. A.: *Electrochim. Acta*, 13, 995 (1968).
26. GURNEY, R. W.: *Ions in Solution*, p. 1–3 (Dover Publications Inc., New York, 1962).
27. PEOVER, M. E.: *Electrochim. Acta*, 13, 1083 (1968).
28. LYONS, L. E.: *Nature, Lond.*, 166, 193 (1950).
29. BUCKINGHAM, A. D.: *Disc. Faraday Soc.*, 24, 151 (1957).
30. KOUTECKY, J.: *Electrochim. Acta*, 13, 1079 (1968).
31. JANO, I.: *Cahiers de Physique*, 20, 1 (1966).
32. PERADEJORDI, F.: *Cahiers de Physique*, 17, 393 (1963).
33. HOIJTINK, G. J., BOER, E. DE, VAN DER MEY, P. H., and WEYLAND, W. P.: *Rec. Trav. Chim.*, 75, 487 (1956).
34. HUSH, N. S., and BLACKLEDGE, J.: *J. Chem. Phys.*, 23, 514 (1955).
35. GOUY, G.: *J. Phys.*, 9, 457 (1910).
36. CHAPMAN, D. L.: *Phil. Mag.*, 25, 475 (1913).
37. DEBYE, P., and HÜCKEL, E.: *Phys. Z.*, 24, 185 (1923).
38. ROBINSON, R. A., and STOKES, R. H.: *Electrolyte Solutions*, 2nd edition, Chapter 4 (Butterworths, London, 1959).
39. FOWLER, R., and GUGGENHEIM, E. A.: *Statistical Thermodynamics*, Chapter 9 (Cambridge University Press, 1960).
40. GOUY, G.: *Ann. Phys.*, *Paris*, 7, 163 (1917).
41. STERN, O.: *Z. Elektrochem.*, 30, 508 (1924).
42. BARLOW, C. A., JR., and MACDONALD, J. R.: *Advances in Electrochemistry and Electrochemical Engineering*, Vol. 6, Chapter 1, Edited by Paul Delahay (Interscience Publishers, Inc., New York, N.Y., 1967).
43. MACDONALD, J. ROSS, and BARLOW, C. A., JR.: *J. Electrochem. Soc.*, 113, 978 (1966).
44. DEVANATHAN, M. A. V.: *Trans. Faraday Soc.*, 50, 373 (1954).
45. MINC, S., JASTRZEBSKA, J., and BRZOSTOWSKA, M.: *J. Electrochem. Soc.*, 108, 1160 (1961).
46. MÜLLER, H.: *Cold Spring Harbor Symposia Quant. Biol.*, 1, 1 (1933).
47. MACDONALD, J. ROSS, and BRACHMANN, M. K.: *J. Chem. Phys.*, 22, 1314 (1954).
48. PARSONS, R.: *Modern Aspects of Electrochemistry*, Vol. 1, Chapter 3, p. 146, Edited by J. O'M. Bockris (Butterworths, London, 1954).
49. WILLIAMS, W. E.: *Proc. Phys. Soc.*, A66, 372 (1953).
50. KIRKWOOD, J. G.: *J. Chem. Phys.*, 2, 351, 767 (1934).
51. GAUTSCHI, W., and CAHILL, W. F.: *Handbook of Mathematical Functions*, p. 227, Edited by M. Abramowitz and I. A. Stegun (Dover Publications Inc., New York, 1965).
52. National Physical Laboratory, *Modern Computing Methods*, 2nd edition, p. 134 (Her Majesty's Stationary Office, 1961).
53. KRYLOV, V. I.: *Approximate Calculation of Integrals*, p. 130–2 and 347–52 (Macmillan, New York and London, 1962).
54. FREISE, V.: *Z. Elektrochem.*, 56, 822 (1952).
55. ANDERSON, T. N., and BOCKRIS, J. O'M.: *Electrochim. Acta*, 9, 347 (1964).
56. MARCUS, R. A.: *J. Chem. Phys.*, 24, 966 (1956).
57. HUSH, N. S.: *Electrochim. Acta*, 13, 1011 (1968).
58. MARCUS, R. A.: *Disc. Faraday Soc.*, 29, 21 (1961).
59. SUTIN, N.: *Ann. Rev. Phys. Chem.*, 17, 119 (1966).

60. REYNOLDS, W. L., and LUMRY, R. W.: *Mechanisms of Electron Transfer*, p. 21 (Ronald Press, New York, 1966).
61. MARCUS, R. A.: *J. Phys. Chem.*, 67, 853 (1963).
62. DAUDEL, R., LEFEBVRE, R., and MOSER, C.: *Quantum Chemistry*, Chapter 5 (Interscience Publishers, Inc., New York, 1959).
63. COULSON, C. A., and GOLĘBIEWSKY, A.: *Proc. Phys. Soc.*, (*London*), 78, 1310 (1961).
64. GORDY, W.: *J. Chem. Phys.*, 15, 305 (1947).
65. RANDLES, J. E. B.: *Can. J. Chem.*, 37, 238 (1959).
66. JOSHI, K. M., MEHL, W., and PARSONS, ROGER: *Trans. Symp. Electrode Processes*, 249–63 (1959), Edited by E. Yeager (John Wiley and Sons, New York, N.Y., 1961).
67. RANDLES, J. E. B.: *Progress in Polarography*, Vol. 1, p. 123, Edited by I. M. Kolthoff and P. Zuman (Interscience, New York, 1962).
68. PARSONS, R., and PASSERON, E.: *J. Electroanal. Chem.*, 12, 524 (1966).
69. OGINO, K., and TANAKA, N.: *Bull. Chem. Soc. Japan*, 40, 1119 (1967).
70. GREEF, R., and AULICH, H.: Personal communication.
71. JAHN, D., and VIELSTICH, W.: *J. Electrochem. Soc.*, 109, 849 (1962).
72. BORISOVA, L. M., and ROTINYAN, A. L.: *Soviet Electrochemistry*, 3, 846, 1163 (1967).
73. GREEF, R., and AULICH, H.: *J. Electroanal. Chem.*, 18, 295 (1968).
74. AZIM, S., and RIDDIFORD, A. C.: *J. Polarographic Soc.*, 12, 20 (1966).
75. JORDAN, J.: *Analyt. Chem.*, 27, 1708 (1955).
76. WIJNEN, M. D., and SMIT, W. M.: *Rec. Trav. Chim.*, *Pays-Bas*, 79, 203, 289 (1960).
77. GALUS, Z., and ADAMS, R. N.: 142nd Am. Chem. Soc., National Meeting (1962), 'Symposium on Mechanism of Electrode Reactions'. See Reference 61.
78. KLATT, L. N., and BLAEDEL, W. J.: *Analyt. Chem.*, 39, 1065 (1967).
79. LAITINEN, H. A., and KIVALO, P.: *J. Am. Chem. Soc.* 75, 2198 (1953).
80. STONEHART, P.: *Anal. Chim. Acta*, 37, 127 (1967).
81. SCHWARZER, O., and LANDSBERG, R.: *J. Electroanal. Chem.*, 14, 339 (1967).
82. PEOVER, M. E., and POWELL, J. S.: *J. Polarographic Soc.*, 12, 106 (1966).
83. SAWYER, D. T., and ROBERTS, J. L.: *J. Electroanal. Chem.*, 12, 90 (1966).
84. DIETZ, R., and PEOVER, M. E.: *Disc. Faraday Soc.*, 45, 154 (1968).
85. ALLENDOERFER, R. D., and RIEGER, P. H.: *J. Am. Chem. Soc.*, 87, 2336 (1965).
86. AYLWARD, G. H., GARNETT, J. L., and SHARP, J. H.: *Analyt. Chem.*, 39, 457 (1967).
87. JAMES, S. D.: *Electrochim. Acta*, 12, 939 (1967).
88. PHELPS, J., SANTHANAM, K. S. V., and BARD, A. J.: *J. Am. Chem. Soc.*, 89, 1752 (1967).
89. PEOVER, M. E., and WHITE, B. S.: *J. Electroanal. Chem.*, 13, 93 (1967).
90. GERISCHER, H.: *Z. Physik. Chem.*, 201, 55 (1952).
91. LAITINEN, H. A., and RANDLES, J. E. B.: *Trans. Farad. Soc.*, 51, 54 (1955).
92. LLOPIS, J., FERNANDEZ-BIARGE, J., and FERNANDEZ, M. PEREZ: *Electrochim. Acta*, 1, 130 (1959).
93. SENDA, M., and DELAHAY, P.: *J. Phys. Chem.*, 65, 1580 (1961).
94. BARKER, G. C.: *Trans. Symp. Electrode Processes*, p. 325 (1959), Edited by E. Yeager (John Wiley and Sons, New York, N.Y., 1961).
95. NAKAGAWA, I., and SHIMANOUCHI, T.: *Spectrochim. Acta*, 20, 429 (1964).
96. KERMARNEC, Y.: *Compt. Rend.*, 258, 5836 (1964).
97. NAKAMOTO, K., FUJITA, J., and MURATA, H.: *J. Am. Chem. Soc.*, 80, 4817 (1958).
98. NAKAGAWA, I., and SHIMANOUCHI, T.: *Spectrochim. Acta*, 18, 101 (1962).
99. KREBS, B., MUELLER, A., and ROESKY, H. W.: *Mol. Phys.*, 12, 469 (1967).
100. NAKAGAWA, I., SHIMANOUCHI, T., and HIRAISHI, J.: *Proc. I.C.C.C. 8*, p. 21 (Vienna, 1964).
101. BARNET, M. T., CRAVEN, B. M., FREEMAN, H. C., LIME, N. E., and IBERS, J. A.: *Chem. Comm.*, 24, 307 (1966).
102. DUCHESNE, J., and NIELSEN, A. H.: *J. Chem. Phys.*, 20, 1968 (1952).
103. Special Publications No. 11, 18 of Chem. Soc., Tables of Interatomic Distances and Configuration in Molecules and Ions.

104. HERZBERG, G.: *Molecular Spectra and Molecular Structure*, Vol. 1, 'Spectra of Diatomic Molecules', 2nd edition, p. 560 (D. van Nostrand Co. Inc., Princeton, New Jersey, 1965).
105. 'The force constant of O_2^- was estimated from its bond length[106] using data given by J. W. Linnett', *J. Chem. Soc.*, 275 (1956).
106. ABRAHAMS, S. C., and KALNAJS, J.: *Acta Cryst.*, **8**, 503 (1955).
107. COULSON, C. A., and STREITWEISER, A.: *Dictionary of π-Electron Calculations* (Pergamon Press, Oxford, 1965).
108. DECIUS, J. C.: *J. Chem. Phys.*, **45**, 1069 (1966).
109. CARRINGTON, A., and TODD, P. F.: *Mol. Phys.*, **7**, 1525 (1964).
110. KATZ, T. J.: *J. Am. Chem. Soc.*, **82**, 3784, 3785 (1960).
111. ANET, F. A. L.: *J. Am. Chem. Soc.*, **84**, 671 (1962).
112. PAULING, L.: *The Nature of the Chemical Bond*, 3rd edition, p. 257 (Cornell University Press, 1960).
113. HALE, J. M., and PARSONS, R.: *Adv. in Polarography*, **1**, 829 (1960).
114. TOSHIMA, S., and OKANIWA, H.: *Denki Kagaku*, **34**, 641 (1966).
115. KUZNETSOV, A. M., and DOGONADZE, R. R.: *Izvestiya Akademii Nauk SSSR Ser. Khim.*, 2140 (1964).
116. VLČEK, A.: *Coll. Czech. Chem. Comm.*, **24**, 181 (1959).
117. PEOVER, M. E.: Personal communication.

OXIDATION AND REDUCTION OF AROMATIC HYDROCARBON MOLECULES AT ELECTRODES

M. E. Peover

I. INTRODUCTION

This survey will be concerned with some of the more recent developments in the study of electrode reactions of conjugated aromatic hydrocarbons and some of their derivatives in non-aqueous solvents. Reviews of earlier work can be found elsewhere.[1,2] Earlier work established that under well chosen conditions a remarkably consistent picture of their electrochemistry emerged, reduction processes being fast one-electron processes forming rather stable radical-ions. These properties have been much exploited to obtain electrochemical series, usually polarographically, for testing calculations of related properties and for obtaining the electron spin resonance spectra of the radical anions. Oxidation behaviour of these compounds is however more complex because of spontaneous reactions of the radical cations formed. Here we shall be concerned with some of the detailed kinetics of the charge-transfer processes relating them to parallel developments in knowledge of

259

double-layer structure and adsorption phenomena in non-aqueous solvents and to homogeneous electron-exchange processes. A particular problem is the location of the pre-electrode site as many of the molecules are rather large compared with solvent and electrolyte ions. Also work under conditions where spontaneous chemical reactions accompany charge-transfer has led to the identification and study of carbanion intermediates which now makes possible a study of the energetics of these and carbonium ion intermediates.

In order to minimize problems arising from coupled chemical reactions the favoured solvents have been the so-called dipolar aprotic solvents[3] in which proton abstraction by the radical anions is usually slow. In this class belong N,N-dimethyl formamide, acetonitrile, dimethylsulphoxide, and sulpholane. Ethereal solvents such as dimethoxyethane and tetrahydrofuran have also been used but even with high concentrations of added quaternary ammonium salts their conductivities remain rather low.

II. EXPERIMENTAL INVESTIGATION OF ELECTRODE KINETICS OF REDUCTION OF CONJUGATED HYDROCARBONS

The earliest kinetic investigations on conjugated hydrocarbons were carried out by Hoytink and co-workers using first AC polarographic techniques[4] and then the impedance bridge techniques[5] first developed by Randles[6] and Ershler.[7]

The electrolytic cell is placed in one arm of an AC impedance bridge and with a few millivolts AC superimposed on the bias potential of the electrode the bridge is balanced. The combination of resistive and capacitive components obtained at balance when an electrode reaction occurs is found to differ from those of the cell resistance and double layer capacity alone. Various means of analysing the measurements lead to the kinetic parameters k_s(app) and α (the transfer coefficient) defined by

$$i = nFAk_s \left\{ C_0 \exp \left[\frac{-\alpha nF}{RT} (E - E^0) \right] - C_R \exp \left[\frac{(1 - \alpha)nF}{RT} (E - E^0) \right] \right\}$$
(1)

where A is the electrode area, C_0 and C_R are the concentrations of the oxidized and reduced forms at the electrode surface and E^0 the standard electrode potential of the couple under investigation.

Aten and Hoytink[5] found that for reduction of a series of polycyclic aromatic hydrocarbons at a dropping mercury electrode in dimethylform-amide solution containing quaternary ammonium salts the experimental results could be analysed satisfactorily to yield rate constants towards the upper limit of the technique. No marked variation with structure was found for the first reduction step but the second addition of one electron was not only apparently a much slower process but showed some marked variations with structure. The authors realized that the second step is complicated by coupled chemical reactions but also remark that apparently the variation in this rate constant is related to the potential at which the addition occurs and this they considered may be related to the structure of the electrical double-layer.

Adams and co-workers[8] made a study of nitrobenzene and perylene reduction on a rotating disc of solid mercury at low temperatures. The experimental problems were formidable and poor agreement with Hoytink's results were obtained. Recent work by Peover and co-workers has returned to the use of the impedance bridge.[9-11]

III. DOUBLE LAYER STRUCTURE AND SPECIFIC ADSORPTION IN NON-AQUEOUS SOLVENTS

It is in general necessary to employ quaternary ammonium salts as background electrolytes because of the negative reduction potentials of the aromatics. In view of the surface activity of these salts in aqueous solution[12] there has always been some reservation, particularly from electrochemists more used to aqueous conditions, on the precise meaning of the kinetic results. Also, the frequency with which aqueous electrochemistry of organic compounds has been beset with problems of specific adsorption[13] suggests a wary approach to any study of their electrode kinetics. However, observations in the DC and AC polarographic behaviour of many aromatic compounds in aprotic solvents strongly suggest qualitatively the absence of strong adsorption phenomena.[2]

Recently, studies of the differential capacity of mercury electrodes, and electrocapillary curves, have shown that solvent molecules such as dimethylformamide,[14,15] dimethylsulphoxide,[14] sulfolane,[16] and acetonitrile[17] are strongly adsorbed on the metal surface. Since specific adsorption of the electroactive material is a displacement process it suggests that organic solutes may not compete very effectively with these solvents. A study of electrocapillary curves of naphthalene and anthracene solutions in dimethylformamide[18] shows much less effect on the surface tension near the electrocapillary maximum than in alcohol. Desorption occurs at slightly negative surface charge and similar results together with the characteristic differential capacity desorption peak have been observed in methylene chloride solutions of pyrene.[19] A direct measure of adsorption can be made by chronocoulometry in which the charge on application of a potential step from below the electrode reaction onto the limiting current region is measured as function of time.[20] Adsorbed electroactive material gives rise to an apparent increase in the double-layer charge at zero time. The results for trans-stilbene reduction in dimethylformamide are shown in Fig. 1 and indicate that neither the

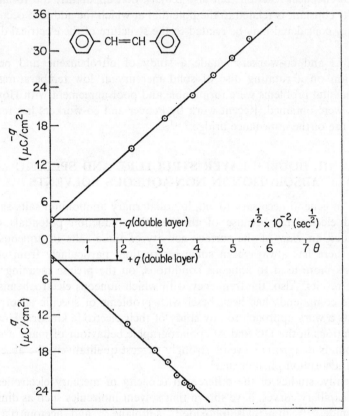

Fig. 1. Chronocoulometry of trans-stilbene in dimethylformamide containing 0·1 M NBu₄ with a double-potential step 0·6 V symmetrical about the redox potential for addition of one electron, illustrating lack of adsorption of neutral molecule and its anion, References 20–21

neutral molecule nor the resulting anion radical is specifically adsorbed under the conditions used.[21] The method is particularly sensitive in non-aqueous solvents as the double-layer capacities are considerably smaller than in water and thus smaller changes in charge due to specifically adsorbed components can be readily detected. On the example given the double layer charging involves about 4 μC/cm² so that with care about 0·4 μC/cm² of adsorbed material could be detected. This result gives support to the view that for fairly high negative values of electrode charge specific adsorption can be neglected for aromatic systems at least in dimethylformamide. This still leaves the problem of the effect of quaternary ammonium ions. Again electrocapillary[15] and differential capacity[9,22] measurements Fig. 2 show that the

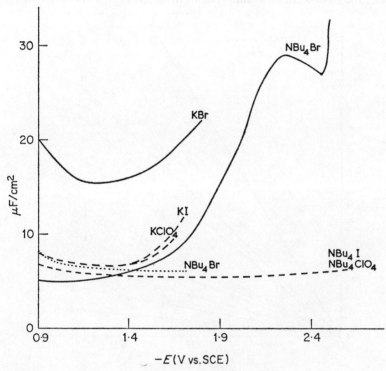

Fig. 2. Comparison of cathodic branches of differential capacities of 0·1 M solutions of inorganic and tetrabutylammonium salts in:
——— water at 25° (schematic curves)
– – – – dimethylformamide at 30°
· · · · · · acetonitrile at 30°
Determined by impedance bridge at a dropping mercury electrode, References 9, 12, 22.

differences in behaviour between the alkali metal cations and organic cations is much less marked in dimethylformamide, dimethylsulphoxide and acetonitrile than in water. It is difficult to make quantitative comparisons because of lack of activity data and problems of uncompensated liquid junction potentials.

A thorough re-study by impedance bridge of reduction of aromatics in dimethylformamide has been carried out by Peover and co-workers[9-11] with particular reference to the internal consistency of the results and the influence of the quaternary ammonium cations. Evaluations of the results via phase angle computation, in which adsorption phenomena show readily, failed to reveal any anomalous behaviour. The behaviour was comparable to that for some of the most ideal inorganic systems in water, Figs. 3 and 4. Results give

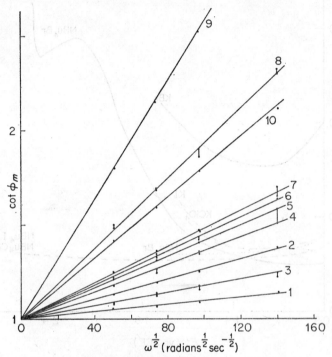

Fig. 3. Dependence of cotangent of phase angle minimum, cot Φ_m, on frequency for one-electron electroreduction in dimethylformamide containing 0·1 M NBu_4^nI at 30°, References 9–10. Key: 1. tetracene; 2. naphthalene; 3. trans-stilbene; 4. mesityl trans-stilbene; 5. α-methyl trans-stilbene; 6. cis-stilbene; 7. α,α′-dideutero cis-stilbene; 8. hexamethyl trans-stilbene; 9. α,α′-diphenyl trans-stilbene; 10. nitro mesitylene.

fair agreement with those of Hoytink, being smaller by a factor of about two.[5] Five fold change in concentration also produced no significant variation in rate constants. A convenient way to examine the influence of the cation on the kinetics is to vary the concentration of base electrolyte. In this way blocking effects can be detected and, if the pre-electrode site is located within the diffuse layer, acceleration of rate may occur through specific adsorption leading to a lowering of the effective electrode charge retardation. Some results are shown in Table 1. The direction of change in apparent rate

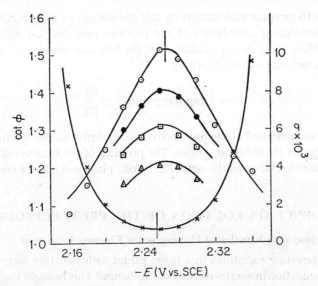

Fig. 4. Dependence of computed phase angle and diffusion polarization resistance (lowest curve) on electrode potential for one-electron reduction of α-methyl trans stilbene in dimethylformamide—0·1 M NBu$_4^n$I at 30°. Curves are theoretical dependence. Frequencies in Hz: O,×, 3125; ●, 1500; □, 870; △, 405; Reference 9. (Reproduced by courtesy of the Faraday Society.)

Table 1. *Effect of inert electrolyte on rate constants k_s (app) for one-electron electro-reduction in dimethylformamide at 30°. The k_s values are calculated at the O.H.P. from diffuse layer theory using concentrations of free ions and assuming no specific adsorption, refs. 5, 9, 29.*

Compound	$-E_{\frac{1}{2}}$ V.vs.SCE	[NBu$_4^n$I] mole/lit.	$-q^M$ μC/cm^2	k_s (app) cm/sec	k_s cm/sec
Tetracene	1·58	0·1	11·5	1·64	20
		0·19		2·19	22
		0·27		2·49	22
		0·43		3·28	25
		0·1(KI)	13·3	∞	∞
cis-Stilbene	2·18	0·1	14·5	0·42	4·2
		0·19		0·65	5·6
		0·43		0·90	6·1
		0·68		1·20	7·2
Hexamethyl-trans stilbene	2·46	0·1	16·1	0·18	2·3
		0·19		0·22	2·2
		0·27		0·245	2·3
		0·43		0·36	2·9
		0·68		0·62	4·2

constant is to increase with increasing salt concentration as expected on the view of diminishing retardation of the reaction predicted on diffuse layer theory. Diffuse layer theory predicts that the true rate constant k_s is related to the experimental value $k_s(\text{app})$ by

$$k_s = k_s(\text{app}) \exp\left(\alpha n - z\right) \frac{F\phi}{RT} \tag{2}$$

where ϕ is the potential at the pre-electrode site relative to the solution and z is the charge on the oxidized species. The problem of the location of the pre-electrode site in relation to the outer Helmholtz plane will now be considered.

IV. EVIDENCE ON LOCATION OF THE PRE-ELECTRODE SITE

1. Comparison with Kinetics of Homogeneous Electron Exchange

Some interesting variations in $k_s(\text{app})$ values with structure were recently found for reduction in a series of hindered stilbenes.[9] This brought the problem of the pre-electrode site into urgent consideration, since if the site is located within the diffuse layer then any variations in location of the site which occur in a series of molecules will give apparent rate variations in addition to those due to reorganization energy. These problems do not enter into considerations of homogeneous electron-exchange between the parent molecule and the anion derived from it. Marcus first pointed out that a simple relation could be expected between the homogeneous rate constant k_{ex} and that for the heterogeneous processes.[23] Table 2 shows values of k_{ex} obtained from esr line-broadening experiments[24] and it is seen that they parallel the heterogeneous rate constants. Thus one can conclude that the pre-electrode site does not vary greatly in the molecules considered. Pursuing the relationship further quantitatively, Marcus predicted one of the form

$$\left(\frac{k_{ex}}{Z_{ex}}\right)^{1/2} = \frac{k_s}{Z_{het}} \tag{3}$$

where Z_{ex} and Z_{het} are the corresponding collision frequencies about 10^{11} mol l.$^{-1}$ sec^{-1} and 10^4 cm sec^{-1} respectively. This relation derives from comparison of the free energies of activation of the two processes[23]

$$\Delta G^*_{het} = -\frac{\alpha^2 e^2}{2}\left(\frac{1}{a_1} - \frac{1}{2\delta}\right)\left(\frac{1}{D_{op}} - \frac{1}{D_s}\right) \tag{4}$$

$$\Delta G^*_{hom} = -\frac{e^2}{4}\left(\frac{1}{2a_1}\right)\left(\frac{1}{D_{op}} - \frac{1}{D_s}\right) \tag{5}$$

Table 2. *Comparison of experimental heterogenous rate constants k_s(app) and k_s values calculated at the O.H.P. with homogeneous electron-exchange rate constants k_{ex} determined by epr experiments, in dimethylformamide containing 0·1 M, NBu_4^nI at room temperature, Refs. 8–10, 24.*

Compound	k_{ex} $\times 10^{-8}$ mol/l./sec	k_s(app) cm/sec	k_s cm/sec	k_{ex}/k_s(app) $\times 10^{-8}$	k_{ex}/k_s $\times 10^{-7}$
trans-Stilbene	10·2 (a)	1·22 (e)	25	8	4
	10·4 (b)			8	4
	7·5 (b, c)	1·84 (c)	23	4	3
Naphthalene	6·0 (a)	1·00	23	6	3
	5·1 (b)			5	2
α-Methyl trans-Stilbene	1·4 (a)	0·43	5	3	3
Hexamethyl trans-Stilbene	0·6 (a)	0·18	2·8	3	2
m-Dinitrobenzene	5·2 (a, d)	4·9	34	1	2

(a) In slow exchange limit.
(b) In fast exchange limit.
(c) In 0·5 M tetrabutyl ammonium iodide.
(d) In 0·01 M tetraethyl ammonium perchlorate.
(e) Aten and Hoytink give k_s(app) 2·5 cm/sec at 25°, Reference 5.

neglecting work terms to bring reactants to the reaction site and remove products therefrom, where α is the transfer coefficient ($\sim$0·5 in all cases we consider here) a_1 is the radius of the reactant in the transition state, δ is the distance from the centre of the molecule to the electrode surface (about equal to a_1), D_{op} and D_s are the optical and local dielectric constants respectively. On the other hand, Hush[25,28] has assumed that the influence of image forces, the term in δ, is negligible, owing to the screening effect of the electrical double layer.

The relation can then be expressed

$$\frac{k_{ex}}{k_s} = \frac{Z_{ex}}{Z_{het}} = \text{constant } 10^7 \tag{6}$$

The fifth column of Table 2 shows that this modified relation is reasonably well obeyed but numerical agreement is better if the rate constants are corrected for double layer effects. The validity of this procedure is considered below. A similar relation will hold if the molecule is specifically adsorbed,[8] but this has been shown not to be the case.[21]

2. Salt Effects on Electrode Kinetics in Non-aqueous Solvents

While comparison of homogenous and heterogenous kinetics is very useful in evaluating the importance of specifically electrochemical influences on kinetics the results quoted above merely serve to show that these influences do not vary greatly in the molecules considered, as would be the case if they lay outside the diffuse layer. Salt effects were mentioned earlier in connection with investigations of the blocking action of quaternary ammonium salts, Table 1; these suggest a double-layer influence of the normal Frumkin type, eqn (2), but another reason why increase in salt concentration could increase the rate constant is by formation of ion-pairs between the aromatic anions and the supportingly electrolyte cations. Measurement of E^0 values with a series of cations of different size will show up ion-associations since the latter shift E^0 values to more positive potentials. The diffusion controlled polarographic half-wave potentials for tetracene reduction in 0·1 M perchlorate solutions of Li^+, Na^+, K^+, and NBu_4^{n+} in dimethylformamide are respectively $-1·564$, $-1·563$, $-1·564$, and $-1·557$ V against a saturated calomel electrode.[26] Neglecting liquid-junction potentials some ion-pairing with NBu_4^{n+} appears to occur. The association constant is about 4 l. mole^{-1} giving about 80 per cent and 60 per cent free ions in 0·1 M and 0·4 M solutions of the salt over which range the value of k_s(app) doubles. It seems unlikely that this degree of ion-association could lead to the observed rate changes. At this point comparison with homogeneous exchange rates can be invoked since ion-association would affect these also. A comparison has been made for trans-stilbene reduction,[24] Table 2, showing a slight decrease in exchange rate with increase in salt concentration, i.e. in a direction opposite to that found in the electrode kinetics. Thus it is reasonably clear that double-layer influences are operative in the kinetic behaviour of the compounds we have so far considered under the conditions employed. A correction, eqn (2), should therefore be applied to the kinetic parameters. If the pre-electrode site is the outer Helmholtz plane then the appropriate equation is

$$k_s(\text{app}) = k_s \, e^{-(\alpha n - z)\phi_2 F/RT} \tag{7}$$

where ϕ_2 values can be calculated from Gouy-Chapman theory on the assumption that no specific adsorption of the quaternary ammonium ion occurs. As can be seen in Table 1 this correction results in a reasonably constant value of k_s, independent of salt concentration, but with such large corrections the method is rather insensitive. It is also possible that specific adsorption of quaternary ammonium ions occurs as the concentration increases leading to a decrease in the effective electrode charge and hence an increase in k_s(app). The abnormally high rates found in 0·68 M tetrabutylammonium solutions may be a reflection of this. The effect of using corrected values of k_s on comparisons with k_{ex} values is also shown in Table 2, with some improvement in constancy of k_{ex}/k_s. Models show that stilbene molecules may be located further out in the diffuse layer than the outer Helmholtz plane at about 9 Å from the electrode surface[27] where the ϕ potential is 60–70 per cent of ϕ_2.

There is some evidence from early work of Hush and Blackledge[28] on polarographic reduction potentials (presumably not effectively E^0 values) in alcoholic solutions of different salt concentrations that when larger rod-like molecules are considered the double-layer influence does disappear, Table 3, since the reduction potential becomes independent of salt concentration.

Table 3. *Effect of change in electrolyte concentration on rate of reduction in alcoholic solution, Ref. 28.*

Compound	$E_{\frac{1}{2}}(1 \cdot 0 \text{ M NEt}_4\text{Br}) - E_{\frac{1}{2}}(0 \cdot 1 \text{ M NEt}_4\text{Br})$ (V)
Distyryl	$+0 \cdot 057$
Anthracene	$+0 \cdot 111$
Diphenyldiacetylene	$-0 \cdot 001$

Also, when the supporting electrolyte cation is made smaller than a quaternary ammonium ion the outer Helmholtz plane will be located nearer the electrode surface plane and the electroactive species may then be even further out in the diffuse layer. This appears to happen with tetracene reduction in dimethylformamide containing potassium iodide[24,29] because a diffusion controlled process is indicated at the highest AC frequencies used, c.f. Table 1, despite higher values of electrode surface charge and hence of ϕ_2. In summary, a much clearer picture of the kinetics of electroreduction of aromatic systems at potentials well removed from the point of zero charge now emerges. What ambiguities remain concern the exact location of the pre-electrode site within the diffuse layer and the extent to which specific adsorption of quaternary ammonium ions modifies calculations of ϕ potentials at the site. A possible approach to this problem may be through a well defined pre-electrode site determined by the stereochemistry of the molecule. The relation of the kinetic data to molecular structure is discussed by Hale in Chapter 4 who assumes no double-layer effects; this as we have seen is an approximation. Also, now that we can see that kinetic behaviour of many aromatics in organic aprotic solvents is likely to be close to ideality the kinetics could be used as a probe to double-layer structure particularly as a wide range of reduction potentials can be covered by suitable small changes in the structure of the probe molecule.

3. Reduction of Organic Anions

If anions are located in the diffuse layer then inhibition of their reduction will be much greater than for neutral molecules. Hoytink found this to be so for polycyclic aromatic hydrocarbon anions but the results were complicated by coupled chemical reactions.[5] Smith[30] has shown that if the AC impedance measurements are evaluated via phase angle computation the chemical reaction has little influence if slow compared with AC frequency. However the AC response was found to be too poor to pursue this line except for dibiphenylene ethylene—already examined by Hoytink—and m-dinitrobenzene anions in which in any case cyclic voltammetry shows the dinegative ion to be reasonably stable.[24] The results in Table 4 show that the negative ion

Table 4. *Comparison of rate constants for electro-reduction of neutral aromatic molecules and their anions, in dimethylformamide containing* 0.1 M NBu_4^+I *at* $30°$ $k_s(app)$ *is the experimental rate constant,* $f(\phi_2)$ *is the retardation at the O.H.P. calculated on diffuse layer theory, Ref. 11.*

Oxidized form of Reactant	$-E_{\frac{1}{2}}$ V.vs.SCE	$-q^M$ $\mu C/cm^2$	$k_s(app)$ cm/sec	$f(\phi_2)$ (a)	$f(\phi_2/2)$
Dibiphenylene ethylene	0·93	7·8	5	—	—
radical anion	1·43	10·6	1·2 (b)	—	—
m-Dinitrobenzene	0·80	6·8	4·9	7	2·6
radical anion	1·23	9·5	0·68	614	29

(a) Retardation factor due to double-layer calculated at the outer Helmholtz plane.
(b) Aten and Hoytink give $k_s(app) = 1.7$ cm/sec.

reductions are about an order of magnitude slower than those of the neutral molecule.

Marcus' theory predicts about the same absolute rate for anion reduction as the neutral molecule reduction and diffuse layer theory predicts about two orders of magnitude difference in retardation. One interpretation of the results is that only about 50 per cent of the ϕ_2 potential is operative, ϕ_2 being calculated assuming no specific adsorption of quaternary ammonium cation. The result is sufficiently similar to that discussed earlier for location of the pre-electrode site for the neutral molecule based on models to suggest that the sites are much the same for the anion reduction as the neutral molecule reduction.

However both the anions are reduced at rather similar potentials and a much wider range needs to be covered to survey this problem adequately. Also, ion-association may also be greater with dinegative ions and influence the exchange rates additionally.

V. CARBANION INTERMEDIATES IN ELECTRO-REDUCTION OF POLYCYCLIC AROMATIC HYDROCARBONS

It was mentioned earlier that the addition of a second electron to an aromatic system is often accompanied by a coupled chemical reaction even in nominally aprotic conditions. The effect of this is most easily seen in cyclic voltammetry. Limitation of the potential sweep to the first electron addition peak usually results in a stable redox system, Fig. 5, but if the second electron

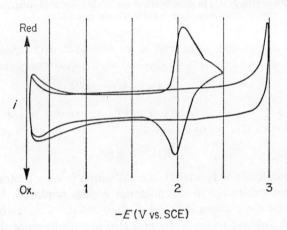

Fig. 5. Steady-state cyclic voltammograms of background electrolyte and first one-electron reduction step of pyrene at a Pt electrode in 0·1 M $NBu_4^+ClO_4$ in dimethylformamide, frequency 50 Hz, Reference 31. (Reproduced by courtesy of the Faraday Society.)

addition is brought in there is little anodic current associated with it, Fig. 6. Interestingly, a new redox system occurs at potentials more positive than that of the original one-electron addition.[31] This behaviour occurs in a wide range of solvents, including normally unreactive examples such as dimethoxyethane, and dihydro derivatives can be isolated. If a proton donor is deliberately added to the solution, then the new redox system disappears. The observations have been rationalized as follows. In nominally aprotic solvents the dinegative ion formed from the aromatic normally abstracts a single proton

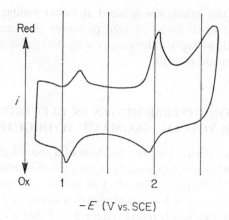

$$-E \text{ (V vs. SCE)}$$

Fig. 6. Steady-state cyclic voltammogram of two-electron reduction of pyrene at a Pt electrode in 0·1 M NBu$_4^n$ClO$_4$ in dimethylformamide showing the formation of a new redox system near $-1·0$ V. Frequency 50 Hz.

either from the solvent, or at least in the case of dimethoxyethane in which dinegative ions are known to be stable from the quaternary ammonium ion:

$$R^{2-} + HX \rightarrow RH^- + X^-$$

The entity RH^- is a carbonium ion and the new redox system is ascribed to the oxidation of this ion to a radical

$$RH^- \leftrightarrow RH^0$$

When a proton donor is present, the carbanion is fully protonated to RH_2 which is not oxidizable in the potential region employed. If these interpretations are correct then the oxidation potentials of the carbanions are of considerable interest theoretically and also practically since the carbanions are intermediates in base catalysed aereal oxidation of hydrocarbons such as fluorene.[32]

When a proton is added to the hydrocarbon dinegative ion a carbon atom is removed from the conjugated system. For alternant aromatic hydrocarbons Hückel theory predicts that the ionization potential of the protonated dianions should be independent of the hydrocarbon since the highest occupied orbital is then the non-bonding molecular orbital of energy independent of the hydrocarbon skeleton. Table 5 shows how closely these predictions are fulfilled in the corresponding oxidation potentials.

The results also show that the oxidation potential is essentially independent of solvent and so the process is unlikely to be due to a solvent fragment, while tributylamine or butene, possible fragments of the tetrabutylammonium

Table 5. *One-electron oxidation-reduction potentials at a platinum electrode of some polycyclic aromatic hydrocarbons and carbanions derived therefrom in various solvents containing* 0·1 M $NBu_4^nClO_4$, *Ref.* 31.

Compound	$-E_{\frac{1}{2}}(R \leftrightarrow R^-)$ V.vs.SCE	$-E_{\frac{1}{2}}(RH^- \leftrightarrow RH)$ V.vs.SCE	Solvent (a)
Pyrene	2·05	1·15	DMF
	2·05 (b)	1·17 (b)	DMF
	2·05	1·13	CH_3CN
	2·08	1·13	DME
Coronene	2·05	1·10	DMF
	2·05 (b)	1·15 (b)	DMF
	2·10	1·18	CH_3CN
1,2 Benzanthracene	1·95	1·12	DMF
Perylene	1·68	1·12	DMF
Tetracene	1·58	1·08	DMF
Fluoranthene	1·72	0·85	DMF
	1·75	0·87	CH_3CN
	1·70 (b)	0·80 (b)	DME
	1·70	0·85	DMSO
Acenaphthylene	1·60	0·78	DMF

(a) DMF = NN — dimethylformamide, DME = dimethoxyethane containing 0·5 M NBu_4ClO_4, DMSO = dimethylsulphoxide.
(b) At a mercury-film electrode.

cation, are not oxidized at this potential. Table 5 also contains results for two non-alternant hydrocarbons fluoranthene and acenaphthylene, in which the structure of the carbanions are thought to be (a) and (b) from considerations of positions of highest charge density in the dinegative ions. It is often held that these are the positions at which proton addition will occur

(a)   (b)

The most loosely bound electrons in (a) and (b) occupy formally bonding orbitals with Hückel coefficients 0·1317 and 0·2240 respectively which means that their oxidation potentials should be more positive than those involving non-bonding orbitals. This is found experimentally, Table 5. The oxidation potentials do of course contain a contribution from the solvation energy of the carbanion but frequently it is found that in a series of similar molecules variations are not sufficient to invalidate comparisons.[33] One can go further and derive directly the gas phase ionization potentials of the carbanions from the simple assumption that the solvation energies of the carbanion RH^- and radical anion R^- are the same so that the difference in the redox potentials for these two systems gives directly the difference between the gas phase electron affinity of R and that of RH^0 (or between the ionization potential of R^- and that of RH^- which is the same thing).

The results are shown in Table 6 and now the similarity in A_{RH} is less striking. This can be ascribed to neglect of electron repulsion in Hückel treatment. Using the self-consistent field methods of Pople,[34] with the repulsion integrals of Hush and Pople,[35] the third column of figures in Table 6 is obtained

Table 6. *Calculated electron affinities* $A_{RH}(SCF)$ *of some protonated hydrocarbon anions compared with values* A_{RH} *derived from gas phase electron affinities* A_R *and oxidation-reduction potentials, Ref.* 31.

Compound	A_R	A_{RH^0}	A_{RH^0} (SCF)
Anthracene	0·56	—	1·31
Pyrene	0·59	1·49	1·80
Coronene	0·60	1·55	1·56
1,2 Benzanthracene	0·63	1·46	1·56
3,4 Benzopyrene	0·63	—	1·80
Perylene	0·93	1·49	1·88
Tetracene	1·01	1·51	1·46

which show fair agreement with experiment, except for perylene and pyrene. A recent equilibrium study of the dinegative ion—carbanion systems by Velthorst and Hoytink[36] has shown that it may be necessary to take into account disproportionation equilibria in assigning the correct structure to the carbanion. These authors made the dinegative ions by electron transfer with alkali metals in tetrahydrofuran and protonation was accomplished by addition of the dihydrocompound or stearyl alcohol. The reaction was followed spectroscopically. The carbanions from pyrene and perylene in particular were found to disproportionate according to the following equilibria

$$2RH^- \rightleftharpoons RH_2 + R^{2-}$$
$$2RH^- \rightleftharpoons RH_2^- + R^-$$
$$2RH^- \rightleftharpoons RH_2^{2-} + R$$

If these equilibria are sufficiently rapid under the electrochemical conditions, then protonation of RH_2^{2-} will certainly occur to give the carbanion RH_3^-. Provided the latter is stable to disproportionation the observed carbanion oxidation potential is then due to RH_3^- and not to RH^-. Dietz[37] has recently calculated the electron affinities of some possible structures for the radicals RH_3^0 from pyrene and perylene. Rather better agreement with experiment is

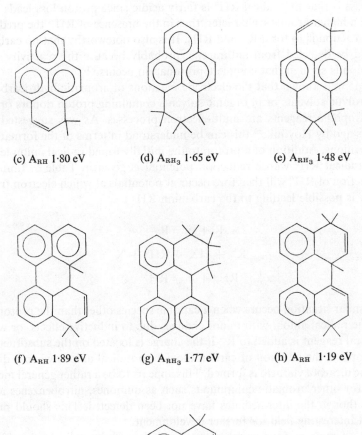

(c) A_{RH} 1·80 eV   (d) A_{RH_3} 1·65 eV   (e) A_{RH_3} 1·48 eV

(f) A_{RH} 1·89 eV   (g) A_{RH_3} 1·77 eV   (h) A_{RH} 1·19 eV

(i) A_{RH_3} 1·46 eV

10

found using structures (e) and (i) but it should be stressed that there is no other positive evidence for disproportionation under the experimental conditions used to determine the solution potentials. An attempt has been made to identify the structure of the carbanions from the electron-spin resonance spectrum of the radicals produced by electrolytic reduction of the carbanions in a flow cell,[38] without success. One possible reason for lack of success is that the radical RH^0 is fairly acidic since proton loss leads to R^- with a higher aromatic character, thus in the presence of RH^- the predominant reaction is to form R^- and RH_2. It is also noteworthy that no carbanion could be detected from anthracene, probably because the reactivity of the dianion is so high that complete protonation occurs.[31]

It is well known that the electrode reactions of aromatic hydrocarbons in hydrolytic solvents or in organic solvents containing proton donors or other electrophilic reagents are multi-electron processes. As was suggested some years ago by Hoytink,[39] this can be understood in terms of the formation of carbanions. Addition of a proton to the initially found radical anion leads to the radical RH^0 whose reduction potential is given in Table 5. Immediate reduction of RH^0 will therefore occur at potentials at which electron transfer to R is possible leading to the carbanion RH^-:

$$R + e \rightleftharpoons R^-$$

$$R^- + HX \rightarrow RH^0 + X^-$$

$$RH^0 + e \rightleftharpoons RH^-$$

A similar situation occurs when a cationic species other than the proton adds to the radical anion, with minor changes due to inductive effects, or when a neutral reagent is added to R^- if the charge is located on the substituent. An example is the addition of carbon dioxide to radical anions when a dianion of the dicarboxylic acid is formed. This appears to be a rather general mechanism for other aromatic compounds such as quinones, nitrobenzenes and so on,[2] though the intermediates have not been detected. This should prove a most interesting field for further development.

In a few cases, for example tetraphenylallene,[40] the radical anions formed at an electrode are so reactive that the above sequence takes place spontaneously even in dimethoxyethane. In these cases the radical RH^0 is not so acidic that proton loss easily occurs and in the case of the tetraphenyl allene carbanion oxidation to a detectable radical was possible, confirming the general mechanism.

The reason why carbanions undergo aerial oxidation readily in organic solvents is easily understood knowing that the reduction potential of oxygen under those conditions is in the region -0.75 to -0.86 V vs. SCE (cf. Table 5). Thus electron transfer to oxygen will occur readily giving the radical RH^0 which can undergo further reaction with oxygen by a radical chain mechanism.[32]

The extent to which carbanions occur as intermediates can be judged from the following scheme for derivatives of tetraphenylallene (j) in which uv and esr spectroscopy were used to confirm carbanion formation by proton addition, base catalysed ionization and elimination reactions:[40]

$$Ph_2C\!\!=\!\!C\!\!=\!\!Ph_2 \qquad\qquad Ph_2C\!\!=\!\!CH\!-\!CPh_2$$

$$\underset{2e+H^+}{\overset{-2.11\,V}{\searrow}} \qquad\qquad \overset{-0.95\,V}{\swarrow}$$

$$Ph_2C\!\!=\!\!CH\!-\!\bar{C}Ph_2$$

$$\underset{H^+}{\overset{Base}{\rightleftharpoons}} \qquad\qquad \underset{-2.2\,V}{\overset{-OEt}{\leftharpoonup}}$$

$$Ph_2C\!\!=\!\!CH\!-\!CHPh_2 \qquad\qquad Ph_2C\!\!=\!\!CH\!-\!CPh_2$$
$$\text{(j)} \qquad\qquad\qquad\qquad\qquad | $$
$$\qquad\qquad\qquad\qquad\qquad\qquad OEt$$

VI. CARBONIUM ION INTERMEDIATES IN THE ELECTRO-OXIDATION OF POLYCYCLIC AROMATIC HYDROCARBONS

The electro oxidation process of aromatic compounds are much less well understood than the corresponding reduction processes, largely because of the high reactivity of the cationic intermediates. Except for compounds in which reactive positions are blocked, such as 9,10-diphenyl anthracene and rubrene, oxidation in solvents such as acetonitrile as nitrobenzene does not stop at the formation of the radical cation.[2,41-42] However because of the symmetry between the filled and unfilled orbitals in unsubstituted alternant hydrocarbons some rationalization of the behaviour is possible. Thus just as when a proton is added to a radical anion the resulting molecule has its odd electron in the non-bonding molecular orbital, of energy independent of carbon skeleton to a first approximation, so when an anion is added to a radical cation the resulting molecule has its odd electron in the same non-bonding molecule orbital (neglecting the small inductive effect of the substituent):

$$R^- + H^+ \rightarrow RH^0$$
$$R^+ + X^- \rightarrow RX^0$$

Similarly, as the subsequent reduction behaviour is determined by the reduction potential of RH^0 leading to the carbanion RH^- we should now consider the oxidation behaviour of RX^0 anticipating it will lead to the formation of the carbonium ion RX^+ at potentials at which R^+ is formed. This can be done by a theoretical study and also by making use of the fact that in strong acids protonation of aromatic hydrocarbons occurs to form the proton complexes RH^+ similar in structure to RX^+. The experimental study of the latter in relation to oxidation behaviour is analogous to the study of RH^- formed by proton addition to R^{2-} in relation to reduction behaviour.

Aalbersberg and Mackor[43] studied the polarographic reduction of species produced from some polycyclic aromatic hydrocarbons in trifluoroacetic acid containing boron trifluoride. Under these conditions both the proton complexes RH^+ and stable radical cations R^+ are formed. Thus both reduction of R^+ and RH^+ could be observed, analogous to oxidation of R^- and RH^-. As anticipated, the radical RH^0 is oxidized more easily than R by 0·4–0·7 V. On the assumption that the solvation energy of R^+ and RH^+ is the same, the difference between their half-wave potentials will be independent of solvent and in this way the values of $E_{1/2}(RH^+ \leftrightarrow RH)$ in Table 7 have been derived

Table 7. *Oxidation-reduction potentials of radicals RH^0 formed from polycyclic aromatic hydrocarbons in acetonitrile containing* 0·1 M $NBu_4^nClO_4$ *derived from reduction potentials of their proton complexes, and derived and theoretical ionization potentials* I_{RH} *and* $I_{RH}(SCF)$, *Refs.* 2, 31, 43.

Compound	$E_{1/2}(R \leftrightarrow R^+)$ V.vs.SCE	$E_{1/2}(RH^+ \leftrightarrow RH^0)$ V.vs.SCE	I_R eV	I_{RH^0} eV	$I_{RH^0}(SCF)$ eV
Anthracene	1·29 (a)	0·78	7·43	6·92	7·05
Pyrene	1·42 (a)	0·45	7·53	6·56 (a)	6·56
1,2 Benzan-thracene	1·39 (a)	0·66	7·75	7·02 (c)	6·80
3,4 Benzo-pyrene	1·10 (b)	0·42	7·20	6·36	6·56
Perylene	1·04 (b)	0·47	7·07	6·33	6·48
Tetracene	0·84 (b)	0·76	6·96	6·54	6·90

(a) At high sweep rates to eliminate chemical reactions.
(b) Some kinetic shift due to chemical reactions.
(c) Using comparison of $E_{1/2}$ values in acetonitrile.

for acetonitrile solution, a solvent commonly used in oxidation studies. There is more scatter in the values for carbonium ion reduction than for the corresponding carbanion oxidation, but the agreement between derived ionization potentials and the calculated[31] values for the radicals RH^0 is similar to that between the electron affinities for the radicals RH^0, Table 6. Again we would expect that addition of a neutral reagent to R^+ will give an intermediate with similar energy to RH^0 provided the charge is localized in the substituent. An example of this is the addition of pyridine to anthracene radical cation which gives the dipyridinium compound of the dication.[44]

Attempts to detect the formation of carbonium ions during electro oxidation of polycyclic aromatic hydrocarbons in non-acidic solvents have not been successful so far, presumably because when reaction of the radical cation with the solvent or nucleophiles occurs reaction does not stop appreciably at the carbonium ion stage but a second addition occurs. This is similar to what happens when proton addition takes place on addition of acids during the reduction process, the carbanion oxidation is eliminated during cyclic voltammetry.[31] However with a substituted aromatic giving a stable radical cation but a reactive dication, 9,10-diphenyl anthracene, in a weakly nucleophilic solvent ethyl methane sulphonate a new cathodic peak cathodic of the first oxidation is detected by cyclic voltammetry[45] when the second oxidation step

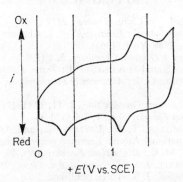

Fig. 7. Steady-state cyclic voltammogram of two-electron oxidation of 9,10-diphenyl anthracene at a Pt electrode in 0·1 M $NBu_4^nClO_4$ in ethyl methane sulphonate, showing the formation of a new reduction peak near +0·6 V. Frequency 40 Hz (compare Fig. 6).

is brought in, Fig. 7. Here it seems probable that the sequence is analogous to that occurring in the reduction process i.e.

$$\text{oxidation} \qquad\qquad \text{reduction}$$

$$R \rightleftharpoons R^+ + e \qquad\qquad R + e \rightleftharpoons R^-$$
$$R^+ \rightleftharpoons R^{2+} + e \qquad\qquad R^- + e \rightleftharpoons R^{2-}$$
$$R^{2+} + X \rightarrow RX^+ \qquad\qquad R^{2-} + HS \rightarrow RH^-$$
$$RX^+ + e \rightarrow RX^0 \qquad\qquad RH^- \rightarrow RH^0 + e$$

The reduction potential of this radical RH^0 occurs near $-1·0$ V in agreement with values in Table 5 and its oxidation near $+0·6$ V (or rather the reduction of RX^+) in agreement with values in Table 6. Further work is necessary to establish the correctness of the assignment of the reduction peak to that for RX^+. Recent work has confirmed this assignment.[45]

During oxidations of polycyclic aromatic hydrocarbons in solvents such as acetonitrile, the reduction can exceed two electrons/molecule. What may be happening here is that the adduct $(RX_2)^{2+}$ in which two carbon atoms are removed from the conjugated system undergoes loss of two protons with consequent aromatization to the disubstituted aromatic RX_2 which may be capable of further oxidation. If the nucleophile X contains p-electrons then the aromatic RX_2 will have a larger π-system than R and will certainly be more easily oxidized. There is experimental evidence to show that a common nucleophile in these reactions is residual water in the solvent which gives rise to oxygen-containing products.[46-47]

REFERENCES

1. GUILLANTON, G. LE: *Bull. soc. Chim.* (*France*), 2359 (1963).
2. PEOVER, M. E.: *Electroanalytical Chemistry*, Vol. 2, Chapter 1, Edited by A. J. Bard (Marcel Dekker, Inc., New York, 1967).
3. PARKER, A. J.: *Advances in Organic Chemistry*, **5**, 1 (1965).
4. ATEN, A. C., BUTHKER, C., and HOYTINK, G. J.: *Trans. Faraday Soc.*, **55**, 324 (1959).
5. ATEN, A. C., and HOYTINK, G. J.: *Advances in Polarography*, 777 (Pergamon, Oxford, 1961).
6. RANDLES, J. E. B.: *Discussions Faraday Soc.*, **1**, 11, 47 (1947).
7. ERSHLER, B.: *Discussions Faraday Soc.*, **1**, 269 (1947).
8. MALACHESKY, P. A., MILLER, T. A., LAYLOFF, T., and ADAMS, R. N.: *Exchange Reactions*, 157 (International Atomic Energy Agency, Vienna, 1965).
9. DIETZ, R., and PEOVER, M. E.: *Discussions Faraday Soc.*, **45**, 154 (1968).
10. PEOVER, M. E., and POWELL, J. S.: *J. Electroanalyt. Chem.*, **20**, 427 (1969).
11. PEOVER, M. E., and POWELL, J. S.: To be published.
12. DAMASKIN, B. B., and NIKOLAEVA-FEDOROVICH, N. V.: *Russ. J. Phys. chem.*, **35**, 627 (1961).
13. DELAHAY, P.: *Double Layer and Electrode Kinetics*, Chapter 11 (Interscience, New York, 1965J.
14. PAYNE, R.: *J. Amer. Chem. Soc.* **89**, 489 (1967).
15. BEZUGLYI, V. D., and KORSHIKOV, L. A.: *Soviet Electrochem.*, **1**, 1279 (1965).
16. PARSONS, R.: *Trans. Faraday Soc.*, **64**, 751 (1968).
17. PAYNE, R.: *J. Phys. Chem.*, **71**, 1548 (1967).
18. BEZUGLYI, V. D., and KORSHIKOV, L. A.: *Soviet Electrochem.*, **4**, 283 (1968).
19. PEOVER, M. E.: *Trans. Faraday Soc.*, **60**, 417 (1964).
20. CHRISTIE, J. H., OSTERYOUNG, R. A., and ANSON, F. C.: *J. Electroanalyt. Chem.*, **13**, 236 (1967).
21. LUND, W., and PEOVER, M. E.: To be published.
22. PEOVER, M. E., and POWELL, J. S.: To be published.
23. MARCUS, R. A.: *J. Chem. Physics*, **43**, 679 (1965).
24. FORNO, A. E. J., PEOVER, M. E., and WILSON, R.: *Trans. Faraday Soc.*, **66**, 1322 (1970).
25. HUSH, N. S.: *Electrochim. Acta*, **13**, 1005 (1968).
26. PEOVER, M. E.: *Discussions Faraday Soc.*, **45**, 186 (1968).
27. PEOVER, M. E.: *Discussions Faraday Soc.*, **45**, 183 (1968).

28. HUSH, N. S.: *Discussions Faraday Soc.*, **45**, 185 (1968).
29. PEOVER, M. E.: *Discussions Faraday Soc.*, **45**, 184 (1968).
30. SMITH, D. E.: *Electroanalytical Chemistry*, Vol. 1, Chapter 1, p. 50, Edited by A. J. Bard (Marcel Dekker, Inc., New York, 1966).
31. DIETZ, R., and PEOVER, M. E.: *Trans. Faraday Soc.*, **62**, 3535 (1966).
32. RUSSELL, G. A., ZANTZEN, E. G., BERNIS, A. G., GEELS, E. J., MOYE, A. J., MAK, S., and STRAN, E. T.: *Advances in Chemistry*, **51**, 112 (1965).
33. HOYTINK, G. J.: *Rec. Trav. Chim.*, **77**, 555 (1958).
34. POPLE, J. A.: *J. Physic. Chem.*, **61**, 6 (1957).
35. HUSH, N. S., and POPLE, J. A.: *Trans. Faraday Soc.*, **51**, 600 (1955).
36. VELTHORST, N. H., and HOYTINK, G. J.: *J. Amer. Chem. Soc.*, **89**, 209 (1967).
37. DIETZ, R.: Personal communication (1968).
38. FORNO, A. E. J.: *Chem. and Ind.*, 1728 (1968).
39. HOYTINK, G. J., SCHOOTEN, J. VAN, BOER, E. DE, and AALBERSBERG, W. Y.: *Rec. Trav. Chim.*, **73**, 355 (1954).
40. DIETZ, R., PEOVER, M. E., and WILSON, R.: *J. Chem. Soc.* (B), 75 (1968).
41. MARCOUX, L. S., FRITSCH, J. M., and ADAMS, R. N.: *J. Amer. Chem. Soc.*, **89**, 5766 (1967).
42. PHELPS, J., SAUTHANAM, K. S. V., and BARD, A. J.: *J. Amer. Chem. Soc.*, **89**, 1752 (1967).
43. AALBERSBERG, W. Y., and MACKOR, E. L.: *Trans. Faraday Soc.*, **56**, 1351 (1960).
44. LUND, H.: *Acta Chim. Scand.*, **11**, 491, 1323 (1957).
45. DIETZ, R., and LARCOMBE, B. E.: *J. Chem. Soc.* (B), 1369 (1970).
46. MAJESKI, E. J., STUART, J. D., and OHNESORGE, W. E.: *J. Amer. Chem. Soc.*, **90**, 633 (1968).
47. SIODA, R. E.: *J. Phys. Chem.*, **72**, 2322 (1968).

REDUCTION POTENTIALS AND ORBITAL ENERGIES OF AZA-HETEROMOLECULES

B. J. Tabner and J. R. Yandle

I. INTRODUCTION

The chemistry of heterocyclic compounds is of great interest in the fields of biochemical and pharmacological processes, particularly the study of vitamins and alkaloids. Consequently, over the last two decades, a very considerable advance in heterocyclic chemistry has occurred, of which the electrochemical behaviour of this group of compounds is by no means the least important.

Originally, it was believed that the polarographic characteristics of heterocyclic compounds followed closely those of aromatic hydrocarbons. Although

283

this concept is still basically true other factors are now also known to be important, such as adsorption of the heterocyclic compound on the mercury surface and the proton-accepting properties of the reduced heterocyclic molecule.

In view of the large number of families of heterocyclics and of the variety of conditions under which their electrochemical and polarographic properties have been measured, a critical assessment of the present literature would be a major task. It was therefore decided to restrict the present discussion to π-deficient heterocyclics and to aza-heteromolecules in particular, as this group appeared to be the least thoroughly investigated.

II. THE NATURE OF THE POLAROGRAPHIC REDUCTION

The polarographic reduction of organic molecules may be either reversible or irreversible and it is therefore important that the nature of the reduction should be established. Consequently, three criteria are usually applied to determine reversibility:

(i) at 20°C a plot of E (the applied potential) vs. log $[(i_d - i)/i]$ should give a straight line of slope $0.058/n$;

(ii) the half-wave potentials of the oxidant and reductant should be identical and the same as that determined potentiometrically;

(iii) the instantaneous current-time curves, for a single drop, should follow the form given by the Ilkovič equation,[1] i.e. $i = kt^a$ with $a = 0.192$.

A knowledge of the number of electrons, n, involved in the reduction is also required. In addition to product analysis, several methods exist to conveniently determine n from a study of the reduction wave itself. For reversible reductions a value may be obtained:

(a) directly from the Ilkovič equation;[1]

(b) from the slope of the logarithmic plot mentioned in (i) above; and finally

(c) from the Toměs criterion[2] which gives a value for $E_{1/4} - E_{3/4}$ of $0.0554/n$ (at 20°C).

The application of the three criteria mentioned above, together with a knowledge of the number of electrons involved in the reduction, is usually sufficient to establish the main features of the reduction mechanism.

III. THE EFFECT OF SOLVENT ON THE REDUCTION MECHANISM

The nature of the solvent is an important factor in determining the mechanism of the reduction. In the absence of proton donors (e.g. reductions studied in dimethylformamide, acetonitrile, and dimethyl-sulphoxide), the reduction follows the simple mechanism:[3-6]

$$R + e \rightleftarrows (R)^- \quad (1)$$

$$(R)^- + e \rightleftarrows (R)^= \quad (2)$$

where R denotes the reactant molecule and $(R)^-$ and $(R)^=$ the successive products of the reduction. In general two reduction waves are observed, corresponding to eqns (1) and (2), and since both steps are one electron reductions, the waves are of equal height.

However, in the presence of a proton donor (e.g. reductions studied in H_2O-dioxan mixtures and 2-methoxyethanol) a different mechanism is observed.[3,7] The product of the reduction given in eqn (1) is immediately protonated to give the corresponding radical and the reduction becomes irreversible:

$$(R)^- + H^+ \rightarrow RH^{\cdot} \quad (3)$$

It is relevant at this stage of the mechanism to consider the electron affinity of $RH^{\cdot}$, i.e.

$$RH^{\cdot} + e \rightarrow RH^- \quad (4)$$

If the electron affinity of $RH^{\cdot}$ is less than that of R, then it will be reduced at a more negative potential. On the other hand, if the electron affinity of $RH^{\cdot}$ is greater than that of R, it will accept a further electron at the same potential (eqn (4)). Molecular orbital calculations have shown that for a large range of aromatic hydrocarbons the electron affinity of $RH^{\cdot}$ was greater than that of R.[8] In such cases, the height of the first reduction wave corresponds to the irreversible addition of two electrons and further protonation usually occurs:

$$RH^- + H^+ \rightarrow RH_2 \quad (5)$$

The change in reduction mechanism has been illustrated rather effectively by the addition of a proton donor, such as phenol, to an aprotic solvent.[7] It was shown that in dimethylformamide the reduction of anthracene followed the mechanism given in eqns (1) and (2). Two waves were observed of equal height. When phenol was added to the solution the height of the first wave increased at the expense of the second wave. Finally, only the first wave remained corresponding to the irreversible addition of two electrons, and the mechanism becomes that given by eqns (1), (3), (4), and (5). In this particular instance, it appears that the reaction of $(R)^-$ may be with the acid itself rather than with solvated protons. In a few cases the product RH_2 may itself be reduced to $(RH_2)^-$ and a new wave may appear at a more negative potential.

IV. ELECTRON SPIN RESONANCE SPECTROSCOPY AS AN AID TO ESTABLISHING THE REDUCTION MECHANISM

The polarographic reduction of an aromatic hydrocarbon in dimethylformamide follows the reversible mechanism given in eqns (1) and (2).[4-7] In particular, the first step for naphthalene is

$$\text{naphthalene} + e \rightleftarrows (\text{naphthalene})^- \tag{6}$$

which clearly parallels the reduction of hydrocarbons with alkali metals in aprotic solvents such as tetrahydrofuran:[9]

$$\text{naphthalene} + Na \rightleftarrows (\text{naphthalene})^- Na^+ \tag{7}$$

In both cases the resulting product should be the radical anion. The relationship between these two methods of reduction has been confirmed not only by product analysis but also by a comparison of the optical spectra and analysis of the electron spin resonance (esr) spectra.[10,11]

Electron spin resonance techniques have been increasingly used to study the above relationship. Nitrobenzene was chosen for an initial study.[12] The polarographic reduction of nitrobenzene in acetonitrile (0·1 M tetra-n-propyl ammonium perchlorate as supporting electrolyte) gave two waves at $-1·15$ V vs. SCE and $\sim-1·9$ V vs. SCE respectively. The first of these two waves was shown to correspond to a reversible one-electron reduction:

$$\underset{NO_2}{\bigodot} + e \rightleftarrows \left(\underset{NO_2}{\bigodot} \right)^- \tag{8}$$

A potential of $-1·3$ V vs. SCE was then applied to a similar solution in an electrochemical cell placed in the cavity of an esr spectrometer. The resulting esr spectrum confirmed the presence of the nitrobenzene radical anion and illustrated the use to which esr spectra can be put as an additional means of interpreting polarographic behaviour.

An esr study of a wide range of electrolytically generated nitrile radicals indicated that in nearly all cases the first observed wave (in dimethylformamide) corresponded to the reversible addition of one electron.[13] One exception, however, was the reduction of phthalonitrile. The first step in this case corresponded to the reversible addition of one electron:

$$\text{(9)}$$

However, the second step ($-2\cdot76$ V vs. Ag-AgClO$_4$ in dimethylformamide with tetra-n-propylammonium perchlorate as supporting electrolyte) corresponded to the addition of two electrons. Electrolysis at $-2\cdot85$ V gave the esr spectrum of the benzonitrile radical anion. From these results it was deduced that a proton had been abstracted from the solvent:

$$\text{(10)}$$

$$\text{(11)}$$

where SH represents a molecule of solvent.

One-electron reversible first waves have been observed for 2,1,3-benzoxadiazole (X = O), 2,1,3-benzothiadiazole (X = S) and 2,1,3-benzoselenadiazole (X = Se) and related compounds.[14] Similarly, the reduction products

of N-phenylphthalamide and of the phthalamide anion gave esr spectra indicating one-electron additions.[15,16]

Unfortunately, this technique was originally applied to only a relatively small number of polynuclear aza-heteromolecules. For example, sym-tetrazine, pyridazine, pyrazine, and phenazine all give the corresponding radical anions by one electron reductions in dimethylsulphoxide.[17] More recently, however, interest has been aroused in further studies of this group using the esr technique not only as an aid to the elucidation of their polarographic behaviour but also because of interest in the interpretation of the resulting esr spectra.

V. THE REDUCTION OF AZA-HETEROCYCLICS

The activity of aza-heterocyclics can be due either to the heterocyclic ring or substituents in that ring. In some cases the distinction is difficult or even impossible, e.g. 1,2-di(4-pyridyl)ethylene.[18] The reduction of aza-heterocyclics has been studied in a wide range of experimental conditions with respect to solvent, reference electrode and supporting electrolyte. These reductions may be either reversible or irreversible depending upon the nature of the heterocyclic and, in many cases, on the conditions used in the experiment. For instance, polarographic data for a limited number of pyridine-like heterocycles in 95 per cent dimethylformamide gives some indication of the influence of annelation on the half-wave reduction potentials.[19] Heterocyclics with two nitrogen atoms in the same ring show similar effects, e.g. pyrazine[20] reduces at some 40 mV more negative than quinoxaline.[21] The position of the nitrogen atom in the ring also has an important effect on the reduction potential: at pH 4·9 pyrazine reduces at -0.74 V while pyrimidine reduces at -1.09 V.[22] In general the reducibility of heterocyclic compounds is eased by the introduction of substituents into the ring. Examples of this are found in pyridine-3-carboxylic acid and its derivatives[23] and methyl-substituted aza-heterocyclics.[24] The interested reader will find the advances made in these areas reviewed periodically.[25,26,27]

The reduction of aza-heterocyclics revises the whole question of reversibility, which has already been discussed. In addition it may be mentioned that although methods such as periodically changed square-wave voltage, single-sweep techniques and constant-current oscillographic polarography may be useful, the critical test for reversibility is measurement of the i-t curves on single drops. This method has been applied in some detail to 1,2-di(4-pyridyl)ethylene[28] and also to pyrazine and 4,4′-bipyridyl.[22]

In the absence of proton donors, it is interesting to note that as a general rule the polarographic reduction of an aza-heterocyclic, parallels that of the corresponding aromatic hydrocarbon, under the same conditions. In all reported cases, the reduction occurs at a more positive potential. For example, the observed difference in $E_{1/2}$ for naphthalene and quinoline, in dimethylformamide is $+0.46$ V.[5]

It is now generally accepted that in dimethylformamide, to which no proton donor has been added, the reduction of aromatic hydrocarbons is reversible and follows the mechanism outlined in eqns (1) and (2). Thirty-five aza-heterocyclics have been studied under similar conditions.[24] Each of these heteromolecules gave one or more well defined reduction waves, and where present, the height of the second was approximately equal to that of the first wave. The first half-wave reduction potentials $E_{1/2}(1)$ and the second half-wave reduction potentials $E_{1/2}(2)$ (where detected) are summarized in Table 1. A comparison of these values with the corresponding aromatic hydrocarbon values indicates that in every case, $E_{1/2}(1)$ is more positive than for the aromatic hydrocarbon.

Table 1. *Experimental half-wave reduction potentials* vs. Ag/AgCl in DMF[24]

Heteromolecule	$-E_{1/2}(1)$	$-E_{1/2}(2)$	Heteromolecule	$-E_{1/2}(1)$	$-E_{1/2}(2)$
1 Pyridine	2·76	—	21 1,2-Bis-(3-		
2 Pyrimidine	2·35	—	pyridyl)ethylene	1·92	2·32
3 Pyrazine	2·17	—	22 Benzylideneaniline	1·94	2·43
4 Pyridazine	2·22	—	23 Diphenylmethyl-		
5 1,3,5-Triazine	2·105	—	eneaniline	1·94	—
6 2,2′-Bipyridyl	2·19	2·76	24 Dibenzylidene-		
7 4,4′-Bipyridyl	1·91	2·47	azine	1·77	2·26
8 4-Phenylpyridine	2·00		25 1,10-Phenan-		
9 5-Phenyl-		2·67	throline	2·12	2·70
pyrimidine	2·24	2·80	26 Acridine	1·62	2·38
10 2-Phenylpyridine	2·30	2·78	27 Benzo(c)cinnoline	1·55	2·40
11 Quinoline	2·175	—	28 5,6-Benzo-		
12 Isoquinoline	2·22	—	quinoline	2·20	2·72
13 Quinoxaline	1·80	—	29 7,8-Benzo-		
14 Phthalazine	2·02	—	quinoline	2·23	2·72
15 1,5-Naphthyri-			30 Phenazine	1·20	2·01
dine	1·86	—	31 Phenanthridine	2·12	2·64
16 Cinnoline	1·68	2·62	32 5,6:7,8-Dibenzo-		
17 Azobenzene	1·39	2·07	quinoxaline	1·78	—
18 2,2′-Azopyridine	1·04	1·65	33 3,4-Benzacridine	1·73	2·38
19 3,3′-Azopyridine	1·21	1·85	34 2,2′-Biquinoline	1·77	2·20
20 1,2-Bis-(2-			35 Dibenzo(a,c)-		
pyridyl)ethylene	1·69	2·11	phenazine	1·35	2·12

The slope of the plot of E vs. log $[(id - i)/i]$ and the value of $E_{1/4} - E_{3/4}$ have theoretical values of 0·058 V and 0·0554 V respectively, for a one-electron reduction at 20°C.[2] The mean values for the reduction waves given in Table 1 lie close to these theoretical values, again showing parallel behaviour to that of the aromatic hydrocarbons. Furthermore, the values of the diffusion current constants, $A(= [id/cm^{2/3}t^{1/6}])$, which is proportional to the number of electrons involved in the reduction,[29] are similar to those obtained for the one-electron reduction of hydrocarbons in dimethylformamide.[6,7] The A values for the heterocyclic series lie in the range two to three whereas A lies in the range 1·8–3·7 for the corresponding hydrocarbons.[6,7]

It was concluded from the above data that the reduction of the series of aza-heterocyclics appears to follow the normal mechanism observed for aromatic hydrocarbons and that $E_{1/2}(2)$, where detected, corresponded to the reduction of R^- to $R^=$.

1. A Correlation with Molecular Orbital Calculations

Many of the studies of the polarographic reduction of organic compounds have been successfully concerned with the correlation of the half-wave reduction potentials with simple Hückel MO type calculations.

The acceptance of a single electron in a reversible reduction results in the formation of the corresponding radical anion (eqn (1)). Electron spin resonance studies indicate that in many of these molecules the electron has entered the lowest unoccupied π-molecular orbital and the potential at which the molecule is reduced should thus be related to the energy of this π-orbital. A correlation of this type was first reported by Maccoll[30] and later by Lyons.[31] The relationship was confirmed shortly afterwards by Hoijtink,[8] who showed that for a series of similar molecules a linear correlation might be expected between the energy of the lowest unoccupied orbital and the half-wave reduction potential. More recently, the relationship has been extended to pyridine-type heteromolecules[19,32] and the correlation expressed by the following equation[19]

$$E_{1/2} = 2·127m_{m+1} - 0·555 \qquad (12)$$

where m_{m+1} is the energy of the lowest unoccupied molecular orbital.

Earlier papers had attempted theoretical correlations between $E_{1/2}$ and the molecular structure of aza-heteromolecules. In some cases the correlation was successful, e.g. aminoacridines,[33] whereas an unsuccessful correlation was found for pyridine-aldehydes in acid media.[34]

One difficulty, however, in correlations of this type for heterocyclics appears to be in the choice of a Coulomb integral for nitrogen, α_N, which may be used throughout a series of related compounds. It is usual, in a simple Hückel calculation, to express the Coulomb integral for nitrogen in terms of the Coulomb integral for carbon, α_C, and a fraction of the resonance integral, β, such that:

$$\alpha_N = \alpha_C + h_N\beta \qquad (13)$$

The problem is then reduced to choosing a suitable value for the parameter h_N. The generally accepted value of h_N varies from $0 \cdot 5-1 \cdot 0^{19,32,35,36,38}$ depending on the nature of the problem to which the calculations have been applied. Although the various values for h_N have been applied with varying success, a unique value for h_N, if indeed one exists, has by no means been determined. For the polarographic reduction of a series of aza-heterocyclics, it would seem appropriate, therefore, to investigate the possibility of a suitable value for h_N that would successfully correlate a wide range of heterocyclics.

A. Estimating h_N by Perturbation Theory. In a first order approximation, the change in energy of the lowest unoccupied molecular orbital of a hydrocarbon, when substituted with a heteroatom, is given by[39]

$$\Delta m_{m+1} = \sum_r C^2_{m+1,r} \cdot h_N \qquad (14)$$

where $C_{m+1,r}$ is the coefficient for the atomic orbital on atom r in the molecular orbital $m + 1$ and h_N as already defined.

From the relationship, eqn (15), derived by Hoijtink and van Schooten,[8] which assumes, for a series of similar heteromolecules, that the difference in free energies of solvation of the heteromolecule and its anion is effectively constant, eqn (16) follows whence a plot of the difference in the $E_{1/2}(1)$'s, $\Delta E_{1/2}(1)$, of the hydrocarbon and the corresponding heteromolecule, against $\sum_r C^2_{m+1,r}$ for

$$E_{1/2}(1) = -m_{m+1} \cdot \beta + C \qquad (15)$$

$$\frac{\Delta E_{1/2}(1)}{\beta} = \sum_r C^2_{m+1,r} \cdot h_N \qquad (16)$$

the 'replaced' carbons, should give a straight line through the origin of gradient h_N. Tabner and Yandle[24] applied the method to 35 aza-heterocyclics. The values of $\Delta E_{1/2}(1)$ and $\sum_r C^2_{m+1,r}$ are shown in columns 1 and 2 respectively of Table 2. The values of $E_{1/2}(1)$ for the hydrocarbons were measured under the same experimental conditions and the coefficients taken from the tables of Coulson and Streitwieser.[40]

The plot is shown in Fig. 1. The largest deviations occur for cinnoline, benzo(c)cinnoline and phthalazine. This might be expected from the presence in these heterocyclics of two adjacent nitrogen atoms, leading to greater stabilization of the lowest vacant energy level, by a change in the resonance integral for a —N=N— bond. Alternatively, since the assumption is made that the solvation energy change from the neutral to the reduced molecule is of the same order for both hydrocarbon and heteromolecule, the less negative reduction potential found (i.e., than predicted by perturbation theory) could possibly imply a greater solvation term for these heteromolecules. Nevertheless, under the conditions of approximation, the application of a single value of h_N appears justified.

Neglecting these three heteromolecules a value of 0·84 was obtained for h_N using the semi-empirical value of $-2·38$ eV for β.[41]

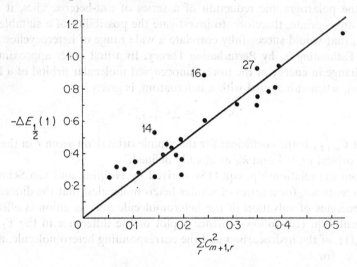

Fig. 1. A plot of $-\Delta E_{1/2}(1)$ against $\sum_r C^2_{m+1,r}$ (for the 'replaced' carbon atoms) for the heteromolecules given in Table 2. The numbers refer to Table 2.

B. The Correlation of $E_{1/2}(1)$ with m_{m+1}. Column 3 of Table 2 shows the calculated Hückel eigenvalues, in units of β, for the lowest unoccupied energy level. These values were obtained using the above-mentioned value for h_N and $\beta_{CN} = \beta_{CC}$. Overlap integrals were neglected and the AIP technique[42] ($\delta = 0·1$) incorporated to allow for the effect of the more electronegative nitrogen atom on the attached carbon atoms. All molecules were assumed planar and the variation of the k parameter ($\beta_{CN} = k\beta_{CC}$) with bond distance was calculated according to Mulliken, Rieke, and Brown.[43] Aromatic C—N and N—N bonds were taken as 1·36 Å and aromatic C—C bonds as 1·39 Å.

The plot of $E_{1/2}(1)$ against $-m_{m+1}$, although reasonable, was not as good as that found for the series of benzenoid hydrocarbons.[39] A least squares correlation gave:

$$E_{1/2}(1) = 1·84m_{m+1} - 1·19 \tag{17}$$

and the slope, the effective value for β, namely $-1·84$ eV or $-42·4$ kcal mole^{-1} is considerably lower than for the hydrocarbon series,[39] viz. $-2·41$ eV or $-55·6$ kcal mole^{-1}.

Table 2. *Simple Hückel energy values for some aza-heterocyclics.*[24]

Heteromolecule	$-\Delta E_{1/2}(1)$	$\Sigma C_{m+1,r}^2$ for 'replaced' atoms	$-m_{m+1}(\beta)$ $h_N=0{\cdot}84$	$\alpha_N = \alpha_C + h_N\beta$ $h_N=0{\cdot}5$	$h_N=0{\cdot}7$	$h_N=1{\cdot}0$
1. Pyridine	—	—	0·760	0·878	0·788	0·712
2. Pyrimidine	—	—	0·654	0·848	0·690	0·592
3. Pyrazine	—	—	0·528	0·805	0·581	0·431
4. Pyridazine	—	—	0·493	0·766	0·626	0·405
5. 1,3,5-Triazine	—	—	0·596	0·769	0·642	0·524
6. 2,2′-Bipyridyl	0·43	0·178	0·542	—	—	—
7. 4,4′-Bipyridyl	0·71	0·316	0·510	0·612	0·536	0·463
8. 4-Phenylpyridine	0·38	0·158	0·606	—	—	—
9. 5-Phenyl-pyrimidine	0·615	0·247	0·520	—	—	—
10. 2-Phenylpyridine	0·315	0·089	0·448	—	—	—
11. Quinoline	0·38	0·180	0·475	0·538	0·492	0·446
12. Isoquinoline	0·335	0·069	0·530	—	—	—
13. Quinoxaline	0·755	0·360	0·287	0·441	0·329	0·201
14. Phthalazine	0·535	0·138	0·519	0·548	0·317	0·500
15. 1,5-Naphthyri-dine	0·69	0·360	0·363	0·496	0·390	0·316
16. Cinnoline	0·875	0·249	0·264	0·471	0·371	0·190
17. Azobenzene	0·77	0·384	0·008	0·232	0·067	−0·101
18. 2,2′-Azopyridine	1·12	0·532	−0·076	—	—	—
19. 3,3′-Azopyridine	0·95	0·390	−0·011	—	—	—
20. 1,2-Bis-(2-pyridyl)ethylene	0·24	0·148	0·425	—	—	—
21. 1,2-Bis-(4-pyridyl)ethylene	0·47	0·196	0·295	—	—	—
22. Benzylidene-aniline	0·22	0·197	0·392	—	—	—
23. Diphenylmethyl-eneaniline	—	—	0·318	—	—	—
24. Dibenzylidene-azine	—	—	0·296	—	—	—
25. 1,10-Phenan-throline	0·365	0·108	0·456	—	—	—
26. Acridine	0·36	0·193	0·284	0·336	0·297	0·262
27. Benzo(c)cinno-line	0·935	0·344	0·170	0·389	0·216	0·087
28. 5,6-Benzoquino-line	0·285	0·116	0·492	0·547	0·507	0·464
29. 7,8-Benzoquino-line	0·255	0·054	0·522	—	—	—
30. Phenazine	0·78	0·386	0·088	0·232	0·127	0·018
31. Phenanthridine	0·365	0·172	0·457	0·522	0·475	0·427
32. 5,6:7,8-Dibenzo-quinoxaline	—	—	0·386	0·543	0·431	0·310
33. 3,4-Benzacridine	—	—	0·334	0·388	0·348	0·312
34. 2,2′-Biquinoline	—	—	0·312	—	—	—
35. Dibenzo(a,c)-phenazine	—	—	0·190	—	—	—

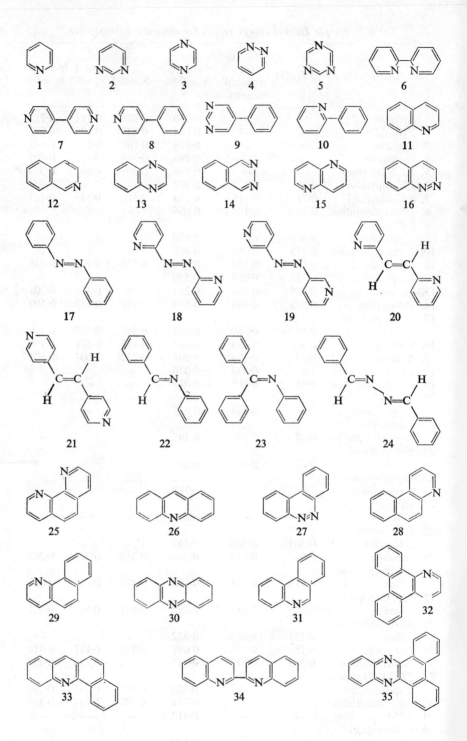

In view of the other values of h_N used in the literature, further calculations were made in order to find the effect of variation of h_N on the correlation. Columns 4, 5, and 6 of Table 2 show the energy values for a cross-section of the heteromolecules calculated with $h_N = 0.5, 0.75$, and 1.0 respectively. It is noteworthy that for $h_N = 1.0$ azobenzene has a bonding lowest unoccupied energy level.

A least-squares determination of the three sets of values gave the following equations:

$$h_N = 0.5; \quad E_{1/2}(1) = 1.63 m_{m+1} - 1.05 \tag{18}$$

$$h_N = 0.75; \quad E_{1/2}(1) = 1.84 m_{m+1} - 1.11 \tag{19}$$

$$h_N = 1.0; \quad E_{1/2}(1) = 1.68 m_{m+1} - 1.35 \tag{20}$$

It was found that the correlation was slightly inferior to that obtained using $h_N = 0.84$, and the standard deviations indicated that the choice of the value for h_N is not too critical. Also, in all cases the derived values of β are considerably less than for the corresponding benzenoid hydrocarbons, although the value depends upon the chosen h_N parameter.

2. The Effect of Substituents on the Polarography of Aza-heterocyclics

It has already been mentioned that the polarographic activity of heterocyclic compounds is due to the ring, substituents in that ring or possibly both. In most cases, however, the reduction takes place in the ring. The effects of substituents on the reduction potential depends on the position of substitution, and on the substituent itself. The positions of the heterocyclic ring, are of course, not equivalent with regard to their deficiency or excess of π-electrons. This is illustrated by some 4-substituted derivatives of pyridine, which reduce at a slightly more positive potential than 2-substituted derivatives and both more positive than substituents at the 3-position.[44] In a qualitative treatment, Volke,[45] suggested that the nitrogen may be looked upon as a strong external substituent. This concept is substantiated by the above-mentioned work on 2-, 3-, and 4-derivatives of pyridine.

It has been predicted that methyl-group substitution in a molecule has the effect of making the half-wave reduction potential more negative.[46] This has been verified for a series of methyl substituted azulenes.[47] This effect has been investigated for a series of 19 methyl substituted aza-heteromolecules.[24] The results are given in Table 3, column 1. In each family of compounds, a more negative $E_{1/2}(1)$ than the corresponding unsubstituted heteromolecule is found (cf. Table 1).

A. The Correlation of $E_{1/2}(1)$ and m_{m+1}. In order to consider such a correlation, the choice of a 'model' for the methyl group is critical. Four models were considered and calculations made in each case.

(a) *Heteroatom Model* (*H*). This model effectively replaces the methyl group by a single heteroatom, which donates two electrons to the π-system. It was first used by Matsen[48] and suitable parameters appear to be $h_x = 2$ and $k_{cx} = 0.7$. The eigenvalues of the lowest unoccupied orbital calculated, using these parameters, are shown in column 5 of Table 3.

(b) *Conjugation Model* (*C*). A theoretical consideration of hyperconjugation was first attempted by Mulliken *et al.*[43] The methyl group is treated as a modified vinyl group in which c_α is the carbon atom of the π-system, C_{sp^2} is the tetrahedral carbon of the methyl group, and X is the pseudo-atom representing the three hydrogen atoms (Fig. 2)

$$C_\alpha \xrightarrow{\beta_1} C_{sp^2} \xrightarrow{\beta_2} X$$
$$\alpha_1 \quad\quad \alpha_2 \quad\quad \alpha_3$$

Fig. 2.

Values chosen for the calculation were $h_2 = -0.1$, $h_3 = -0.3$, $k_1 = 0.8$, and $k_2 = 2.8$. The α_1 term was unaltered. The resulting energy values are shown in column 6 of Table 3.

(c) *Inductive-Conjugation Model* (*IC*). The model was first suggested by Streitwieser *et al.*[49] and consists of using the conjugation model for the methyl group, but also making the Coulomb integral for the attached carbon, more positive. The parameters suggested by Streitwieser, $h_1 = -0.3$, $h_2 = -0.3$, $h_3 = -0.6$, $k_1 = 0.8$, and $k_2 = 2.8$, were used and gave the results shown in column 7, Table 3.

(d) *Inductive Model* (*I*). Wheland and Pauling[42] introduced a simple model for the methyl group where conjugation was completely neglected and the inductive effect only considered. It is incorporated by altering α for the attached carbon of the π-system by a negative value for h. Streitwieser *et al.*[49] and Mackor *et al.*[50] have used this model with reasonable success for values of $h = -0.3$ to -0.5.

The model offers the opportunity of estimating the value of h by perturbation theory (cf. unsubstituted aza-heterocyclics discussed earlier). A plot of $\delta E_{1/2}$ (now the difference between the parent and methyl-substituted heteromolecule) against $\sum_r C^2_{m+1,r}$ for the attached carbon yields a value of h from the gradient. The values of $\delta E_{1/2}$ and $\sum_r C^2_{m+1,r}$ are given in columns 2 and 3 of Table 3, respectively. The coefficients for the parent heteromolecules were taken from Coulson and Streitwieser.[40] A plot of $\delta E_{1/2}$ against $\sum_r C^2_{m+1,r}$ is shown in Fig. 3 and yields a value of -0.14 for h (neglecting 2,6-dimethylquinoline and using the same value of β as earlier).

Table 3. *Experimental half-wave reduction potentials and simple Hückel energy values for methyl-substituted heteromolecules*[24]

Substituted Heteromolecules	$-E_{1/2}$ (1)	$\delta E_{1/2}$	$-\Sigma C^2_{m+1,r}$ for substituted atoms	Eigenvalues using various models for the methyl group (β)			
				I	H	C	IC
1. 2-methylpyridine	2·80	0·04	0·142	−0·779	−0·783	−0·754	−0·793
2. 3-methylpyridine	2·77	0·01	0·016	−0·765	−0·775	−0·758	−0·769
3. 4-methylpyridine	2·86	0·10	0·206	−0·805	−0·815	−0·748	−0·847
4. 2,5-Dimethylpyridine	2·82	0·06	0·158	−0·781	−0·785	−0·765	−0·793
5. 2,6-Dimethylpyridine	2·85	0·09	0·284	−0·804	−0·814	−0·748	−0·845
6. 2,4,6-Trimethylpyridine	2·91	0·15	0·490	−0·851	−0·871	−0·736	−0·940
7. 2-Methylpyrazine	2·23	0·06	0·103	−0·543	−0·547	−0·526	−0·557
8. 2,6-Dimethylpyrazine	2·28	0·11	0·206	−0·559	−0·578	−0·523	−0·593
9. 2,3,5,6-Tetramethylpyrazine	2·50	0·33	0·412	−0·593	−0·616	−0·519	−0·670
10. 2-Methylquinoline	2·20	0·025	0·125	−0·497	−0·504	−0·473	−0·519
11. 4-Methylquinoline	2·25	0·075	0·228	−0·509	−0·523	−0·474	−0·549
12. 6-Methylquinoline	2·19	0·015	0·049	−0·481	−0·483	−0·472	−0·487
13. 8-Methylquinoline	2·22	0·045	0·132	−0·491	−0·497	−0·476	−0·508
14. 2,4-Dimethylquinoline	2·285	0·11	0·353	−0·532	−0·551	−0·470	−0·493
15. 2,6-Dimethylquinoline	2·31	0·135	0·174	−0·502	−0·511	−0·485	−0·538
16. 2-Methylquinoxaline	1·85	0·05	0·127	−0·304	−0·312	−0·287	−0·325
17. 2,3-Dimethylquinoxaline	1·90	0·10	0·254(a)	−0·325	−0·363	−2·287	−0·377
18. 6,7-Dimethylquinoxaline	1·82	0·02	0·104(a)	−0·302	−0·310	−0·284	−0·324
19. 3-Methylisoquinoline	2·28	0·06	0·161(a)	−0·531	−0·531	−0·531	−0·533

(a) Calculated by the authors.

Column 4, of Table 3 gives the calculated values for the lowest unoccupied energy level using $h = -0.14$.

A least squares determination of the plots of $E_{1/2}$ against m_{m+1} calculated for each of the four models gave the following equations:

$$\text{Model } (H): \quad E_{1/2}(1) = 2.02m_{m+1} - 1.19 \tag{21}$$

$$\text{Model } (C): \quad E_{1/2}(1) = 2.11m_{m+1} - 1.24 \tag{22}$$

$$\text{Model } (IC): \quad E_{1/2}(1) = 1.94m_{m+1} - 1.20 \tag{23}$$

$$\text{Model } (I): \quad E_{1/2}(1) = 2.02m_{m+1} - 1.22 \tag{24}$$

The conjugation model gave the poorest correlation and the heteroatom and inductive models appeared to best represent the methyl group in the present investigation, yielding a value of -2.02 eV or -46.6 kcal mole^{-1} for β.

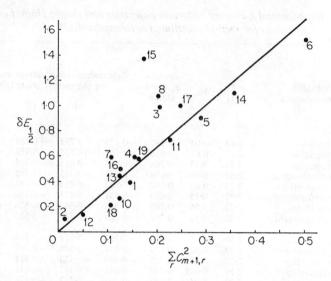

Fig. 3. A plot of $\delta E_{1/2}$ against $\sum_r C^2_{m+1,r}$ (for the carbon atoms substituted with methy groups) for some heteromolecules. The numbers refer to Table 3.

3. Some Electron Affinities and Free Energies of Solvation

The reduction potential, relative to the standard calomel electrode, can be expressed[51] as eqn (25) if ion association is assumed to be negligible, where

$$E_{1/2} = EA + \Delta F_{solv} - K \tag{25}$$

EA is the electron affinity of the neutral molecule, ΔF_{solv} is the difference in free energy of solvation of the neutral and the mono-negative ion, and K is a constant.

In a very detailed study Hush, Peover and their co-workers[52] have obtained a value of 4·70 eV for K, with acetonitrile as solvent. Through lack of other data, this value will be used in the following discussion.

A. Calculated Electron Affinities. Roothaan[53] has shown that, in a more rigorous theoretical treatment, the ionization potential of a molecule is given by eqn (26)

$$-I = \sum_{\mu\nu} \chi_{N\nu} F_{\mu\nu} \chi_{N\mu} \tag{26}$$

where $\chi_{N\nu}$ and $\chi_{N\mu}$ are the coefficients of atoms ν and μ for the highest occupied energy level and $F_{\mu\nu}$ is the matrix element of the Hartree-Fock Hamiltonian.

By a completely analogous method Hush and Pople[54] have deduced the electron affinity as eqn (27)

$$-A = \sum_{\mu\nu} \chi_{N+1,\mu} F_{\mu\nu} \chi_{N+1,\nu} \tag{27}$$

where $\chi_{N+1,\mu}$ and $\chi_{N+1,\nu}$ are now coefficients for the lowest unoccupied orbital.

In Table 4, columns 3 and 4, respectively are shown the calculated electron affinities for some 20 neutral aza-heterocyclics and of 13 negative ions where second half-wave potentials have been detected.[24] These values have been calculated by making the following assumptions and approximations:

(a) simple Hückel coefficients have been employed instead of self-consistent coefficients;

(b) all molecules have been assumed to be planar and rigid;

(c) aromatic C—N and N—N bonds were taken as 1·36 Å and aromatic C—C bonds as 1·39 Å unless more accurate data was available;

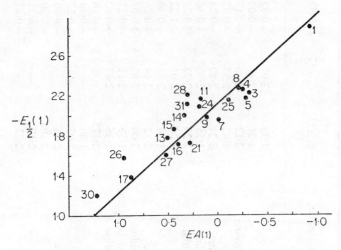

Fig. 4. A plot of $-E_{1/2}(1)$ against electron affinity, $EA(1)$, for the heteromolecules given in Table 4. The numbers refer to Tables 2 and 4.

(d) one- and two-centre repulsion integrals were obtained by the method of Pariser and Parr[41] as were the resonance integrals for β_{CC} and β_{CN};

(e) the variation of resonance integrals with bond length was estimated by the method of Mulliken, Rieke, and Brown;[43]

Table 4. *Experimental half-wave reduction potential, electron affinities and solvation energies of some unsubstituted heteromolecules*

Ref. Table 1	Heteromolecule	$-E_{1/2}$ (1)	$-E_{1/2}$ (2)	EA (1)	EA (2)	R_{REP}	$\Delta E_{1/2}$	ΔE_{SOLV} (a)	ΔE_{SOLV} (b)
17	Azobenzene	1·39	2·07	+0·83	−4·69	5·52	0·68	2·48	2·59
7	4,4'-Bipyridyl	1·91	2·47	−0·07	−5·34	5·27	0·56	2·86	2·35
25	1,10-Phenanthroline	2·12	2·70	+0·17	−4·88	5·05	0·58	2·41	2·24
26	Acridine	1·62	2·38	+0·92	−4·29	5·21	0·76	2·16	2·23
8	4-Phenylpyridine	2·24	2·80	−0·24	−5·70	5·46	0·56	2·70	2·45
27	Benzo(c)cinnoline	1·55	2·40	+0·56	−5·48	6·04	0·85	2·59	2·60
28	5,6-Benzoquinoline	2·20	2·72	+0·23	−4·97	5·20	0·52	2·27	2·34
30	Phenazine	1·20	2·01	+1·22	−4·20	5·42	0·81	2·28	2·31
16	Cinnoline	1·68	2·62	+0·40	−5·85	6·25	0·94	2·62	2·66
31	Phenanthridine	2·12	2·64	+0·26	−5·10	5·36	0·52	2·32	2·42
9	5-Phenylpyrimidine	2·00	2·67	+0·04	−5·49	5·53	0·67	2·66	2·43
21	1,2-Bis-(4-pyridyl)ethylene	1·69	2·11	+0·27	−4·78	5·05	0·42	2·74	2·32
15	1,5-Naphthyridine	1·86	2·67	+0·40	−5·34	5·74	0·81	2·44	2·47
11	Quinoline	2·175	—	+0·16	—	—	—	2·37	—
13	Quinoxaline	1·80	—	+0·49	—	—	—	2·41	—
14	Phthalazine	2·02	—	+0·27	—	—	—	2·41	—
1	Pyridine	2·76	—	−0·90	—	—	—	2·84	—
5	1,3,5-Triazine	2·105	—	−0·30	—	—	—	2·90	—
3	Pyrazine	2·17	—	−0·33	—	—	—	2·86	—
4	Pyridazine	2·22	—	−0·27	—	—	—	2·75	—

(a) Calculated from eqn (25).   (b) Calculated from eqn (29).

(f) the valence state ionization potential and penetration integral for nitrogen were those used by Brown and Heffernan.[55] Only nearest-neighbour penetration integrals were considered;

(g) a value for α_{CC} of -9.5 eV was used.[54]

Discussion of individual values will not be given here, although the relative values are comparable with the values for corresponding hydrocarbons.

The plot of $-E_{1/2}(1)$ against $EA(1)$, the first electron affinity, is shown in Fig. 4 and a least squares determination gave eqn (28):

$$E_{1/2}(1) = 0.86EA(1) - 2.08 \tag{28}$$

In view of the approximations made, the correlation is good and suggests that variations in the solvation-energy term are not critical. More rigorous calculations or experimental electron affinities may quite possibly result in a poorer correlation and hence indicate a greater dependence on solvation energy. However, for the present treatment, the solvation term may well be considered as effectively constant.

B. Solvation Energies. With no experimental electron affinity (EA) values available at present for the above considered aza-heterocyclics, substitution of the calculated $EA(1)$ values in eqn (25) yields a measure of ΔF_{solv}. These values are shown in column 7, Table 4 and if zero entropy change is assumed, may be considered as solvation energy differences.

ΔE_{solv} can also be expressed by eqn (29):[56]

$$2\Delta E_{solv} = R - \Delta E_{1/2} \tag{29}$$

where R is the disproportionation energy of the reaction:[57]

$$2A^-(\text{gas}) \rightarrow A^0(\text{gas}) + A^{2-}(\text{gas}) \tag{30}$$

and $\Delta E_{1/2}$ is the difference in the first and second half-wave reduction potentials. The values of R (difference in the electron affinities of A^0 and A^-) and $\Delta E_{1/2}$ are given in columns 5 and 6, respectively, of Table 4 and ΔE_{solv} calculated from eqn (29) in column 8. Although these values tend to be smaller, they compare favourably with the ΔE_{solv} values of column 7.

Since R is the difference in electron affinities, the ΔE_{solv} values calculated from eqn (29) are likely to be more accurate. They are approximately 0.5 eV greater than the most recent values obtained for some aromatic hydrocarbons.[52] The variation is small and the 'average' value from column 8 of Table 4 is 2.42 eV. Substitution of this value in eqn (25) yields eqn (31):

$$E_{1/2}(1) = EA(1) - 2.38 \tag{31}$$

which agrees reasonably well with the experimentally determined eqn (28).

The assumption that ΔE_{solv} is constant suggests that the major reason for the difference in solvation energy is the presence of the more electronegative nitrogen atoms. From these considerations, we conclude that correlation of the reduction potentials with the energy of the lowest vacant orbital, without detailed consideration of the variations of solvation terms, is generally as successful in the aza-aromatic series as for the parent hydrocarbon molecules.

REFERENCES

1. KÜTA, J., and SMOLER, I.: *Progress in Polarography*, Vol. 1, p. 43 (Interscience, N.Y., 1962).
2. TOMĚS, J.: *Coll. Trav. chim. tchécoslov.*, 9, 12 (1937).
3. HOIJTINK, G. J., SCHOOTEN, J. VAN, BOER, E. DE, and AALBERSBERG, W. I.: *Rec. Trav. chim.*, 73, 355 (1954).
4. WAWZONEK, S., BLAHA, E. W., BERKEY, R., and RUNNER, M. E.: *J. Electrochem. Soc.*, 102, 235 (1955).
5. GIVEN, P. H.: *J. Chem. Soc.*, 2684 (1958).
6. ATEN, A. C., BÜTHKER, C., and HOIJTINK, G. J.: *Trans. Faraday Soc.*, 55, 324 (1959).
7. GIVEN, P. H., and PEOVER, M. E.: *J. Chem. Soc.*, 385 (1960).
8. HOIJTINK, G. J., and SCHOOTEN, J. VAN, *Rec. Trav. chim.*, 71, 1089 (1952).
9. TUTTLE, T. R., WARD, R. L., and WEISSMAN, S. I.: *J. Chem. Phys.*, 25, 189 (1956).
10. AUSTEN, D. E. G., GIVEN, P. H., INGRAM, D. J. E., and PEOVER, M. E.: *Nature*, 182, 1784 (1958).
11. HOIJTINK, G. J., MEIJ, P. H. VAN DER: *Z. physik. Chem. N.F.*, 20, 1 (1959).
12. GESKE, D. H., and MAKI, A. H.: *J.A.C.S.*, 82, 2611 (1960).
13. RIEGER, P. H., BERNAL, I., REINMUTH, W. H., and FRAENKEL, G. K.: *J.A.C.S.*, 85, 683 (1963).
14. ATHERTON, N. M., OCKWELL, J. N., and DIETZ, R.: *J. Chem. Soc.*, A, 771 (1967).
15. NELSON, S. F.: *J.A.C.S.*, 89, 5256 (1967).
16. SISDA, R. E., and KOSKI, W. S.: *J.A.C.S.*, 89, 475 (1967).
17. STONE, E. W., and MAKI, A. H.: *J. Chem. Phys.*, 39, 1635 (1963).
18. VOLKE, J., and HOLUBEK, J.: *Coll. Czech. Chem. Comm.*, 27, 1777 (1962).
19. PÁRKÁNYI, C., and ZAHRADNIK, R.: *Bull. Soc. chim. belges*, 73, 57 (1964).
20. VOLKE, J., DUMANOVIĆ, D., and VOLKOVÁ, V.: *Coll. Czech. Chem. Comm.*, 30, 246 (1965).
21. SARTORI, G., and FURLANI, C.: *Ann. Chim. (Italy)*, 45, 251 (1955).
22. VOLKE, J.: *Talanta*, Vol. 12, 1081 (1965).
23. SORM, F.: *Rozpravy Ceské akademie II*, tr. 53, No. 27 (1942).
24. TABNER, B. J., and YANDLE, J. R.: *J. Chem. Soc.*, A, 381 (1968).
25. WAWZONEK, S.: *Analyt. Chem.*, 34, 182R (1962).
26. WAWZONEK, S., and PIETRZYLE, D. J.: *Analyt. Chem.*, 36, 220R (1964).
27. PIETRZYLE, D. J.: *Analyt. Chem.*, 38, 278R (1966).
28. VOLKE, J.: 'Die Polarographie in der Chemotherapie, Biochemie and Biologie' Abhandl. der DAW, Berlin, 1964, Jona 1962, p. 70.
29. KOLTHOFF, I. M., and LINGANE, J. J.: *Polarography* (Interscience, New York, 1941).
30. MACCOLL, A.: *Nature*, 163, 178 (1949).
31. LYONS, L. E.: *Nature*, 166, 193 (1950).
32. BERG, H., and KÖNIG, K. M.: *Analyt. Chim. Acta*, 18, 140 (1958).
33. HUSH, N. S.: *J. Chem. Phys.*, 20, 1660 (1952).
34. FORNASARI, E., GIACOMETTI, G., and RIGATTI, G.: *Advances in Polarography*, Vol. 3, p. 895 (Pergamon Press, London, 1960).
35. MATAGA, N.: *Bull. Chem. Soc., Japan*, 31, 463 (1958).

36. MURRELL, J. N.: *Mol. Phys.*, **1**, 384 (1958).
37. BROWN, R. D.: *J. Chem. Soc.*, 272 (1956).
38. CARRINGTON, A., and SANTOS-VEIGA, J. DOS: *Mol. Phys.*, **5**, 21 (1962).
39. STREITWIESER, A.: *Molecular Orbital Theory*, p. 180–2 (John Wiley and Sons, New York, 1961).
40. COULSON, C. A., and STREITWIESER, A.: *Dictionary of π-Electron Calculations* (Pergamon Press, Oxford, 1965).
41. PARISER, R., and PARR, R. G.: *J. Chem. Phys.*, **21**, 466, 767 (1953).
42. WHELAND, G. W., and PAULING, L.: *J.A.C.S.*, **57**, 2086 (1935).
43. MULLIKEN, R. S., RIEKE, C. A., and BROWN, W. G.: *J.A.C.S.*, **63**, 41 (1941).
44. HOLUBEK, J., and VOLKE, J.: *Coll. Czech. Chem. Comm.*, **27**, 680 (1962).
45. VOLKE, J.: *Physical Methods in Heterocyclic Chemistry*, Vol. 1, p. 260 (Academic Press, 1963).
46. LYONS, L. E.: *Research*, **2**, 587 (1949).
47. CHOPARD-DIT-JEAN, L. H., and HEILBRONNER, E.: *Hllv. Chim. Acta*, **36**, 144 (1953).
48. MATSEN, F. A.: *J.A.C.S.*, **72**, 5243 (1950).
49. STREITWIESER, A., and NAIR, P. M.: *Tetrahedron*, **5**, 149 (1959).
50. MACKOR, E. L., HOFSTRA, A., and WAALS, J. H. VAN DER: *Trans. Faraday Soc.*, **54**, 186 (1958).
51. MATSEN, F. A.: *J. Chem. Phys.*, **24**, 602 (1956).
52. CASE, B., HUSH, N. S., PARSONS, R., and PEOVER, M. E.: *J. Electro-analyt. Chem.*, **10**, 360 (1965).
53. ROOTHAAN, C. C. J.: *Rev. Mod. Physics*, **23**, 61 (1951).
54. HUSH, N. S., and POPLE, J. A.: *Trans. Faraday Soc.*, **51**, 600 (1955).
55. BROWN, R. D., and HEFFERNAN, M. L.: *Austral. J. Chem.*, **12**, 319, 543 (1959).
56. HEDGES, R. M., and MATSEN, F. A.: *J. Chem. Phys.*, **28**, 950 (1958).
57. HUSH, N. S., and BLACKLEDGE, J.: *J. Chem. Phys.*, **23**, 514 (1955).

REACTIONS AT ORGANIC SEMICONDUCTOR ELECTRODES

W. Mehl*

* Based on Habilitationsschrift W. Mehl, Technische Hochschule München, November 1968.

I. INTRODUCTION

This chapter deals with charge transfer reactions between organic molecules which are lattice elements of molecular crystals and redox systems in electrolytes which are in contact with such crystals. The molecular crystals will be referred to as either 'organic semiconductors' or, because their intrinsic carrier density at room temperature and in the dark is extremely small, as 'insulators'. Condensed aromatic hydrocarbons form a group of compounds for which such investigations are possible. As model substance mainly anthracene has been used. For experimental details a previous publication may be consulted.[1]

A discussion of electron transfer between a semiconductor and an electrolyte containing a redox system requires consideration of three effects which are absent in the case of electron transfer between a metal and a redox electrolyte.

1. It becomes necessary to account for the absence of allowed energy levels for electrons at the surface of the solid near the Fermi level of the system. This property of semiconductor introduces a term for electronic activation which must be supplied in order to raise the charge from its energy in the initial state in the Fermi level to the lowest allowed level in the surface which for an electron is the bottom of the conduction band and for a hole the top of the valence band. For a metal electrode there are many occupied and unoccupied states at the Fermi level, so that in a wide current range the activation energy is purely environmental.

2. The intrinsic carrier density for a semiconductor is so small that an electric field can be supported in the electrode surface so that variations in the potential difference applied between the semiconductor and the electrolyte may be shared between the surface space charge layer of the semiconductor and the inner Helmholtz layer of the solid-electrolyte interface while for a metal electrode this variation is restricted to the latter region.

3. On charge transfer vibrational reorganization may occur around a localized site in the semiconductor surface (lattice polarization), leading to a contribution to the activation energy which is absent for metal electrodes.

These effects are most pronounced for intrinsic wide gap semiconductors which experience no thermally activated intrinsic carrier generation and thus are very poor conductors (insulators). For such materials 'concentration polarization' of charge carriers opposing current flow can be so high, that the interfacial reaction can be assumed in equilibrium and the current voltage curve is controlled by the solid (space charge limited current). In this situation the interfacial reaction controls the current voltage curve when the applied voltage is high enough so that the rate of carrier transport through the solid equals the rate of the interfacial reaction. A limited (saturation) current is then observed, on the anodic side when electrons in the conduction band are involved, and on the cathodic side when holes in the valence band are the charge carrier.

II. THE EQUILIBRIUM CONDITION

1. General Remarks

The redistribution of charge at the interface between a semiconductor and an electrolyte results in the formation of an electrical double layer, an electronic charge in the solid is compensated by an excess of ionic charge in the electrolyte. In contrast to a metal for which the density of carriers is so high that the injected charge presents only a minor perturbation, insulators have at room temperature a small concentration of intrinsic carriers so that the injected carriers are in large excess. Upon contact with an electrolyte equilibrium can thus be established only by modifying the carrier distribution in valence or conduction band to display accumulation of electrons in the former and holes in the latter.

2. The Space Charge in the Semiconductor

We choose the vacuum level at infinity as reference for the electrochemical potential of electrons. The electrochemical potential is then identical with the Fermi level and the electrostatic potential difference $\Delta\phi$ at the interface between insulator and electrolyte may then be written as

$$\Delta\phi = F_i - F_s \qquad (1)$$

F_i and F_s denote the Fermi level of the insulator electrode and the electrolyte before contact, the sign of $\Delta\phi$ refers to the sign of the charge on the semiconductor surface using the electrolyte as reference. All potentials refer to the charge free bulk of semiconductor and electrolyte, i.e. for electron injection $\phi > 0$ in the semiconductor surface and $\phi < 0$ in the electrolyte while for hole injection $\phi < \theta$ in the semiconductor surface and $\phi > 0$ in the electrolyte.

The differential equations describing the space charge variations of the electrostatic field E and the electrostatic potential ϕ at equilibrium are

$$\frac{dE}{dx} = \frac{4\pi q e_0 n(x)}{\kappa_i} \qquad (2)$$

$$\frac{d\phi}{dx} = -E \qquad (3)$$

$$\frac{dn}{dx} = \frac{\mu n E}{q D} \qquad (4)$$

II

For simplicity a one-dimensional system was chosen. $n(x)$ is the density of carriers at the point x, e_0 is the magnitude of the electronic charge, qe_0 is the actual charge on each carrier and κ_i is the permittivity of the organic material. The first equation is Poisson's equation in which the volume charge is due to injected carriers following the assumption that these are in large excess over the thermally generated carriers. Equation (3) expresses the field as a gradient of potential and eqn (4) states that in the equilibrium case the diffusion of electrons away from the interface is balanced by the drift of electrons toward the contact. Solutions to the system of differential eqns (2)–(4) with various boundary conditions have been given by several authors[2,3,4,5] who may be consulted for details of the calculations. Figure 1 shows the variation of the electrostatic potential through a 10μ thick insulator film for various combinations of contacts. Curve a represents the case of injection of carriers of one sign only into one face while at the other face the contact is blocking for both electrons and holes. This is a condition which is frequently realized experimentally and the complete solution for electrolytic contacts has recently been given.[1] A combination of ohmic contacts, typified by the symmetrical situation (Fig. 1, curve c), is of interest for the experimental investigation of the equilibrium condition at insulator electrodes.

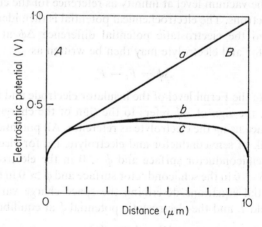

Fig. 1. Variation of the electrostatic potential through a 10 μm thick insulator film sandwiched between two electrolytic contacts. At interface A the potential step is $e_0 V = 0.1$ eV. At interface B the potential step is: 1 eV for curve a (monotonic case); 0.472 eV for curve b (intermediate case); 0.1 eV for curve c (symmetrical case) (based on Reference 2).

We shall present here in some detail the simpler case of an infinite solid with zero field at infinity.[5] To a very high degree of accuracy this case may be used as an approximation to the other more complicated solutions in many practical situations.[3] It has recently been used for a detailed discussion of the potential distribution at an insulator electrode.[6]

$$E \to 0 \quad \text{for} \quad x \to \infty \tag{5}$$

It leads to the solutions:[6]

$$\frac{d\phi(x)}{dx} = \frac{q}{L_D} \exp \frac{-qe_0}{kT} \tag{6}$$

$$qe_0\phi(x) = 2kT \log \frac{x+1}{L_D} \tag{7}$$

$$L_D = \left(\frac{\kappa_i kT}{4\pi ne_0^2}\right)^{1/2} \tag{8}$$

Assuming the carriers to obey the Boltzmann distribution law their bulk concentration is

$$n = N \exp \frac{-\Delta}{kT} \tag{9}$$

N is the density of states in the carrier band and Δ is the distance between Fermi level and carrier band. As we assume the insulator to be an intrinsic semiconductor, the bulk Fermi level rests at all temperatures halfway between conduction and valence band. If the width of the carrier band $\Delta E \simeq kT$, the effective density of states approaches a maximum value of twice the density of molecules in the crystal (for anthracene, e.g. $2 \cdot 4 \times 10^{19}$ cm^{-3}). The density of carriers n_s in the insulator immediately at the electrolyte boundary is thus

$$n_s = N \exp \frac{-\Delta_{n,p}}{kT} \tag{10}$$

$\Delta_{n,p}$ represents the distance between carrier band and Fermi level at the semiconductor surface, i.e.

$$\Delta_n = (E_c - F_i)_{x=0}$$
$$\Delta_p = (E_v - F_i)_{x=0}$$

From eqns (8) and (9)

$$L_D = L_D^0 \exp \frac{\Delta}{2kT} \tag{11}$$

L_D^0 is a formal Debye length which would be observed for an insulator when all electrons in the valence band are released to the conduction band assuming the density of states in both bands to be equal.[6]

From eqns (11) and (6)

$$\left(\frac{d\phi}{dx}\right)_{x=0} = \left(\frac{q}{L_D^0}\right)^{-1} \exp\left(\frac{-\bar{\Delta}_{n,p}}{2kT}\right) \tag{12}$$

The classical approach outlined here assumes a continuous distribution of states over potential energy and therefore must fail when the potential gradient becomes large. For the calculation of the local concentrations of electrons and holes the Fermi distribution function has then to be combined with the Poisson equation. This point has been discussed in detail by various authors.[1,6,7,8]

3. The Potential Distribution in the Electrolyte

The interfacial potential distribution may be represented (to a good approximation) by the sum of three contributions

$$\Delta\phi_T = \Delta\phi_i + \Delta\phi_H + \Delta\phi_e \tag{13}$$

where $\Delta\phi_i$ represents the space charge inside the semiconductor, $\Delta\phi_H$ the potential difference across the inner Helmholtz layer, the region between the solid and the electrolyte into which ions do not penetrate unless there are specific chemisorption forces, and $\Delta\phi_e$, the potential difference across the diffuse double layer, the ionic space charge in the electrolyte formed by counter ions. Neglecting specific adsorption the potential gradient in the Helmholtz layer will be approximately constant, i.e. the potential difference $\Delta\phi_H$ across the layer is determined by the field E_H multiplied by the thickness of this layer.

Because of the continuity of the normal component of the displacement at the interface between insulator and electrolyte

$$\kappa_i \left(\frac{d\phi}{dx}\right)_{x=0} = \kappa_H \left(\frac{d\phi}{dx}\right)_{x_H < x < 0} \tag{14}$$

and the potential drop across the Helmholtz layer is thus

$$\Delta\phi_H = \frac{\kappa_i}{\kappa_H} x_H \left(\frac{d\phi}{dx}\right)_{x=0} \tag{15}$$

From eqns (12) and (15)

$$\Delta\phi_H = \frac{\kappa_i}{\kappa_H} \frac{qx_H}{L_D^0} \exp\left(\frac{-\bar{\Delta}_{n,p}}{2kT}\right) \tag{16}$$

From eqn (16) the potential drop across the Helmholtz layer has been calculated as a function of the distance between Fermi level and carrier band at the insulator surface at equilibrium (ref. 6) (Fig. 2), verifying the previously made[1] assumption (supported by experimental evidence) that only for the degenerate insulator surface can a sizeable potential drop across the Helmholtz layer be expected.

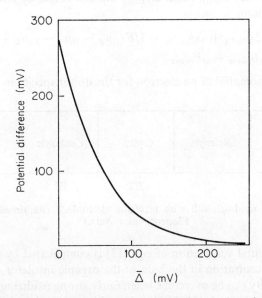

Fig. 2. Potential difference at equilibrium across the Helmholtz layer in the electrolyte at the anthracene/electrolyte interface as function of the difference between the Fermi level and the band edge (diffuse layer in the electrolyte neglected).[6] (Reprinted by permission Surface Science.)

4. Experimental Results

An experimental verification of the results of the preceding discussion would involve a direct measurement of the capacitance of the space charge in the insulator surface which has not yet been accomplished. Equilibrium potentials at insulator surfaces have however been measured.[9,10] For such investigations a galvanic cell is used consisting of a series of contacts between different materials ending in terminals of the same material. The cell voltage represents the difference in free energy of electrons at the two ends. Figure 3 presents the principle of a galvanic cell of this type with two hydrogen electrodes. The cell voltage comprises the sum over the potential differences at all interfaces:

$$V = \Delta\phi_{\text{I/II}} + \Delta\phi_{\text{II/III}} + \Delta\phi_{\text{III/IV}} + \Delta\phi_{\text{IV/V}} = \phi_{\text{I}} - \phi_{\text{V}}$$

Assuming the acidity of the two electrolytes to be equal:

$$\Delta\phi_{I/II} + \Delta\phi_{IV/V} = 0$$

When equilibrium is established at the insulator/electrolyte interface and the potential differences $\Delta\phi_{II/III}$ and $\Delta\phi_{III/IV}$ are controlled by electron transfer reactions we obtain

$$V = \Delta\phi_{II/III} + \Delta\phi_{III/IV} = 1/F \left(_{II}\mu_D - _{II}\mu_A - _{IV}\mu_D + _{II}\mu_A\right)$$
$$= _I E_{redox} - _{II} E_{redox} \tag{17}$$

$_{II}\mu_D$: chemical potential of an electron for the donor species in electrolyte II.

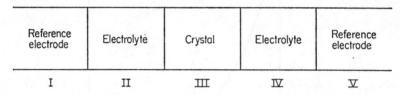

Reference electrode	Electrolyte	Crystal	Electrolyte	Reference electrode
I	II	III	IV	V

Fig. 3. Galvanic sandwich cell with insulator electrode.[10] (Reprinted by permission Electrochimica Acta.)

The experimental verification of eqn (17) is complicated by the extremely low carrier concentration in the bulk of the organic insulator. It was found that this difficulty can be overcome when fairly strong oxidizing (or reducing) redox systems and extremely thin insulator films were chosen. The situation can then be realized that the thickness of the crystal is small compared to the Debye lengths of the two space charge layers which thus overlap so that the low intrinsic 'bulk' concentration of carriers nowhere materializes.

For simplicity we assume the contact at each insulator face to be established by the same redox electrolyte. The electrostatic potential then has a maximum at the centre of the crystal and a single solution to the boundary problem can be given by moving the origin of the coordinate system to the centre:[11]

$$n(x) = n_c \sec^2 \left(\left\{ \frac{2\pi e_0^2 n_c}{kT} \right\}^{1/2} x \right) \tag{18}$$

$$\frac{e_0\phi(x)}{kT} = -2 \log \cos \left(\left\{ \frac{2\pi e_0^2 n_c}{kT} \right\}^{1/2} x \right) \tag{19}$$

n_c is the carrier density at the centre of the crystal (Fig. 1).

Unfortunately eqns (18) and (19) describe the general principle only as for a real crystal the free carrier density is due to trapping effects many orders of magnitude smaller than eqn (18) would suggest. Nevertheless it has been possible to measure reversible redox potentials at very thin anthracene membranes.[10]

III. ELECTRODE REACTIONS WITH CHARGE TRANSFER BY ELECTRONS

1. General Kinetics

The charge transfer reaction may be described by

$$D \rightleftharpoons A \pm C \tag{20}$$

where D and A stand for donor and acceptor species (of electrons or holes) both of which are assumed present in solution and C for a charge carrier in the electrode surface. The discussion shall be restricted to those simple reactions for which the only sources of activation energy are the reorganization processes of the solvent in the immediate vicinity of the species D and A which are concurrent with the charge transfer process. It has been shown previously[12] that the energy of the system during this reaction may be represented as a parabolic function of a reaction coordinate z, which can be interpreted as the virtual charge on the oxidized species with which the orientational polarization of the solvent would be in equilibrium. The minima of the parabolae for oxidized and reduced species occur at values of z equal to the charges of these ions, and therefore separated by unity in the case of a single electron transfer. The simultaneous applicability of the Franck-Condon principle and of conservation of energy to the process under consideration, specifies that the electron transfer event must occur at the crossing point of the parabolae for oxidized and reduced species. Introduction of the significance of the equilibrium potential, that the difference in the total free energies of activation for forward and backward processes is zero, allows the determination of the magnitude of z appropriate to the crossing point and hence of the individual energies of activation. The value of z found differs from that appropriate to a metal electrode, since here we account for the energy expended in the excitation of the charge carrier in the electrode, from the Fermi level to the lowest available level in the surface.

As discussed by various authors[8,12,13] the current density should be formulated over the distance x from the interface and over energy. Assuming first order kinetics in the integrand the partial forward and backward currents are expressed as a triple product of the concentration of solvated species, the density of states in the electrode and a transmission coefficient for electrons through the barrier which lies between donor and acceptor states.

As has been detailed previously this operation leads to the following expressions for the exchange currents at the equilibrium potential of a redox reaction at an insulator electrode:

$$i_0 = \mathscr{F} v C \left\{ \frac{-\lambda}{4RT} \left[1 + \frac{1}{\lambda} \bar{\Delta}_{n,p} \right]^2 \right\} \tag{21}$$

Here i_0 stands for the exchange current due to either electrons or holes; v is independent of the parameters of interest in this application containing a product of the transmission coefficient, the frequency of vibration of the activated complex and the faraday; C_0 is the concentration of the oxidized species and C_R that of the reduced species at the outer Helmholtz plane at the equilibrium potential; λ is an energy parameter which in magnitude equals the energy required to reorient the solvation atmosphere around the oxidized species until it becomes identical with that around the reduced species in equilibrium, it can include contributions from the stretching of bounds in the inner solvation shell λ_i, from the rotations of solvent molecules in the outer solvation shell λ_0 and finally from the change in the polarization of the semi-conductor lattice λ_c. The magnitude of this term can be estimated from continuum theory in the Born approximation

$$\lambda_c \cong \frac{e_0^2}{2} \left(\frac{1}{a} - \frac{1}{R} \right) \left(\frac{1}{\varepsilon_{op}} - \frac{1}{\varepsilon} \right) \tag{22}$$

where a is the radius of a spherical region of the lattice within which the dielectric saturation by an electron or hole is assumed, and ε_{op} is the high frequency dielectric constant of the crystal. Noting that $-e_0^2/\varepsilon a = P$ is the total energy of polarization of the lattice by a carrier, and that $e_0^2/\varepsilon R = W$ is the energy of coulombic interaction of the electron and hole at the separation R, we may write

$$\lambda_c = \tfrac{1}{2} (|P| - |W|) \left(\frac{\varepsilon}{\varepsilon_{op}} - 1 \right) \tag{23}$$

For anthracene we use $\varepsilon = 3\cdot4$ and for ε_{op} the value of $2\cdot765$ obtained by extrapolation of the refractive index to infinite wavelength, for P the value $-1\cdot74$ eV,[17] and for W the value $-0\cdot4$ eV being the difference between the energy of the charge transfer exciton and the band gap of anthracene. These data yield $\lambda_c = 0\cdot17$ eV.

In addition to the 'outer' environmental contribution to the free energy of activation there must be an 'inner' contribution arising from the changes in equilibrium bond lengths as an electron is added. This is approximately:[14]

$$\Delta F^{+}_{\text{inner}} = \tfrac{1}{8} \sum_{\mu v} k_{\mu v} \delta^2_{r\mu v}$$

where $|\delta_{r\mu\nu}|$ is the change in equilibrium length of the bond between atoms μ and ν on electron addition, and $k_{\mu\nu}$ is the force constant, assumed to be identical for molecule and anion. It is sufficiently accurate here to consider only the carbon framework, and to assume that the stretching constant for all C—C bonds in a molecule has a constant average value $\bar{k}$. Then, as $\bar{k}$ is high enough for the difference between $\Delta F^{\ddagger}$ and $\Delta G^{\ddagger}$ to be neglected we have

$$\Delta G^{\ddagger}_{\text{inner}} = \tfrac{1}{8} \sum_{\mu\nu}^{\text{carbon}} \bar{k}_{\mu\nu}\delta^2_{r\mu\nu}$$

It is possible to relate the changes $\delta_{r\mu\nu}$ to the change in π-bond order of the μ—ν bond.

The simplest such relationship is that proposed by Coulson and Poole[15]

$$l = 1\cdot517 - 0\cdot18p$$

where l is the bond length (Å) and p is the bond order. Hence

$$|l^0_{\mu\nu} - l^-_{\mu\nu}| = 0\cdot18 \times c^i_\mu c^i_\nu$$

where the additional electron occupies orbital i (the lowest antibonding orbital in this case). Hence

$$\Delta G^{\ddagger}_{\text{inner}} = \tfrac{1}{8} \times (0\cdot18)^2 \sum_{\mu<\nu}^{\text{carbon}} \bar{k}(c^i_\mu c^i_\nu)^2$$

For anthracene, the coefficients c^i for the various bonds (Fig. 4) are given in Table 1.

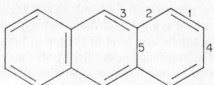

Fig. 4. Numbering of bonds in anthracene.

Table 1. *Hückel coefficients for first antibonding orbital*

bond	number	$c^i_\mu\mu c^i_\nu$
1	4	0·31094 × 0·21987
2	4	0·31094 × 0·09108
3	4	0·43973 × 0·09108
4	2	0·21987 × 0·43973
5	2	(0·09108)²

Thus $\Delta G^{\pm}_{\text{inner}} = \frac{1}{8} \times 3 \cdot 24 \times 10^{-2} \times \bar{k} \times 0 \cdot 047$ ergs-molecule^{-1} (approx.) or

$$\Delta G^{\pm}_{\text{inner}} = 2 \cdot 37 \times 10^{-8} \bar{k} \text{ (eV)} \quad (\bar{k} \text{ in dyne cm}^{-1})$$

Taking the value 5×10^{-5} dynes/cm as a reasonable average for $\bar{k}$, we get

$$\Delta G^{\pm}_{\text{inner}} = 0 \cdot 01 \text{ eV}$$

The 'inner' contribution, although not negligible, is thus quite small, and certainly less than the 'outer' one, for anthracene.

$\bar{\Delta}_{n,p}$ is the magnitude of the forbidden energy gap between the Fermi level and the energy state for the carrier in the surface (the bottom of the conduction band for an electron and the top of the valence band for a hole) at the equilibrium potential.

If the whole equilibrium interfacial potential is located within the solid $\bar{\Delta}$ is determined by the relative position of the Fermi levels in the solid and in the electrolyte and we can make use of the Nernst equation:

$$\bar{\Delta}_n = \Delta + e_0(E_R - E_{FB}) = \Delta + e_0(E_s - E_{FB}) + kT \ln \frac{C_0}{C_R} \quad (24a)$$

$$\bar{\Delta}_p = \Delta - e_0(E_s - E_{FB}) - kT \ln \frac{C_0}{C_R} \quad (24b)$$

E_{FB} (the 'flat band' potential) represents the Fermi level of the insulator in electrochemical units.

2. The Net Current Passing through the Interface

Insulators have very narrow carrier bands so that the density of states for carriers may be presented as a delta function of energy centred on the energy of the bottom of the conduction band or the top of the valence band at the insulator surface. The transmission coefficient also may be presented as a delta function centred on the distance separating the acceptor and donor orbitals. For simplicity the density of states and the transmission coefficient have been treated as true delta functions and the integrations were performed to yield:[1]

$$qi = \nu(N_c C_{D*} - n(o)C_{A*})$$

$n(o)$ is the carrier density at the insulator surface when the current density is i A/cm^2, N_c is the density of unoccupied states in the carrier band at the surface and C_{D*} and C_{A*} are the concentrations of the donor and acceptor molecules in the transition state.

If we can assume that the total applied voltage drops across the insulator only the concentration of carriers in the solid phase will vary with current so that we obtain in the limiting case

$$n(o) \to 0 \quad \text{and} \quad qi \to \nu N_c C_{D*} = i_0$$

We thus come to the important conclusion that when the assumption about the potential distribution is applicable the current through the insulator/electrolyte phase boundary approaches with increasing applied voltage a limiting value which is equal in magnitude to the exchange current at the equilibrium potential. This property of insulator electrodes has been used to determine the exchange current at such electrodes, in most cases anthracene, for various redox systems.[1,16]

3. The Magnitude of Exchange Currents at Insulator Electrodes

From eqns (21), (24a), and (24b) it follows that the relative magnitude of $e_0(E_s - E_{FB})$ and Δ directly control the possibility of observable current flow through the insulator/electrolyte interface. For these materials easily measurable currents can only then be expected when $|E_s - E_{FB}| \gg 0$ so that it will be impossible to inject carriers of both signs from one particular redox system. A further prediction of these formulae which deserves specific mention is that for $e_0(E_s - E_{FB}) \cong \Delta$ the exchange current becomes of the order of magnitude that it takes at a metal.

From Marcus' treatment of electron transfer reactions of the outer sphere type the following expression for the exchange rate of a redox reaction at a metal electrode can be obtained when all work terms are neglected:[12]

$$i_0^M = \frac{kT}{h}\mu(C_0 C_R)^{1/2}\mathscr{F}\exp\frac{-\lambda}{4RT} \tag{25}$$

where μ is the reaction layer thickness.

According to Marcus[18]

$$\frac{kT}{h} \cong 10^4 \text{ cm sec}^{-1} \tag{26}$$

Comparison of eqns (21) and (25) shows that in a very rough approximation the following relationship exists between the exchange current for a particular redox system at an insulator i_0^s and a metal i_0^M:

$$i_0^s \cong i_0^M \exp\left\{\frac{-\lambda}{4RT}\left[\left(1 + \frac{\bar\Delta_{n,p}}{\lambda}\right)^2 - 1\right]\right\} \tag{27a}$$

For $\bar\Delta_{n,p} \ll \lambda$, eqn (27a) can be reduced to[19]

$$i_0^s \cong i_0^M \exp\frac{-\bar\Delta}{2RT} \tag{27b}$$

Differences are caused by two partially compensating factors:

1. the very high density of states which may possibly be found in the insulator bands;

2. the polarization of the crystal lattice associated with the charge transfer process.

For an order of magnitude estimate of the contribution of the increased density of states for an insulator to the exchange current we assume the bandwidth in a metal to be $\Delta E_M \cong 5$ eV and we estimate the width of the carrier band in, e.g. anthracene, to be $\Delta E_A \cong 0.05$ eV.[25] Each metal atom should contribute roughly one electron state to the band, each anthracene molecule likewise. For platinum the density of atoms is about twenty times higher than the density of molecules for anthracene, thus we obtain for the ratio of the effective densities of states $N_M/N_A \cong 0.2$.

The comparison of exchange currents seemed particularly interesting for the system Ce^{4+}/Ce^{3+} on anthracene and platinum. With $E_s \cong 1.4$ V in sulphuric acid and the following parameters for anthracene $E_{FB} = -0.35$ V,[20] $2\Delta = 3.8$ V,[21] using the normal hydrogen electrode as reference we obtain $\bar{\Delta}_p = 0$, i.e. the Fermi energy of the Ce^{4+}/Ce^{3+} system is about equal in magnitude to the ionization energy of crystalline anthracene and surface degeneracy occurs. Differences between the exchange currents at a platinum and an anthracene electrode will thus be due to the polarization of the anthracene lattice and the pre-exponential factors. An experimental determination of the exchange current at an anthracene electrode for this case is unfortunately not possible as the potential drop across the Helmholtz layer is no longer independent of the applied voltage. Experimentally it was found with a rotating anthracene disc electrode[22] that the rate of hole injection from Ce^{4+}/Ce^{3+} into anthracene is determined by the rate of transport of ceric ions to the electrode surface. From pulse measurements[23] the highest measurable injection current was determined as $i_{max} = 2$ amp/cm^2 which compares with $i_0^0 = 0.13$ amp/cm^2 for the standard exchange current at a platinum electrode.[24] If we assume the exchange rate at the degenerate anthracene electrode to be equal to that at a platinum electrode then this result indicates an overvoltage in the Helmholtz layer of about 160 mV if the transfer coefficient $\alpha = 0.5$.

Another useful correlation has been given[11,18] between the rate constant of a (chemical) electron exchange reaction, k_{ex}, and the electrochemical rate constant at zero activation overpotential $k_{el} = i_0/C\mathscr{F}$ again when the work terms are negligible:

$$(k_{ex}Z_{soln})^{1/2} \approx k_{el}Z_{el} \tag{28}$$

Here Z_{soln} and Z_{el} are collision frequencies, namely about 10^3 l./mol/sec and 10^4 cm/sec. In eqn (28) $\approx$ should be replaced by $\geqq$ when the ion-electrode distance in k_{el} exceeds one-half the ion-ion distance in k_{ex}.[18]

4. Energetic Considerations on Carrier Generation

For molecular solids, the lattice forces are of the van der Waals type and there will be only small overlap between the orbitals of adjacent molecules. For anthracene, e.g. the calculated intermolecular resonance integrals range in value from 5 to 30×10^{-16} erg for nearest-neighbour pairs, depending on orientation.[26] Thus the energetics of carrier formation may be described in terms of localized electrons and holes,[27] although the mobilities are probably best described in terms of the band approximation. Localized electrons and holes can be considered as negative and positive hydrocarbon ions respectively. The properties of these ions have been well investigated,[28] and, provided that the molecule is alternant, the spin and charge density distribution apart from the sign of the latter are identical for cations and ions. The frequencies of the excited electronic states, and excitation probabilities, are also independent of the sign of the charge. This simplifies the discussion of the properties of these ions.

According to Hush[27] the energy distance between the lower edge of the lowest band for free electrons (the conduction band) and the upper edge of the highest band for free holes (the valence band) may be determined from the energy which is required to form a pair of carriers. We denote a positive carrier by M^+ and a negative carrier by M^-, it being understood that these refer to holes and electrons trapped on individual molecules. The energy for formation of a hole-electron pair at infinite separation in the crystal can be calculated from the ionization potential I, the electron affinity A and the polarization energies P^+ and P^- of the ions. As values for I and A have been determined experimentally,[29] the remaining unknown quantities are the polarization terms. Some calculations of these have been made by Lyons and Mackie,[17] who approximated the ions fields as resulting from classical point charges. As Choi and Rice have noted,[30] this will tend to overestimate the polarization terms, since the π-charge distribution is in fact delocalized over the molecular framework.

The analogous polarization energies in liquid solution (differential real potentials) have recently been reported for a series of hydrocarbon positive and negative ions.[31] The solvent used was methyl cyanide. The difference between P^+ and P^- for a given molecule was found to be small and nearly constant. The average value was

$$P^+ - P^- = -0.20 \text{ eV}$$

This difference was attributed to the effect of the surface potential χ of the solvent, and it is calculated from this result, that χ for acetonitrile is -0.10 V. Apart from this, the polarization energies are identical for positive and negative ions of the same hydrocarbon. This implies that the solvent has the role simply of a dielectric medium for these large ions, and that there are no specific chemical solvation effects. It follows that a very good approximation to the polarization terms for these ions in another medium will be obtained by the use of the Born relationship to correct for the change in dielectric constant. Applying this to the case of ions in hydrocarbon crystals we have[27]

$$P_c^{\pm} = P_s^{\pm} \frac{(\varepsilon_c - 1)\varepsilon_s}{(\varepsilon_s - 1)\varepsilon_c} \tag{29}$$

where the subscripts c and s refer to the crystal and solution respectively, and ε is the dielectric constant.

In Table 2, the ionization potentials, electron affinities and ion polarization energies $P_s^{\pm}$ for CH_3CN solution are listed for a series of alternant hydrocarbons. The dielectric constants of the crystals and the corresponding

Table 2. *Based on Ref.* (27)

Hydrocarbon	I (a)	A (b)	$-P_s^{\pm}$ (c)	ε_c (d)	$-P_c^{\pm}$	I_c	A_c	2Δ	F_c
naphthalene	8·20	0·15			1·28(17)	6·92	1·43	5·49	4·18
anthracene					1·74(17)	5·86	2·29	3·57	
	7·61	0·55	1·94	3·38	1·41	6·20	1·96	4·24	4·08
chrysene	8·01	0·33	1·83	2·97	1·25	6·76	1·58	5·18	4·17
pyrene	7·72	0·39	1·86	3·36	1·35	6·37	1·74	4·63	4·06
phenanthrene	8·06	0·20	1·76	2·54	1·10	6·96	1·30	5·66	4·13
1·2 benzanthracene	7·74	0·46	1·86	3·41	1·35	6·39	1·81	4·58	4·10
triphenylene	8·19	0·14	1·84	3·41	1·34	6·85	1·48	5·37	4·17

Ionization potentials and electron affinities (I, A) of hydrocarbons in the gas phase and in the crystal (I_c, A_c), the energy gap $2\Delta = I_c - A_c$. $P_s^{\pm}$ and $P_c^{\pm}$ are the polarization energies of ions in methyl cyanide solution and in the hydrocarbon crystal respectively. ε_c is the dielectric constant of the crystal, F_c is the energy of the Fermi level. The data for naphthalene crystal and for the first line for anthracene crystals were taken from (17). All energies are in eV.

(a) For sources of data see Reference 31.
(b) See Reference 33.
(c) See Reference 31.
(d) Calculated from molecular polarizabilities using the method of Denbigh, *Trans. Faraday Soc.*, **36**, 936 (1940). Crystal densities are calculated from unit cell data of A. I. Kitaigorodskii, *Organic Chemical Crystallography* (translated from Russian by Consultant Bureau, New York, 1965).

polarization energies $P_c^{\pm}$ calculated using eqn (29) are also shown. Comparison with the calculation of Reference (17) for $P_c^{\pm}$ can be made only for anthracene. As anticipated, the value obtained by Hush ($-1\cdot41$ eV) is smaller than that calculated with the point charge model ($-1\cdot74$ eV). In the last three columns of the table, the crystal ionization potential I_c ($=I + P_c^{\pm}$), the crystal electron affinity $A_c(=A - P_c^{\pm})$ and the difference $2\Delta = I_c - A_c$ are given. The quantity 2Δ is the difference between the energies of the valence and conduction bands on this approximation.

We consider these crystals as intrinsic semiconductors and thus determine the energy of the Fermi level to be

$$F_c = \frac{I_c + A_c}{2} = 4\cdot15 \text{ eV} \tag{30}$$

as the mean value for the hydrocarbons listed in Table 2. The constancy of the Fermi level is a consequence of the Hush-Pople rule of the constancy of the sum of electron affinity and ionization energy for alternant hydrocarbons.[32] In Table 2 an electron in vacuum at infinity has been used as the reference energy state. Conversion between energy levels measured on this reference system and those measured on the electrochemical scale may be achieved through use of the energy on the vacuum scale of the origin of the hydrogen electrode scale which has been determined[25] to

$$E_{H^+H_2} = -4\cdot5 \text{ eV}$$

so that

$$E^H = -4\cdot5 - E \tag{31}$$

The band structure at the interface between an aromatic hydrocarbon crystal and an electrolyte which consists of a solution of the radical ions of the hydrocarbon is determined by the difference between the Fermi levels of the crystal and the electrolyte before contact, eqn (22)

$$F_c - F_s = \frac{I_c + A_c}{2} - A_s \tag{32a}$$

$$F_c - F_s = \frac{I_c + A_c}{2} + I_s \tag{32b}$$

Equation (32a) applies for a solution of radical anions and eqn (32b) for a solution of radical cations. As the distance between the carrier band and the Fermi level at the solid-electrolyte interface at equilibrium is determined by (see eqns (24a) and (24b))

$$\bar{\Delta} = \Delta - e_0(E_{FB} - E_s) \tag{33}$$

we obtain with $\Delta = \frac{1}{2}(I_c + A_c)$:

$$\bar{\Delta} = P_s^{\pm} - P_K \tag{34}$$

i.e. the distance between the carrier band and the Fermi level at the surface of an aromatic hydrocarbon crystal which is in contact with a solution of its radical ions is determined by the difference of the polarization energies of the crystal and the solution.[22]

For anthracene we obtain with $P_c = 1 \cdot 41$ eV and $P_s^{\pm} = 1 \cdot 94$ eV for acetonitrile $\bar{\Delta} \cong 0 \cdot 5$ V.

IV. CHARGE TRANSFER WITH EXCITONS

1. The Mechanism of Photoinjection

It is well known from studies of homogenous systems that the rate of the reaction

$$M + D \underset{k_b}{\overset{k_f}{\rightleftarrows}} M^{+(-)} + A \tag{35}$$

where M is a hydrocarbon molecule, $M^{+(-)}$ a hydrocarbon cation or anion and D,A donors and acceptors of position or negative charges can be accelerated by irradiation with light of the proper wavelength which leads to the formation of excited states of the hydrocarbon molecules, e.g. the first excited singlet state:[34]

$$^1M^* + D \underset{k_b^*}{\overset{k_f^*}{\rightleftarrows}} M^{+(-)} + A \tag{36}$$

We expect a similar increase of the rate of electron transfer if the hydrocarbon molecules are part of the surface of a molecular crystal electrode.

Recent theoretical developments[35] make it possible to quantitatively predict the increase of the electron transfer rate on illumination.

For adiabatic reactions of the outer sphere type, in which the activation process involves only reorganization of the solvation atmosphere about the reactant, together with symmetrical stretching or compression of bonds in the inner coordination or solvation shell, the rate constants may be expressed in the form

$$k_f = Z \exp \frac{-m^2 \lambda^*}{RT} \tag{37a}$$

$$k_b = Z \exp \frac{-(1 - m)^2 \lambda^*}{RT} \tag{37b}$$

$$m = \frac{1}{2} \left[1 + \frac{\Delta F^*}{\lambda} \right] \tag{37c}$$

Z is a collision number, λ^* is the reorganization term, ΔF^* is the standard free energy difference between the reactants and the products in the states immediately prior and after electron transfer, respectively.

λ^* will be left a semiempirical parameter, to be fitted by inspection of the rates of transfer reactions. ΔF^*, however, will be estimated from more fundamental quantities. For the dark reaction the state prior to electron transfer consists of the donor species, in its equilibrium solvation environment, adjacent to the insulator surface in which there is an unoccupied level available in the carrier band. Similarly in the state immediately after electron transfer the acceptor species exist in an equilibrium solvation environment adjacent to the surface, which now contains a carrier in a state at the bottom of the band.

The bands are bent in the region near the surface, but our assumption about the distribution of potential in the interface implies that in this post electron-transfer state the plane of the crystal in which the carrier is to be found is equipotential with the solution. For simplicity we regard the energies of donor and acceptor species as being independent of distance from the solid. Thus, the standard free energy change ΔF may be written as the difference between the standard chemical potential of the donor species in solution μ_A^0, and the sum of those of the carriers in the solid μ_C^0 and the acceptor species in solution $\mu_{A\pm}^0$

$$\Delta F = \mu_{A\pm}^0 + \mu_C^0 - \mu_A^0 \tag{38}$$

μ_C^0 is represented by the energy of the edge of the carrier band in the uncharged regions of the solid, which we relate to the Fermi level F_C and the gap between the carrier band edge and the Fermi level in the bulk of the solid:

$$\mu_C^0 = F_C + \Delta \tag{39}$$

The difference in chemical potentials $\mu_{A\pm}^0 - \mu_A^0$ may be rewritten after making use of the condition for equilibrium between the solid and the redox system in the absence of irradiation. This may be expressed by:

$$\bar{\mu}_{A\pm}^0 + \bar{\mu}_C = \bar{\mu}_A^0 \tag{40}$$

that is:

$$\mu_{A\pm}^0 - \mu_A^0 = -F_C - q\mathscr{F}V_d^0 \tag{41}$$

The superscript 0 on V_d^0 indicates the standard state of equal activities of donor and acceptor species in solution. $\bar{\mu}_i$ represents the electrochemical potential of species i, which in eqn (41) has been divided into chemical and electrical terms. Substitution of eqns (39) and (41) into eqn (38) results in:

$$\Delta F = \Delta - q\mathscr{F}V_d^0 \tag{42}$$

ΔF, k_f, and k_b are independent of irradiation because the assumed potential distribution causes the relative positions of the energy levels involved in electron transfer to be unalterable.

The photoinjection reaction may now be dealt with summarily, by consideration of the difference in the pre-electron transfer states of this reaction and the dark reaction. The system now includes one excited molecule in the transition state, which has energy E_X, say, with respect to a ground state molecule.* In general terms, this energy difference is given by the energy of the absorption which gives rise to the first excited singlet in the solid. Hence

$$\Delta F^* = \Delta - q\mathscr{F}V_d^0 - E_X \tag{43}$$

The relative positions of the various states are illustrated in Fig. 5. Although

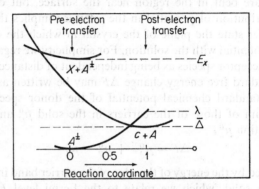

Fig. 5. Cross-section through the potential energy surface for dark and photoinjection into a molecular crystal. $Z = 10^7$ cm^4/mol/sec; $D = 1\cdot25 \times 10^3$ cm^2/sec; $\tau = 2 \times 10^{-8}$ sec; $\lambda = 2$ eV.[42] (Reprinted by permission Electrochim. Acta.)

the band gaps of some insulators are known, on the whole information from which it might be estimated is scarce. V_d^0 may be replaced by $E_S - E_{FB}$ where E_S is the standard potential of the redox system and E_{FB}, the flat band potential of the solid, may be calculated as described by Lohmann.[20]

2. The Transport Problem for Excitons

In order to obtain an expression for the magnitude of the photostimulated electron transfer rate we have to consider the density of excited state ('excitons') at the crystal surface.

The singlet exciton created by absorption of light in the first allowed adsorption band of a molecular solid is mobile.[36] It is localized on

---

* Due to the trapping of excitons by imperfections in the crystal surface, there is in a real system a spectrum of values of E_X. In such a case, estimation of the rate constant involves an integration over transition states.

particular lattice sites (unlike the charge transfer exciton), and its transport obeys the diffusion equation.[37] It is found that the exciton has a well defined lifetime, due to radiative decay or radiationless transition to a lower energy state.[38] If during its lifetime an exciton arrives at the boundary between the solid and the electrolyte, one of four things might happen. It might decay radiatively, the emitted photon being detectable outside the solid as fluorescence. Secondly, it might decay non-radiatively by interaction with charges trapped in bulk- or surface-states. Thirdly, it may loose its excitation energy as a result of resonance transfer to an appropriate acceptor state. We are here concerned mainly with a fourth possibility, which involves reaction between the exciton and a reactant in solution, a carrier being generated inside the solid where it can be accelerated by an applied electric field.

For irradiation times much longer than the exciton lifetime τ_X, a steady state concentration of excitons exists described by the diffusion equation:

$$D_X \frac{d^2 D_X}{dx^2} - \frac{C_X}{\tau_X} + \varepsilon I_0 \, e^{-\varepsilon x} = 0 \tag{44}$$

D_X is the diffusion coefficient and C_X the concentration of the excitons. ε is the extinction coefficient of the solid, wavelength dependent, and I_0 the intensity of the incident radiation. The case analysed is that in which the exciting radiation is normally incident (along the x-axis) upon the interface between the solid and the electrolyte; x has its origin at this interface. In eqn (44) conservation of excitons exists inside each volume element, since the first term gives the net rate of transport into the slab through its faces, the second gives the rate of decay by all mechanisms, and the third the rate of formation by absorption of light. Equation (44) has the solution:[37]

$$C_X = \tau_X \varepsilon I_0 \left\{ \alpha \, e^{x/\sqrt{(D_X \tau_X)}} + \beta \, e^{x/\sqrt{(D_X \tau_X)}} + \frac{e^{-\varepsilon x}}{1 - \varepsilon^2 D_X \tau_X} \right\} \tag{45}$$

α and β are arbitrary constants which are chosen to satisfy boundary conditions.

The extinction coefficient of molecular crystals is often so high ($\varepsilon \sim 3.2 \times 10^4$ at 390 mμ in anthracene) that virtually all of the incident light is absorbed inside the crystals, with thicknesses $d > 20 \, \mu$m, normally used in these experiments. Hence we justify the simplifying assumption that the crystal is semifinite, $C_X = 0$ at $x \to \infty$. Evidently this leads to $\beta = 0$. As a condition upon C_X at the electrolyte interface, we equate the net flux of excitons to the surface to the net rate of their disappearance through the agency of the photoinjection reaction:

$$D_X \frac{dC_X}{dx} = k_f^* C_X C_{A^\pm} - k_b^* C_c C_A \tag{46}$$

at $x = 0$. C_C, $C_{A^\pm}$, and C_A are the surface concentrations of carriers, donors and acceptors respectively; we can choose to express these, and C_X, as moles/cm³.

For further calculation we require only the concentration of excitons in the transition state of the electron transfer reaction, that is, at $x = 0$. By incorporation of eqn (46) into eqn (45) this is found to be:

$$C_X(0) = \frac{I_0/k_f^* C_{A^\pm}}{1 + [k_f^* C_{A^\pm} \sqrt{(D_X/\tau_X)}]^{-1}} \left\{ \left[1 + \frac{1}{\varepsilon\sqrt{(D_X\tau_X)}} \right]^{-1} + \frac{k_b^* C_A C_C(0)}{I_0} \right\} \quad (47)$$

3. The Current Flowing through the Interface

Let the charge on the carrier be qe_0, where e_0 is the magnitude of the electronic charge. Then if $A^\pm$ is an oxidizing species, a cathodic current flows, holes carry the current through the crystal and $q = +1$. Similarly, an electron current flows with $q = -1$, when $A^\pm$ is reducing.

The total current density i through the interface has a dark, and a photo-component. It may be expressed:

$$\frac{i}{q} = k_f^* C_X(0) C_{A^\pm} - k_b^* C_C(0) C_A + k_f \rho C_{A^\pm} - k_b C_C(0) C_A \quad (48)$$

$\mathscr{F}$ is the Faraday. ρ, the molar density of states in the carrier band, is introduced in order that all rate constants have the same dimensions (cm⁴ mole⁻¹ sec⁻¹).

4. The Concentration of Carriers in the Surface

Equations (47) and (48) are simultaneous equations involving two unknowns, namely the surface concentrations of excitons and of carriers. Solving them for C_C, we find

$$C_C(0)$$
$$= \frac{I_0\{[1 + 1/\varepsilon\sqrt{(D_X\tau_X)}][1 + 1/k_f^* C_{A^\pm} \sqrt{(D_X/\tau_X)}]\}^{-1} + k_f \rho C_{A^\pm} - (i/q\mathscr{F})}{C_A\{k_b + k_b^*[1 + k_f^* C_{A^\pm} \sqrt{(\tau_X/D_X)}]^{-1}\}}$$
$$(49)$$

In particular the surface concentration of carriers at open circuit, $C_C^0(0)$, is obtained from this equation by putting $i = 0$. Evidently:

$$\frac{C_C(0)}{C_C^0(0)}$$
$$= 1 - \frac{i}{q} \left\{ k_f \rho C_{A^\pm} + I_0 \left[\left(1 + \frac{1}{\varepsilon\sqrt{(D_X\tau_X)}} \right) \left(1 + \frac{1}{k_f^* C_{A^\pm} \sqrt{(\tau_X/D_X)}} \right) \right]^{-1} \right\}^{-1}$$
$$(50)$$

$C_C(0)$ and $C_C^0(0)$ increase linearly with the intensity of irradiation, I_0. $C_C(0)$ decreases linearly with the current density.

5. The Transport Problem for Carriers

At the steady state the flux of carriers through all planes parallel to the interface must be constant, determining the current. Migration is the dominant means of transport throughout most of the crystal, diffusion being appreciable only within the reservoir of carriers generated by charge separation through the interface. A familiar approximation therefore consists of equating the total current to the migration current everywhere

$$i = \mu_C \mathscr{F} C_C E \tag{51}$$

The error made is least serious at high applied voltages, which happens to be the region of most interest to electrochemistry as will become apparent in the following. The previously undefined symbols in eqn (51) are the carrier mobility μ_C and the local electric field E. The latter is determined by the distribution of carriers within the insulator, according to Poisson's equation:

$$\frac{dE}{dx} = \frac{4\pi q \mathscr{F} C_C}{\kappa} \tag{52}$$

κ is the dielectric constant of the insulator. Equations (51) and (52) may be integrated simultaneously, using $C_C = C_C(0)$ as boundary condition at $x = 0$, in order to determine the field and carrier distributions inside the crystal. The current-voltage relationship follows from a further integration of the field over the insulator, using

$$V = \int_0^d E dx \tag{53}$$

where d represents the thickness of the solid. V, here, is the total applied voltage. In its most convenient form, the result of this is:

$$\left[\frac{q\mu_C\kappa}{32\pi qid^3}\right]^{1/2} V = \left[1 + \frac{i}{i_1}\left(\frac{C_C^0(0)}{C_C(0)}\right)^2\right]^{3/2} - \left[\frac{i}{i_1}\left(\frac{C_C^0(0)}{C_C(0)}\right)^2\right]^{3/2} \tag{54}$$

where

$$i_1 = \frac{8\pi d\mu_C}{\kappa}[\mathscr{F} C_C^0(0)]^2 \tag{55}$$

i_1 is the current which would flow if a uniform volume charge of $\mathscr{F} C_C(0)$ C cm^{-3} were subjected to the field $2Q/C_G d$ where C_G is the geometrical capacitance, and $Q = C_C^0(0)\mathscr{F} d$. i/i_1 is a small quantity, because the average concentration of carriers is much less than $C_C^0(0)$. Whilst $(i/i_1)(C_C^0(0)/C_C(0))^2$ is also small in comparison with unity, the right hand side of eqn (54) is practically unity and Child's law results. This inequality in practice is obeyed over most of the range of variation of the current, i.e. the illuminated surface of the crystal makes an 'ohmic contact' to the bulk of the crystal into which a space charge limited current is injected.

6. The Maximum Photoinjection Current

It was shown in Section 4 that the current through the interface can increase only if the concentration of carriers at the surface decreases linearly with the current. This is so, because the only means of variation of the net rate of charge transfer through the interface at constant light intensity is through a change in the rate of the carrier capture reaction.

Writing eqn (50) in the form:

$$\frac{C_C(0)}{C_C^0(0)} = 1 - \frac{i}{i_{11m}} \tag{56}$$

where

$$i_{11m} = q\mathscr{F}\left\{k_{f}\rho C_A^{\pm} + I_0\left[\left(1 + \frac{1}{\varepsilon\sqrt{(D_X\tau_X)}}\right)\left(1 + \frac{1}{k_f^*C_{A^{\pm}}\sqrt{(\tau_X/D_X)}}\right)\right]^{-1}\right\} \tag{57}$$

we see that the carrier concentration can continue to decrease only if $i < i_{11m}$; i_{11m} therefore is a limiting value for the rate of injection. Reference to eqn (54) reveals that an infinite voltage is required to raise i to i_{11m}. Combination of eqns (54) and (56), therefore, yields a single equation describing the variation of the current, with voltage over the range $0 < V < \infty$, and with the extinction coefficient.

The limiting current is seen to have additive contributions from the maximum dark injection current, equal to the exchange current i_0 of the redox system at the insulator electrode ($i_0 = q\mathscr{F}k_f\rho C_{A^{\pm}}$), and from the photoinjection. The latter contribution is linearly proportional to the illumination intensity, and shows the structure of the absorption spectrum due to the influence of the extinction coefficient (Fig. 6).

According to eqn (57) the quantum efficiency for this extrinsic charge separation process is

$$\frac{i_{11m} - i_0}{q\mathscr{F}I_0} = \Phi = f_1 f_2 \tag{58a}$$

where

$$f_1 = \frac{1}{1 + 1/\varepsilon\sqrt{(D_X\tau_X)}} \tag{58b}$$

and

$$f_2 = \frac{1}{1 + 1/k_f^*C_{A^{\pm}}(\sqrt{\tau/D_X})} \tag{58c}$$

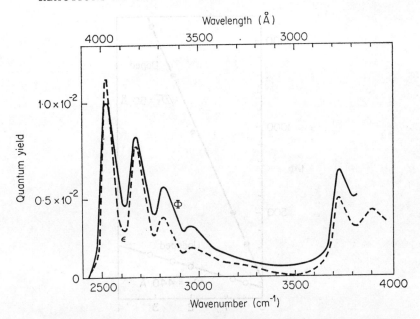

Fig. 6. Comparison of photoconduction spectrum with absorption spectrum of anthracene. Light polarized parallel to b-axis. Electrolytic contact: 1 N NaOH.[50] (Reprinted by permission Philips Res. Repts.)

For convenience of discussion Φ has been factorized into the efficiencies of collection of excitons by the surface, f_1, and the efficiency of the charge separation reaction, f_2. Previous theories of photoinjection[35,36] which did not include a treatment of the kinetics of the surface reaction, yielded the same formula for f_1 but left f_2 undetermined. From eqns (58a) and (58b) the mean free path for the diffusion of excitons, $\sqrt{(D_X \tau_X)}$, can be determined. Conveniently a graphical procedure can be used based on a plot of $1/(i_{11m} - i_0)$ vs. $1/\varepsilon$. Figure 7 shows the results of measurements on an undoped and a tetracene-doped anthracene crystals, illustrating the reduction of the exciton mean free path due to the activity of tetracene as a trap for excitons.

The present kinetic treatment of the exciton reaction enables us to make certain predictions about the efficiency of the charge separation reaction. f_2 approaches unity:

(a) if the rate constant for the heterogeneous reaction tends to infinity;

(b) if the solution is very concentrated in the donor species;

(c) if the velocity of the exciton $\sqrt{(D/\tau)}$ is small so that the maximum rate of annihilation of excitons by reaction exceeds their maximum rate of arrival at the surface.

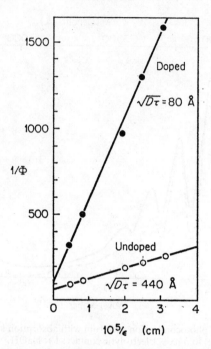

Fig. 7. The ratio of the quantum efficiencies of the photoinjection process in an undoped anthracene crystal and a crystal doped with 2×10^{-4} M/M tetracene increases with increasing excitation depth $1/\varepsilon$.[50] (Reprinted by permission Philips Res. Repts.)

7. The Photovoltage

In the absence of illumination, eqn (49) yields for the carrier concentration at open circuit, $C_C^d(0)$

$$C_C^d(0) = \frac{k_f \rho C_{A^\pm}}{C_A\{k_b + k_b^*[1 + k_f^* C_A \sqrt{(\tau_X/D_X)}]^{-1}\}} \tag{59}$$

The presence of the rate constants for reaction eqn (36) in this expression is accounted for by the possibility that injection of a carrier might create an exciton in the solid. Normally one can assume $k_b^* \sim 0$.

The expression derived from eqn (49) for $C_C^0(0)$ may be rewritten, after using eqns (57) and (59), in the form:

$$C_C^0(0) = C_C^d(0) \frac{i_{1\text{lim}}}{i_0} \tag{60}$$

which underlines the proportionality between i_{lim}, and the capacity of the carrier reservoir at open circuit. These carrier concentrations, being equilibrium quantities, may also be written in the following manner, however:

$$C_C^0(0) = \rho \exp\left[-\frac{\Delta - q\mathscr{F}V_p}{RT}\right] \tag{61}$$

$$C_C^d(0) = \rho \exp\left[-\frac{\Delta - q\mathscr{F}V_d}{RT}\right] \tag{62}$$

where V_p and V_d are the potential differences within the solid in the presence of illumination and in the absence of illumination, respectively; V_d and V_p are defined to have the sign of the charge of the carrier in the space charge layer. Elimination of the concentrations between eqns (60), (61), and (62) yields a simple, direct relationship between the photovoltage, measured in the absence of current flow, and the maximum photocurrent i_{lim} which is measured at large applied voltages,

$$V_p - V_d = \frac{RT}{q\mathscr{F}} \ln\frac{i_{lim}}{i_0} \tag{63}$$

Evidently the spectral dependence of the photovoltage must resemble that of the absorbance by the crystal, peaking in the same wavelength regions. Extrapolation of a V_p, $\ln i_{lim}$ plot to V_d determines the exchange current. This procedure can even be carried out for systems which give measurable photocurrents but dark currents less than the leakage current, since the potential V_d should be identical at insulating and metallic membranes provided that secondary effects, e.g. corrosion, can be neglected.

Figures 8(a) and 8(b) present experimental verifications of the relationship eqn (63) between limiting photocurrent i_{lim} and photovoltage. In a galvanic sandwich cell (Fig. 3) one side of the anthracene crystal was illuminated through the electrolyte. The non-illuminated half of the cell contained Ce^{4+}/Ce^{3+} in $15n$-H_2SO_4 which quickly established a well defined potential at the anthracene electrode. The photovoltage was plotted against the light-intensity I_0 on which i_{lim} depends linearly (Fig. 10). Notice that

$$\frac{d(V_p - V_d)}{d\ln i_{lim}} = 29\ mV$$

for Tl^{3+}/Tl^+, and

$$\frac{d(V_p - V_d)}{d\ln i_{lim}} = 58\ mV$$

for $IrCl_6^{3-}/IrCl_6^{2-}$.

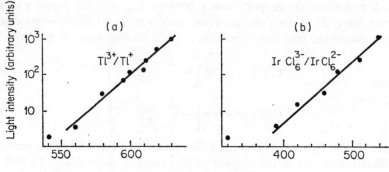

Photovoltage (mV)

Fig. 8(a). Dependence of photovoltage upon light intensity at interface between anthracene and Tl^{3+}/Tl^+ solution.[10] (Reprinted by permission Electrochimica Acta.)

Fig. 8(b). Dependence of photovoltage upon light intensity at interface between anthracene and $IrCl_6^{2-}/IrCl_6^{3-}$ solution.[10] (Reprinted by permission Electrochimica Acta.)

8. Quenching of Excitons at the Crystal/Electrolyte Interface by Energy Transfer

Following Förster's theory of resonance energy transfer between a donor and an acceptor when the interaction between them is of the dipole-dipole type, the rate of quenching of the excitation energy in species A by an energy absorber B situated the distance R away is given by[48]

$$k_{DA} = \frac{K^2 e_0^4}{16\pi^2 rm^2 R^6} \int_0^\infty F_D(\nu)F_A(\nu)\frac{d\nu}{\nu^2} \ \sec^{-1} \tag{64}$$

Here K is a dimensionless constant of order unity, r is the refractive index of the medium, m is the mass of the electron, ν is the wave number of the energy transferred and $F_A(\nu)$ $F_D(\nu)$ is the oscillator strength for the donor and acceptor at the frequency ν.

Equation (64) may be rewritten as:

$$k_{DA} = \frac{1}{\tau}\left(\frac{R_0}{R}\right)^6 \tag{65}$$

where τ is the lifetime of the exciton and R_0 is the critical separation distance at which normal decay and energy transfer are equally likely.

In order to compute the total rate of quenching in the interfacial region we have to integrate over R which leads to:[48]

$$k_Q = \frac{\pi R_0^6}{12\tau R_e^2} \ cm^4 \ \sec^{-1} \tag{66a}$$

or

$$k_Q = \frac{N\pi R_0^6}{12\tau R_e^2} \text{ cm}^4 \text{ mole}^{-1} \text{ sec}^{-1} \tag{66b}$$

where R_e is the minimum distance of separation between energy donor and acceptor $\sim$5 Å.

9. The Photocurrent at an Insulator/Electrolyte Interface in Presence of Quenching of Excitons by Energy Acceptors in Solution

We shall now discuss the case when the energy of the exciton is dissipated simultaneously by a charge transfer process and an energy transfer process. The boundary value problem defining this system is as follows:

$$D\frac{d^2C_x}{dx^2} - \frac{C_x}{\tau} + \varepsilon I_0 \, e^{-\varepsilon x} = 0 \tag{67a}$$

at $x = 0$:

$$D\frac{dC_x}{dx} = k_Q C_x C_A + k C_x C_e \tag{67b}$$

at $x \to \infty$:

$$C_x = 0 \tag{67c}$$

Here C_A represents the concentration in solution of an energy acceptor and C_e the concentration of a species involved in an electrode reaction with excitons.

A solution of this boundary value problem is (see Sections 1–7):

$$C_x = I_0 \varepsilon \tau \left\{ \alpha \left[\exp \frac{-X}{\sqrt{(D\tau)}} \right] + \frac{\exp(-\varepsilon X)}{1 - \varepsilon^2 D\tau} \right\} \tag{68}$$

From the boundary condition at $X = 0$ we obtain

$$\alpha = - \frac{\varepsilon/(1 - \varepsilon^2 D\tau) + [(k_Q C_A + k C_e)/D][1/(1 - \varepsilon^2 D\tau)]}{[1/\sqrt{(D\tau)}] + (k_Q C_A + k C_e)/D} \tag{69}$$

The maximum photocurrent must now be calculated from

$$i_p = i_0 + e_0 k C_x C_e \tag{70}$$

where i_0 is the dark current. Equation (70) leads after substituting eqns (68) and (69) for C_x to

$$\frac{i_\infty}{i_p - i_0} = 1 + \frac{1}{kC_e} + \frac{k_q C_A}{k C_e} \tag{71}$$

where $i_\infty = e_0 I_0$.

$(i_p - i_0)/i_\infty$ is the experimental quantum efficiency for charge separation by the electrode reaction, and $[1 + 1/kC_e\sqrt{(\tau/D)}]^{-1}$ is the theoretical efficiency for $k_Q = 0$. We denote these f_{ex}^* and f_{th}^* respectively and thus obtain from eqn (71)

$$\frac{1}{f_{ex}^*} = \frac{1}{f_{th}^*} + \frac{k_Q C_A}{kC_e} \tag{72a}$$

and

$$k_Q = \frac{kC_e}{C_A}\left[\frac{1}{f_{ex}^*} - \frac{1}{f_{th}^*}\right] \tag{72b}$$

This formula may be used for evaluation of quenching rate constants from experiment.

10. Experimental Determination of the Quantum Efficiency for the Extrinsic Charge Separation Process

The formulae derived in Sections 1–9 have been tested in an experimental arrangement which permitted illumination of an anthracene crystal through a 1 mm thick volume of electrolyte. 403 mμ was chosen for the wavelength of the exciting light because the absorption coefficient of anthracene has near this wavelength its maximum. Other experimental details have been discussed in Reference 38. Figure 9 shows a current voltage curve obtained for a 10^{-2}

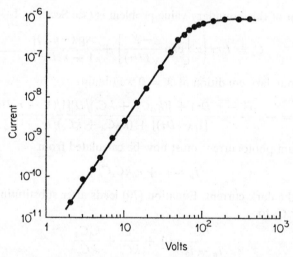

Fig. 9. Typical current (amps/cm²) voltage curve for photo-injection experiment. 10^{-2} M Tl³⁺ in 0·5 N HCl.³⁸ (Reprinted by permission Discussions Faraday Soc.)

molar solution of Tl^{3+}/Tl^+ in $0.5n$-HCl representing the general type of curve observed in these experiments. Below a critical voltage the current is controlled by the bulk properties of the crystal giving rise to an increase of the current with the third power of the applied voltage as observed previously for the positive space charge limited current in anthracene.[1] For higher voltages current saturation is observed. The current is now controlled by the rate of carrier generation at the anthracene surface which is controlled by eqn (58). Figure 10 shows that in the intensity range investigated the limiting current

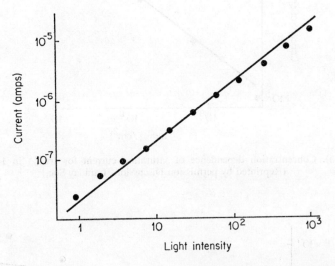

Fig. 10. Variation of saturation current with relative light intensity. 5×10^{-2} M Tl^{3+} in 0.5 N HCl.[38] (Reprinted by permission Discussions Faraday Soc.)

increases linearly with light intensity. The dependence of the photocurrent on the concentration of the electron acceptor in the electrolyte is shown for three different systems in Figs. 11(a), (b), (c). Only for the $IrCl_6^{3-}/IrCl_6^{2-}$ solution does the system show the linear increase of current with the acceptor concentration as predicted by eqn (58). For Fe^{3+}/Fe^{2+} in 1 M H_2SO_4 and Tl^{3+}/Tl^+ in $0.5n$ HCl an appreciable variation of the photocurrent with concentration was observed for low concentrations of the redox systems only while no concentration dependence was observed for higher concentrations. In each case, however, an appreciable increase of the photocurrent over that observed with the supporting electrolyte alone was detected. Nothing is known as yet about the nature of the interaction of Fe^{3+} and Tl^{3+} ions with the anthracene surface which gives rise to this saturation.

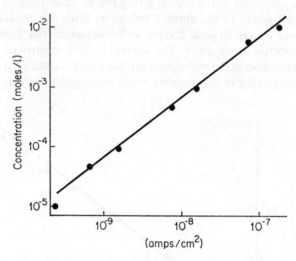

Fig. 11(a). Concentration dependence of saturation current for $IrCl_6^{2-}$ in 1 N HCl.[38] (Reprinted by permission Discussions Faraday Soc.)

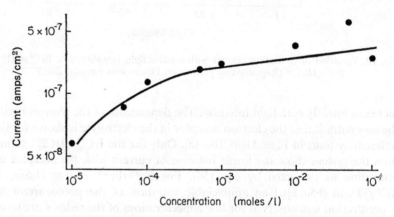

Fig. 11(b). Concentration dependence of saturation current for Fe^{3+} in 1 N H_2SO_4.[38] (Reprinted by permission Discussions Faraday Soc.)

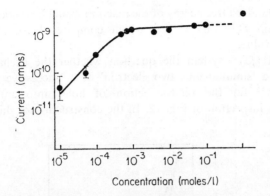

Concentration (moles/l)

Fig. 11(c). Concentration dependence of saturation current for Tl^{3+} in 1 N HCl.[38] (Reprinted by permission Discussions Faraday Soc.)

For the determination of the efficiency of photogeneration of charge carriers the following data for the photocurrents were used:

$$10^{-2}\ \text{M}\ \text{IrCl}_6^{2-} \text{ in } 1n\text{-HCl } 4\cdot5 \times 10^{-8}\ \text{A}$$
$$10^{-5}\ \text{M}\ \text{Tl}^{3+} \text{ in } 0\cdot5n\text{-HCl } 1\cdot8 \times 10^{-9}\ \text{A}$$
$$10^{-5}\ \text{M}\ \text{Fe}^{3+} \text{ in } 1n\text{-H}_2\text{SO}_4\ 4\cdot0 \times 10^{-9}\ \text{A}$$

If each photon reaching the electrode surface were converted into an electron we would obtain a current $i_\infty = I_0 e_0 = 2\cdot83 \times 10^{-5}$ A. This value is reduced due to the finite exciton collection efficiency of the electrode to

$$i_{1\text{lim}} = f_1 I_0 e_0 = 9\cdot6 \times 10^{-6}\ \text{A}$$

(see eqn (58b)) using $\varepsilon = 10^5$ and $\sqrt{(D\tau)} = 5 \times 10^{-6}$ cm.[37] For the prediction of the charge separation efficiency f_2 we need according to eqn (58c) the values of $\bar{\Delta}$ and λ for the redox systems investigated. $\bar{\Delta}$ is determined from the flatband potential for anthracene ($E_{FB} = -0\cdot35$ V) and the standard redox potential to be $\bar{\Delta} = 0\cdot8$ V for Fe^{3+}/Fe^{2+}. For this system $\lambda = 1\cdot16$ eV may be derived from homogeneous exchange studies eqn (39) to which we add $0\cdot17$ eV representing polarization of the anthracene lattice. Inserting these data in eqn (58) and using $\sqrt{(\tau/D)} = 4 \times 10^{-3}$ sec cm^{-1}, we obtain $f_2 = 8\cdot4 \times 10^{-7}$ which compares with the experimental value of $4\cdot7 \times 10^{-4}$. There is a serious discrepancy between the theoretically predicted and the experimentally observed values for f_2.

From the absorption spectrum of Fe^{2+},[40] we estimate the energy of the electronic excitation from the ground state to the first excited state $3d_s^5 d_j^1$ to be $1\cdot29$ eV. Hence, if this were the product of the electron transfer reaction the standard free energy of the reaction would be reduced by 1·29 eV and we obtain $f_2 = 1·65 \times 10^{-4}$ which compares favourably with the experimental value.

For the Tl^{3+}/Tl^+ system the question whether the exchange reaction proceeds via a 'simultaneous two electron' or a 'successive one electron' transfer path,[41] for the photoinjection of holes into anthracene may be decided by inspection of Fig. 12. In the construction of this diagram we

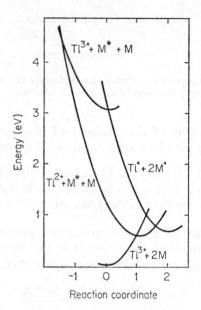

Fig. 12. Cross-section through the potential energy surface for dark and photo-injection into an anthracene crystal from Tl^{3+}/Tl^+ in 0·5 N HCl.[38] (Reprinted by permission Discussions Faraday Soc.)

used the estimate of the free energy change in the reaction $Tl^{3+} + Tl^+ = 2Tl^{2+}$ made by Hush,[43] viz. 0·5 eV, in order to place the energy of the $Tl^{2+} + M^+$ state relative to that of $Tl^{3+} + M$. The reaction coordinate varies from 0–2 in the states between Tl^{3+} and Tl^+. λ may be determined from the homogeneous exchange rate, eqn (44), to 1·75 eV to which we have to add 0·68 eV for the polarization of the anthracene lattice due to two positive charges. Finally we estimate the energy of one radical cation (hole) in the surface of the anthracene crystal to be $\bar{\Delta} = 0·35$ eV. Evidently the transition state for transfer of two electrons is more accessible to the system Tl^{2+} plus exciton, than is that for a single electron transfer. Indeed, the ratio of the two pathways is computed to be 10^{25}. It is predicted, however, that Tl^{2+} should be involved as an intermediate in the 'dark injection' of holes by Tl^{3+}:

$$Tl^{3+} + 2M \rightarrow Tl^+ + 2M^+$$

The activation free energy of this process, seen from Fig. 12 to be 0·85 eV is so high, that the injection rate is very small at room temperature.

We now find for the charge separation efficiency $f_2 = 7\cdot85 \times 10^{-5}$ which is in satisfactory agreement with the experimentally determined value $f_2 = 2 \times 10^{-4}$. Excited states of Tl^+ have not to be considered in this case because the energies involved are too high.

For $IrCl_6^{2-}/IrCl_6^{3-}$ we derive from the redox potential $\bar{\Delta} = 0\cdot6$ eV. For this system the value $\lambda = 0\cdot67$ was previously[16] determined from the homogeneous rate constant[45] and an estimate of the lattice polarization. For the formation of the ground state of $IrCl_6^{3-}$ we obtain $f_2 = 10^{-20\cdot6}$ and for formation of the first excited state $(4d_e^3 d_j^3)$ at 2·02 eV above the ground state,[46] $f_2 = 1\cdot3 \times 10^{-1}$ which compares with the experimental value of 4×10^{-3}. The discrepancy remaining between the theoretically predicted and the experimentally observed quantum efficiencies is due to the quenching of excitons by the solution.

$IrCl_6^{2-}$ has an absorption peak[46] at 490 mμ. With

$$f_{\text{ex}}^* = 0\cdot004 \quad \text{and} \quad f_{\text{th}}^* = 0\cdot1316$$

$$k = 9\cdot25 \times 10^5 \text{ cm}^4 \text{ sec}^{-1} \text{ mole}^{-1} \quad \text{(Reference 45)}$$

and because in this case $C_A = C_e$, eqn (72b) yields

$$k_Q = 2\cdot24 \times 10^8 \text{ cm}^4 \text{ mole}^{-1} \text{ sec}^{-1}$$

This value may be compared with a theoretical one obtained from eqn (66b) with $R_0 = 50$ Å, $R_e = 5$ Å, and $\tau = 2 \times 10^{-8}$ sec. This yields $k_Q = 5\cdot2 \times 10^7$ cm^4 mole^{-1} sec^{-1} in fairly good agreement with the experimental value.

When R_0 is evaluated for $IrCl_6^{2-}$ using Förster's expression[48]

$$R_0^6 = \frac{9 \times 10^6 (\ln 10)^2 K^2 C \tau}{16\pi^4 n N^2 v_0^2} \int_0^\infty \varepsilon_A(v)\varepsilon_D(2v_0 - v)dv \tag{73}$$

using $K^2 = \frac{2}{3}$, $\tau = 2 \times 10^{-8}$, $n = 1\cdot85$, and $v_0 = 24\,725$ cm^{-1} we find after numerical integration $R_0 = 48$ Å which leads to $k_Q = 3\cdot9 \times 10^7$ cm^4 mole^{-1} sec^{-1}.

12

V. CHARGE TRANSFER REACTIONS WITH ELECTRONICALLY EXCITED DONORS AND ACCEPTORS IN THE ELECTROLYTE

1. The Mechanism of Photoinjection

In the preceding paragraph we have discussed the charge transfer reaction between excited hydrocarbon molecules which are located in the crystal surface with electron donors and acceptors which are located in the solution near the crystal/electrolyte interface. We now turn to the case of excited donors or acceptors in the electrolyte which are in contact with an unexcited organic molecular crystal. Experimentally this means illumination of the system with a wavelength which is absorbed by the ions in the electrolyte but not by the crystal.

The charge transfer reaction can now be written as:

$$M + D^* \underset{k_{b*}}{\overset{k_{f*}}{\rightleftharpoons}} M^{+(-)} + A \qquad (74)$$

For adiabatic reactions of the outer sphere type the rate constants may be expressed in the form:

$$k_{f*} = Z \exp \left[\frac{-m^2 \lambda^*}{RT} \right] \qquad (75a)$$

$$k_{b*} = Z \exp \left[\frac{-(1-m)^2 \lambda^*}{RT} \right] \qquad (75b)$$

$$m = \tfrac{1}{2} \left[1 + \frac{\Delta F^*}{\lambda^*} \right] \qquad (75c)$$

We approximate the semiempirical parameter λ^* with the equivalent parameter λ for the dark reaction.

A more detailed discussion of the change ΔF^* in the standard free energy of the charge transfer reaction is possible. As previously detailed,[33] the standard free energy change of the dark reaction may be written as $\Delta F = \bar{\Delta}$, where $\bar{\Delta}$ is the distance between the carrier band and the Fermi level at the crystal surface after equilibrium has been established. On illumination the system contains an excited donor ion in the transition state, which may have the energy $\Delta\varepsilon$ with respect to the ground state ion. $\Delta\varepsilon$ may be determined from the energy of absorption which is necessary for obtaining the excited state. Thus

$$\Delta F^* = \bar{\Delta} - \Delta\varepsilon \qquad (76)$$

i.e. by optical excitation energy can be supplied to a redox system so that donor or acceptor states in molecular crystals may be reached which are otherwise unaccessible. The question whether for a particular system such a process is feasible or not can be answered only after the transport problem for excited states in the electrolyte has been solved.

2. The Transport Problem for Excited States in the Electrolyte

During illumination of the electrolyte e.g., through the molecular crystal the steady state concentration of excited states can be described by the equivalent of eqn (44):

$$D_x \frac{d^2 C_x}{dx^2} - \frac{C_x}{\tau x} + \varepsilon I_0 \exp(-\varepsilon x) = 0 \tag{77}$$

We assume all of the incident light to be absorbed inside the electrolyte so that $C_x = 0$ at $x \to \infty$ which is equivalent to the boundary condition in (Section IV.2). Also for this case we assume the net flux of excited states to the interface to be equal to the net rate of their disappearance through the agency of the photoinjection reaction:

$$D_x \frac{dC_x}{dx} = k^* C_x C_{A^\pm} - k_{b^*} C_C C_{A^\pm} \tag{78}$$

at $x = 0$. C_C, $C_{A^\pm}$, and C_A are the surface concentration of carriers, donors, and acceptors respectively. In the manner outlined in Section IV we obtain with the equivalent assumptions a solution for the transport problem which may be formulated as the following expression for the quantum efficiency of the injection current with electronically excited redox systems:

$$\frac{i}{i_{11m} - i_0} = \left(1 + \frac{1}{\varepsilon \sqrt{(D_x \tau_x)}}\right) \alpha \tag{79}$$

where i_{11m} is the actually observable limiting photocurrent and i_0 is the dark current.

From a discussion of the quantum yield of the charge separations reaction we obtain (see eqn (58c)):

$$\alpha = \frac{1}{1 + 1/k_{f_*} C_{A^\pm} \sqrt{(\tau_x/D_x)}}$$

From eqn (79) the mean free path of excited states in electrolytes can be determined.

3. Quenching of Excited States by Energy Transfer at the Electrode Surface

As pointed out in Section V.1 the generation of excited states in the electrolyte at the insulator/electrolyte interface asks for an insulator which is transparent in the wavelength region in which species in the electrolyte are absorbing which has as a consequence that resonance transfer between excited species in the solution and energy acceptor states in the insulator may be neglected because of the lack of overlapping energy levels.

Thus the paucity of states lends an insulator more easily to the investigation of charge transfer processes with excited species in solution than, e.g. a metal electrode.

4. Experimental Results

Figure 13 shows the spectral dependence of the photocurrent i_p across the interface between an electrolyte and a crystal of perylene and the photo-

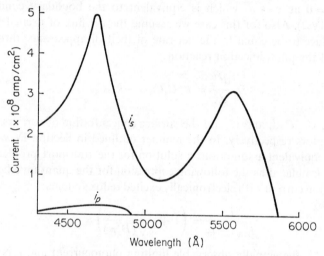

Fig. 13. Spectral dependence of the photocurrent i_p and the photosensitized current at a perylene electrode. The dye was rhodamine B (ca 10^{-5} M).[51] (Reprinted by permission Electrochimica Acta.)

current i_s which can be observed after addition of the dye rhodamine B to the electrolyte. While i_p is the result of the excitation of the molecular crystal and the reactions of excitons with electron acceptors in the electrolyte (see Section IV), the peak at 550 mμ is caused by excited dye molecules which are adsorbed at the perylene surface.[50,51]

A similar process has recently been reported[55] for a solution of I_2 in carbon tetrachloride, which does not form a complex with iodine so that the absorption spectrum of the solution in the visible closely resembles that of iodine vapour.[52]

The visible absorption of the iodine molecule (near 520 mμ) arises mostly from transitions between the ground state $^1\Sigma_0^+$ (electron configuration $\sigma_g^2\pi_u^4\pi_g^4$) and the excited state $^3\Pi_0^+$ ($\sigma_g^2\pi_u^4\pi_g^3\sigma_u$).[53]

Because of the large amount of spin-orbit coupling in the iodine atoms the radiative lifetime of the triplet is quite short; it has recently been determined to be (7·2 ± 1) × 10⁻⁷ sec.[54]

Figure 14 shows the spectral dependence of the photocurrent across the

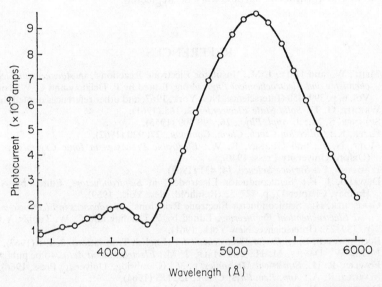

Fig. 14. Spectral dependence of the photosensitized current across the interface between anthracene and a solution of I_2 in CCl_4.[55] (Reprinted by permission Physics Letters.)

anthracene/iodine solution interface, which is illuminated through the anthracene crystal.

The peaks around 400 mμ are caused by reactions with anthracene excitons (Section IV), the peak at 520 mμ closely resembles the absorption spectrum of iodine. A plot of $1/i_{11m}$ against $1/\varepsilon$ (ε: extinction coefficient for I_2 in CCl_4),[52] gives indeed a straight line as predicted by eqn (79). Because the value of ε for I_2/CCl_4 is at the absorption peak more than two orders of magnitude smaller than the value of ε for anthracene at its main absorption peak, for a determination of i_∞ the graphical extrapolation method employed for anthracene,[50] is in this case very inaccurate, instead a direct determination of I_0 is necessary. From the data given in Fig. 10 and eqn (79) for I_2/CCl_4 for the value of the mean free path $\sqrt{(D_x\tau_x)} = (3 \pm 2) \times 10^{-6}$ cm was obtained from which $D_x = 1\cdot25 \times 10^{-5}$ cm²/sec was determined with the gas phase value for τ_x. This value for D_x may be compared with the diffusion coefficient in CCl_4 of ground state molecular iodine of $1 \cdot 36 \times 10^{-5}$ cm^2/sec.

We thus conclude that for this system the intermolecular transfer of excitation energy in the solution can be neglected so that energy transfer is closely coupled to mass transfer. It was found that i_{11m} increased linearly with the concentration of iodine so that complications due to adsorption of iodine at the anthracene surface may be neglected.

REFERENCES

1. MEHL, W., and HALE, J. M.: 'Insulator Electrode Reactions', in *Advances in Electrochemistry and Electrochemical Engineering*, Edited by P. Delahay and C. W. Tobias, Vol. 6, p. 399–458 (Interscience, New York, 1967, and other references quoted there).
2. WRIGHT, G. T.: *Solid State Electronics*, 2, 165 (1961).
3. SKINNER, S. M.: *J. Appl. Phys.*, 26, 498 509 (1955).
4. KLIER, K.: *Collection Czech. Chem. Commun.*, 27, 920 (1962).
5. MOTT, N. F., and GURNEY, R. W.: *Electronic Processes in Ionic Crystals*, p. 170 (Oxford University Press, 1940).
6. LOHMANN, F.: *Surface Science*, 14, 431 (1969).
7. DEWALD, J. F.: 'Semiconductor Electrodes', in *Semiconductors*, Edited by N. B. Hannay, Chapter 17, p. 727–52 (Reinhold, New York, 1959).
8. GERISCHER, H.: 'Semiconductor Electrode Reactions', in *Advances in Electrochemistry and Electrochemical Engineering*, Edited by P. Delahay and C. W. Tobias, Vol. 1, p. 139–232 (Interscience, New York, 1961).
9. BOGUSLAVSKII, L. I.: *Soviet Electrochemistry* (English Translation), 3, 191 (1967).
10. MEHL, W., DAVIES, M. H., and HALE, J. M.: *Electrochimica Acta.*, to be published.
11. FOWLER, R. H.: *Statistical Mechanics*, p. 364 (Cambridge University Press, 1966).
12. MARCUS, R. A.: *Ann. Rev. Phys. Chem.*, 15, 155 (1964).
13. DOGONADZE, R. R., KUZNETSOW, A. M., and CHIZMADZHEV, YU. A.: *Russ. J. Phys. Chem.* (English Translation), 38, 652 (1964).
14. HUSH, N. S.: Personal communication.
15. COULSON, C. A., and POOLE, M. D.: *Tetrahedron*, 20, 1859 (1964).
16. MEHL, W., and LOHMANN, F.: *Electrochimica Acta*, 13, 1459 (1968).
17. LYONS, L. E., and MACKIE, J. C.: *Proc. Chem. Soc. (London)*, 71 (1962).
18. MARCUS, R. A.: *Electrochimica Acta*, 13, 995 (1968).
19. DOGONADZE, R. R., and CHIZMADZHEV, YU. A.: *Proc. Acad. Sci. U.S.S.R., Phys. Chem. Sect.* (English Translation), 150, 333 (1963).
20. LOHMANN, F.: *Z. Naturf.*, 22A, 843 (1967).
21. GEACINTOV, N., and POPE, M.: *J. Chem. Phys.*, 45, 3884 (1966).
22. LOHMANN, F., and MEHL, W.: *Ber. Ber. Bunsenges. Physik. Chem.*, 77, 493 (1967).
23. LOHMANN, F., and MEHL, W.: *Electrochimica Acta*, 13, 1449 (1968).
24. GREEF, R., and AULICH, H.: *J. Electroanal. Chem.*, 18, 295 (1968).
25. DRESNER, J.: *Phys. Rev. Letters*, 21, 356 (1968).
26. BLANC, O. H. LE: *J. Chem. Phys.*, 35, 1275 (1961).
27. HUSH, N. S.: Unpublished manuscript.
28. BALK, P., BRUIJN, S. DE, and HOIJTINK, G. J.: *Mol. Phys.*, 1, 151 (1958), and further references quoted there.
29. BECKER, R. S., and CHEN, E.: *J. Chem. Phys.*, 45, 2403 (1966).
30. CHOI, SANG-IL, and RICE, S. A.: *J. Chem. Phys.*, 38, 366 (1963).

31. CASE, B., HUSH, N. S., PARSONS, R., and PEOVER, M. E.: *J. Electroanal. Chem.*, **10**, 360 (1965).
32. HUSH, N. S., and POPLE, J. A.: *Trans. Faraday Soc.*, **51**, 600 (1955).
33. WENTWORTH, W. E., and BECKER, R. S.: *J. Amer. Chem. Soc.*, **85**, 2210 (1963).
34. See for example LEONHARDT, H., and WELLER, A.: *Ber. Bunsenges. Physik. Chem.*, **67**, 791 (1963).
35. MARCUS, R. A.: *J. Chem. Phys.*, **43**, 2654 (1965).
36. LYONS, L. E.: *J. Chem. Phys.*, **23**, 220 (1955).
37. EREMENKO, V. V., and MEDVEDEV, V. S.: *Fizika tverd. Tela*, **2**, 1572 (1960).
38. MEHL, W., and HALE, J. M.: 'Disc. Faraday Soc., Newcastle' (1968), No. 45.
39. SILVERMAN, J., and DODSON, R. W.: *J. Physic. Chem.*, **56**, 846 (1952).
40. WEISS, J.: *Trans. Faraday Soc.*, **37**, 463 (1941).
41. JAMES, S. D.: *Electrochimica Acta*, **12**, 939 (1967); see also J. Jordan, R. Greef, 'Disc. Faraday Soc., Newcastle' (1968), No. 45.
42. HALE, J. M., and MEHL, W.: *Electrochimica Acta*, **13**, 1483 (1968).
43. HUSH, N. S.: *Trans. Faraday Soc.*, **57**, 557 (1961).
44. ROIG, E., and DODSON, R. W.: *J. Physic. Chem.*, **65**, 2175 (1961).
45. HURWITZ, P., and KUSTIN, K.: *Trans. Faraday Soc.*, **62**, 427 (1966).
46. JORGENSEN, C. K.: *Acta Chem.*, **10**, 500 (1956).
47. HALE, J. M., and MEHL, W.: To be published.
48. FÖRSTER, TH.: *Ann. Physik.*, **2**, 55 (1947).
49. WRIGHT, G. T.: *Proc. Phys. Soc.*, **68B**, 241 (1955).
50. MULDER, B. J.: Philips Res. Repts. Suppl. No. 4 (1968).
51. GERISCHER, H., MICHEL-BEYERLE, M. E., REBENTROST, F., and TRIBUTSCH, H.: *Electrochimica Acta*, **13**, 1509 (1968).
52. MAINE, P. A. D. DE: *J. Chem. Phys.*, **26**, 1192 (1957).
53. See for example HAM, J.: *J. Am. Chem. Soc.*, **76**, 3886 (1954).
54. BREWER, L., BERG, R. A., and ROSENBLATT, G. M.: *J. Chem. Phys.*, **38**, 1381 (1963).
55. MEHL, W., DRURY, J. S., and HALE, J. M.: *Physics Letters*, **28A**, 205 (1968).

THE ELECTRODE REACTIONS OF ORGANIC MOLECULES

M. Fleischmann and D. Pletcher

347

In recent years there have been considerable advances in the control of electrolysis conditions (for example, of the potential of the working electrode, and of solvent media) and this control has also been accompanied by an increase in the understanding of the fundamental steps of the electrode reactions of organic molecules. As a consequence, it seems likely that electrosynthesis will play an increasing role in preparative organic chemistry, both in the laboratory and on an industrial scale. Furthermore, when electrolytic processes are compared with competitive industrial processes, it should be remembered that due to absence of chemical oxidants or reductants the extraction and work up costs of products from electrolysis are likely to be less and that, with the advent of nuclear power, electricity will become cheaper compared with chemical reagents. At the present time, many of the powerful oxidants and reductants used in synthesis (e.g. Na, MnO_2, Cl_2) are prepared electrolytically and there are also obvious advantages in avoiding such intermediate steps. Moreover, as the requisite knowledge and suitable equipment, particularly potentiostats and electrochemical cells, become more widely available, electrochemical methods should become standard techniques for carrying out oxidations and reductions in the laboratory.

This chapter is divided into four sections; the first describes the intermediates found in the electrode processes of organic compounds. The second illustrates the variables which may be used in the control of electrode reactions, while the final two sections will enumerate some of the electrode reactions which may prove useful in the synthesis of organic molecules.

A number of reviews has appeared on various aspects of the subject matter of this chapter. The early work on the application of electrochemistry to organic synthesis was summarized in two books.[1,2] Later reviews have been written covering organic polarography,[3-5] electrode mechanisms,[6,7] synthesis,[8-13] reductions,[14,15] oxidations,[16-20] substitution,[21,22,23] and the Kolbe reaction.[24-26]

I. ELECTRODE INTERMEDIATES

It is possible to formulate the electrode reactions of organic compounds as proceeding via intermediates which are very similar to those postulated for homogeneous chemical reactions, for example carbonium ions, carbanions, and radicals. This has become clear with the widespread use of aprotic solvents such as acetonitrile, dimethylformamide, dimethylsulphoxide, propylene carbonate, and pyridine in which the lifetime of these intermediates is much longer than in aqueous solutions. During the last decade a number of electrochemical and spectroscopic techniques have also been devised for the study and characterization of reactive intermediates; of the electrochemical techniques, cyclic voltammetry is by far the most important although a number of other linear sweep and pulse methods for the study of reactive intermediates has also been described. Similarly, of the spectroscopic techniques, esr for the study of radical intermediates is of outstanding importance, although more recently IR, UV and visible spectroscopy using optically transparent electrodes, total reflectance spectroscopy and electrochemiluminescence have been used for radical and non-radical intermediates. Thus, in some cases there is direct evidence for the existence of the intermediates described in this section. However, in many other cases the existence of the intermediate is inferred from the products of the reaction and by analogy to known reaction mechanisms. In these cases, the development of improved analytical techniques which allow the identification of minor as well as major products is of great assistance.

The main reason for differences in behaviour between intermediates formed in electrode processes and in homogeneous reaction is that in the former case intermediates can be adsorbed on the electrode. Adsorption is particularly strong for uncharged radicals both because of the unpaired electron which is available for bonding with the electrode and because of the low solvation energy of uncharged species. Thus esr spectra can seldom be observed for uncharged radicals during electrode reactions. Adsorption is probably responsible for the high activity of catalytic electrodes such as platinum blacks, the stereospecific nature of some electrode reactions and the variation of products with electrode materials. For example, the oxidation of carboxylic acids on platinum electrodes gives products derived from alkyl radicals while on carbon electrodes the formation of alkyl carbonium ions is favoured.

Clearly, a complete understanding of the intermediates produced during electrode reactions is necessary for an intelligent approach to electrosynthesis, since only with this knowledge can any systematic attempt be made to build up complex products from relatively simple starting materials. Therefore, this section will outline the preparation and behaviour of some types of common electrode intermediates. Furthermore, in this context it is convenient to consider some further reactive species which may be produced at electrodes and used *in situ* for chemical reactions (e.g. cobalt(III), solvated electrons, inorganic radicals etc.) and the reactions of these species will also be described.

1. Uncharged Radicals

Radicals are produced as intermediates in many electrode reactions, for example, the oxidation of carbanions,[27,28] aliphatic carboxylic acids,[29] and certain phenols[30] and the reduction of carbonium ions,[31] acetone,[32] quinone,[33] tetraarylammonium ions,[34] and pyrimidine.[35]

$$\bar{C}H(COOEt)_2 \rightarrow e + \dot{C}H(COOEt)_2$$

$$AlMe_4^- \rightarrow Me^\cdot + AlMe_3 + e$$

$$RCOO^- \rightarrow e + R^\cdot + CO_2$$

$$\varnothing_3 C^+ + e \rightarrow \varnothing_3 C^\cdot$$

$$CH_3COCH_3 + H^+ + e \rightarrow CH_3 \underset{\underset{OH}{|}}{-}C^\cdot -CH_3$$

$$\varnothing_4 N^+ + e \rightarrow \varnothing_3 N + \varnothing^\cdot$$

However, with a few exceptions, radicals are very reactive species and can react in a number of ways, the most important of which are summarized in Fig. 1. The commonest reaction is dimerization as illustrated by the normal Kolbe reaction, the formation of pinacol during the reduction of acetone and the production of biphenyl during the reduction of tetraphenylammonium ions. A reaction which is of importance for the preparation of large molecules is the addition of radicals to olefin compounds as, for example, in the oxidation of monoethyl oxalate in the presence of butadiene which leads to 2,6-octadiene-1,8-dicarboxylic acid[36]

$$\underset{\text{COOEt}}{\overset{\text{COO}^-}{|}} \rightarrow e + CO_2 + \overset{\cdot}{C}OOEt \xrightarrow{CH_2=CH-CH=CH_2}$$

$$\overset{\cdot}{C}H_2-CH=CH-CH_2COOEt \rightarrow$$
$$EtOOCCH_2-CH=CH-CH_2-CH_2-CH=CH-CH_2COOEt$$

and under suitable conditions reactions which produce radicals as intermediates can be used to initiate polymerization.[37]

The preparation of organometallic compounds is of growing importance and several metal alkyls can be prepared by the generation of radicals on attackable metal electrodes.[38,39]

$$RMgCl \rightarrow e + MgCl^+ + R^{\cdot} \xrightarrow{\text{Pb anode}} PbR_4$$

However, in most cases, byproducts formed by competing radical reactions will also be obtained. Moreover, it is reasonable to expect that the products

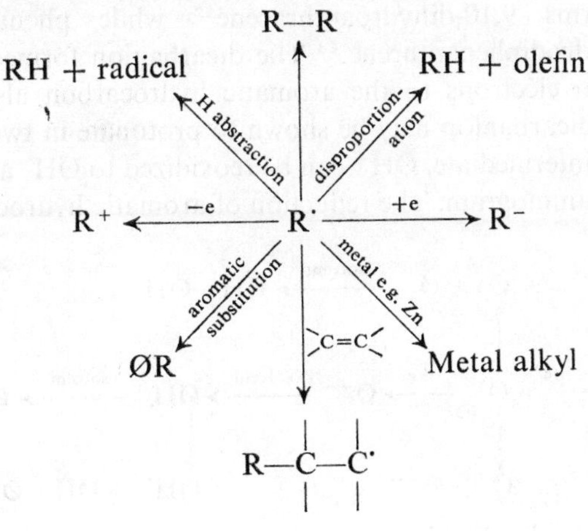

Fig. 1

will be dependent on the concentration of radicals present, i.e. electrode potential and concentration of electroactive species, and a high concentration of radicals will favour dimerization.

Another typical radical process which has been observed at an electrode is a chain reaction.[40] Several times the coulombic yield of product, benzyl mercuric iodide, was obtained from the reduction of benzyl iodide at a mercury cathode and the mechanism postulated for this reaction is

$$C_6H_5CH_2I + e \rightarrow C_6H_5CH_2^- + I^-$$
$$C_6H_5CH_2^- + Hg \rightarrow C_6H_5CH_2Hg^{\cdot}$$
$$C_6H_5CH_2Hg^{\cdot} + C_6H_5CH_2I \rightarrow C_6H_5CH_2HgI + C_6H_5CH_2^-$$

since the direct chemical reaction

$$C_6H_5CH_2I + Hg \rightarrow C_6H_5CH_2HgI$$

is very slow.

2. Carbanions and Carbanion Radicals

In aprotic solvents, aromatic hydrocarbons may be shown to reduce in two, one electron steps to the carbanion radical and the dicarbanion.[41] In the absence of proton donors or other added reagents, the carbanion radicals are relatively stable and on the time scale of an electrochemical experiment, the addition of the first electron to an aromatic hydrocarbon is reversible. However, on a longer timescale, dimerization or disproportionation will take place; the dianions thus formed will abstract protons relatively easily even in an aprotic solvent so that dihydroaromatic or dihydrodiaromatic compounds will be formed at the potential at which only one electron is added. Thus anthracene forms 9,10-dihydroanthracene[42] while phenanthrene gives 9,9′,10,10′-dihydrodiphenanthrene.[43] The dicarbanion formed by the direct addition of two electrons to the aromatic hydrocarbon also readily protonates.[41] The dicarbanion may be shown to protonate in two steps since in some cases the intermediate, $\emptyset H^-$ can be reoxidized to $\emptyset H^{\cdot}$ as can be shown on a cyclic voltammogram. The reduction of aromatic hydrocarbons may be summarized

$$
\begin{array}{ccc}
& \emptyset^- \!-\! \emptyset^- \xrightarrow{\text{solvent}} \emptyset H \!-\! \emptyset H & \\
& \uparrow & \\
\emptyset \xrightarrow{+e} & \emptyset^{\cdot -} \xrightarrow{+e} \emptyset^{--} \xrightarrow{\text{solvent}} \emptyset H^- \xrightarrow{\text{solvent}} \emptyset H_2 & \\
\downarrow & & \downarrow{\scriptstyle -e} \\
\emptyset + \emptyset^{--} & & \emptyset H^{\cdot} \rightarrow \emptyset H \!-\! \emptyset H
\end{array}
$$

In certain cases the dihydro aromatic compound reduces further.

The polarography of aromatic hydrocarbons has also been thoroughly studied in aprotic solvents in the presence of a proton donor (e.g. water, benzoic acid). As indicated above, in the absence of protons aromatic hydrocarbons give two waves of equal height. However, in the presence of added protons the first wave increases in height at the expense of the second wave; when sufficient proton is added a single, two electron wave is obtained.[41] The reduction scheme

$$\emptyset + e \rightarrow \emptyset^{\cdot -}$$
$$\emptyset^{\cdot -} + H^+ \rightarrow \emptyset H^{\cdot}$$
$$\emptyset H^{\cdot} + e \rightarrow \emptyset H^-$$
$$\emptyset H^- + H^+ \rightarrow \emptyset H_2$$

has been postulated, the single wave being obtained since the species $\Phi H^{\cdot}$ has a higher electron affinity (i.e. it is more readily reduced) than the parent hydrocarbon. A similar behaviour is found for the reduction of the hydrocarbons in the presence of carbon dioxide. For example, naphthalene is reduced to 1,4-dihydro-1,4-dicarboxy-naphthalene at the potential at which only one electron would be added in the absence of carbon dioxide.[42]

Several classes of aromatic compounds such as quinones,[44] nitrocompounds,[45] and acid anhydrides[46] show reduction patterns similar to the hydrocarbons in that they are reduced in two one electron steps to give carbanion radicals and dicarbanions. In each case the carbanion radicals have been the subject of esr studies.

The addition of carbon dioxide is a general reaction of carbanions and it has been used as a diagnostic test for the presence of carbanion intermediates. Thus the carbanions formed by the reduction of stilbene,[43] benzyl halides,[47] and benzophenone[48] react to give 1,2-diphenylsuccinic acid, phenylacetic acid and benzilic acid while in the absence of carbon dioxide the carbanions would proton abstract to form 1,2-diphenylethane, toluene, and diphenylmethanol respectively.

Aliphatic carbanions are also intermediates in electrode reactions as, for example, in the reduction of alkyl halides[49] and activated olefins[43,50]

$$RCl + 2e \rightarrow R^- + Cl^-$$

$$ØCH{=}CHØ + 2e \rightarrow Ø\bar{C}H{-}\bar{C}ØH$$

$$CH_2{=}CHCN + 2e \rightarrow \bar{C}H_2{-}\bar{C}HCN$$

and the products normally arise by proton abstraction. For example, the reduction of ethyl halides yields ethane. Under suitable conditions, however, the dicarbanions formed by the reduction of activated olefins will react with a further molecule of olefin to form the hydrodimer[50]

$$\bar{C}H_2{-}\bar{C}H{-}CN + CH_2{=}CH{-}CN + 2H^+ \rightarrow \begin{array}{l} CH_2{-}CH_2{-}CN \\ | \\ CH_2{-}CH_2{-}CN \end{array}$$

This reaction has important applications in synthesis and will be discussed later.

3. Carbonium Ions and Carbonium Ion Radicals

It might be expected that in an aprotic solvent, aromatic hydrocarbons would oxidize in two, one electron steps to give a carbonium ion radical and a dication or a carbonium ion.

$$ØH \rightarrow e + ØH^{\cdot +}$$

$$ØH^{\cdot +} \rightarrow ØH^{+ +} + e$$

or

$$ØH^{\cdot +} \rightarrow e + H^+ + Ø^+$$

However, the cation species are much less stable than the corresponding anion species and this simple behaviour is only observed for relatively few hydrocarbons such as 9,10-diphenylanthracene[51] and rubrene.[52] In most cases multi-electron oxidations are observed. The removal of the first electron has been shown to be reversible using rapid cyclic techniques and it therefore seems likely that this first electron transfer is followed by a very fast chemical reaction to give a product which is electroactive at the same potential.[41] The products of the overall process have not been definitely elucidated although in the presence of pyridine, anthracene has been found to give the 9,10-dihydro anthraquinyl pyridinium ion.[53] However, the presence of pyridine is believed to effect a basic change in the electrode mechanism since in the absence of pyridine the number of electrons transferred is greater than two.

In the case of hydrocarbons containing side chains, oxidation in acetonitrile gives nitrilium salts which on hydrolysis yield substituted acetamides. Examples of such hydrocarbons are toluene[54] and hexamethylbenzene.[55]

Furthermore, the electrochemical substitution of aromatic hydrocarbons is now believed to take place via such carbonium ion intermediates[56] since substitution has been found only at potentials at which the hydrocarbon is itself oxidized even if the substituting anion is oxidized at less positive potentials. Conversely, substitution has been observed at potentials below the discharge potential of the anion in the cases where the hydrocarbon is oxidized at lower potentials.[57] The distribution of product isomers also corresponds more closely to a reaction between a carbonium ion and an anion than attack on an aromatic species by a radical.[26] Some examples of substitution reactions will be discussed later.

Aliphatic carbonium ions are also intermediates in electrode reactions such as the oxidation of alkyl halides,[58] aliphatic hydrocarbons,[59] and certain carboxylic acids.[60]

$$RI \rightarrow e + R^+ + \tfrac{1}{2}I_2$$
$$RH \rightarrow 2e + R^+ + H^+$$
$$RCOO^- \rightarrow R^+ + 2e + CO_2$$

A summary of the probable reactions of aliphatic carbonium ions in common solvents is shown in Fig. 2. Since these species are extremely reactive,

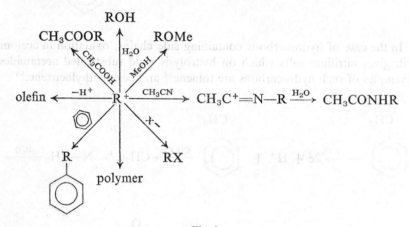

Fig. 2

the products usually arise from reaction with the solvent. Carbonium ions which occur in electrode reactions show the usual carbonium ion rearrangements. For example, the neopentyl rearrangement has been observed during the oxidation of neopentyl iodide[58] and cyclizations,[60] ring contractions[61] and ring expansions[62] have all been observed. For example,

$$CH_3-CH_2-CH_2COO^- \xrightarrow{-2e} CO_2 + CH_3-CH_2-CH_2^+ \xrightarrow{-H^+} CH_2-CH_2$$

Non-classical carbonium ions may also be prepared electrochemically as in the oxidation of exo-norbornene-2-carboxylic acid[62]

$$\text{(structure) } \xrightarrow{-2e} CO_2 + \text{(structure)} \longrightarrow \text{(structure)}$$

4. Charged and Uncharged Nitrogen Intermediates

In the electrode reactions of nitrogen containing organic compounds, intermediates with the charge or unpaired electron situated on the nitrogen atom will commonly be observed.

The negatively charged species occur during the reduction of heterocyclic compounds such as quinoline in aprotic solvents.[63]

$$\text{(quinoline)} \xrightarrow{e} \left[\text{(radical anion)} \longleftrightarrow \text{(radical anion)} \longleftrightarrow etc. \right]$$

After proton abstraction from the solvent and a further electron transfer the final product will be the dihydrocompound.

$$\text{(structure)} \xrightarrow{solvent} \text{(structure)} \xrightarrow{+e} \text{(structure)} \xrightarrow{solvent} \text{(structure)}$$

The same product would be obtained in aqueous solution, but in this case it is not clear whether the mechanism is the same or involves preprotonation, i.e.

$$\text{(structure)} \xrightarrow{H^+} \text{(structure)} \xrightarrow{e} \text{(structure)} \quad etc.$$

In many cases, it can be proved that the reduction of nitrogen derivatives in aqueous solutions occurs via protonation and the reduction of the protonated form. Examples are the reduction of oximes[64] and hydroxamic acids[10] to amines and amides respectively.

$$RR'C{=}N{-}OH \overset{H^+}{\rightleftharpoons} RR'C{=}\overset{+}{N}H{-}OH$$

$$\xrightarrow{4e+4H^+} RR'CH{-}\overset{+}{N}H_3 + H_2O$$

$$RCONHOH \overset{H^+}{\rightleftharpoons} R\overset{\overset{\displaystyle OH}{|}}{C}{=}N^+HOH \xrightarrow{2e+2H^+} RCONH_2 + H_2O$$

Nitrogen cation radicals are produced during the oxidation of amines[65] and amides[66] in aprotic solvents and the final products are usually formed by hydrogen atom abstraction:

$$R_3N \xrightarrow{-e} R_3N^{\cdot+} \xrightarrow{solvent} R_3N^+H$$

$$CH_3CONHR \xrightarrow{-e} CH_3CON^{\cdot+}HR \xrightarrow{solvent} CH_3CON^+H_2R$$

and an example of the formation of a nitrogen radical occurs during the oxidation of p-nitroaniline[67] in an aprotic medium:

As can be seen from the examples quoted, the reactions of the nitrogen intermediates are very similar to those of their carbon analogues. That is, the products are generally formed from the intermediates by proton or hydrogen abstraction or by dimerization.

5. Biradicals

The biradicals, dichlorocarbene[68] and benzyne[69] have been prepared electrochemically. Naturally, since these intermediates are extremely reactive, their presence is inferred by products from reactions with trapping agents. Dichlorocarbene and benzyne are prepared by the reduction of carbon tetrachloride and dibromobenzene respectively in acetonitrile or dimethylformamide and the trapping agents used were tetramethylethylene and furan respectively.

6. Inorganic Radicals

Inorganic radicals have been postulated as intermediates in a number of electrode processes. In acetonitrile, the anodic limit has been found to be strongly dependent on the base electrolyte present.[56] Thus it seems likely that the electrode reaction is the oxidation of the anion present. In a perchlorate solution, the mechanism

$$ClO_4^- \rightarrow ClO_4^{\cdot} + e$$

$$ClO_4^{\cdot} + CH_3CN \rightarrow HClO_4 + {}^{\cdot}CH_2CN$$

has been proposed but while there is good evidence for the production of perchloric acid there is no conclusive proof that succinonitrile is the other product.[70]

The methoxylation of dimethylformamide in a nitrate medium is also believed to proceed via the production of nitrate radicals.[71]

$$NO_3^- \rightarrow NO_3^{\cdot} + e$$

A number of attempts have been reported to initiate polymerization with inorganic radicals.[37]

7. Transition Metal Ions

Metal ions in lower or higher oxidation states may be used as intermediates in reductions and oxidations. A number of powerful oxidizing agents such as silver(II), cobalt(III), and manganese(III) may be generated at the anode and may be used *in situ* to carry out highly specific reactions. For example,

$$Co^{2+} \rightarrow Co^{3+} + e$$

The electroactive lower oxidation state is regenerated during the chemical reaction and the actual electrode reaction serves to drive the homogeneous catalytic reaction.[72]

While there are probably lower oxidation states, for example, samarium(II) and chromium(II), which could be used in the same way to carry out reductions, the only examples reported to the present time are the reduction of 2-methoxyphenyl mesityl ketone and nitrosobenzene, azoxybenzene and azobenzene by magnesium(I)[73] and aluminium(I)[74] respectively. These oxidation states may be produced by the anodic dissolution of the respective metals.

$$Al \rightarrow Al^+ + e$$

$$3Al^+ + 2C_6H_5NO + 4H_2O \rightarrow 6OH^- + 3Al^{3+} + C_6H_5{-}NH{-}NHC_6H_5$$

However, these reactions are not catalytic since the reducing agent cannot be regenerated.

An example of such an indirect reaction in a molten salt medium is the preparation of tetramethylsilane in a sodium chloride/aluminium chloride melt.[75] A suspension of aluminium metal is produced at the cathode

$$Al^{3+} + 3e \rightarrow Al$$

and is used in the following reaction sequence

$$2Al + 3CH_3Cl \rightarrow Al(CH_3)_3 + AlCl_3$$
$$3SiCl_4 + 4Al(CH_3)_3 \rightarrow 3Si(CH_3)_4 + 4AlCl_3$$

a product being aluminium trichloride which can be used further in the electrode reaction. The overall equation, with power being consumed is

$$3SiCl_4 + 12CH_3Cl \xrightarrow[\text{Al cathode}]{\text{NaCl/AlCl}_3} 3Si(CH_3)_4 + 12Cl_2$$

Some further examples of electrode reactions coupled to chemical reactions will be given later.

8. Other Intermediates

There are several other intermediates which are of importance in organic electrosynthesis.

The superoxide ion, generated by the reduction of oxygen in aprotic solvents, has basic properties which allow its use as in electrochemical autocatalysis.[76] For example, in the presence of excess oxygen a 400 per cent current efficiency is obtained for the oxidation of diphenylmethane to benzophenone and the reaction occurs at the cathode. The mechanism postulated is

$$\emptyset_2CH_2 + O_2^- \to \emptyset_2CH^- + HO_2 \qquad (1a)$$

$$\emptyset_2CH_2 + OH^- \to \emptyset_2CH^- + H_2O \qquad (1b)$$

$$\emptyset_2CH^- + O_2 \to \emptyset_2CH^{\cdot} + O_2^- \qquad (2a)$$

$$\emptyset_2CH^- + HO_2 \to \emptyset_2CH^{\cdot} + HO_2^- \qquad (2b)$$

$$\emptyset_2CH^{\cdot} + O_2 \to \emptyset_2CHOO^{\cdot} \qquad (3)$$

$$\emptyset_2CHOO^{\cdot} + \emptyset_2CH^- \to \emptyset_2CHOO^- + \emptyset_2CH^{\cdot} \qquad (4a)$$

$$\emptyset_2CHOO^{\cdot} + O_2^- \to \emptyset_2CHOO^- + O_2 \qquad (4b)$$

$$\emptyset_2CHOO^- \to \emptyset_2CO + OH^- \qquad (5)$$

$$2HO_2^- \to 2OH^- + O_2 \qquad (6)$$

where reactions (1a) or (1b) is a proton transfer, steps (2a) or (2b) and (4a) and (4b) are electron transfers and reactions (2a), (5), and (6) produce species to continue the base catalysis. The superoxide ion may also be used as a nucleophilic substituting agent as in the formation of alkyl peroxides during the reduction of alkyl halides

$$O_2 + e \rightarrow O_2^-$$
$$O_2^- + RX \rightarrow RO_2^{\cdot} + X^-$$
$$RO_2^{\cdot} + O_2^- \rightarrow RO_2^- + O_2$$
$$RO_2^- + RX \rightarrow ROOR + X^-$$

The solvated electron has been prepared electrochemically in solvents such as ammonia, amines, and hexamethylphosphoramide and this intermediate is clearly an extremely powerful reducing agent. It has been shown to be a likely intermediate in the reduction of benzene, since the reduction of benzene is only possible in solvents in which solvated electrons are formed.[77]

$$Li^+(sol) + e \rightarrow Li^+(sol) + e(sol) \xrightarrow{\text{solvent}}$$

It has also been used in the reduction of olefins.[78]

A number of electrophilic reagents, for example NO_2^+, NO^+, and I^+ are capable of preparation in aprotic solvents and their application to electrochemical aromatic substitution can be expected in the near future.

Finally, two of the earliest recognized intermediates, the adsorbed hydrogen atom and the adsorbed hydroxyl radical are worthy of mention and are of particular importance on catalytic electrodes such as platinum black. Thus, hydrogen atoms are intermediates in the electrochemical hydrogenation of olefins[79] and hydroxyl radicals are likely intermediates in the formation of methanol during the oxidation of acetate in basic solution[80] and the complete oxidation of hydrocarbons to carbon dioxide on platinum catalysed fuel cell electrodes.[81] The hydroxyl radical is also likely to be of importance on transition metal oxide electrodes.[72]

II. THE CONTROL OF ELECTRODE REACTIONS

While the possibilities of electrosynthesis were realized before the end of the last century, the early work was marred by an inability to achieve selective reactions. This failure may largely be attributed to the lack of control of the conditions under which the electrolyses were carried out. The variables which exist are now more fully understood and control of these variables allows a measure of selectivity to be introduced. This section will deal, in turn, with the main electrolysis parameters, indicate the effect of varying these parameters and show how a fuller understanding allows some more sophisticated ways of controlling electrode reactions.

1. Electrode Potential

The well known relationship between the driving force for any reaction, ΔG^0, and the cell potential E^0

$$\Delta G^0 = -zFE^0$$

indicates immediately that for a reversible reaction, the energy that may be introduced during an electrode reaction, and therefore the nature of that reaction, is dependent on the cell potential.

Similarly, for an irreversible process the approximate expression

$$i = zFk_i \prod_i c_i^{\gamma_i} \exp - \left(\frac{E^{\ddagger}}{RT} - \frac{\alpha z \phi F}{RT} \right)$$

shows that by varying the electrode potential, ϕ, it is possible to modify the energy of activation, $E^{\ddagger}$, of any electrode reaction so that the reaction takes place at room temperature, and again, the nature of the electrode reaction is clearly controlled by the potential of the electrode.

It is of interest to note here, that since the potential range in many solvents is from about $+3 \cdot 0$ V to $-3 \cdot 0$ V, the driving force for an oxidation or reduction is of the order of 3 eV or 60–70 kcals/mole. This energy is, in fact, comparable to the strength of chemical bonds and therefore it is not surprising that electrochemical reactions take place and that many 'high energy' oxidants and reductants are usually prepared electrochemically.

It is, therefore, obviously necessary to control the potential of the working electrode if selective reactions are to be attempted. This is best achieved electronically using a potentiostat and the development of this equipment during the last decade is of prime importance to electrosynthesis.

The importance of the electrode potential was demonstrated as early as 1898, when it was shown that at low negative potentials nitrobenzene could be reduced to phenylhydroxylamine while at more negative potentials the product was aniline.[82]

With the development of potentiostats examples of this idea have become fairly common. Thus the products from the reduction of p-dinitrobenzene,[83] benzene diazonium salts,[84] arylketones,[85] and phthalazines[86] and the oxidation of o-toluidine[87] have all been shown to depend on the electrode potential

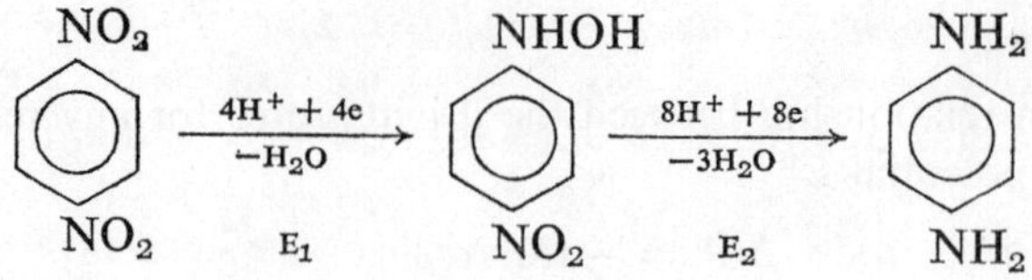
NO2
NO2
4H+ + 4e
−H2O
E1
NHOH
NO2
8H+ + 8e
−3H2O
E2
NH2
NH2

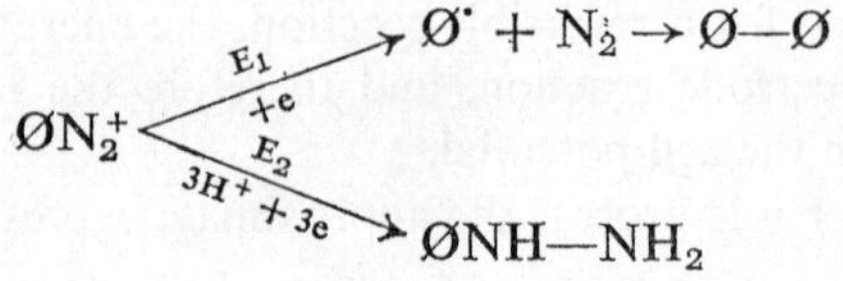
ØN2+
E1
+e
E2
3H+ + 3e
Ø˙ + N2 → Ø—Ø
ØNH—NH2

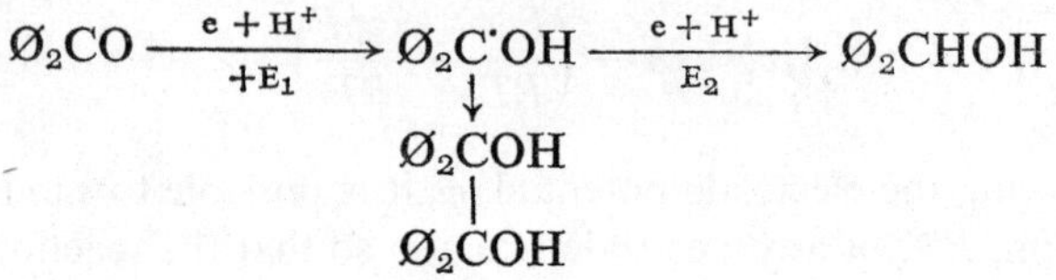
Ø2CO
e + H+
+E1
Ø2C˙OH
e + H+
E2
Ø2CHOH
Ø2COH
Ø2COH

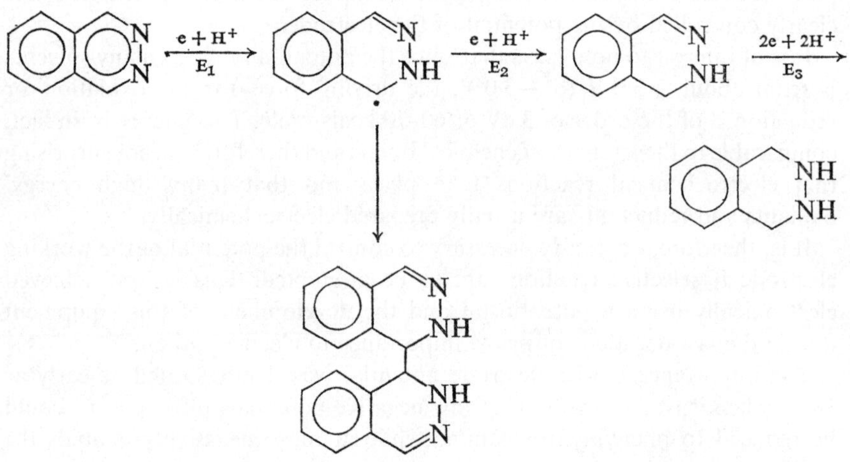
N
N
e + H+
E1
N
NH
e + H+
E2
N
NH
2e + 2H+
E3
NH
NH
N
NH
NH
N

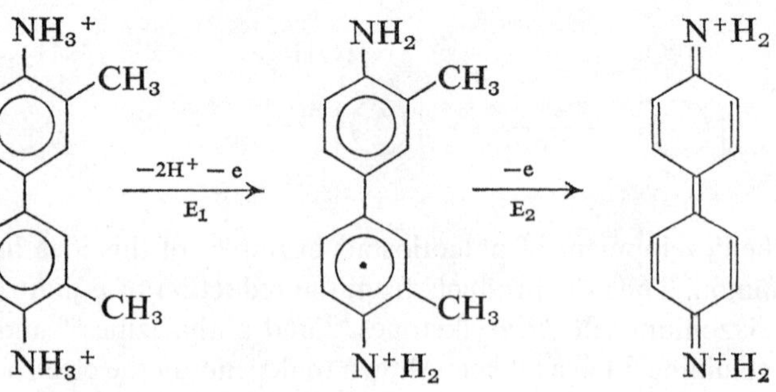
NH3+
CH3
CH3
NH3+
−2H+ − e
E1
NH2
CH3
CH3
N+H2
−e
E2
N+H2
N+H2

It has also been shown that the stereochemistry of an electrode reaction can depend on the electrode potential.[88] In acid or neutral solution benzil is reduced to stilbene diol, and the ratio of cis to trans stilbene diol formed is potential dependent.

The control of an electrode reaction by the application of a steady potential is now well known. However, it is less often realized that an electrode reaction may also be controlled by applying a pulse profile, for example a square wave potential-time profile, so that several different reactions take place successively and repetitively and that the products of an electrode reaction can be changed completely in this way. An example is the variation in the reduction mechanism of nitrobenzene caused by reoxidizing the intermediate phenylhydroxylamine before it has time to diffuse away from the electrode. Thus, by choosing the potential, E_1, of the reduction step it is possible to accumulate the coupling product, azoxybenzene, or its reduction product.[89] In effect, by applying a suitable square wave potential/time profile, benzidene may be prepared directly from nitrobenzene and it is clear that other products could be prepared from nitrobenzene if suitable pulse profiles are chosen.

It will be apparent that in such pulse electrolyses the duration of the pulse is an additional control variable. For example, it has been shown that during the pulse electrolysis of acetate in alkaline solution, the ratio of products oxygen, methanol, and ethane can be varied by changing the pulse lengths.[90] The variation of the yields of the products with pulse length and amplitude gives detailed information on the kinetics of the reaction as well as indicating that suitable electrocatalytic conditions must be built up at the surface before ethane or methanol can be formed. A further application of pulse electrolysis is likely to be the control of electrode activity by control of the electrode history. Thus the oxidation of high concentrations of aromatic hydrocarbons is hampered by steadily decreasing currents. However, the current can be maintained by applying very short pulses at a potential below the oxidation potential of the hydrocarbon.[72]

A question of key importance for achieving a selective reaction is whether the desired product is itself electroactive at the potential at which the reaction is carried out. The half wave potentials for the substrate and product under the experimental conditions will be a useful guide. For example, it is clear that it will always be difficult to find selective routes for the formation of phenols and acetates since they are usually more easily oxidized than the parent hydrocarbon. On the other hand, cyanation of aromatic compounds is a favourable reaction since the products have more positive half wave potentials. However, polarographic data will not necessarily be an accurate guide for the course of a reaction since large scale preparations are carried out on a longer time scale and can, therefore, be affected by disproportionations. An example is the reduction of aromatic hydrocarbons in aprotic solvents when the dihydroaromatic may be formed at the potential at which only one electron has been added to the hydrocarbon[42]

$$\varnothing + e \to \varnothing^{\cdot -} \xrightarrow{\text{solvent}} \varnothing H^{\cdot} \xrightarrow{\varnothing^{\cdot -}} \varnothing + \varnothing H^{-} \xrightarrow{\text{solvent}} \varnothing H_2$$

2. Choice of Solvent

In many cases the choice of solvent for an electrochemical reaction will be governed by physical data such as solubilities and dielectric constant and in other cases the reasons have been implied earlier. For example, if the intermediates which occur have much greater stability in aprotic solvents than in water and other protic solvents, and if this stability is required in order to use the intermediate in a constructive reaction such solvents must be used. It is for these reasons that aprotic solvents have been widely used in the laboratory. However, on an industrial scale the use of these solvents poses some difficult problems, particularly the ohmic drop across the cell, and this factor leads to attempts to carry out the reactions in aqueous media by the use of solubilizing agents such as p-toluene sulphonates (McKee's salts), emulsifying agents and tetraalkylammonium salts which also adsorb strongly on the electrode and can lead to a nominally aprotic environment at the surface.

The solvent can also play a role in the kinetics of an electrode reaction by virtue of its solvating properties. Thus the electron transfer reactions of aromatic hydrocarbons are known to be slow in aqueous solution while in the aprotic solvents the addition and removal of the first electron has been shown to be reversible.

3. pH

Clearly in both aqueous and non-aqueous solvents the supply of hydrogen ion will be of critical importance in controlling the mechanism of the electrode reaction, both by determining whether it is the protonated or unprotonated form which is oxidized or reduced at the electrode and by determining whether the intermediates become protonated. An example is the reduction of aryl alkyl ketones at various pHs in aqueous solution for which the following three schemes may be written.[85]

Low pH

$$\emptyset COR + H^+ \rightleftharpoons \emptyset C^+ OHR$$

$$\emptyset C^+ OHR + e \overset{E_1}{\rightleftharpoons} \emptyset C^\cdot OHR \rightarrow products$$

$$\emptyset C^\cdot OHR + e \overset{E_2}{\longrightarrow} \emptyset C^- OHR \overset{H^+}{\longrightarrow} \emptyset CHOHR$$

Two one electron waves observed.

Medium pH

$$\emptyset COR + e \overset{E_1}{\longrightarrow} \emptyset C^{\cdot -} OR$$

$$\emptyset C^{\cdot -} OR + H^+ \rightleftharpoons \emptyset C^\cdot OHR$$

$$\emptyset C^\cdot OHR + e \rightarrow \emptyset C^- OHR$$

$$\emptyset C^- OHR + H^+ \overset{E_2}{\longrightarrow} \emptyset CHOHR$$

Since E_2 is more positive than E_1 one wave is observed.

High pH

$$\emptyset COR + e \overset{E_1}{\longrightarrow} \emptyset C^{\cdot -} OR$$

$$\emptyset C^{\cdot -} OR + e \overset{E_2}{\longrightarrow} \emptyset C^{--} OR$$

$$\emptyset C^{--} OR + 2H^+ \rightleftharpoons \emptyset CHOHR$$

Two one electron waves observed.

Thus the highest yield of alcohol can be expected at medium pH while pinacol can be formed by controlled potential electrolysis at low and high pH.

A second example is the reduction of 2-cyanopyridine where at high pH the cyanide group is split out to give pyridine and at low pH the product is 2-picolylamine. At intermediate pH values the ratio of the two products is dependent on the potential.[91]

These 'pH' effects are also apparent in aprotic solvents. It has already been pointed out that the presence of pyridine totally changes the oxidation mechanism of anthracene in acetonitrile. Similarly, the products from the oxidation of 3,4-dimethoxypropenylbenzene in acetonitrile are totally different in the presence and absence of pyridine.[92]

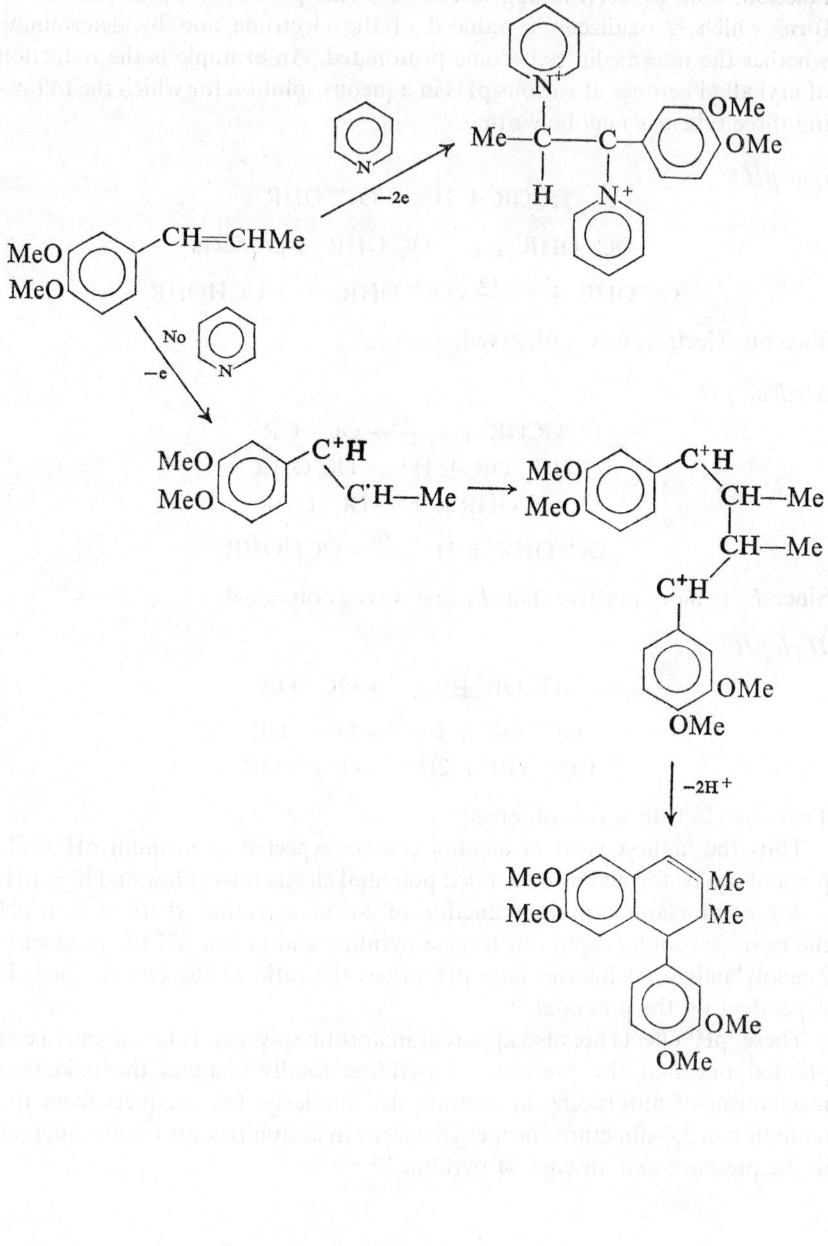

N⁺
OMe
Me—C—C—
OMe
N
−2e
H
N⁺
CH=CHMe
MeO
MeO
No
N
−e
C⁺H
MeO
MeO
C·H—Me
MeO
MeO
C⁺H
CH—Me
CH—Me
C⁺H
OMe
OMe
−2H⁺
MeO
MeO
Me
Me
OMe
OMe

4. Electrode Material

The choice of electrode material can have considerable influence on the products of an electrode reaction. Thus, during the reduction of acetone in 20 per cent sulphuric acid at a mercury electrode the main product (95 per cent) is isopropanol, at a lead electrode the products were isopropanol (68 per cent), pinacol (7 per cent), and propane (25 per cent) while at cadmium, zinc, aluminium, nickel, and copper cathodes the products were almost 100 per cent propane.[93] Again, the oxidation of carboxylic acids at platinum electrodes leads to products derived from radical intermediates in most cases while at carbon anodes the same acids give rise to products formed by carbonium ion reactions,[60] and other examples of this variation of products with electrode material could be cited. In fact, the products of some electrolyses have been shown to be dependent on the casting treatment of the metal electrode as well as the actual metal used.

The reasons for this behaviour are not fully understood but possible factors include the nature and degree of adsorption of organic compounds on the different metals, the nature of the surface (e.g. whether the surface is covered by a surface oxide) and the degree to which the electrode materials can form organometallic compounds with intermediates. In the general case, the first factor is likely to be the most important. The varying degrees to which metals will adsorb organic compounds is strikingly shown by the case of olefins at platinum and gold electrodes. At platinum anodes, the olefins are strongly adsorbed and can completely hinder oxygen evolution while at gold electrodes the presence of olefins has almost no effect on the evolution of oxygen.[72]

III. ELECTRODE REACTIONS OF FUNCTIONAL GROUPS

This section summarizes the products which may be expected from the electrode reactions of functional groups. As is stressed in other sections of this chapter, the electrode reactions usually take place via reactive intermediates and in synthesis these intermediates are often used in coupled chemical reactions with other substrates in order to produce different products. However, it is also useful to summarize the reactions according to the nature of functional groups. No indication of the electrolysis conditions (pH, electrodes, etc.) will be given here. Owing to the large number of references that would be required, this section is not referenced. Relevant references may be found in the list of reviews and elsewhere in this chapter.

1. Reductions

A. Oxygen Containing Groups. Aldehydes and ketones can be reduced by a one electron route to give the glycol or pinacol, a two electron route to give an alcohol or a four electron route to give the hydrocarbon. However, these routes appear to be independent since generally alcohols are not reduced at electrodes.

Aliphatic carboxylic acids are not reducible unless they contain an activating group, but aromatic acids have been reported to reduce to the corresponding alcohol. Similarly, esters of benzoic acids are reported to reduce to ethers. Amides can be reduced to the amine and the reactivity increases with substitution on the nitrogen atom.

Quinones are reducible to give the quinol but phenol groups cannot be reduced electrochemically. Peroxides reduce with splitting of the oxygen-oxygen bond and simple peroxides give the alcohols.

Sulphur derivatives usually behave similarly to the oxygen species. For example, sulphur-sulphur bonds can be reductively fractured to give a mixture of thiols.

B. Halogen Groups. Alkyl halides are reduced by either a one electron or a two electron route. In each case the product at inactive electrodes will be a saturated hydrocarbon. More fully halogenated hydrocarbons may also be reduced, the halogen atoms are replaced by hydrogen in a stepwise fashion. In fact, the replacement of halogen by hydrogen is a general cathodic reaction and can be carried out with many aliphatic compounds. The reduction of dihalides can lead to olefins or cyclic derivatives. Similarly, aryl halides are generally reducible to give the aromatic hydrocarbon.

C. Hydrocarbons. Aromatic hydrocarbons are only reduced very irreversibly in aqueous solution. However, in aprotic media they can be reduced to give either the dihydroderivative or the dihydrodimer of the parent hydrocarbon.

In the aliphatic series it is possible to reduce both olefin and acetylene groups. However, the partial electrochemical hydrogenation of acetylenes to olefins is often difficult since the olefin is more readily reduced than the acetylene.

D. Nitrogen Containing Groups. The reduction of aromatic nitrocompounds can give rise to a large number of different products. The phenyl hydroxylamine and the aniline are the most common products but p-aminophenols, azoxycompounds, azocompounds, and hydrazobenzenes, have all been reported. Nitrosocompounds can be reduced to similar products.

Diazonium salts can be reduced either in a one electron route to give the dimer of the hydrocarbon fragment and nitrogen or by a four electron route to the hydrazine. Nitramines and nitrosamines also reduce to the hydrazine while triazenes, azides and oximes reduce to give the amine.

2. Oxidations

During recent years, the complete oxidation at low overpotentials of organic compounds (e.g. hydrocarbons, alcohols, aldehydes, and acids) to carbon dioxide has been studied at a variety of catalytic electrodes in connection with the development of electrodes for fuel cells. However, these reactions are of no direct interest in synthesis and will not be considered here.

A. Oxygen Containing Groups. Carboxylic acids can be oxidized by a 1e route or a 2e route. In the first case we have the Kolbe reaction or the related Brown-Walker and Hofer-Moest reactions. In the Kolbe reaction the acid, RCOOH, is oxidized to carbon dioxide plus the dimer R—R. This is a very general reaction and occurs readily with a few exceptions such as the benzoic acids and α,β-unsaturated acids. The Hofer-Moest reaction gives rise to the alcohol ROH, probably by a radical reaction involving hydroxyl radicals. In the second case, the acid gives rise to a carbonium ion, R^+, and the products are very dependent on the solvent. For example, in water the product will be an alcohol and in alcohol the product will be an ether.

The oxidation of alcohols, aldehydes and ketones has not been widely studied. However, there is evidence that oxidation to the corresponding acid is possible and that in the case of secondary alcohols, selective oxidation to the ketone can be carried out. The only esters which have been studied are those such as diethyl malonate which are acidic in nature and, in these cases, the carbanion may be oxidized to the corresponding radical followed by dimerization.

Phenols are very readily oxidized but their behaviour is difficult to generalize since a great variety of products is obtained. However, the initial reaction is normally a one electron oxidation to give a radical in basic media and a two electron oxidation in acid media. Quinones are formed by the oxidation of quinols, p-substituted phenols and certain phenols.

B. Halogen Compounds. The oxidation of alkyl iodides and aryl iodides has been studied in acetonitrile. In the case of alkyl iodides the products are a carbonium ion and iodine. The carbonium ion reacts with the solvent. The aryl iodide, $\emptyset I$, oxidizes to give the cation $\emptyset - I^+ - \emptyset I$.

C. Hydrocarbons. The oxidation of aromatic hydrocarbons has been studied in several non-aqueous solvents. The products are generally derived from the chemical reactions of carbonium ion radicals or carbonium ions. Recently it has also been shown that certain aliphatic hydrocarbons in acetonitrile can be oxidized to carbonium ions.

13

In aqueous solution, few oxidations of hydrocarbons have been attempted. However, the oxidation of benzene to quinone and oxidation of the side chain in alkyl benzenes have been reported.

D. Nitrogen Containing Compounds. As in the case of phenols, aromatic amines are readily oxidized, but the wide variety of products does not encourage generalizations.

In acetonitrile, amides and amines are oxidized to give quaternary ammonium salts. In aqueous solution amines can be oxidized with breakage of the carbon nitrogen bond. The products are a lower amine and an aldehyde.

E. Sulphur Containing Compounds. The electrochemical oxidation of thiols gives the corresponding disulphides while the thioethers have been shown to oxidize to the sulphoxide and the sulphone. This oxidation of compounds to the corresponding oxide should be a general reaction.

IV. EXAMPLES OF USEFUL SYNTHETIC ELECTRODE REACTIONS

1. Hydrodimerizations

A hydrodimerization reaction may be defined by the general equation

$$2 \quad \underset{R_2}{\overset{R_1}{>}}C{=}C\underset{E}{\overset{R_3}{<}} + 2H^+ + 2e \rightarrow \quad \begin{array}{c} \underset{R_2}{\overset{R_1}{>}}C{-}CH\underset{E}{\overset{R_3}{<}} \\ | \\ \underset{R_2}{\overset{R_1}{>}}C{-}CH\underset{E}{\overset{R_3}{<}} \end{array}$$

where E is an electron withdrawing group such as a nitrile, amide, ester, aryl, pyridyl, vinyl, or acetyl group.[94,98] The importance of these reactions is that they give rise to bifunctional compounds.

The reactions are carried out under conditions which optimize the yield of hydrodimer as opposed to the simple hydrogenated product formed by the competing reaction

$$\underset{R_2}{\overset{R_1}{>}}C{=}C\underset{E}{\overset{R_3}{<}} + 2H^+ + 2e \rightarrow \underset{R_2}{\overset{R_1}{>}}CH{-}CH\underset{E}{\overset{R_3}{<}}$$

Thus the electrolyses are carried out using concentrated, aqueous solutions of the olefin at high hydrogen overpotential cathodes such as mercury, lead, cadmium, and tin. The base electrolyte used is a concentrated solution of a salt such as tetraalkylammonium p-toluenesulphonate which also acts as a solubilizing agent for the olefin and the tetraalkylammonium cation further serves to suppress the hydrogen evolution reaction even at the very cathodic potentials required for these reactions. A low concentration of olefin, a low pH and the presence of metal ions all lower the yield of the hydrodimer.[94]

The hydrodimerization of acrylonitrile under these conditions is the basis of a new American industrial plant for the production of adiponitrile[50]

$$2CH_2{=}CH{-}CN + 2H^+ + 2e \rightarrow \begin{array}{l} CH_2{-}CH_2{-}CN \\ | \\ CH_2{-}CH_2{-}CN \end{array}$$

Other examples of this reaction which may be cited include the hydrodimerization of ethylacrylate, diethylmaleate, N,N-dimethylcrotonamide, tiglonitrile, 1-cyano-1,3-butadiene, 4-vinyl-pyridine, and maleonitrile.[95,96,97,98,99,100]

$$2CH_2{=}CHCOOEt + 2H^+ + 2e \rightarrow \begin{array}{l} CH_2{-}CH_2{-}COOEt \\ | \\ CH_2{-}CH_2{-}COOEt \end{array}$$

$$2\begin{array}{l} CH{-}COOEt \\ \| \\ CH{-}COOEt \end{array} + 2H^+ + 2e \rightarrow \begin{array}{l} EtO{-}\underset{\underset{O}{\|}}{C}{-}CH{-}CH_2COOEt \\ | \\ EtO{-}\underset{\underset{O}{\|}}{C}{-}CH{-}CH_2COOEt \end{array}$$

$$2CH_3{-}CH{=}CH{-}CON\begin{array}{l} {}^{\displaystyle Me} \\ {}_{\displaystyle Me} \end{array} + 2H^+ + 2e$$

$$\rightarrow \begin{array}{l} CH_3{-}CH{-}CH_2CON\begin{array}{l}{}^{\displaystyle Me}\\{}_{\displaystyle Me}\end{array} \\ | \\ CH_3{-}CH{-}CH_2CON\begin{array}{l}{}^{\displaystyle Me}\\{}_{\displaystyle Me}\end{array} \end{array}$$

$$2\text{MeCH}=C\begin{array}{c}\text{Me}\\ \diagup\\ \diagdown\\ \text{CN}\end{array} + 2\text{H}^+ + 2\text{e} \rightarrow \begin{array}{c}\text{Me}\\ \diagup\\ \text{Me}-\text{CH}-\text{CH}-\text{CN}\\ |\\ \text{Me}-\text{CH}-\text{CH}-\text{CN}\\ \diagdown\\ \text{Me}\end{array}$$

$$2\text{CH}_2=\text{CH}-\text{CH}=\text{CHCN} + 2\text{H}^+ + 2\text{e} \rightarrow \begin{array}{c}\text{CH}_2=\text{CH}-\text{CH}-\text{CH}_2\text{CN}\\ |\\ \text{CH}_2=\text{CH}-\text{CH}-\text{CH}_2\text{CN}\end{array}$$

$$2\ \begin{array}{c}\text{CH}=\text{CH}_2\\ \bigcirc\\ \text{N}\end{array} + 2\text{H}^+ + 2\text{e} \rightarrow \begin{array}{c}\text{CH}_2-\text{CH}_2-\bigcirc\text{N}\\ |\\ \text{CH}_2-\text{CH}_2-\bigcirc\text{N}\end{array}$$

$$2\ \begin{array}{c}\text{CH}=\text{CH}\\ |\quad\ |\\ \text{CN}\ \ \text{CN}\end{array} + 2\text{H}^+ + 2\text{e} \rightarrow \begin{array}{c}\text{CH}_2-\text{CH}-\text{CH}-\text{CH}_2\\ |\quad\ |\quad\ |\quad\ |\\ \text{CN}\ \ \text{CN}\ \ \text{CN}\ \ \text{CN}\end{array}$$

As can be seen from the above examples, coupling is always head to head; that is, the coupling always occurs at the position β to the functional group.

It is also possible to obtain crossed coupled products from mixtures of olefins. Selective results can only be obtained if there is a difference in the reduction potentials of the two olefins; then if the electrolysis is carried out at a potential where only one of the olefins is reduced and the non-reducible olefin is in large excess a single product is obtained. For example, if a mixture of mesityl oxide and acrylonitrile is reduced at a potential below that required for the reduction of acrylonitrile, high yields of the crossed product are obtained.[101]

$$\begin{array}{c}\text{Me}\\ \diagdown\\ \quad\ \ \text{C}=\text{CHCOCH}_3 + \text{CH}_2=\text{CHCN} + 2\text{H}^+ + 2\text{e}\\ \diagup\\ \text{Me}\end{array}$$

$$\rightarrow \begin{array}{c}\text{Me}\\ \diagdown\\ \quad\ \ \text{C}-\text{CH}_2\text{COCH}_3\\ \diagup\ |\\ \text{Me}\ \ \text{CH}_2-\text{CH}_2\text{CN}\end{array}$$

If the reduction potentials of the two olefins are similar or the cathode potential is not carefully controlled, a mixture of the crossed and the two uncrossed hydrodimers is obtained.

The mechanism of hydrodimerization reactions is believed to be carbanionic rather than radical, i.e.

$$CH_2{=}CH{-}CN + 2e \rightarrow \bar{C}H_2{-}\bar{C}H{-}CN$$

$$\bar{C}H_2{-}\bar{C}H{-}CN + CH_2{=}CH{-}CN \rightarrow \begin{array}{l} CH_2{-}\bar{C}H{-}CN \\ | \\ CH_2{-}\bar{C}H{-}CN \end{array}$$

$$\begin{array}{l} CH_2{-}\bar{C}H{-}CN \\ | \\ CH_2{-}\bar{C}H{-}CN \end{array} + 2H^+ \rightarrow \begin{array}{l} CH_2{-}CH_2CN \\ | \\ CH_2{-}CH_2CN \end{array}$$

rather than

$$CH_2{=}CH{-}CN + e \rightarrow C\dot{H}_2{-}\bar{C}H{-}CN$$

$$\rightarrow \tfrac{1}{2} \begin{array}{l} CH_2{-}\bar{C}HCN \\ | \\ CH_2{-}\bar{C}HCN \end{array} \xrightarrow{2H^+} \begin{array}{l} CH_2{-}CH_2{-}CN \\ | \\ CH_2{-}CH_2{-}CN \end{array}$$

because of the absence of polymer under the working conditions and because the mechanism of the hydrodimerization of a pair of different olefins at a potential below that required for the reduction of one of the olefins cannot be radical in nature. Here the mechanism must be

$$\begin{array}{c} Me \\ \diagdown \\ C{=}CHCOCH_3 \\ \diagup \\ Me \end{array} + 2e \rightarrow \begin{array}{c} Me \\ \diagdown \\ \bar{C}{-}\bar{C}HCOCH_3 \\ \diagup \\ Me \end{array}$$

$$\begin{array}{c} Me \\ \diagdown \\ \bar{C}{-}\bar{C}HCOCH_3 \\ \diagup \\ Me \end{array} + CH_2{=}CHCN$$

$$\rightarrow \begin{array}{c} Me \\ \diagdown \\ \bar{C}{-}\bar{C}HCOCH_3 \\ \diagup \quad | \\ Me \quad CH_2{-}\bar{C}H{-}CN \end{array} \xrightarrow{2H^+} \begin{array}{c} Me \\ \diagdown \\ C{-}CH_2COCH_2 \\ \diagup \quad | \\ Me \quad CH_2{-}CH_2CN \end{array}$$

Another reaction carried out under similar conditions is the reduction

$$L(CH_2)_nE + 2e \rightarrow L^- + {}^-(CH_2)_nE$$

where L is a leaving group and E is again an electron withdrawing group.[102,103] An example is the reduction

$$C_6H_5CH_2S^+ \Big\backslash^{Me}_{Me} + 2e \to C_6H_5CH_2^- + \ ^{Me}\!\diagup S \diagdown_{Me}$$

which when carried in the presence of an activated olefin leads to a coupled product.

$$C_6H_5CH_2^- + CH_2{=}CH{-}CN \longrightarrow C_6H_5CH_2{-}CH_2{-}\bar{C}H{-}CN$$
$$\xrightarrow{H^+} C_6H_5(CH_2)_3CN$$

2. Radical Reactions

A. Radical-Radical Reactions. The simple Kolbe reaction, that is the anodic formation of ethane and carbon dioxide from the electrolysis of acetate ion was reported as early as 1849 and it has since been shown

$$2CH_3COO^- \to 2CO_2 + C_2H_6 + 2e$$

that this reaction may be extended to acids containing up to twenty carbon atoms and that branching of the carbon chain (except at the α position) has no effect on the reaction. Examples are[104,105]

$$2CH_3(CH_2)_4COO^- \to 2CO_2 + 2e + CH_3(CH_2)_8CH_3$$

$$2 \ ^{CH_3}\!\diagdown_{CH_3}\!\!\diagup CH{-}(CH_2)_3CH{-}CH_2COO^- \to 2CO_2 + 2e$$
$$\underset{\underset{CH_3}{|}}{}$$

$$+ \ ^{CH_3}\!\diagdown_{CH_3}\!\!\diagup CH{-}(CH_2)_3{-}\underset{\underset{CH_3}{|}}{CH}{-}(CH_2)_2{-}\underset{\underset{CH_3}{|}}{CH}{-}(CH_2)_3{-}CH\diagup^{CH_3}_{\diagdown CH_3}$$

The reaction with the unsubstituted acids is seldom of interest in synthetic work. However, the Kolbe synthesis is a general reaction for substituted carboxylic acids except for α substituted acids, α,β-unsaturated acids, benzoic acids and dicarboxylic acids where little success has been achieved in obtaining the Kolbe dimer. Thus, the reaction using substituted acids gives rise to bifunctional compounds which are useful intermediates for synthesis. For example[106,107,108]

$$2CH_3COCH_2-CH_2COO^- \rightarrow 2CO_2 + 2e + CH_3CO-(CH_2)_4-COCH_3$$

$$2CH_3CONH-(CH_2)_5COO^-$$
$$\rightarrow 2CO_2 + 2e + CH_3CONH-(CH_2)_{10}-NHCOCH_3$$

$$2CH_2-(CH_2)_9COO^- \rightarrow 2CO_2 + 2e + CH_2-(CH_2)_{18}-CH_2$$
$$\quad | \qquad\qquad\qquad\qquad\qquad\qquad\qquad | \qquad\qquad\qquad |$$
$$\quad OH \qquad\qquad\qquad\qquad\qquad\qquad\quad OH \qquad\qquad\quad OH$$

Although it is not possible to carry out the reaction with dicarboxylic acids, it is possible with their half esters. The synthesis is then known as the Brown-Walker reaction. Examples of this reaction include[109,110]

$$2EtOOC-(CH_2)_4COO^- \rightarrow 2CO_2 + 2e + EtOOC-(CH_2)_8COOEt$$

$$2MeOOC-(CH_2)_2COO^- \rightarrow 2CO_2 + 2e + MeOOC-(CH_2)_4COOMe$$

and it should be noted that the reaction is also possible with substituted and unsaturated half esters.[111]

$$\qquad\qquad\qquad CH_3$$
$$\qquad\qquad\qquad |$$
$$2EtOOC-CH=C-(CH_2)_2COO^-$$
$$\qquad\qquad\qquad\qquad\qquad\qquad CH_3 \qquad\qquad CH_3$$
$$\qquad\qquad\qquad\qquad\qquad\qquad | \qquad\qquad\qquad |$$
$$\rightarrow 2CO_2 + 2e + EtOOC-CH=C-(CH_2)_4-C=CH-COOEt$$

Also important in synthetic work is the Kolbe reaction carried out on a mixture of acids, half esters or an acid and half ester, since it is possible to obtain the crossed product. Although the product is a mixture of the uncrossed dimers as well as the crossed product the reaction can be important for the synthesis of products which are otherwise difficult to prepare. Again, examples include[112,113,114]

$$CH_3CO(CH_2)_2COO^- + CH_3COO^-$$
$$\qquad\qquad\qquad\qquad \rightarrow CH_3CO(CH_2)_2-CH_3 + 2CO_2 + 2e$$

$$CH_3O(CH_2)_{10}COO^- + MeOOCCH_2-CH-CH_2COO^-$$
$$\qquad\qquad\qquad\qquad\qquad\qquad\qquad\qquad |$$
$$\qquad\qquad\qquad\qquad\qquad\qquad\qquad\qquad CH_3$$
$$\qquad\qquad \rightarrow CH_3O(CH_2)_{11}-CH-CH_2COOMe + 2CO_2$$
$$\qquad\qquad\qquad\qquad\qquad\qquad\qquad |$$
$$\qquad\qquad\qquad\qquad\qquad\qquad\qquad CH_3 \qquad\qquad + 2e$$

$$CH_3(CH_2)_7 \diagdown \diagup (CH_2)_7COO^-$$
$$C=C$$
$$H \diagup \diagdown H$$

$$+ MeOOC(CH_2)_4COO^-$$

$$\rightarrow \quad CH_3(CH_2)_7 \diagdown \diagup (CH_2)_{11}COOMe$$
$$C=C$$
$$H \diagup \diagdown H$$

and several such crossed products have been used in the synthesis of large natural products.[24]

The Kolbe and Brown-Walker reactions studied early in the century were carried out in aqueous solutions, but in recent years anhydrous alcohols and, in some cases, the acids themselves have been shown to be more suitable solvents. The reactions are usually carried out in acidic or neutral solution at room temperature or below and at a smooth platinum or other platinum group metal anode. As is indicated elsewhere in this chapter, certain other electrodes, particularly carbon, are poor materials for this reaction.

Although other mechanisms for the Kolbe reaction have been suggested, it is now believed to occur via the mechanism

$$CH_3COO^- \rightarrow CH_3COO^{\cdot} + e$$

$$CH_3COO^{\cdot} \rightarrow CO_2 + CH_3^{\cdot}$$

$$2CH_3^{\cdot} \rightarrow C_2H_6$$

The Hofer-Moest reaction occurs during the oxidation of carboxylate anions in solutions of high pH and it gives rise to an alcohol with one carbon less than the original acid. The mechanism of this reaction is believed to involve reaction between an alkyl and a hydroxyl radical,[80] i.e.

$$CH_3COO^- \rightarrow CH_3COO^{\cdot} + e$$

$$CH_3COO^{\cdot} \rightarrow CO_2 + CH_3^{\cdot}$$

$$H_2O \rightarrow H^+ + OH^{\cdot} + e$$

$$CH_3^{\cdot} + OH^{\cdot} \rightarrow CH_3OH$$

Esters such as diethyl malonate,[27] ethyl acetoacetate,[27] and ethyl phenylacetate,[27] diketones such as acetylacetone[116] and nitrocompounds[115] which contain an acidic hydrogen can be oxidized at a platinum electrode to give a radical which then dimerizes so that these reactions are further examples of dimerization reactions

$$2\overset{-}{C}H\begin{array}{c}COOEt\\\\COOEt\end{array}\rightarrow 2e + \begin{array}{c}EtOOC\\\\EtOOC\end{array}CH\!-\!CH\begin{array}{c}COOEt\\\\COOEt\end{array}$$

$$2CH_3CO\overset{-}{C}HCOCH_3 \rightarrow 2e + \underset{\underset{CH_3COCHCOCH_3}{|}}{CH_3COCHCOCH_3}$$

$$2\;\;\begin{array}{c}CH_3\\\\ \overset{-}{C}\!-\!NO_2\\\\CH_3\end{array}\rightarrow 2e + CH_3\!-\!\underset{\underset{NO_2}{|}}{\overset{\overset{CH_3}{|}}{C}}\!-\!-\!-\!\underset{\underset{NO_2}{|}}{\overset{\overset{CH_3}{|}}{C}}\!-\!CH_3$$

A further radical-radical reaction which is useful in building large molecules is the oxidation of certain thiols at a mercury anode to give a sulphur-sulphur bond. An example of such a compound is L-cysteine.[117]

$$2HS\!-\!CH_2\!-\!\underset{\underset{NH_2}{|}}{CH}\!-\!COOH$$

$$\rightarrow 2H^+ + 2e + HOOC\!-\!\underset{\underset{NH_2}{|}}{CH}\!-\!CH_2\!-\!S\!-\!S\!-\!CH_2\!-\!\underset{\underset{NH_2}{|}}{CH}COOH$$

The oxidation of the boron tetraphenyl anion has been shown to give biphenyl by a radical dimerization process.[118] Furthermore

$$B(C_6H_5)_4^- \rightarrow 2e + B^+(C_6H_5)_2 + C_6H_5\!-\!C_6H_5$$

has been shown to be an intramolecular reaction since when a mixture of $B(C_6H_5)_4^-$ and $B(C_6D_5)_4^-$ is electrolysed, the product was shown to be $C_6H_5\!-\!C_6H_5$ and $C_6D_5\!-\!C_6D_5$ with no crossed product.[119]

The reactions discussed above are all oxidations but radicals are also formed in reductions at cathodes. For example, biphenyl is formed by a radical-radical reaction following the reduction of tetraarylammonium ions.[44]

$$\varnothing_4N^+ + e \rightarrow \varnothing_3N + \varnothing^{\cdot}$$

$$2\varnothing^{\cdot} \rightarrow \varnothing\!-\!\varnothing$$

Dimerization of the carbanion radicals has been observed during the reduction of aromatic hydrocarbons in aprotic solvents. For example, phenanthrene can be reduced in a one electron step to give the dihydrodiaromatic derivative[43]

However, from a synthetic point of view the most important reduction involving radicals is the one electron reduction of an aldehyde or ketone and the subsequent dimerization of the radicals in the pinacol reaction.

This reaction has been widely studied for aromatic aldehydes and ketones and some examples are[120,121,122]

The reaction is more difficult to carry out in the aliphatic series because of the competing reactions. However the reaction has been studied with acetone and a few other compounds.[2,32,123] For example

Few examples are to be found of the reaction with substituents in the aliphatic chains. However the coupling of glyoxylic acid has been reported.[124]

No attempt to obtain cross coupled products by the reduction of a mixture of a pair of aldehyde or ketones seems to have been reported. The solution conditions and cathode recommended for the pinacol reaction seems to vary from compound to compound. However, the reaction is usually performed in either acidic or basic conditions and particular metals are advised for electrodes in order to obtain the maximum yield of the pinacol in each case.[2]

B. Radical-Olefin Reactions. The reactions of electrochemically generated radicals with olefins[125] present in solution show great promise for the synthesis of large molecules although as yet these reactions have not been studied in great numbers. There have been reports that olefin polymerization can be initiated in this way.

Methyl radicals generated by the oxidation of sodium acetate in acetic acid have been shown to react with acrylonitrile,[126] isoprene,[126] butadiene,[128] and styrene[127] and, in some cases, products from the acetate radical have also been isolated. Thus the scheme

can be written for the reaction between acrylonitrile and methyl radicals and all products have been identified. In the other cases the main products are shown by the equations

$$CH_3^{\cdot} + CH_2{=}CH{-}CH{=}CH_2 \rightarrow CH_3{-}CH_2{-}CH{=}CH{-}CH_2^{\cdot}$$
$$\rightarrow (C_2H_5{-}CH{=}CH{-}CH_2)_2$$

$$CH_3^{\cdot} + CH_2{=}CHC_6H_5 \rightarrow CH_3{-}CH_2{-}\overset{\cdot}{C}HC_6H_5$$

In each case, other products from the various radical-radical combinations are found and products derived from the carboxyl radical are also found. Moreover, it is clear that by suitable choice of conditions these reactions could be modified to yield polymers.

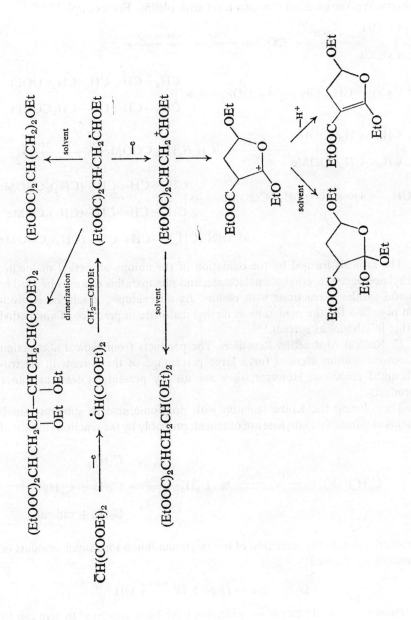

Similarly, it has been shown that radicals which are intermediates in the Brown-Walker reaction can also react with olefins. For example,[36,129]

$$\underset{\overset{|}{COOEt}}{\overset{COOH}{\underset{-CO_2}{\xrightarrow{-2e-H^+}}}} \cdot COOEt \xrightarrow{CH_2=CH-CH=CH_2}$$

$$\overset{\cdot}{C}H_2-CH=CH-CH_2COOEt \longrightarrow \underset{CH_2-CH=CH-CH_2COOEt}{\overset{CH_2-CH=CH-CH_2COOEt}{\underset{|}{}}}$$

$$\underset{\overset{|}{CH_2-CH_2COOMe}}{\overset{CH_2-CH_2COOH}{\underset{-CO_2}{\xrightarrow{-2e-H^+}}}} \cdot CH_2(CH_2)_3COOMe \xrightarrow{CH_2=CH-CH=CH_2}$$

$$\cdot CH_2-CH=CH-(CH_2)_5COOMe \longrightarrow \underset{CH_2-CH=CH-(CH_2)_5COMe}{\overset{CH_2-CH=CH-(CH_2)_5COOMe}{\underset{|}{}}}$$

$$+$$
$$MeOOC(CH_2)_6-CH=CH-(CH_2)_5COOMe$$

The radicals formed by the oxidation of the anions of diethyl malonate, ethyl acetoacetate, ethyl phenylacetate, and nitroparafins can also be used in useful synthetic reactions with olefins. As an example, a scheme is given on page 383 for the oxidation of diethyl malonate in presence of vinyl ethyl ether in ethanol as solvent.[130]

C. Radical Abstraction Reactions. The products from radical abstraction reactions seldom account for a large percentage of the current in electrochemical reactions. However, they are usually present, at least as minor products.

Thus during the Kolbe reaction with propionic acid as substrate small yields of ethane and ethylene are obtained, probably by the reactions indicated

$$C_2H_5COOH \xrightarrow{-H^+-CO_2-e} C_2H_5 \cdot \begin{array}{l} \overset{C_2H_5}{\nearrow} C_4H_{10} \\ \xrightarrow{C_2H_5} C_2H_6 + C_2H_4 \\ \underset{solvent}{\searrow} C_2H_6 + \text{radical} \end{array}$$

Similarly, during the reduction of tetraarylammonium ions small amounts of benzene are formed[34]

$$\varnothing_4N^+ + e \rightarrow \varnothing_3N + \varnothing\cdot \xrightarrow{solvent} \varnothing H$$

However, radical abstraction reactions have been suggested to give rise to the major products in the side chain substitution reactions of alkyl benzenes under suitable conditions. Thus in acetic acid containing toluene and nitrate ions the mechanism for the oxidation is suggested to be[71]

$$NO_3^- \rightarrow NO_3^{\cdot} + e$$

$$NO_3^{\cdot} + CH_3 \text{⟨ring⟩} \rightarrow HNO_3 + \dot{C}H_2 \text{⟨ring⟩}$$

$$CH_2^{\cdot} \text{⟨ring⟩} \quad -e \rightarrow \quad CH_2^+ \text{⟨ring⟩} \xrightarrow{\text{HOAc}} \quad CH_2OAc \text{⟨ring⟩} + H^+$$

since current/potential curves suggest that the nitrate ion is the most readily oxidized component of the system and the major product isolated is benzyl acetate. The methoxy radical is postulated to play a similar role in the methoxylation of toluene in methanol containing sodium methoxide.[131]

D. Formation of Metal Alkyls. The electrochemical methods of preparation of metal alkyls may all be considered as the generation of radicals at an electrode followed by reaction between the radicals and the electrode. The radicals are generally made by the reduction of alkyl halides in an aprotic solvent or by the oxidation of a carbanion species such as another organometallic compound in an aprotic solvent or a molten salt medium. The Kolbe reaction is not suitable for the preparation of metal alkyls because of the dissolution of the suitable metal anodes at the potentials required for the Kolbe reaction. Electrochemical preparations of mercury, lead, tin, zinc, cadmium, aluminium, magnesium, berylium, calcium manganese, gold, indium, and antimony alkyls have all been reported.[39] However, the bulk of the literature covers preparations of lead tetraethyl and lead tetramethyl because of their importance as anti-knock agents in the petroleum industry.

Lead tetraalkyls have been prepared by the oxidation of Grignard reagents in ether at lead anodes and this method is the basis of a new American plant for the preparation of lead tetramethyl and lead tetraethyl[38]

$$RMgCl \xrightarrow[\text{anode}]{Pb} MgCl^+ + R^{\cdot} + e$$

$$Pb + 4R^{\cdot} \rightarrow PbR_4$$

Similar reactions have been reported using aluminium triethyl, zinc diethyl or sodium aluminium tetraethyl as the electroactive species at a lead anode. In the case of sodium aluminium tetraethyl, the molten complex can be used as electrolysis medium.[28]

$$AlEt_4^- \rightarrow e + AlEt_3 + Et^{\cdot} \xrightarrow{Pb} PbEt_4$$

The reduction of alkyl halides in propylene carbonate at a lead cathode is reported to give high yields of lead tetraalkyls.[39]

$$\text{EtBr} + e \rightarrow \text{Br}^- + \text{Et}^{\cdot} \xrightarrow{\text{Pb}} \text{PbEt}_4$$

Metal alkyls have also been found as products in the reduction of ketones. For example, the reduction of acetone at a mercury electrode gives rise to some disopropylmercury and the same reaction at a lead electrode gives disopropyl lead as the product.[32,132]

3. Cyclization Reactions

The study of the electrochemical preparation of cyclic compounds from non-cyclic starting materials is still in its early stages and many possibilities have still to be investigated. However, at the present stage it seems worthwhile to collect together the reports of cyclization reactions in one section.

A. Carbanionic Cyclizations

Using conditions similar to those discussed earlier for intermolecular hydrodimerizations, it has been found possible to bring about the intramolecular hydrodimerization of an activated diolefin to form a cyclic product.[133] The reaction

$$Z\begin{array}{c} \text{CH=CHCOOEt} \\ \diagup \\ \diagdown \\ \text{CH=CHCOOEt} \end{array} + 2\text{H}^+ + 2e \rightarrow Z\begin{array}{c} \text{CH—CH}_2\text{COOEt} \\ \diagup \\ | \\ \text{CH—CH}_2\text{COOEt} \end{array}$$

has been carried out where $Z = (\text{CH}_2)_n$, $n = 2, 3, 4, 6, 8, 12,$ and 16 and

$Z = \bigcirc$ and $\text{O—CH}_2\text{—CH}_2\text{—O}$. The best yields are obtained for the compounds $Z = (\text{CH}_2)_n$, $n = 2\text{–}4$. Examples are

$$\begin{array}{c} \text{CH}_2 \\ \diagup \quad \diagdown \\ \text{CH}_2 \qquad \text{CH=CHCOOEt} \\ | \qquad\qquad\quad + 2\text{H}^+ + 2e \\ \text{CH}_2 \qquad \text{CH=CHCOOEt} \\ \diagdown \quad \diagup \\ \text{CH}_2 \end{array}$$

$$\rightarrow \begin{array}{c} \text{CH}_2 \\ \diagup \quad \diagdown \\ \text{CH}_2 \qquad \text{CH—CH}_2\text{COOEt} \\ | \qquad\qquad | \\ \text{CH}_2 \qquad \text{CH—CH}_2\text{COOEt} \\ \diagdown \quad \diagup \\ \text{CH}_2 \end{array}$$

The activating group can also be nitrile in certain cases.

The related leaving group reaction has also been used in a cyclization reaction but the yields are poor.

It has recently been shown that it is possible to form cyclopropane and cyclobutane by the reduction of 1,3-dibromopropane and 1,4-bibromobutane respectively.[134,135] The reaction was not possible for the formation of larger rings.

$$BrCH_2-CH_2-CH_2Br + 2e \rightarrow \underset{CH_2}{CH_2-CH_2} + 2Br^-$$

Similar reactions were possible for the formation of bicyclo compounds

These reactions are believed to proceed via the formation of carbanions, i.e.

rather than by the formation of a biradical intermediate since the reduction of chlorobromocyclobutane took place in a single two electron step at the potential required to break the carbon-bromine bond [the C—Br bond is known to be weaker than the C—Cl bond and the potential required for their fissures will be different] and not in two one electron waves as would be expected for the radical-radical mechanism. Therefore it was concluded that the mechanism was

This mechanism is not surprising since the normal reduction mechanism of alkyl halides involves a two electron step to give a carbanion

$$RCl + 2e \rightarrow Cl^- + R^- \xrightarrow{\text{solvent}} RH$$

B. Radical Mechanisms

Generally it is not possible to carry out the Kolbé reaction on dicarboxylic acids, but in the case of the cyclobutane derivative shown it was possible to decarboxylate and form the bicyclobutane derivative.[136]

It is also possible to perform an intramolecular pinacol reaction of the type[137]

$$(CH_2)_n \underset{COC_6H_5}{\overset{COC_6H_5}{\diagdown}} + 2H^+ + 2e \rightarrow (CH_2)_n \begin{array}{c} C_6H_5 \\ | \\ C\!-\!OH \\ | \\ C\!-\!OH \\ | \\ C_6H_5 \end{array}$$

C. Carbonium Ion Mechanisms

The recent reports that at carbon anodes the oxidation of carboxylic acids gives carbonium ions as does the oxidation at platinum anodes of aliphatic hydrocarbons, alkyl iodides and aliphatic amines, gives rise to many possibilities for the preparation of cyclic compounds. It has already been reported that at a carbon anode, butyric acid forms cyclopropane,[60]

$$CH_3\!-\!CH_2\!-\!CH_2\!-\!COO^- \xrightarrow{-2e-CO_2} CH_3\!-\!CH_2\!-\!CH_2^+ \xrightarrow{-H^+} CH_2\!-\!CH_2 \diagdown CH_2$$

and several ring expansions and contractions have been noted earlier. Modification of the mechanism of these reactions to non-cyclic compounds suggest that further carbonium ion cyclizations will be possible.

The reaction of anthracene at a platinum anode in the presence of pyridine was stated in an earlier section. A modified form of this reaction has been shown to be possible intramolecularly,[137] thus leading to a cyclic compound

D. Heterocyclic Compounds

The examples of ring closure as well as ring contraction and ring expansion are more numerous in the heterocyclic series.

Thus ring closures have been noted by:

(a) reduction of a nitro or nitrosogroup to a suitable nucleophilic reagent followed by its attack on an electrophilic centre;

Reactions of type (a)

Reactions of type (b)

Reactions of type (c)

(b) reduction of an azogroup to a hydrazine followed by attack on an electrophilic centre; and

(c) reduction of an azomethine to an amine followed by attack on an electrophilic centre.

Examples of each type of reaction are given in the equations above.[137]

4. Aromatic Substitution

The aromatic substitution reactions which have been studied include fluorination, chlorination, bromination, iodination, hydroxylation, cyanation, thiocyanation, alkoxylation, acyloxylation, and acetamidation.

These substitution reactions are believed to occur by a similar mechanism involving discharge of the aromatic hydrocarbon to form a cationic species followed by reaction between the cation and the substituting anion or a solvent molecule. An earlier suggestion[139] that the reactions took place by a mechanism involving the discharge of the anion or the solvent to give a radical, followed by radical substitution is now discounted for the following reasons:

(a) substitution is found to occur only at potentials at which the aromatic hydrocarbon is oxidized, even when the solvent or substituting anion is discharged at less anodic potential;[138]

(b) substitution can take place at less positive potentials than that required to oxidize the solvent or anion if the hydrocarbon is more readily oxidized;[140]

(c) the distribution of product isomers is not that predicted for radical substitution;[26]

(d) some of the minor products expected from a radical mechanism are absent.[56]

However in certain special circumstances it is possible that the mechanism of substitution is

$$X^- \to X^{\cdot} + e$$

$$C_6H_5CH_3 + X^{\cdot} \to HX + C_6H_5CH_2^{\cdot}$$

$$C_6H_5CH_2^{\cdot} \to e + C_6H_5CH_2^+ \to products$$

In these cases, for example, the oxidation of toluene in the presence of methoxide ion,[131] minor products such as the dimer, dibenzyl, can be identified. Generally these minor products are absent.

In order to study the mechanism of substitution it is necessary to choose reactions where the products are stable. Thus hydroxylation reactions are very difficult to study because of the ease with which the primary products are oxidized and total disintegration of the hydrocarbons is common. Similarly, acetoxylation reactions lead to a complicated mixture of multi-substituted compounds and isomers. On the other hand the primary products of cyanation reactions are stable to further oxidation. Thus in many cases the products of substitution reactions are found to be time dependent and simple behaviour with one time independent product is not common.

In the case of the alkyl substituted aromatic hydrocarbons, there is competition between nuclear and side chain substitution. In fact nuclear substitution is found when the substituting anion is present in solution while side chain substitution is found when the product depends on a solvolysis reaction.[56] For example, in the presence of acetate ion, toluene gives tolyl acetate while if the reaction is carried out in acetic acid without free acetate ion, the product is benzyl acetate. Thus, clearly the presence of acetate ion has a profound effect on the reaction mechanism and it has been suggested that nuclear substitution takes place by a concerted mechanism, i.e.

while in the absence of acetate ion, the mechanism is

Substitution reactions have been carried out using the polynuclear aromatic hydrocarbons, substituted benzenes, and heterocyclic compounds but seldom with benzene itself because of its positive oxidation potential.

The reactions considered above all occur at the anode but it should be remembered that it is possible to carboxylate and alkylate aromatic hydrocarbons at the cathode using the reaction between the carbanion and carbon dioxide or methyl iodide (see earlier section). The products of these reactions are the dihydro compounds but they are readily reoxidized under mild conditions to give the substituted aromatic species.

Finally, it should be noted that the species NO_2^+, NO^+, I^+, and Br^+, widely used for the substitution of organic compounds by chemical methods, have recently been prepared by the anodic oxidation of NO_2, NO, I_2, and Br_2 respectively, and it is to be expected that their use for electrochemical substitution will not be far in the future.

5. Aliphatic Substitution

The halogenation of saturated and unsaturated aliphatic compounds has been studied. For example, the chlorination and bromination of acetone, the oxidation of alcohols to haloforms and the chlorination of ethylene.[17,21] However, the products of these reactions seem typical of those expected from the chemical reactions of the compounds with halogen, and it seems likely that the electrode reaction is the generation of the halogen.

The only reactions of commercial significance are the fluorination reactions.[21,22] It is not clear whether these reactions proceed via anodic generation of fluorine or whether the reactions take place via carbonium ion intermediates. However, there has been considerable work reported on the complete and partial fluorination of organic compounds in anhydrous hydrogen fluoride. Typical reactions are

$$n\text{-}C_8H_{18} \rightarrow n\text{-}C_8F_{18}$$
$$C_2H_5\text{—}O\text{—}C_2H_5 \rightarrow C_2F_5\text{—}O\text{—}C_2F_5$$
$$CH_3COOH \rightarrow CF_3COF$$

although yields are sometimes poor due to carbon-carbon bond fragmentation and a multitude of smaller molecules are usually obtained.

6. Elimination Reactions

The removal of certain substituent groups is also possible electrochemically. Of the cathodic reactions, dehalogenation is by far the best documented and it has been shown that the removal of halogen atoms is possible both from aromatic rings and aliphatic chains.[141,142,143,144] For example, it is possible to remove one, two or all three chlorine atoms from trichloroacetic acid by control of the electrode potential,[143]

$$CCl_3COOH \xrightarrow[-Cl^-]{H^+ + 2e} CHCl_2COOH \xrightarrow[-Cl^-]{H^+ + 2e} CH_2ClCOOH$$
$$\xrightarrow[-Cl^-]{H^+ + 2e} CH_3COOH$$

and it is possible to remove the iodine atom from iodobenzoic acid, iodo-phthalates and iodophthalic anhydride[143]

while bromobenzene reduces to benzene[69]

These reactions are believed to proceed via a carbanionic intermediate since typical carbanion reactions with carbon dioxide, etc., are possible in many cases.

The reduction of vicinal dibromides to the olefin has been studied. For example[145]

and the variation of two isomeric products with pH, potential and stereo-chemistry of the starting materials (in the case of optically active starting materials[146]) has been examined.

Similar elimination reactions have been studied for the nitro group. Thus the reduction of dinitrobenzene[147] and 2,2-dinitropropane[148] in dimethyl-formamide both lead to elimination of the nitrite ion

$$CH_3-\underset{\underset{NO_2}{|}}{\overset{\overset{NO_2}{|}}{C}}-CH_3 + 2e + H^+ \rightarrow CH_3-\underset{NO_2}{\overset{|}{C}}H-CH_3 + NO_2^-$$

Elimination reactions also occur during oxidations, for example during the oxidation of p-substituted phenols.[150,151]

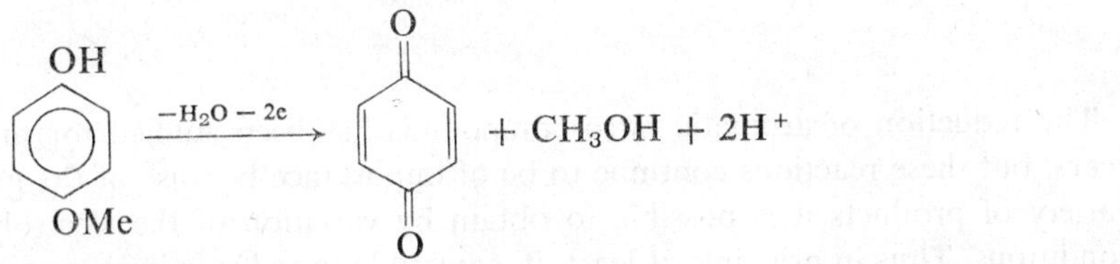

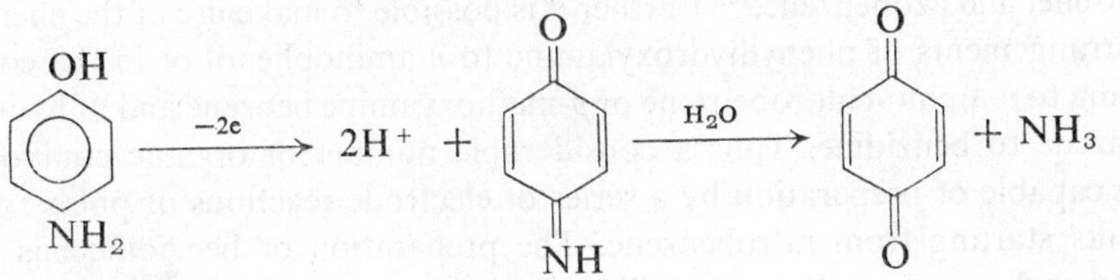

7. Further Reductions

The electrochemical hydrogenation of olefinic and acetylenic bonds (both the selective reduction to the olefin and complete reduction to the saturated molecule) have been studied at low hydrogen overpotential electrodes such as platinum where the reaction is postulated to proceed via electrochemically generated, active hydrogen, and at high hydrogen overpotential electrode where the reaction is believed to proceed via electron transfer and a following chemical reaction between an intermediate and the solvent. In the latter case organometallic intermediates have also been postulated particularly at silver and copper cathodes.[151] Certain hydrogenation reactions which are very slow under normal conditions have been shown to proceed smoothly at platinum black electrodes.[79] Furthermore, if the cell is arranged

Anode

$$H_2 \rightarrow 2H^+ + 2e$$

Cathode

$$CH_3\text{—}CH\text{=}CH_2 + 2H^+ + 2e \xrightarrow{\text{Pt black}} CH_3\text{—}CH_2\text{—}CH_3$$

it is possible to obtain power from the cell. An example of the second type of reaction is the reduction of maleic and fumaric acids to succinic acid at a lead cathode in aqueous solutions.[152] The hydrogenation of olefins is also possible in aprotic solvents, for example the reduction of styrene in dimethylformamide[43]

The reduction of aromatic nitro compounds has been studied for many years, but these reactions continue to be of importance because of the great variety of products it is possible to obtain by variation of the electrolysis conditions. Thus in principle at least, it is possible to reduce nitrobenzene to nitrosobenzene, phenylhydroxylamine, aniline, azoxybenzene, hydrazo-benzene, and azobenzene.[1,2] Further it is possible to make use of the chemical rearrangements of phenylhydroxylamine to p-aminophenol or in the correct media to p-amino-chlorobenzene or p-methoxyamine benzene and of hydrazo-benzene to benzidine. Thus a considerable number of organic compounds are capable of preparation by a series of electrode reactions or pulse experiments starting from nitrobenzene. The preparation of benzidine has been cited earlier and another possibility is the preparation of quinone.

In view of the current interest in dipyridyls it is worth noting the coupling of a substituted pyridinium ion to give a dihydro derivative of a N-methyl dipyridyl which can readily be reoxidized to the aromatic species,[153]

8. Other Oxidations

The study of electrochemical oxidations was largely neglected until recent years and for this reason our knowledge of anodic reactions still lags behind our knowledge of cathodic reductions, particularly with respect to the systematic study of reaction products. Moreover, the electrode intermediates found during oxidation reactions in aprotic solvents are very sensitive to changes of solvent and to trace impurities such as water and this has led to considerable difficulties in identifying products. For these reasons, there are many new anodic reactions still to be discovered and considerable advances can be expected in the next few years. These comments are likely to be particularly true in the aliphatic series where electrochemical carbonium ion chemistry is only in its very early stages.

Another field which is likely to be of growing importance in the next decade is the 'destructive' electrode reaction of metal complexes. The simple electron transfer properties of metal complexes of the type

$$(\pi\text{-}C_5H_5)_2Fe \xrightarrow{-e} (\pi\text{-}C_5H_5)_2Fe^+$$

have been widely studied in recent years, but no attempt seems to have been made to use the electrode reactions of metal complexes in organic syntheses. Perhaps the first example is the oxidation of the mercury(II) propene complex which in aqueous acid solution at a platinum anode leads to fracture of the double bond and acetic and formic acids as products.[72]

A rather unusual example of the fracture of a carbon-carbon single bond has been reported during the oxidation of the substituted glycol[154]

9. Indirect Electrode Reactions

A number of systems has been investigated where strong oxidizing or reducing agents are produced at an electrode and used in a chemical reaction with an organic substrate either *in situ* or in a circulating system. In many cases, the electroactive species is regenerated in the chemical reaction and then the actual electrode reaction serves to 'drive' a homogeneous catalytic reaction. Air or oxygen have been widely used for such synthetic procedures, but the use of electricity allows the synthetic use of more powerful reagents. These methods have the advantages that the oxidizing or reducing agent need not be very stable (e.g. silver(II)) and the handling of large quantities of unpleasant reagents (e.g. bromine, sodium) can be avoided.

In passing, it is worth noting that electrochemical techniques, particularly cyclic voltammetry, the ring-disc electrode and multiple pulse experiments, can be used for the study of the kinetics of the reaction between the oxidizing or reducing agent and an organic substrate, and could be used for the study of many catalytic reactions. At the present time there is considerable interest in the reactions of the powerful oxidizing ions such as silver(II), cobalt(III), manganese(III) and lead(IV). Since these ions are often fairly unstable and difficult to prepare by other methods, the electrochemical preparation and *in situ* use described in this section offers many advantages.

An early example of an indirect electrolytic oxidation was the oxidation of xylene to tolualdehyde by manganese(III) in a sulphuric acid medium,[155] the manganese(III) being generated by the electrolytic oxidation of manganese(II)

$$Mn^{2+} \rightarrow Mn^{3+} + e$$

$$4Mn^{3+} + \underset{CH_3}{\overset{CH_3}{\bigcirc}} + H_2O \rightarrow 4Mn^{2+} + \underset{CH_3}{\overset{CHO}{\bigcirc}} + 4H^+$$

Other examples of such reactions include the electrochemical modification of the Clauson-Kaas methoxylation of furans where the bromine required for the reaction is generated at the anode,[156]

$$2Br^- \rightarrow Br_2 + 2e$$

the oxidation of propylene to acrolein by mercury(II) in a nitric acid medium[157] where the mercuric ions are regenerated electrochemically

$$CH_2{=}CH{-}CH_3 + 4Hg^{2+} + H_2O$$
$$\rightarrow CH_2{=}CH{-}CHO + 4H^+ + 2Hg_2^{2+}$$
$$Hg_2^{2+} \rightarrow 2Hg^{2+} + 2e$$

and the oxidation of propylene to propylene oxide by electrochemically generated hypochlorite.[158]

$$Cl^- + H_2O \rightarrow ClO^- + 2H^+ + 2e$$

$$CH_2{=}CH{-}CH_3 + ClO^- \rightarrow CH_2{-}CH{-}CH_3 + Cl^-$$
$$\underset{O}{\diagdown\diagup}$$

Two examples believed to be commercially exploited are the chromic acid oxidation of oleic acid to pelargonic and azelaic acids with electrolytic regeneration of the chromic acid[159] and the periodate oxidation of sugars where the periodic acid is regenerated electrochemically.[160]

REFERENCES

1. FICHTER, F.: *Organische Electrochemie* (Steinkopf, Dresden, 1942).
2. ALLEN, M. J.: *Organic Electrode Processes* (Reinhold Pub. Co., New York, 1958).
3. WAWZONEK, S.: *Anal. Chem.* (Reviews Section, 1958, 1960, 1962, 1964).
4. PIETRZYK, D. J.: *Anal. Chem.* (Reviews Section, 1966, 1968).
5. MÜLLER, O. H.: *Chem. Revs.*, **24**, 93 (1963).
6. PERRIN, C. L.: *Progress in Phys. Org. Chem.*, **3**, 165 (1965).
7. ZUMAN, P.: *Progress in Phys. Org. Chem.*, **5**, 1 (1967).
8. ZUMAN, P.: *J. Polarog. Soc.*, **13**, 53 (1967).
9. WAWZONEK, S.: *Science*, **155**, 39 (1967).
10. LUND, H.: *Öster. Chem. Zeitung*, **68**, 43 (1967); **68**, 152 (1967).
11. SWANN, S.: *Techniques of Organic Chem.*, Edited by Weissberger (Interscience, New York, 1956).
12. SWANN, S.: *Electrochemical Technology*, **1**, 308 (1961); **4**, 550 (1966); **5**, 53, 101, 393, 467, 547 (1967); **6**, 59 (1968).
13. TOMILOV, A. P., and FIOSHIN, M. YA.: *Russ. Chem. Revs.*, **32**, 30 (1967).
14. POPP, F. D., and SCHULTZ, H. P.: *Chem. Revs.*, **62**, 19 (1962).
15. ELVING, P. J., and PULLMAN, B.: *Advances Chem. Phys.*, **3**, 1 (1961).
16. ELMING, N.: *Advances in Organic Chem.*, **2**, 67 (1960).
17. WEINBERG, N. L., and WEINBERG, H. R.: *Chem. Revs.*, **68**, 449 (1968).
18. PAPON-DESBUQUOIS, C.: *Rev. Inst. franc. Petrole*, **19**, 627 (1964).
19. SASAKI, K., and NEWBY, W.: *J. Electroanal. Chem.*, **20**, 137 (1969).
20. ROSS, S. D.: *Trans. of New York Academy Sciences*, **30**, 901 (1968).
21. TOMILOV, A. P.: *Russ. Chem. Revs.*, **26**, 639 (1961).
22. BURDON, J., and TATLOW, J. C.: *Advances in Fluorine Chem.*, **1**, 129 (1960).
23. NAGASE, S.: *Fluorine Chem. Revs.*, **1**, 77 (1967).
24. WEEDON, B. C. L.: *Advances in Org. Chem.*, **1**, 1 (1960).
25. CONWAY, B. E., and VIJH, A. K.: *Chem. Revs.*, **67**, 623 (1967).
26. EBERSON, L.: *Chem. of the Carboxyl Group*, Edited by Patai (Interscience, New York, 1968).
27. OKUBO, T., and TSUTSUMI, S.: *Bull. Chem. Soc. Japan*, **37**, 1794 (1964).
28. LEHMKUHL, H., SCHAEFER, R., and ZIEGLER, K.: *Chem. Ing. Tech.*, **36**, 612 (1964).
29. GLASSTONE, S., and HICKLING, A.: *Chem. Revs.*, **25**, 407 (1939).
30. VERMILLION, F. J., and PEARL, I. A.: *J. Electrochem. Soc.*, **111**, 1392 (1964).
31. WAWZONEK, S., BERKEY, R., and THOMSON, D.: *J. Electrochem. Soc.*, **103**, 513 (1956).
32. SEKINE, T., YAMURA, A., and SUGINO, K.: *J. Electrochem. Soc.*, **112**, 439 (1965).

33. BERKEY, R., WAWZONEK, S., BLAHA, E. W., and RUNNER, M. E.: *J. Electrochem. Soc.*, 103, 456 (1956).
34. FINKELSTEIN, M., PETERSON, R. C., and ROSS, S. D.: *Electrochim. Acta*, 10, 465 (1965).
35. SMITH, D. L., and ELVING, P. J.: *J. Am. Chem. Soc.*, 84, 2741 (1962).
36. FIOSHIN, M. YA., MIRKIN, L. A., SALMIN, L. A., and KORNIENKO, A.: *Zh. Vses. Obstichestva im D. I. Mendeleeva*, 10, 238 (1965).
37. BREITENBACH, J. W., and SRNA, CH.: *Macromolecular Chem., Montreal, 1961* (Butterworths, London, 1962).
38. MARTLETT, E. M.: *Ann. New York Academy of Science*, 12, 125 (1965).
39. GALLI, R.: *Chem. and Ind.* (Milan), 50, 977 (1968).
40. HUSH, N. S., and OLDHAM, K. B.: *J. Electroanal. Chem.*, 6, 34 (1963).
41. PEOVER, M.: *Electroanal. Chem.*, Vol. II (Edward Arnold, London, 1967).
42. WAWZONEK, S., BLAHA, E. W., BERKEY, R., and RUNNER, M. E.: *J. Electrochem. Soc.*, 102, 235 (1955).
43. WAWZONEK, S., and WEARRING, D.: *J. Am. Chem. Soc.*, 81, 2067 (1959).
44. PEOVER, M.: *J. Chem. Soc.*, 4540 (1962).
45. MAKI, J., and GESKE, D.: *J. Am. Chem. Soc.*, 83, 1852 (1961).
46. PEOVER, M.: *Trans. Faraday Soc.*, 58, 2370 (1962).
47. WAWZONEK, S., DUTY, R. C., and WAGENKNECHT, J. H.: *J. Electrochem. Soc.*, 111, 74 (1964).
48. WAWZONEK, S., and GUNDERSON, A.: *J. Electrochem. Soc.*, 107, 537 (1960).
49. LANTELME, F., and CHEMLA, M.: *Electrochim. Acta*, 10, 657 (1965).
50. BAIZER, M. M.: *J. Electrochem. Soc.*, 111, 215 (1964).
51. VISCO, R. E., and CHANDROSS, E. A.: *J. Am. Chem. Soc.*, 86, 5350 (1964).
52. PEOVER, M. E., and WHITE, B. S.: *J. Electroanal. Chem.*, 13, 93 (1967).
53. LUND, H.: *Acta Chem. Scand.*, 11, 1323 (1957).
54. PARKER, V. D., and BURGERT, R. E.: *Tetrahedron Letters*, 2411 (1968).
55. EBERSON, L., and NYBERG, K.: *Tetrahedron Letters*, 2389 (1966).
56. EBERSON, L., and NILSSON, S.: *Disc. Faraday Soc.*, 45, 242 (1968).
57. EBERSON, L.: *J. Am. Chem. Soc.*, 89, 4669 (1967).
58. MILLER, L. L., and HOFFMANN, A. K.: *J. Am. Chem. Soc.*, 89, 593 (1967).
59. FLEISCHMANN, M., and PLETCHER, D.: *Tetrahedron Letters*, 6255 (1968).
60. KOEHL, W. J.: *J. Am. Chem. Soc.*, 86, 4686 (1964).
61. BAGGALEY, A. J., and BRETTLE, R.: *J. Chem. Soc.*, C, 2055 (1968).
62. COREY, E. J., BAULD, N. L., LONDE, R. L. LA., CASANOVA, J., and KAISER, E. T.: *J. Am. Chem. Soc.*, 82, 2645 (1960).
63. FUJINAGA, T., IZUTSU, K., and TAKAOKA, K.: *J. Electroanal. Chem.*, 12, 203 (1966).
64. LUND, H.: *Acta Chem. Scand.*, 13, 249 (1959).
65. RUSSELL, C. D.: *Anal. Chem.*, 35, 1293 (1963).
66. O'DONNELL, J. F., and MANN, C. K.: *J. Electroanal. Chem.*, 13, 157 (1967).
67. WAWZONEK, S., and MCINTYRE, T. W.: *J. Electrochem. Soc.*, 114, 1025 (1967).
68. WAWZONEK, S., and DUTY, R. C.: *J. Electrochem. Soc.*, 108, 1135 (1961).
69. WAWZONEK, S., and WAGENKNECHT, J. H.: *J. Electrochem. Soc.*, 110, 420 (1963).
70. SCHMIDT, H., and NOACK, J.: *Z. anorg. u. allgem. Chem.*, 296, 262 (1958).
71. ROSS, S. D., FINKELSTEIN, M., and PETERSON, R. C.: *J. Am. Chem. Soc.*, 89, 4088 (1967).
72. FLEISCHMANN, M. *et al.*, unpublished work.
73. RAUSCH, M. D., POPP, F. D., MCEWEN, W. E., and KLEINBERG, J.: *J. Org. Chem.*, 21, 212 (1956).
74. TSAI, T. T., MCEWAN, W. E., and KLEINBERG, J.: *J. Org. Chem.*, 25, 1186 (1960).
75. SUNDERMEYER, W. E.: *Angew. Chem. Int. Ed.*, 5, 1 (1966).
76. DIETZ, R., PEOVER, M. E., and ROTHBAUM, H. P.: Durham U.S.A. Conference, 137 (1968).
77. STERNBERG, H. W., MARKBY, R. E., and WENDER, I.: *J. Electrochem. Soc.*, 113, 1060 (1966).

78. BENKESER, R. A., and KAISER, E. M.: *J. Am. Chem. Soc.*, **85**, 2858 (1963).
79. LANGER, S. H., and LANDI, H. P.: *J. Am. Chem. Soc.*, **85**, 3043 (1963).
80. ATHERTON, G., FLEISCHMANN, M., and GOODRIDGE, F.: *Trans. Faraday Soc.*, **63**, 1468 (1967).
81. GILEADI, E., and PIERSMA, B.: *Modern Aspects of Electrochemistry*, Vol. IV, 47 (Butterworths, London, 1966).
82. HABER, F.: *Zeit. Electrochem.*, **4**, 506 (1898).
83. HOLLECK, L., and SCHMIDT, H.: *Zeit. Electrochem.*, **59**, 56 and 1039 (1955).
84. RUETSCHI, P., and TRUMPLER, G.: *Helv. Chim. Acta*, **36**, 1649 (1953).
85. ZUMAN, P., BARNES, D., and RYVOLOVA-KEJHAROVA, A.: *Disc. Faraday Soc.*, **45**, 205 (1968).
86. LUND, H.: 'Proceedings Prague Conference on Polarography', 1966.
87. KUWANA, T., and STROJEK, J. W.: *Disc. Faraday Soc.*, **45**, 134 (1968).
88. VINCENZ-CHODKOWSKA, A., and GRABOWSKI, Z. R.: *Electrochim. Acta*, **9**, 789 (1964).
89. FLEISCHMANN, M., PETROV, I. N., and WYNNE-JONES, W. F. K.: *Electrochemistry— Proceedings of 1st Australian Conference*, 500 (Pergamon Press, London, 1964).
90. FLEISCHMANN, M., and GOODRIDGE, F.: *Disc. Faraday Soc.*, **45**, 254 (1968).
91. KARDOS, A. M., VALENTA, P., and VOLKE, J.: *J. Electroanal. Chem.*, **12**, 84 (1966).
92. O'CONNOR, J. J., and PEARL, I. A.: *J. Electrochem. Soc.*, **111**, 335 (1964).
93. SEKINE, T., YAMURA, A., and SUGINO, K.: *J. Electrochem. Soc.*, **112**, 439 (1965).
94. Monsanto Co., British Patent 967, 956.
95. BAIZER, M. M.: *J. Electrochem. Soc.*, **111**, 223 (1964).
96. BAIZER, M. M., and ANDERSON, J. D.: *J. Electrochem. Soc.*, **111**, 226 (1964).
97. BAIZER, M. M., and PETROVICH, J. P.: *J. Electrochem. Soc.*, **114**, 1023 (1967).
98. BAIZER, M. M., ANDERSON, J. D., WAGENKNECHT, J. H., ORT, M. R., and PETROVICH, J. P.: *Electrochim. Acta*, **12**, 1377 (1967).
99. BAIZER, M. M., and ANDERSON, J. D.: *J. Org. Chem.*, **30**, 1348, 1357 (1965).
100. BAIZER, M. M., ANDERSON, J. D., and PRILL, E. J.: *J. Org. Chem.*, **30**, 1645 (1965).
101. BAIZER, M. M., and ANDERSON, J. D.: *J. Org. Chem.*, **30**, 3138 (1965).
102. WAGENKNECHT, J. H., and BAIZER, M. M.: *J. Org. Chem.*, **31**, 3885 (1966).
103. WAGENKNECHT, J. H., and BAIZER, M. M.: *J. Electrochem. Soc.*, **114**, 1095 (1967).
104. PETERSEN, J.: *Z. Physik. Chem.*, **33**, 294 (1900).
105. MOROE, T.: *J. Pharm. Soc. Japan*, **71**, 121 (1951).
106. HOFER, H.: *Ber.*, **33**, 655 (1900).
107. SVADKOVSKAYA, G. E., VOITKEVICH, S. A., SMOL'YANINOVA, Y. K., and BELOV, V. N.: *J. Gen. Chem. (U.S.S.R.)*, **27**, 2146 (1957).
108. LINSTEAD, R. P., SHEPHARD, B. R., and WEEDON, B. C. L.: *J. Chem. Soc.*, 2854 (1951).
109. BOUVEAULT, L.: *Bull. Soc. Chim. France Mém.*, **29**, 1042 (1903).
110. FIOSHIN, M. YA., KAMNEVA, A. J., ITENBURG, SH. M., KAZAKOVA, L. I., and ERSHOV, YR. A.: *Khim. Prom.*, 263 (1963).
111. LINSTEAD, R. P., LUNT, J. C., WEEDON, B. C. L., and SHEPHARD, B. R.: *J. Chem. Soc.*, 3621 (1952).
112. MOTOKI, S., and ODAKA, T.: *J. Chem. Soc. Japan*, **77**, 163 (1956).
113. HUNSDIECKER, H.: *Ber.*, **75**, 1197 (1942).
114. BOUNDS, D. G., LINSTEAD, R. P., and WEEDON, B. C. L.: *J. Chem. Soc.*, 448 (1954).
115. BAHNER, C. T.: *Ind. Eng. Chem.*, **44**, 317 (1952).
116. JOHNSTON, K. M., and STRIDE, J. D.: *Chem. Comm.*, 325 (1966).
117. DAVIS, R. G., and BIANCO, E.: *J. Electroanal. Chem.*, **12**, 254 (1966).
118. GESKE, D.: *J. Phys. Chem.*, **63**, 1062 (1959).
119. GESKE, D.: *J. Phys. Chem.*, **66**, 1743 (1962).
120. ALBERT, W. C., and LOWY, A.: *Trans. Electrochem. Soc.*, **75**, 367 (1939).
121. ALLEN, M. J.: *J. Org. Chem.*, **15**, 435 (1950).
122. PEARL, I. A.: *J. Am. Chem. Soc.*, **74**, 4260 (1952).
123. ALLEN, M. J., and LEVINE, D.: *J. Chem. Soc.*, 254, 2220 (1952).
124. BAUR, E.: *Z. Electrochem.*, **37**, 255 (1931).

125. TOMILOV, A. P., and FIOSHIN, M. YA.: Russ. Chem. Revs., 32, 30 (1963).
126. GOLDSCHMIDT, S., and STOCKL, E.: Chem. Ber., 85, 630 (1952).
127. GOLDSCHMIDT, S.: Angew Chem., 89, 132 (1957).
128. LINDSAY, R., and PETERSON, M.: J. Am. Chem. Soc., 81, 2073 (1959).
129. SALMIN, L. A., and MIRKIN, L. A.: Izv. Vysshikh. Uchebn. Zavedenii Khim i Khim Tekhol, 7, 607 (1964).
130. SCHAFER, H.: Meeting of G. D. Ch. Fachgruppe, Angew. Electrochem. (Lindau, 1968).
131. TSUTSUMI, S., and KOYAMA, K.: Disc. Faraday Soc., 45, 247 (1968).
132. BROWN, O. R., and LISTER, K.: Disc. Faraday Soc., 45, 106 (1968).
133. ANDERSON, J. D., BAIZER, M. M., and PETROVICH, J. P.: J. Org. Chem., 31, 3890, 3897 (1966).
134. RIFI, M. R.: J. Am. Chem. Soc., 89, 4442 (1967).
135. RIFI, M. R.: Tetrahedron Letters, 1043 (1969).
136. VELLTURO, A. F., and GRIFFIN, G. W.: J. Am. Chem. Soc., 87, 3021 (1965).
137. LUND, H.: Durham, U.S.A. Conference, 197 (1968).
138. PARKER, V. D., and BERGERT, R. E.: Tetrahedron Letters, 4065 (1965).
139. KOYAMA, K., SUSUKI, T., and TSUTSUMI, S.: Tetrahedron Letters, 627 (1965).
140. EBERSON, L., and NYBERG, K.: J. Am. Chem. Soc., 88, 1686 (1966).
141. ELVING, P. J., and TANG, CHING SIANG: J. Am. Chem. Soc., 74, 6109 (1952).
142. ROSENTHAL, I., ELVING, P. J., and TANG, CHING SIANG: J. Am. Chem. Soc., 74, 6112 (1952).
143. ELVING, P. J., and HILTON, C. L.: J. Am. Chem. Soc., 74, 3368 (1952).
144. ELVING, P. J., ROSENTHAL, I., and KRAMER, M. K.: J. Am. Chem. Soc., 73, 1717 (1951).
145. ROSENTHAL, I., and ELVING, P. J.: J. Am. Chem. Soc., 73, 1880 (1951).
146. ELVING, P. J., ROSENTHAL, I., and MARTIN, A. J.: J. Am. Chem. Soc., 77, 5218 (1955).
147. CHAMBERS, J., and ADAMS, R. N.: J. Electroanal. Chem., 9, 400 (1965).
148. WASSERMAN, R. A., and PURDY, W. C.: J. Electroanal. Chem., 9, 51 (1965).
149. HAWLEY, M. D., and ADAMS, R. N.: J. Electroanal. Chem., 8, 163 (1964).
150. SNEAD, W. K., and REMICK, A. E.: J. Am. Chem. Soc., 79, 6021 (1957).
151. TOMILOV, A. P.: Russ. Chem. Rev., 31, 569 (1962).
152. KANAKAM, R., PATHY, M. S. V., and UDUPA, H. V. K.: Electrochim. Acta, 12, 329 (1967).
153. BURNETT, J. N., and UNDERWOOD, A. L.: J. Org. Chem., 30, 1154 (1965).
154. ROSE, H. A.: Anal. Chem., 31, 2103 (1959).
155. RAMASWAMY, R., VENKATACHALPATHY, M., and UDUPA, H. U. K.: J. Electrochem. Soc., 110, 1086 (1963).
156. CLAUSON-KAAS, N., LIMBORG, F., and GLENS, K.: Acta Chem. Scand., 6, 531 (1952).
157. GOODRIDGE, F., PLIMLEY, R. E., ROBB, I. D., and COULSON, J. M.: British Patent Application, 16364/67.
158. LEDUC, J. A. M.: U.S. Patent, 3 288 692 (1966).
159. MANTELL, C. L.: Chem. Eng., 74, 128 (1967).
160. CONWAY, H. F., LANCASTER, E. G., and SOHNS, V. E.: Electrochem. Technol., 2, 43 (1964).

REDOX REACTIONS AND ELECTRON TRANSFER CHAINS OF INERT TRANSITION METAL COMPLEXES

J. A. McCleverty

There seems to be a slow, but definite, revival of interest among inorganic chemists in polarography and related electrochemical techniques. This may be due, in part, to the recent activity in two very specific areas of inorganic endeavour, namely transition metal sulphur chemistry and organometallic electrochemistry. In the former, electrochemical techniques have played a particularly important and sustaining part. This review is devoted to a discussion of these two areas of current activity, mainly because they represent important points in the growth of electrochemistry as a tool for inorganic, rather than analytical chemists.

The first area, the chemistry of transition metal dithiolene complexes, has been extensively reviewed by, and for, inorganic chemists[1,2,3] and, in consequence, this article does not contain details of syntheses and characterization of transition metal dithiolenes, except where pertinent, nor does it

403

14

present a rigorous account of the electrochemistry of these compounds. Instead, I have tried to indicate how electrochemical techniques have led to the development of the field, particularly in the design of synthetic routes to dithiolene complexes, and to the better understanding of their interesting electronic properties.

The second area of review, that of transition metal organometallic electro-chemistry, has not been reviewed specifically, but excellent background material relating to the usefulness and relevance of organometallic chemistry is available.[4]

By selection of these two topics, this discussion is limited to the study of the electron transfer reactions of inert transition metal complexes in organic (i.e. non-aqueous) media.

I. TRANSITION METAL 1,2-DITHIOLENE COMPLEXES

The reasons for the interest in transition metal dithiolene chemistry are many, not the least of which are the novel electronic structures possessed by these species. The facility with which dithiolene complexes can undergo electron exchange reactions, either by oxidation or reduction, and the interesting structural and spectral changes in the molecules which accompany these electron transfer reactions, have attracted the attention of inorganic chemists for nearly ten years.

The ligands in these dithiolene complexes are derived from cis-1,2-disub-stituted ethylene-1,2-dithiolates or the closely related benzene-1,2-dithiols, and the types of ligands normally encountered in a discussion of dithiolene chemistry are shown in Fig. 1. Transition metal ions may form complexes incorporating these ligands alone or, as we shall see later, in conjunction with other ligand systems (e.g. CO, NO, π-C_5H_5, PR_3), and most of the complexes seem to have the property of engaging in extremely facile one-electron transfer reactions. It is from this property that difficulties in the description

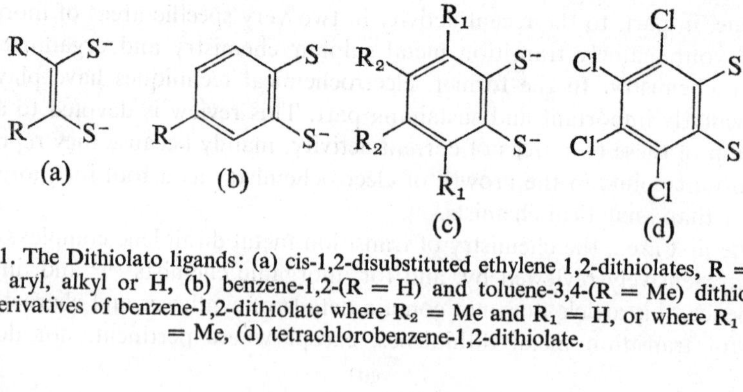

Fig. 1. The Dithiolato ligands: (a) cis-1,2-disubstituted ethylene 1,2-dithiolates, R = CN, CF_3, aryl, alkyl or H, (b) benzene-1,2-(R = H) and toluene-3,4-(R = Me) dithiolates, (c) derivatives of benzene-1,2-dithiolate where R_2 = Me and R_1 = H, or where R_1 = R_2 = Me, (d) tetrachlorobenzene-1,2-dithiolate.

of the complexes, using formal oxidation states for the metals and valence bond representations for the ligands, arise. A special nomenclature has been devised[1] which circumvents this problem. Thus, when a metal complex containing one, or more, cis-1,2-disubstituted ethylene-1,2-dithiolate or benzene-1,2-dithiolate ligand undergoes, or is capable of undergoing, electron transfer reactions, then the complex is referred to by the generic term '1,2-dithiolene'. Accordingly, $[Et_4N][NiS_4C_4(CN)_4]$ would be called 'tetraethylammonium nickel bis-dicyano-1,2-dithiolene, $[VS_6C_6(CF_3)_6]^0$ 'vanadium tri(bis(trifluoromethyl)-1,2-dithiolene)', and $[Bu_4^nN]_2[Fe(NO)(S_2C_6Cl_4)_2]$ 'tetra-n-butylammonium iron nitrosyl bis(tetrachlorobenzene-1,2-dithiolene)'. This terminology specifically avoids the use of particular ligand and metal valence formalisms which may be ambiguous or meaningless. Thus, the description of $[NiS_4C_4(CN)_4]^-$, which is paramagnetic (one unpaired electron), as containing two bidentate dinegative dithiolate ligands and Ni^{III}, is ambiguous, since detailed esr spectral measurements[5,6] and molecular orbital calculations[7,8] have shown that the orbital of the unpaired electron is very largely sulphur ligand in character (being, at the most, 25 per cent metal-based). The ground-state electronic structure of this species cannot therefore be described adequately in classical valence bond terms, and must be regarded as highly delocalized over both the metal and sulphur ligands. The valence bond formalism becomes meaningless in such compounds as $[VS_6C_6(CF_3)_6]^0$, which, at face value, appears to contain V^{VI}. No self-respecting inorganic chemist would believe in sexavalent vanadium (it implies the promotion of a $3p$ electron!), especially when 'stabilized' by six highly polarizable sulphur atoms. Complexes which cannot, as a whole, engage in electron transfer reactions of the type discussed herein are referred to by the normal 'dithiolate' nomenclature and by the use of formal oxidation numbers. For example, $[Zn(S_2C_6H_3Me)_2]^{2-}$ is referred to as the 'bis(toluene-3,4-dithiolato) zinc(II) dianion'. It might be argued that the most reduced species of an electron transfer series, such as $[NiS_4C_4Ph_4]^{2-}$ in the series

$$[NiS_4C_4Ph_4]^+ \rightleftharpoons [NiS_4C_4Ph_4]^0 \rightleftharpoons [NiS_4C_4Ph_4]^- \rightleftharpoons [NiS_4C_4Ph_4]^{2-}$$

could be described as containing a divalent metal ion and two dianionic dithiolates, $\{^-SC(Ph)C=C(Ph)CS^-\}$, and this would not be inconsistent with spectral properties and molecular orbital predictions. However, these reduced species are regarded as members of a general class of molecules which are termed generically as 'metal-1,2-dithiolenes' or '1,2-dithiolene metal complexes'.

Transition metal complexes containing sulphur ligands of the type $\{S_2C=X\}^{2-}$, where X may be $C(CN)_2$, $N(CN)$, $C(NO_2)_2$, etc., and which can undergo electron transfer reactions similar to those of the 1,2-dithiolene complexes, are referred to as 'metal 1,1-dithiolene complexes'. These species are comparatively rare.

1. Simple Metal 1,2-Dithiolene Complexes

In 1962, G. N. Schrauzer and V. P. Mayweg reported[9] that they had prepared an unusual nickel complex, $NiS_4C_4Ph_4$, in the reaction between $Ni(CO)_4$, sulphur and diphenylacetylene. Their original purpose in carrying out this unusual reaction was to investigate the role of transition metal catalysts in the formation of thioaromatics, particularly thiophenes, from acetylenes and sulphur. The new nickel complex was intensely coloured (green), volatile, diamagnetic and quite unlike any other nickel complex known at that time. It was found that in solution in pyridine or piperidine, the green colour was gradually discharged and was replaced by a red-brown colour, the resulting solution being paramagnetic, with one, and not two, unpaired electrons. At the same time, H. B. Gray and his colleagues reported[10] the formation of metal complexes of $\{S_2C_2(CN)_2\}^{2-}$, of the type $[MS_4C_4(CN)_4]^{2-}$ where M was Cu, Ni, Pd, and Co. Shortly thereafter, Davison, Edelstein, Maki, and Holm[11] pointed out that there was only a two-electron difference between the neutral complexes of Schrauzer, abbreviated to $[Ni—(S—S)_2]^0$, and the dianionic species of Gray, abbreviated to $[Ni—(S—S)_2]^{2-}$ (considering only the valence electrons and ignoring the ligand substituents). They reasoned that it should be possible to effect an oxidation of the dianion to the neutral species, or reduction of the neutral compound to a dianion, possibly via an intermediate monoanionic species. They discovered that this was possible chemically, and isolated several different paramagnetic monoanions, $[Ni—(S—S)_2]^-$, which was very exciting at that time because of the relative rarity of paramagnetic, planar nickel complexes. However, perhaps the most significant contribution that Davison et al. made was to demonstrate that the one-electron transfer reactions linking the dianions, monoanions and neutral species could be detected polarographically and voltammetrically. Not long after the publication of this work, Gray and his co-workers announced[12] that the most stable complexes formed between nickel, cobalt, and iron and toluene-3,4-dithiol were monoanionic, e.g. $[M(S_2C_6H_3Me)_2]^-$, the nickel complex having a spin-doublet ground state and the other two reputedly having spin-quartet ground states. From this point, it was recognized that there existed an extensive body of compounds containing cis-1,2-disubstituted ethylene-1,2-dithiolates and related ligands, having unusual electron transfer capabilities and interesting electronic properties.

Before discussing the electrochemical and electronic properties of the 1,2-dithiolene complexes, it is appropriate to mention the methods by which the electrochemical data were obtained. Normally, after synthesis of a particular dithiolene complex, the electron transfer capability of this complex was investigated by polarography or voltammetry, the latter almost exclusively employing a rotating platinum electrode. The cells were generally of the conventional two- or three-electrode type, and the reference electrodes were of the saturated calomel, $Ag/AgClO_4$, $Ag/AgCl$ or Ag/AgI types.[13] The solvents used in this work varied considerably, depending on the solubility and stability of the electro-active species. Thus, use was made of dichloromethane, acetone, dimethylformamide, and dimethylsulphoxide, but the first was generally most popular, perhaps because of its low dielectric constant and minimal tendency to coordinate to the complexes under study. Concentrations of metal complexes were almost invariably 10^{-3} M. The base electrolytes were almost always tetra-alkyl-ammonium salts, e.g. $[Et_4N][ClO_4]$, $[n\text{-}Pr_4N]$ $[ClO_4]$, $[n\text{-}Bu_4N][PF_6]$. The concentrations were generally 0·05 M (Et_4N^+, $n\text{-}Pr_4N^+$), 0·1 M ($n\text{-}Pr_4N^+$, Li^+) or 0·5 M ($[n\text{-}Bu_4N][PF_6]$), as indicated in the tables. Evidence of reversibility was collected in three ways—by measurement of $E_{3/4} - E_{1/4}$, by graphical methods (plotting E_{obsd} vs. $\log i(i_d - i)^{-1}$), or less commonly, by cyclic voltammetry. In a very few cases the number of electrons involved in the electron transfer process was determined directly by coulometry, reliance on the knowledge of 'n' being placed more often on a correlation between wave heights of a particular electrode reaction and the synthetic chemistry associated with that reaction. In a few instances, controlled potential electrolysis (cpe) was used to generate oxidized or reduced species, but this was only carried out in conjunction with esr spectral examination of the desired species; as yet, cpe has not been used to prepare dithiolene complexes. All electrochemical data were measured at, or very near, room temperature. For further details, the reader is referred to the original papers as indicated in the Tables.

A systematization of the half-wave potential data for the one-electron transfer reactions which the dithiolene complexes undergo has been made by Davison and Holm.[14] The generalizations about the synthetic and chemical behaviour of such complexes are made below, with the addition of modifications made by other workers. These are as follows:

(i) reduced species in couples less positive than ca 0·00 V (vs. SCE) are susceptible to aerial oxidation in solution, whereas all reduced species in couples more positive than this value are air-stable;

(ii) oxidized species in couples more positive than ca +0·20 V are unstable to reduction by weakly basic solvents such as ketones or alcohols;

(iii) oxidized species in couples within the very approximate range +0·20 and −0·12 V can be reduced by stronger bases such as aromatic amines (e.g. o- or p-phenylenediamine) or by sulphite ion;

(iv) oxidized species in couples more negative than ca −0·10 V are readily reduced by strong reducing agents such as hydrazine, sodium amalgam, Zn in pyridine, sodium borohydride, or alkali metal alkoxides;

(v) reduced forms in couples less positive than ca $+0.20$ V can be oxidized by iodine, whereas those in couples between $+0.20$ and $+0.50$ V may be oxidized by bromine, or, occasionally, by silver ion. Otherwise, one-electron oxidation must be effected by reagents such as $[NiS_4C_4(CF_3)_4]^0$ and its analogues.

Several points arise from these generalizations. Firstly, it must be emphasized that the limits quoted are very approximate, depending on the nature of the species to be oxidized or reduced, and on the solvent in which the one-electron change is being effected. Secondly, in synthetic oxidations or reductions, care must be exercised in the choice of oxidizing or reducing agent, since contact time between the reactants must be minimal in order to reduce the possibility of ligand exchange reactions (see later), or to avoid the possibility that certain reducing agents (e.g. hydrazine) can function as coordinating ligands thereby upsetting the course of the electron transfer reaction. The chemical reduction of oxidized species in couples more negative than ca -0.95 V, or the oxidation of reduced species in couples more positive than ca $+0.95$ V, do not seem to be practically possible, particularly since reagents capable of effecting these one-electron changes usually cause complete disruption of the complexes. Thus, it is necessary to employ cpe to achieve these electron transfers.

The simple bis-1,2-dithiolene complexes can, in general, exist as part of a four-membered electron transfer series:

$$[M-(S-S)_2]^{2-} \rightleftharpoons [M-(S-S)_2]^- \rightleftharpoons [M-(S-S)_2]^0 \rightleftharpoons [M-(S-S)_2]^+$$

Monocationic species have not yet been isolated in a pure state, but several have been detected voltammetrically.[15,16] In these complexes, the order of decreasing oxidative stability of the dianions, $[M-(S-S)_2]^{2-}$, that is, the order of increasingly negative potentials for the reactions[17]

$$[MS_4C_4R_4]^- + e^- \rightleftharpoons [MS_4C_4R_4]^{2-}$$

is

$$R = CN > CF_3 > Ph > H > Me > Et > n\text{-}Pr > i\text{-}Pr > n\text{-}Bu$$

In other words, the dianions were the most unstable to oxidation when the ligand substituent, R, in $[MS_4C_4R_4]^{2-}$, was an electron-releasing group. A similar ordering exists for the half-wave potentials for the couple

$$[MS_4C_4R_4]^0 + e^- \rightleftharpoons [MS_4C_4R_4]^-,$$

and it was shown[17] that a linear relationship existed between the half-wave potentials for these reactions, where M = Ni, Pd, and Pt, and an inductive substituent constant, viz. Taft's σ^* constant.[18] These results were believed to be in agreement with the proposed electronic structure of the complexes. Unfortunately, many of the free ligands could not be investigated polarographically, but those which were (Table 1) showed the expected parallel

Table 1. *Polarographic and Voltammetric Data Obtained For Free Ligands: Couple* $[L]^z + e^- \rightleftharpoons [L]^{z-1}$ *(volts)**

Ligand	CH₃CN (a)		CH₂Cl₂ (b)		DMF (c)	
	$-2 \leftrightarrow -1$	$-1 \leftrightarrow 0$	$-2 \leftrightarrow -1$	$-1 \leftrightarrow 0$	$-2 \leftrightarrow -1$	$-1 \leftrightarrow 0$
Phenanthroquinone	-1.21	-0.65	-1.3	-0.51	—	—
Benzoquinone	-0.90	-0.31	(-0.15)	(-0.74)	—	—
Chloranil	-0.61	$+0.14$	-0.70	$+0.30$	—	—
$S_2C_2(CN)_2{}^{2-}$	—	—	—	—	-0.31 (d)	0.00 (d)
$S_2C_2H_2{}^{2-}$	—	—	—	—	-0.81 (d)	-0.21 (d)
$[CH_3CNC_6H_5]_2$	-1.82 (e)		—	—	—	—

(a) Versus SCE using dropping mercury electrode.
(b) Versus Ag/AgI using rotating platinum electrode with $[n\text{-Bu}_4\text{N}][\text{PF}_6]$ (0·5 M) as supporting electrolyte.
(c) Versus Ag/AgCl using dropping mercury electrode, with LiClO_4 (0·1 M) as supporting electrolyte.
(d) Irreversible waves.
(e) Two-electron reduction wave.
* Data from References 13 and 17.

substituent effects. This is in agreement with Vlček's observation[19] that the influences exerted by given substituents in ligands (bound to various metal ions) on the oxidation or reduction potentials of their metal complexes were parallel, but not necessarily proportional to, the effects observed in the redox behaviour of the free ligands themselves.

Similar substituent effects were observed on the half-wave potentials of the substituted bis(arene-1,2-dithiolene) complexes,[20] where the monoanionic species, $[M(S_2C_6X_2Y_2)_2]^-$, were stabilized with respect to the dianionic complexes in proportion to the number of electron-releasing groups around the benzene ring. Thus, the order of decreasing potentials for the reaction

$$[M(S_2C_6X_2Y_2)_2]^- + e^- \rightleftharpoons [M(S_2C_6X_2Y_2)_2]^{2-},$$

where $X = Y = H$, Me or Cl, $X = H$, $Y = $ Me, or one $X = H$, one $X = $ Me and $Y = H$, was

$$S_2C_6Cl_4 > S_2C_6H_4 > S_2C_6H_3Me > S_2C_6H_2Me_2 > S_2C_6Me_4$$

The half-wave potentials for the tetrachloro- and tetramethyl-substituted complexes differed by about 0·7 V (vs. Ag/AgClO₄, in DMF),[20] and the introduction of only one Me group into the benzene ring was sufficient to stabilize the respective monoanion by about 0·05 V. By comparison of the data in Tables 2 and 3, and from recent synthetic and voltammetric studies of

Table 2. *Polarographic and Voltammetric Data Obtained from Bis-1,2-Dithiolenes* [*Data from References 20, 24, and 91*]

Complex	Solvent (a)	Elec. (b)	Ref. Cell (c)	$E_{1/2}$ (d)	D (e)	Process
$ZnS_4C_4(CN)_4$	DMF	DME	Ag/AgClO$_4$ (g)	—	—	
$CuS_4C_4(CN)_4$	DMF	DME	Ag/AgClO$_4$ (g)	-0.201	—	$-2 \rightarrow -1$
	DMF	DME	Ag/AgCl (h)	$+0.37$	—	$-2 \rightarrow -1$
	CH$_2$Cl$_2$	RPE	Ag/AgI (i)	$+0.41$	20	$-2 \rightarrow -1$
	CH$_2$Cl$_2$	RPE	Ag/AgI (i)	$+1.37$	24	$-1 \rightarrow \ \ 0$
	CH$_3$CN	RPE	SCE (j)	$+0.33$	—	$-2 \rightarrow -1$
$CuS_4C_4(CF_3)_4$	CH$_3$CN	RPE	SCE (j)	-0.01	—	$-2 \rightarrow -1$
$Cu(S_2C_6Cl_4)_2$	DMF	DME	Ag/AgClO$_4$ (g)	-0.752	—	$-1 \rightarrow -2$
$Cu(S_2C_6H_4)_2$	DMF	DME	Ag/AgClO$_4$ (g)	-1.14	—	$-1 \rightarrow -2$
$Cu(S_2C_6H_3Me)_2$	DMF	DME	Ag/AgClO$_4$ (g)	-1.15	—	$-1 \rightarrow -2$
$Cu(S_2C_6H_2Me_2)_2$	DMF	DME	Ag/AgClO$_4$ (g)	-1.21	—	$-1 \rightarrow -2$
$Cu(S_2C_6Me_4)_2$	DMF	DME	Ag/AgClO$_4$ (g)	-1.41	—	$-1 \rightarrow -2$
$CuS_4C_4H_4$	DMSO	DME	SCE (j)	-0.74	—	$-1 \rightarrow -2$
$AuS_4C_4(CN)_4$	DMF	DME	Ag/AgClO$_4$ (g)	-0.961	—	$-1 \rightarrow -2$
	CH$_3$CN	RPE	SCE (j)	-0.419	—	$-1 \rightarrow -2$
$AuS_4C_4(CF_3)_4$	CH$_3$CN	RPE	SCE (j)	$+1.20$ (f)	—	$-1 \rightarrow \ \ 0$
	CH$_2$Cl$_2$	RPE	SCE (j)	-0.97	26	$-1 \rightarrow -2$
	CH$_2$Cl$_2$	RPE	SCE (j)	$+1.32$	25	$-1 \rightarrow \ \ 0$
$Au(S_2C_6H_3Me)_2$	DMF	DME	Ag/AgClO$_4$ (g)	-1.95	—	$-1 \rightarrow -2$
$NiS_4C_4(CN)_4$	DMF	DME	Ag/AgClO$_4$ (g)	-0.218	—	$-2 \rightarrow -1$
	CH$_2$Cl$_2$	RPE	Ag/AgI (i)	$+0.37$	21	$-2 \rightarrow -1$
	CH$_2$Cl$_2$	RPE	Ag/AgI (i)	$+1.38$	24	$-1 \rightarrow \ \ 0$
	CH$_2$Cl$_2$	RPE	SCE (k)	$+0.12$	—	$-2 \rightarrow -1$
	CH$_3$CN	RPE	SCE (j)	$+0.226$	—	$-2 \rightarrow -1$
	CH$_3$CN	RPE	SCE (j)	$+1.02$	—	$-1 \rightarrow \ \ 0$
	DMF	DME	Ag/AgCl (h)	$+0.26$	—	$-2 \rightarrow -1$
	DMF	DME	Ag/AgCl (h)	$+1.05$	—	$-1 \rightarrow \ \ 0$
$NiS_4C_4(CF_3)_4$	CH$_3$CN	RPE	SCE	-0.121	—	$-1 \rightarrow -2$
	CH$_3$CN	RPE	SCE	$+0.997$	—	$0 \rightarrow -1$
	CH$_2$Cl$_2$	RPE	SCE	-0.01	—	$-1 \rightarrow -2$
	CH$_2$Cl$_2$	RPE	SCE	$+0.80$	—	$0 \rightarrow -1$
	CH$_2$Cl$_2$	RPE	Ag/AgI	-0.05	21	$-1 \rightarrow -2$
	CH$_2$Cl$_2$	RPE	Ag/AgI	$+1.01$	23	$0 \rightarrow -1$
$Ni(S_2C_6Cl_4)_2$	DMF	DME	Ag/AgClO$_4$	-0.532	—	$-1 \rightarrow -2$
$Ni(S_2C_6H_4)_2$	DMF	DME	Ag/AgClO$_4$	-1.05	—	$-1 \rightarrow -2$
$Ni(S_2C_6H_3Me)_2$	DMF	DME	Ag/AgClO$_4$	-1.07	—	$-1 \rightarrow -2$
	CH$_3$CN	RPE	SCE	-0.58	—	$-1 \rightarrow -2$
	DMSO	RPE	SCE	-0.52	—	$-1 \rightarrow -2$
	DMSO	RPE	SCE	$+0.45$	—	$-1 \rightarrow \ \ 0$
$Ni(S_2C_6H_2Me_2)_2$	DMF	DME	Ag/AgClO$_4$	-1.14	—	$-1 \rightarrow -2$
$Ni(S_2C_6Me_4)_2$	DMF	DME	Ag/AgClO$_4$	-1.24	—	$-1 \rightarrow -2$
$NiS_4C_4(p\text{-}ClC_6H_4)_4$	DMF	DME	Ag/AgCl	-0.757	3.64	$-1 \rightarrow -2$
	DMF	DME	Ag/AgCl	$+0.218$	—	$0 \rightarrow -2$
$NiS_4C_4Ph_4$	DMF	DME	Ag/AgCl	-0.881	5.15	$-1 \rightarrow -2$
	DMF	DME	Ag/AgCl	$+0.134$	—	$0 \rightarrow -1$

(a) DMF = dimethylformamide; DMSO = dimethylsulphoxide.
(b) Test electrode: DME = dropping mercury electrode; RPE = rotating platinum electrode.
(c) Reference Cell (SCE = saturated calomel electrode).
(d) Half-wave potential in volts.
(e) D = diffusion current criterion, i_d/c in μamp/mmole.
(f) No wave from 0 to -1.5 V.
(g) [n-Pr$_4$N][ClO$_4$] (0·1 M) as supporting electrolyte.
(h) LiClO$_4$ (0·1 M) as supporting electrolyte.
(i) [n-Bu$_4$N][PF$_6$] (0·5 M) as supporting electrolyte.
(j) [n-Pr$_4$N][ClO$_4$] (0·05 M) as supporting electrolyte.
(k) [Et$_4$N][ClO$_4$] (0·05 M) as supporting electrolyte.

Table 3. *Normalized Potential Data for Bis-*1,2-*Dithiolene Complexes,* $[ML_2]^z$
[*Data from Reference* 1]

M	Ligands, L_2	Electrode Processes (a)	
		$-2 \to -1$	$-1 \to 0$
Cu	$S_4C_4(CN)_4$	$+0.34$	
	$S_4C_4(CF_3)_4$	-0.01	
	$(S_2C_6Cl_4)_2$	-0.29	
	$(S_2C_6H_4)_2$	-0.68	
	$(S_2C_6H_3Me)_2$	-0.69	
	$S_4C_4H_4$	-0.74	
	$(S_2C_6H_2Me_2)_2$	-0.75	
	$(S_2C_6Me_4)_2$	-0.95	
Au	$S_4C_4(CN)_4$	-0.42	
	$(S_2C_6Cl_4)_2$	-1.21	
NiS$_4$	$S_4C_4(CN)_4$	$+0.23$	$+1.02$
	$S_4C_4(CN)_2(CF_3)_2$	$+0.06$	
	$S_4C_4(CF_3)_4$	-0.12	$+0.92$
	$(S_2C_6Cl_4)_2$	-0.07	
	$S_4C_4(CN)_2Ph_2$	-0.35	$+0.57$
	$S_4C_4(CF_3)_2Ph_2$	-0.55	$+0.43$
	$(S_2C_6H_4)_2$	-0.59	
	$(S_2C_6H_3Me)_2$	-0.58	
	$(S_2C_6H_2Me_2)_2$	-0.68	
	$(S_2C_6Me_4)_2$	-0.78	
	$S_4C_4(p\text{-}ClC_6H_4)_4$	-0.79	$+0.19$
	$S_4C_4Ph_4$	-0.82	$+0.12$
	$S_4C_4Ph_2H_2$	-0.91	$+0.09$
	$S_4C_4H_4$	-0.95	$+0.09$
	$S_4C_4(p\text{-}MeC_6H_4)_4$	-0.99	$+0.05$
	$S_4C_4Ph_2Me_2$	-1.02	00.00
	$S_4C_4(p\text{-}MeOC_6H_4)_4$	-0.98	0.00
	$S_4C_4Me_4$	-1.14	-0.14
	$S_4C_4Et_4$	-1.17	-0.15
	$S_4C_4(n\text{-Pr})_4$	-1.18	-0.15
	$S_4C_4(i\text{-Pr})_4$	-1.23	-0.18
Pd	$S_4C_4(CN)_4$	$+0.46$	—
	$S_4C_4(CF_3)_4$	$+0.08$	$+0.96$
	$S_4C_4Ph_4$	-0.51	$+0.15$
	$S_4C_4H_4$	-0.75	$+0.14$
	$S_4C_4(p\text{-}MeOC_6H_4)_4$	-0.75	$+0.06$
	$S_4C_4Me_4$	-0.90	-0.09

Table 3. (contd.)

M	Ligands, L_2	Electrode Processes (a)	
		$-2 \to -1$	$-1 \to 0$
Pt	$S_4C_4(CN)_4$	$+0.21$	—
	$S_4C_4(CF_3)_4$	-0.27	$+0.82$
	$(S_2C_6H_3Me)_2$	-0.59	
	$S_4C_4Ph_4$	-0.81	$+0.06$
	$S_4C_4(p\text{-}MeC_6H_4)_4$	-0.93	$+0.01$
	$S_4C_4(p\text{-}MeOC_6H_4)_4$	-0.95	-0.03
	$S_4C_4Me_4$	-1.10	-0.16
Co	$S_4C_4(CN)_4$	$+0.05$	(b)
	$S_4C_4(CF_3)_4$	-0.24	(b)
	$(S_2C_6Cl_4)_2$	-0.39	
	$(S_2C_6H_4)_2$	-0.92	
	$S_4C_4H_4$	-0.93	
	$(S_2C_6H_3Me)_2$	-0.95	
	$(S_2C_6H_2Me_2)_2$	-1.00	
	$(S_2C_6Me_4)_2$	-1.11	
Fe	$S_4C_4(CN)_4$	-0.40	(b)
	$S_4C_4(CF_3)_4$	-0.44	(b)
	$(S_2C_6H_3Me)_2$	-1.00	(b)

(a) Converted, arbitrarily, to values equivalent to those in CH_3CN, vs. SCE; conversion factors: from DMF, $Ag/AgClO_4$, $E_{1/2} + 0.44$, from DMF, $Ag/AgCl$, $E_{1/2} - 0.03$; estimated values in italics.

(b) Electrode process corresponds to $2[ML_2]^z + 2e^- \rightleftharpoons [ML_2]_2^{2-}$.

tris-tetrachlorobenzene-dithiolene complexes,[21] it may be seen that the potentials of the couples of these compounds are approximately similar to those of the analogous di(bis(trifluoromethyl)-dithiolenes), and therefore that the overall order of stability of reduced complexes (i.e. $[M-(S-S)_2]^{z-1}$) vs. their oxidized counterparts (i.e. $[M-(S-S)_2]^z$), in terms of the sulphur ligands, would seem to be

$$S_2C_2(CN)_2 > S_2C_2(CF_3)_2 \geqslant S_2C_6Cl_4 > S_2C_6H_4 > S_2C_6H_3Me$$
$$> S_2C_6H_2Me_2 > S_2C_6Me_4 > S_2C_2(4\text{-}ClC_6H_4)_2 > S_2C_2Ph_2 >$$
$$S_2C_2H_2$$
$$> S_2C_2(4\text{-}MeC_6H_4)_2 > S_2C_2(4\text{-}MeOC_6H_4)_2 > S_2C_2Me_2$$
$$> S_2C_2Et_2 > S_2C_2(n\text{-}Pr)_2 > S_2C_2(i\text{-}Pr)_2 > S_2C_2(n\text{-}Bu)_2$$

An inspection of this order led to the prediction that if monocationic species could be observed in the voltammograms of $[NiS_4C_4Ph_4]^0$,[15] then they should be readily detected in the voltammograms of all similar species to the right of the phenyl substituent in the above order, and this has been confirmed.[16]

The synthetic implications of the half-wave potential data presented in Tables 2 and 3 are, of course, of paramount importance to the inorganic chemist. A few examples of how the 'rules' for using the electrochemical data have been successfully applied in the syntheses of simple bis-1,2-dithiolene complexes are given in Table 4.

Table 4. *Synthetic Applications of Potential Data Obtained from Bis*-1,2-*Dithiolene Complexes*

Oxidized or Reduced Species	$E_{1/2}$ (a)	Starting Material	Redox Reagent	Solvent
$[NiS_4C_4(CN)_4]^-$	$+0.23$	$[NiS_4C_4(CN)_4]^{2-}$	I_2	DMSO
$[PdS_4C_4(CN)_4]^-$	$+0.44$	$[PdS_4C_4(CN)_4]^{2-}$	$[NiS_4C_4(CF_3)_4]^0$	CH_2Cl_2
$[CoS_4C_4(CN)_4]_2^{2-}$	$+0.05$	$[CoS_4C_4(CN)_4]^{2-}$	I_2 or O_2	acetone
$[NiS_4C_4(CF_3)_4]^-$	$+0.997$	$[NiS_4C_4(CF_3)_4]^0$	basic solvent	
$[CoS_4C_4(CF_3)_4]^{2-}$	-0.398 (b)	$[CoS_4C_4(CF_3)_4]_2^0$	Na/Hg	THF
$[NiS_4C_4(CN)_4]^{2-}$	$+0.23$	$[NiS_4C_4)(CN)_4]^-$	SO_3^{2-}	H_2O acetone
$[NiS_4C_4Ph_4]^-$	-0.824	$[NiS_4C_4Ph_4]^0$	p-phenylene-diamine	DMSO
$[NiS_4C_4Ph_4]^{2-}$	-0.74 (c)	$[NiS_4C_4Ph_4]^0$	N_2H_4	ethanol
$[Ni(S_2C_6H_3Me)_2]^{2-}$	-0.58	$[Ni(S_2C_6H_3Me)_2]^-$	KOEt	ethanol
$[AuS_4C_4(CN)_4]^{2-}$	-0.419	$[AuS_4C_4(CN)_4]^-$	BH_4^-	THF

(a) In volts; in CH_3CN vs. SCE with $[Et_4N][ClO_4]$ (0.05 M), $[n\text{-}Pr_4N][ClO_4]$ (0.05 M), or $[n\text{-}Bu_4N][ClO_4]$ (0.05 M) as supporting electrolytes; using rotating Pt electrode.

(b) Value for couple $[CoS_4C_4(CF_3)_4]_2^{2-} + 2e^- \rightleftharpoons 2[CoS_4C_4(CF_3)_4]^{2-}$.

(c) Value for couple $[NiS_4C_4Ph_4]^- + e^- \rightleftharpoons [NiS_4C_4Ph_4]^{2-}$.

It has been established that all of the species so far isolated in the nickel and copper series of bis-1,2-dithiolenes have planar structures,[1] this being confirmed crystallographically by studies of the salts of $[NiS_4C_4(CN)_4]^{-,2-}$, $[NiS_4C_4Ph_4]^0$, $[NiS_4C_4H_4]^0$, and $[CuS_4C_4(CN)_4]^-$; $[Co(S_2C_6H_3Me)_2]^-$ and $[CoS_4C_4(CN)_4]^{2-}$ are also planar. The effect of reducing charge in the series of nickel complexes, i.e. in going from $[Ni-(S-S)_2]^{2-}$ through $[Ni-(S-S)_2]^-$

to $[Ni—(S—S)_2]^0$, is to lengthen the 'olefinic' $C{=}C$ bond and shorten the C—S bond in the sulphur ligands thereby conferring, in valence bond terms, more dithioketonic character (and less dithiolate character) in the ligands as a whole. In other words, gradual oxidation results in the progressive alteration of the sulphur ligand from a dithiolate to a dithioketone.

Before discussing the electronic structure of these planar complexes, it is pertinent to summarize their spectral properties. All those complexes having spin-doublet ground states (e.g. $[Ni—(S—S)_2]^-$, its Pd and Pt analogues, $[Cu—(S—S)_2]^{2-}$, and $[Co—(S—S)_2]^{2-}$) exhibited electron spin resonance signals in solution and in frozen glasses. The g-factors, and metal hyperfine splittings when observed, were characteristic of the various metal complexes, and frequently served as diagnostic evidence for the formation of the relevant charged species when isolation proved difficult or impossible. The electronic spectra of the oxidized members of the electron transfer series, i.e. those species having charge $z = 0$ or -1, were characterized by very intense absorptions in the visible and near infrared regions. These bands were clearly not of 'd–d' type, being, more probably, charge transfer absorptions. They were absent in the spectra of the reduced species, and 'd–d' bands could be detected.

There have been three major attempts to understand and explain the electronic structures of the planar bis-1,2-dithiolenes. Two of these[7,8] used simple modified Hückel and Wolfsberg-Helmholtz theories and applied the resultant molecular orbital schemes to an interpretation of the spectral, electrochemical and chemical properties of this class of molecules. The third approach, which was perhaps the most rigorous, was developed[5] from a prima facie examination and interpretation of the esr spectra of a number of complexes having spin-doublet ground states. The result of these three different approaches was to produce, for $[Ni—(S—S)_2]^-$, three superficially different ground state assignments, but all of these indicated that the orbital of the unpaired electron in this nickel system must contain significant sulphur ligand character. It now appears that two of the calculations provided, essentially, the correct 'answer', namely that agreement was reached as to the symmetry of the ground state half-filled orbital, and this was confirmed by an independent esr spectral study of ^{33}S hyperfine splittings in $[NiS_4C_4(CN)_4]^-$.[6] It is now conceded that the remaining calculation contained some ambiguities.

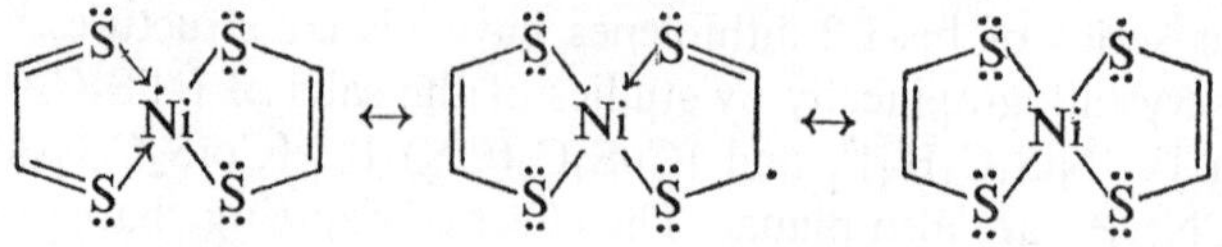

Fig. 2. Kékulé resonance forms for $[NiS_4C_4R_4]^-$ (R omitted for simplicity).

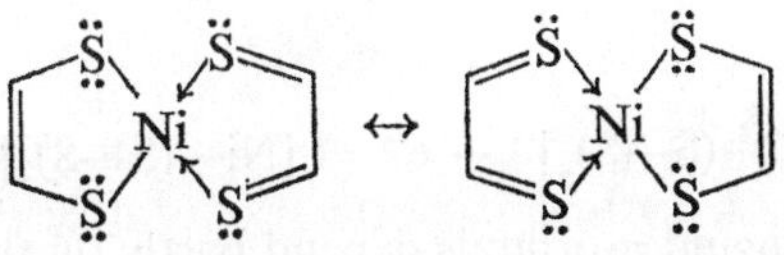

Fig. 3. Kékulé resonance forms for $[NiS_4C_4R_4]^0$ (R omitted for simplicity).

The monoanionic and neutral complexes, particularly of the nickel group, which have highly delocalized ground states, may be visualized in valence bond forms as shown in Figs. 2 and 3. The reduced species (i.e. $[NiS_4C_4R_4]^{2-}$) can be correctly regarded as containing a divalent metal ion complexed by two dithiolato dianions.

Little is known of the electronic structure of the monocationic species observed in the voltammograms of some of the nickel bis-dithiolenes. However, their undoubted existence leads us to believe that the nickel bis-dithiolenes may belong to a five-membered electron transfer series (Fig. 4),

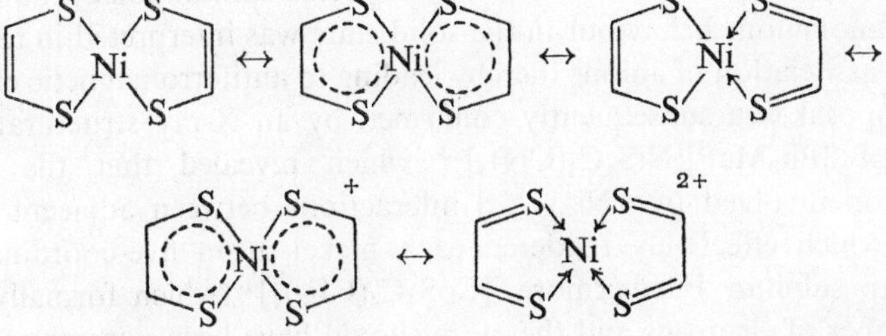

Fig. 4. The postulated electron transfer series for $[NiS_4C_4R_4]^z$ (R omitted for simplicity).

in which the most oxidized species (as yet undetected) would be represented by a divalent nickel ion complexed by two dithioketonic ligands and the most reduced species by Ni^{2+} complexed by two dithiolato ligands.

The dependence on the sulphur ligand substituents of the half-wave potentials for the various one-electron transfer processes which the planar complexes undergo is entirely consistent with the molecular orbital proposals. The molecular orbitals in the metal complexes which are involved in the electron transfer reactions are predominantly sulphur ligand π-orbital in character, and therefore any alterations in the relative energy of these sulphur ligand orbitals would have a direct, and measureable, effect on the metal complex orbitals and consequently on the values of the potentials for such reactions as

$$[Ni-(S-S)_2]^0 + e^- \rightleftharpoons [Ni-(S-S)_2]^-$$

and

$$[Ni\text{---}(S\text{---}S)_2]^- + e^- \rightleftharpoons [Ni\text{---}(S\text{---}S)_2]^{2-}$$

The energies of the ligand π-orbitals depend partly on the inductive effect of the substituents on the ligand, and, of course, this applies also to ring substituents in arene-1,2-dithiolates, and therefore, $+I$ substituents would cause an increase in the energy of these ligand orbitals and consequently in the energy of the orbitals of the metal complex. Substituents exerting a $-I$ effect would cause a decrease in the energy of the metal complex orbitals.

The discovery that many of the apparently monanionic cobalt and iron bis-dithiolene complexes are in fact dimeric dianions is relatively recent in the history of the development of dithiolene chemistry. An early indication of interionic or intermolecular interaction of dithiolene complexes was discovered[22] in the observation of reduced magnetic moments of some crystalline monoanionic dicyano-dithiolene complexes of nickel, palladium and platinum, $[MS_4C_4(CN)_4]^-$. These anions exhibited magnetic moments in solution entirely consistent with their expected spin doublet ground states, and the anomalous behaviour in the solid state was interpreted in terms of a pairwise association of anions thereby leading to antiferromagnetic exchange. This proposal was subsequently confirmed by an X-ray structural investigation of $[Ph_3MeP][NiS_4C_4(CN)_4]^{23}$ which revealed that the pairwise association involved two Ni . . . S interactions between adjacent pairs of anions, which effectively rendered each nickel atom five-coordinate with respect to sulphur. Furthermore, $[CoS_4C_4(CF_3)_4]^0$ (which formally has an odd number of electrons and therefore should have been paramagnetic) was found to be diamagnetic in the solid state[24] and to be dimeric in CCl_4; the structure of the compound in the solid state was shown to involve dimeric units, i.e. $[CoS_4C_4(CF_3)_4]_2^{0,25}$ involving direct Co—S bonding (Fig. 5) some-

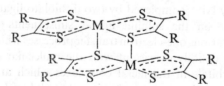

Fig. 5. Idealized dimeric structure of $[MS_4C_4R_4]_2^{2-}$.

what similar to that found in the nickel monoanions described above. Subsequent conductivity measurements on some other cobalt monoanionic species suggested[15,26] that they too were dimeric and dianionic in solution. Dimeric structures have now been conclusively established for certain salts of

$[FeS_4C_4(CN)_4]_2^{2-}$ and $[Co(S_2C_6Cl_4)_2]_2^{2-}$ (although the latter is monomeric in solution).

It has been clearly and elegantly demonstrated by polarography and voltammetry[15] that these discrete dimeric dianions, and the dimeric neutral species, can undergo several one-electron transfer reactions, and one two-electron transfer process. Thus, $[CoS_4C_4(CF_3)_4]_2^{2-}$ exhibits two anodic waves of equal diffusion current and one cathodic wave whose diffusion current is nearly twice those of the anodic waves (Table 5). This behaviour is

Table 5. *Dimeric Dithiolene Complexes*
[*Data from Reference* 15]

	Electrode Processes					
Complex	$2[\ ]^{2-} \rightleftharpoons [\ ]_2^{2-}$		$[\ ]_2^{2-} \rightleftharpoons [\ ]_2^{-}$		$[\ ]_2^{-} \rightleftharpoons [\ ]_2^{0}$	
	$E_{1/2}$ (c)	I (d)	$E_{1/2}$ (c)	I (d)	$E_{1/2}$ (c)	I (d)
$[CoS_4C_4(CF_3)_4]_2^{0}$	−0·14	+34	+0·52	+18	+1·19	+20
$[CoS_4C_4(CF_3)_4]_2^{-}$ (a)	−0·11	+41	+0·54	+22	+1·23	−21
$[CoS_4C_4(CF_3)_4]_2^{2-}$ (b)	−0·09 (i)	+40	+0·56	−17	+1·24	−20
$[CoS_4C_4(CF_3)_4]_2^{2-}$ (a)	−0·05	− (e)	+0·60	− (e)	+1·29	− (e)
$[CoS_4C_4(CN)_4]_2^{2-}$ (b)	+0·17	+35	+1·03	−18	No wave	
$[CoS_4C_4(CN)_4]_2^{2-}$ (b)	+0·24	−17	+1·06	−9	No wave	
$[Co(S_2C_6H_3CH_3)_2]^{-}$ (b)	−0·71	+17	No wave		No wave	
$[FeS_4C_4(CF_3)_4]_2^{0}$	(f)		+0·67	+19	+1·24	+19
$[FeS_4C_4(CF_3)_4]_2^{-}$ (a)	(f)		+0·69	+19	+1·27	−18
$[FeS_4C_4(CF_3)_4]_2^{2-}$ (b)	−0·60 (g)	+9·5 (g)	+0·71	−16	+1·27	−17
$[FeS_4C_4(CN)_4]_2^{2-}$ (b)	(h)		+1·06	−17	No wave	
$[Fe(S_2C_6H_3CH_3)_2]_2^{-}$ (b)	(f)		+0·16	+17	+0·48	−17
$[Fe(S_2C_6H_3CH_3)_2]^{-}$ (b)	−0·95 (g)	+15 (g)	+0·18	−7	+0·48	−7

(a) $[(C_2H_5)_4N]^+$ salts.

(b) $[(n\text{-}C_4H_9)_4N]^+$ salts.

(c) In volts, using $[(n\text{-}C_4H_9)_4N][ClO_4]$ (0·50 M) as supporting electrolyte, in CH_2Cl_2 using rotating platinum electrode and Ag/AgI reference cell.

(d) i_d/c, in μamp/mmole, concentrations calculated on the basis of the formulae shown in column one.

(e) The instability of this compound to oxidation prevented determination of I.

(f) Very drawn out waves which may correspond to the reduction, observed both in CH_2Cl_2 and CH_3CN.

(g) Obtained in CH_3CN solution using dropping mercury electrode, and $[n\text{-}Pr_4N][ClO_4]$ (0·1 M) supporting electrolyte. Reference is SCE.

Table 6. *Applications of $E_{1/2}$-data to the Synthesis of Dimeric Dithiolene Complexes*

Oxidized or Reduced Species	$E_{1/2}$ (a)	Starting Material	Redox Reagent	Solvent
$[FeS_4C_4(CF_3)_4]_2^-$	+0·71 (+0·47)	$[FeS_4C_4(CF_3)_4]_2^{2-}$	$[FeS_4C_4(CF_3)_4]_2^0$	CH_2Cl_2
$[CoS_4C_4(CF_3)_4]_2^-$	+0·56 (+0·34)	$[CoS_4C_4(CF_3)_4]_2^{2-}$	$[CoS_4C_4(CF_3)_4]_2^0$	CH_2Cl_2
$[Fe(S_2C_6H_3Me)_2]_2^-$	+0·18 (+0·01)	$[Fe(S_2C_6H_3Me)_2]^-$	I_2	CH_2Cl_2
$[CoS_4C_4(CN)_4]_2^{2-}$	+0·17 (+0·00)	$[CoS_4C_4(CN)_4]^{2-}$	I_2	DMF
$[CoS_4C_4(CF_3)_4]_2^{2-}$	+1·24 (+0·84) (b)	$[CoS_4C_4(CF_3)_4]_2^0$	acetone	acetone
$[FeS_4C_4(CF_3)_4]_2^{2-}$	+1·24 (+0·84) (b)	$[FeS_4C_4(CF_3)_4]_2^0$	acetone	acetone

(a) In volts: using rotating Pt electrode in CH_2Cl_2 vs. Ag/AgI electrode with $[n\text{-}Bu_4N][PF_6]$ (0·5 M) as supporting electrolyte. Values in parentheses are those calculated for CH_3CN vs. SCE using the relationship $E_{1/2}(CH_3CN) = E_{1/2}(CH_2Cl_2) \times 0\cdot 88 - 0\cdot 15$.

(b) Potential for couples $[MS_4C_4(CF_3)_4]_2^0 + e^- \rightleftharpoons [MS_4C_4(CF_3)_4]_2^-$; for second reduction potential see above.

consistent with the occurrence of two oxidative processes which leave the dimeric unit intact, thereby producing consecutively $[CoS_4C_4(CF_3)_4]_2^-$ and $[CoS_4C_4(CF_3)_4]_2^0$. The reduction process involves a two-electron transfer yielding two dianions

$$[CoS_4C_4(CF_3)_4]_2^{2-} + 2e^- \rightleftharpoons 2[CoS_4C_4(CF_3)_4]^{2-}$$

The other complexes investigated generally exhibited similar behaviour, and it is evident that these dimeric systems exist as part of at least a four-membered electron transfer series

$$2[M\text{—}(S\text{—}S)_2]^{2-} \rightleftharpoons [M\text{—}(S\text{—}S)_2]_2^{2-} \rightleftharpoons [M\text{—}(S\text{—}S)_2]_2^- \rightleftharpoons [M\text{—}(S\text{—}S)_2]_2^0$$

As with the monomeric bis-dithiolenes, the $E_{1/2}$-values follow a dependence on the nature of the ligand substituents such that the order of oxidative stability is $S_2C_6H_3Me < S_2C_2(CF_3)_2 < S_2C_2(CN)_2$. The large positive potentials required to produce the neutral dicyano-dithiolene dimers has prevented their isolation. Thus, to date, the only completely characterized series are those of the cobalt and iron bis-perfluoromethyl-dithiolenes. The bis-diphenyl-dithiolene complexes of iron and cobalt are too insoluble in suitable solvents to permit voltammetric examination, but a study of $[FeS_4C_4(4\text{-}MeOC_6H_4)_4]_n^0$ ($n \geqslant 2$),[16] which is soluble in dichloromethane, has shown that several oxidative and reductive processes do occur, although their nature is not yet understood. It is possible, however, that the electron transfer series for dimeric species may well involve cationic as well as anionic and neutral species.

The formation and isolation of the various charged dimers follow the same general rules as those used to produce bis-dithiolene complexes, and these are summarized in Table 6. It is worthy of note, however, that $[FeS_4C_4(CF_3)_4]_2^-$ could only be produced by reaction of $[FeS_4C_4(CF_3)_4]_2^{2-}$ with $[NiS_4C_4(CF_3)_4]^0$, and that the monoanionic cobalt analogue was formed in the electron-exchange reaction

$$[CoS_4C_4(CF_3)_4]_2^{2-} + [CoS_4C_4(CF_3)_4]_2^0 \rightarrow 2[CoS_4C_4(CF_3)_4]_2^-$$

It is important to recognize that the voltammetric and polarographic characterization of the dimeric species was executed principally in dichloromethane solution. Previous and comparative studies in acetonitrile[15,24] had led to less conclusive and often confusing results. Thus, while the potentials of reduction waves observed in acetonitrile compared well with those measured in dichloromethane, in only two reported cases were the potentials for oxidation waves similar. The difference in electrochemical behaviour in dichloromethane in comparison with other solvents may be due to the ability of basic solvents to cause fission of the dimeric species, and also possibly due to the facility with which basic solvents reduce the oxidized members of couples with $E_{1/2}$ values more positive than ca $+0.2$ V.

Molecular orbital calculations on these iron and cobalt dimeric dithiolenes have not been reported, but the complexes have been described in terms of valence-bond resonance forms developed[27] along lines similar to those used in the pictorial description of the ground states of the planar bis-1,2-dithiolenes. Thus, there appear to be five limiting structures (Fig. 6) which may

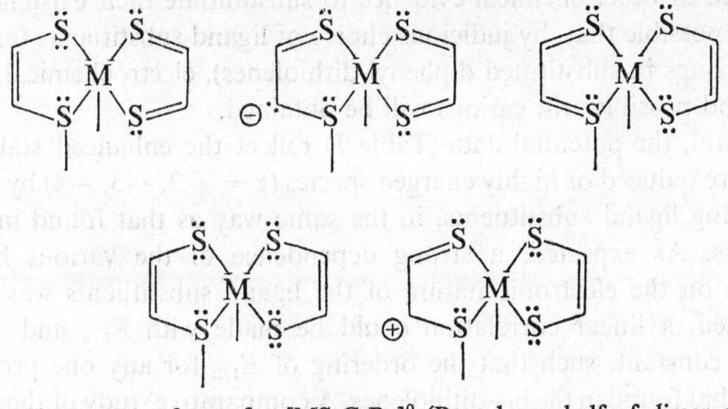

Fig. 6. Kékulé resonance forms for $[MS_4C_4R_4]_2^0$ (R and one half of dimer omitted for simplicity).

be regarded as the principal contributions to the ground states of the neutral dimers, although no account of M—S multiple bonding is taken. Here, as in the nickel group complexes, the metal oxidation number is assumed to be $+2$. Assuming that these five resonance structures contributed equally to the overall electronic structures of the dimeric molecules, the bond distances calculated[27] from the estimated bond multiplicities were in fair agreement with the observed bond lengths.

Following shortly on the discovery of the extensive series of bis-1,2-dithiolene metal complexes, King reported in 1963[28] that $Mo(CO)_6$ reacted with $S_2C_2(CF_3)_2$ in refluxing methylcyclohexane giving $[MoS_6C_6(CF_3)_6]^0$. This observation led to the successful search for other six-coordinate dithiolene complexes[29] and to the development of a chemistry of these systems which is largely similar, in electrochemical terms, to that of their bis-substituted analogues. However, the discovery that some of the highly oxidized members of this six-coordinate group have trigonal prismatic structures (i.e. the $M—S_6$ coordination unit has idealized D_{3h} symmetry) is one of the most remarkable and significant features of dithiolene chemistry.

It is now established that the tris-substituted complexes belong to at least a six-membered electron-transfer series[21]

$$[M—(S—S)_3]^{4-} \rightleftharpoons [M—(S—S)_3]^{3-} \rightleftharpoons [M—(S—S)_3]^{2-} \rightleftharpoons [M—(S—S)_3]^-$$
$$\rightleftharpoons [M—(S—S)_3]^0 \rightleftharpoons [M—(S—S)_3]^+$$

Of course, not every voltammogram or polarogram obtained from a particular tris-dithiolene showed waves corresponding to the generation of all of these species, but it is likely that this may be attributed, as in the bis-substituted species, to limitations in voltage scan. Certainly, at this time, the tetra-anionic and monocationic species must be viewed with caution, since there is no spectral or other chemical evidence to substantiate their existence. However, it is possible that, by judicious choice of ligand substituents (e.g. o- and p-MeO groups in substituted diphenyl-dithiolenes), electrochemical evidence for di-, and possibly tri-, cations will be obtained.

In general, the potential data (Table 7) reflect the enhanced stabilization of the more reduced or highly charged species ($z = -2, -3, -4$) by electron-withdrawing ligand substituents, in the same way as that found in the bis-dithiolenes. As expected, a strong dependence of the various half-wave potentials on the electronic nature of the ligand substituents was found,[17] and, indeed, a linear correlation could be made with $E_{1/2}$ and Taft's σ^* inductive constant, such that the ordering of $E_{1/2}$ for any one process was virtually that found in the bis-dithiolenes. A comparative study of the electron-transfer reactions of tris-dicyano-, perfluoromethyl-, tetrachlorobenzene- and toluene-dithiolenes in dichloromethane[21] confirmed the predicted positions of the arene-dithiolenes in the half-wave potential series (see p. 9).

The relationships between the half-wave oxidation/reduction potentials and the syntheses and chemical stabilities of the tris-substituted dithiolenes were the same as those discussed earlier in the bis-dithiolene series. Generally

Table 7. *Polarographic and Voltammetric Data Obtained from Tris-1,2-Dithiolenes*

[*Data from References* 17, 21, *and* 92]

Complex	Solvent	Ref. Cell	$E_{1/2}$ (a)	$i_{a/}c$ (b)	$E_{3/4} - E_{1/4}$ (c)	Slope (d)	Process
Ti(S$_2$C$_6$Cl$_4$)$_3$	CH$_2$Cl$_2$ (h)	SCE	+0·56	−6·6	+66		−2 → −1
			−1·05	+5·3	−69		−2 → −3
VS$_6$C$_6$(CN)$_6$	CH$_2$Cl$_2$ (h)	SCE	+0·48	+30 (e)	+72	69	−2 → −1
			−0·61	+33 (e)	−66	60	−2 → −3
	CH$_3$CN (i)	SCE	+0·66				−2 → −1
			−0·49				−2 → −3
VS$_6$C$_6$(CF$_3$)$_6$	CH$_2$Cl$_2$ (h)	SCE	+1·20				−1 → 0
			−0·01		+60		−1 → −2
			−1·06		−59		−2 → −3
V(S$_2$C$_6$Cl$_4$)$_3$	CH$_2$Cl$_2$ (h)	SCE	+0·17	−6·5	+68		−2 → −1
			−0·97	+6·2	−65		−2 → −3
VS$_6$C$_6$H$_6$	DMF (j)	Ag/AgCl	+0·25	8·12 (f)			−0 → −1
			−0·722				−1 → −2
VS$_6$C$_6$(p-MeC$_6$H$_4$)$_6$	DMF (j)	Ag/AgCl	+0·323	3·46 (f)			0 → −1
			−0·745				−1 → −2
VS$_6$C$_6$(p-MeOC$_6$H$_4$)$_6$	DMF (j)	Ag/AgCl	+0·260	2·83 (f)			0 → −1
			−0·783				−1 → −2
CrS$_6$C$_6$(CN)$_6$	CH$_2$Cl$_2$ (h)	SCE	+0·76	−33 (f)	+66	74	−2 → −1
			+0·05	−32 (f)	+66	64	−2 → −3
	CH$_3$CN (i)	SCE	+0·69	−27 (f)	+76		−2 → −1
			+0·16	−23 (f)	+71		−2 → −3
CrS$_6$C$_6$(CF$_3$)$_6$	CH$_2$Cl$_2$ (h)	SCE	+0·65		−80		0 → −1
			−0·09		−70		−1 → −2
	CH$_3$CN (i)	SCE	+1·14				0 → −1
Cr(S$_2$C$_6$Cl$_4$)$_2$	CH$_2$Cl$_2$ (h)	SCE	+0·85	−6·3	+70		−1 → 0
			+0·11	−5·7	+54		−2 → −1
			−0·36	+5·6	−56		−2 → −3
MoS$_6$C$_6$(CN)$_6$	CH$_2$Cl$_2$ (h)	SCE	+0·59	+36 (e)	+50		−2 → −1
			−1·15	+33 (e)	−50		−2 → −3
	CH$_3$CN (i)	SCE	+0·66	−31	+61		−2 → −1
			−1·12	+27	−132		−2 → −3
MoS$_6$C$_6$(CF$_3$)$_6$	CH$_2$Cl$_2$ (h)	SCE	+0·58 (g)		−58		0 → −1
			+0·05		−59		−1 → −2
	CH$_3$CN (i)	SCE	+0·95				0 → −1
			+0·36				−1 → −2
Mo(S$_2$C$_6$H$_3$Me)$_3$	DMF (k)	Ag/AgClO$_4$	−0·219				0 → −1
			−0·895				−1 → −2
			−2·62				−2 → −3
MoS$_6$C$_6$Ph$_6$	CH$_2$Cl$_2$ (h)	SCE	+1·10		+56		0 → +1
			−0·30		−75		0 → −1
			−0·87		−67		−1 → −2
	DMF (j)	Ag/AgCl	+0·009	4·42 (f)			0 → −1
			−0·617				−1 → −2
	DMF (k)	Ag/AgClO$_4$	−0·489				0 → −1
			−1·095				−1 → −2
			−2·92				−2 → −3
MoS$_6$C$_6$H$_6$	DMF (j)	Ag/AgCl	−0·090	7·52 (f)			0 → −1
			−0·617				−1 → −2
MoS$_6$C$_6$Me$_6$	DMF (j)	Ag/AgCl	−0·307	5·86			0 → −1
			−0·936				−1 → −2
WS$_6$C$_6$(CN)$_6$	CH$_2$Cl$_2$ (h)	SCE	+0·43	−33 (e)	+56	58	−2 → −1
			−1·52	+31 (e)	−68	60	−2 → −3
	CH$_3$CN (i)	SCE	+0·76	−26	+67		−2 → −1
WS$_6$C$_6$(CF$_3$)$_6$	CH$_3$CN (i)	SCE	+0·88				0 → −1
			+0·32				−1 → −2
W(S$_2$C$_6$Cl$_4$)$_3$	CH$_2$Cl$_2$ (h)	SCE	+0·54	−5·4	+57		−1 → 0
			−0·05	−5·7	+57		−2 → −1
W(S$_2$C$_6$H$_3$Me)$_3$	CH$_2$Cl$_2$ (h)	SCE	+0·02	−5·8	+60		−1 → 0
			−0·54	+7·4	−56		−1 → −2
	DMF (k)	Ag/AgClO$_4$	−0·247				0 → −1
			−1·075				−1 → −2
WS$_6$C$_6$Ph$_6$	CH$_2$Cl$_2$ (h)	SCE	+1·10				0 → +1
			−0·34		−73		0 → −1
			−0·87		−69		−1 → −2
	DMF (j)	Ag/AgCl	−0·041	3·78 (f)			0 → −1
			−0·684				−1 → −2
	DMF (k)	Ag/AgClO$_4$	−0·542				0 → −1
			−1·135				−1 → −2

Table 7 (contd.)

Complex	Solvent	Ref. Cell	$E_{1/2}$ (a)	i_d/c (b)	$E_{3/4} - E_{1/4}$ (c)	Slope (d)	Process
$WS_6C_6H_6$	DMF (j)	Ag/AgCl	$-0\cdot133$	$6\cdot86$ (f)			$0 \rightarrow -1$
			$-0\cdot845$				$-1 \rightarrow -2$
$WS_6C_6(p\text{-}MeC_6H_4)_6$	DMF (j)	Ag/AgCl	$-0\cdot091$	$2\cdot41$ (f)			$0 \rightarrow -1$
			$-0\cdot681$				$-1 \rightarrow -2$
$WS_6C_6(p\text{-}MeOC_6H_4)_6$	DMF (j)	Ag/AgCl	$-0\cdot138$	$2\cdot32$ (f)			$0 \rightarrow -1$
			$-0\cdot751$				$-1 \rightarrow -2$
$WS_6C_6Me_6$	DMF (j)	Ag/AgCl	$-0\cdot333$	$5\cdot30$			$0 \rightarrow -1$
			$-0\cdot994$				$-1 \rightarrow -2$
	DMF (k)	Ag/AgClO$_4$	$-0\cdot839$				$0 \rightarrow -1$
			$-1\cdot405$				$-1 \rightarrow -2$
$MnS_6C_6(CN)_6$	CH_2Cl_2 (h)	SCE	$-0\cdot35$	$+31$ (e)	-63	67	$-2 \rightarrow -3$
			$-0\cdot72$	$+17$ (e)	-60	70	$(-3 \rightarrow -4?)$
	CH_3CN (i)	SCE	$-0\cdot36$	$+22$	-165		$-2 \rightarrow -3$
			$-0\cdot97$	$+13$	-197		$(-3 \rightarrow -4?)$
$Mn(S_2C_6Cl_4)_3$	CH_2Cl_2 (h)	SCE	$+0\cdot61$	$-2\cdot8$	$+70$		?
			$+0\cdot40$	$-7\cdot0$	$+53$		$-2 \rightarrow -1$
			$-0\cdot92$	$+6\cdot0$	-50		$-2 \rightarrow -3$
			$-1\cdot12$	$+6\cdot0$	-100		$-3 \rightarrow -4$
$Re(S_2C_6Cl_4)_3$	CH_2Cl_2 (h)	SCE	$+1\cdot34$	$-11\cdot0$	$+90$		$0 \rightarrow +1$
			$+0\cdot71$	$-8\cdot8$	$+65$		$-1 \rightarrow 0$
			$-0\cdot68$	$+9\cdot6$	-57		$-1 \rightarrow -2$
			$-1\cdot35$	$+8\cdot0$	-64		$-2 \rightarrow -3$
$Re(S_2C_6H_3Me)_3$	DMF (k)	Ag/AgClO$_4$	$+0\cdot387$				$0 \rightarrow +1$
			$-0\cdot065$				$0 \rightarrow -1$
			$-1\cdot577$				$-1 \rightarrow -2$
			$-2\cdot375$				$-2 \rightarrow -3$
$ReS_6C_6Ph_6$	DMF (k)	Ag/AgClO$_4$	$+0\cdot163$				$0 \rightarrow +1$
			$-0\cdot340$				$0 \rightarrow -1$
			$-1\cdot812$				$-1 \rightarrow -2$
			$-2\cdot591$				$-2 \rightarrow -3$
$FeS_6C_6(CN)$	CH_2Cl_2 (h)	SCE	$+0\cdot53$	-25 (e)	$+117$	118	$-2 \rightarrow -1$
			$-0\cdot38$	$+28$ (e)	-52	55	$-2 \rightarrow -3$
			$-1\cdot18$	$+16$	-76	-119	$(-3 \rightarrow -4?)$
	CH_3CN (i)	SCE	$+0\cdot70$	-31	$+61$		$-2 \rightarrow -1$
			$-0\cdot29$	$+26$	-84		$-2 \rightarrow -3$
			$-1\cdot25$	$+16$	-200		$(-3 \rightarrow -4?)$
$Fe(S_2C_6Cl_4)_3$	CH_2Cl_2 (h)	SCE	$+0\cdot27$	$-6\cdot2$	$+54$		$-2 \rightarrow -1$
			$-0\cdot77$	$+6\cdot2$	-52		$-2 \rightarrow -3$
			$-1\cdot51$	$+5\cdot0$	-57		$-3 \rightarrow -4$
$Fe(O_2C_6Cl_4)_3$	CH_3CN (i)	SCE	$-1\cdot66$				$-3 \rightarrow -4$
	$CH_2Cl_2^l$	Ag/AgI	$+0\cdot37$	16			$-3 \rightarrow -2$
			$+0\cdot76$	20			$-2 \rightarrow -1$
			$+1\cdot10$	16			$-1 \rightarrow 0$
$CoS_6C_6(CN)_6$	CH_2Cl_2 (h)	SCE	$+0\cdot03$	-28 (e)	$+72$	75	$-2 \rightarrow -3$
	CH_3CN (i)	SCE	$+0\cdot12$	-27	$+98$		$-3 \rightarrow -2$

(a) In volts.

(b) In μamp/mmole, except where stated.

(c) For reversible wave, $E_{3/4} - E_{1/4}$ should be ±56 mV, for $n = 1$.

(d) Slope of plot of E_{obs} vs. $\log i(i_d - i)^{-1}$ for anodic waves, and of $-E_{obs}$ vs. $\log i(i_d - i)^{-1}$ for cathodic waves. For a reversible one-electron wave in either case, slope is theoretically equal to 59 mV; in mV.

(e) Obtained with a platinum electrode different in length from that used in the series including $[Ti(S_2C_6Cl_3)_3]^{2-}$, $[V(S_2C_6Cl_4)_3]^{2-}$, $[Cr(S_2C_6Cl_4)_3]^{2-}$, etc.

(f) Diffusion coefficient, in cm^2 sec^{-1}.

(g) No anodic wave observed.

(h) Using rotating Pt electrode and $[Et_4N][ClO_4]$ ($0\cdot05$ M) as supporting electrolyte.

(i) Using dropping mercury electrode and $[n\text{-}Pr_4N][ClO_4]$ ($0\cdot1$ M) as supporting electrolyte.

(j) Using dropping mercury electrode and $LiClO_4$ ($0\cdot1$ M) as supporting electrolyte.

(k) Using dropping mercury electrode and $[Et_4N][ClO_4]$ ($0\cdot1$ M) or $[n\text{-}Pr_4N][ClO_4]$ ($0\cdot1$ M) as supporting electrolytes.

(l) Using rotating Pt electrode and $[n\text{-}Bu_4N][PF_6]$ ($0\cdot5$ M) as supporting electrolyte.

speaking, it was possible to isolate neutral, mono-, di-, and tri-anionic species using conventional oxidizing or reducing agents, and, as in the bis-substituted series, the perfluoromethyl-dithiolene species found favour as powerful one-electron oxidizing agents. Some examples of how the 'rules' for using the potential data in the design of syntheses of tris-dithiolene complexes have been applied are given in Table 8.

Table 8. *Applications of $E_{1/2}$-data to the Synthesis of Tris-1,2-Dithiolene Complexes*

Oxidized or Reduced Species	$E_{1/2}$ (a)	Starting Material	Redox Reagent	Solvent
$[VS_6C_6(CN)_6]^-$	+0·48 (b)	$[VS_6C_6(CN)_6]^{2-}$	$[MoS_6C_6(CF_3)_6]^0$	CH_2Cl_2
$[CrS_6C_6(CN)_6]^-$	+0·76	$[CrS_6C_6(CN)_6]^{2-}$	$[CrS_6C_6(CF_3)_6]^0$	CH_2Cl_2
$[CrS_6C_6(CN)_6]^{2-}$	+0·05	$[CrS_6C_6(CN)_6]^{3-}$	I_2	CH_2Cl_2
$[W(S_2C_6Cl_4)_3]^0$	+0·54	$[W(S_2C_6Cl_4)_3]^-$	Br_2	CH_2Cl_2
$[W(S_2C_6Cl_4)_3]^-$	−0·05	$[W(S_2C_6Cl_4)_3]^{2-}$	I_2	acetone
$[MoS_6C_6(CF_3)_6]^{2-}$	+0·05 (b)	$[MoS_6C_6(CF_3)_6]^{-1,0}$	N_2H_4	ethanol
$[MoS_6C_6(CF_3)_6]^-$	+0·54	$[MoS_6C_6(CF_3)_6]^{2-}$	acetone/DMF	acetone/DMF
$[FeS_6C_6(CN)_6]^{3-}$	−0·38	$[FeS_6C_6(CN)_6]^{2-}$	SO_3^{2-}	acetone/water
$[CoS_6C_6(CN)_6]^{2-}$	+0·03	$[CoS_6C_6(CN)_6]^{3-}$	I_2	CH_2Cl_2

(a) In CH_2Cl_2 vs. SCE with $[Et_4N][ClO_4]$ (0·05 M) as base electrolyte, using rotating Pt electrode.

(b) $E_{1/2}$ for couple $[MoS_6C_6(CF_3)_6]^- + e^- \rightleftharpoons [MoS_6C_6(CF_3)_6]^{2-}$, but reaction involved reduction of $[MoS_6C_6(CF_3)_6]^0$.

In general, the magnetic properties of the six-coordinate species were not strictly similar to those of the bis-1,2-dithiolenes. Among the complexes containing first-row metals, those species having an even number of electrons were found to be diamagnetic if the charge on the complexes was $z = 0$ or -1, or to have two unpaired spins if $z = -2$ or -3, with the sole exception of $[Ti-(S-S)_3]^{2-}$ which was diamagnetic. Those species having an odd number of electrons were found to have one or three unpaired spins, the latter predominating with first row metals having $z = -2$. Species having one unpaired electron gave rise to characteristic esr spectra which proved useful in the elucidation of the electronic structures of these complexes. The infrared data obtained from the perfluoromethyl-[29] and diaryl-dithiolenes[30] were interpreted in terms of the gradual oxidation of the sulphur ligand system as the charge on the complexes changed from -2 to 0.

It is now known that many, if not all, neutral vanadium, Group VI metal and rhenium tris-dithiolene complexes have trigonal prismatic geometries (Fig. 7), and, further, it seems highly probable that at least some mono-anionic species, e.g. $[VS_6C_6Ph_6]^{-}$,[31] also adopt this configuration both in the

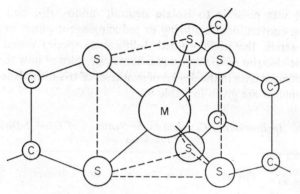

Fig. 7. Idealized trigonal prismatic structure of $[MS_6C_6R_6]^0$ (R omitted for simplicity).

solid state and in solution. The situation for the more reduced members of this group is less clear. However, it is likely that as any one metal dithiolene species becomes more reduced (or relatively more electron-rich) trigonal prismatic symmetry (D_{3h}) gradually disappears and trigonally-distorted octahedral symmetry (D_3) takes its place. It is tempting to visualize this as a gradual process, and, indeed, the structure of $[VS_6C_6(CN)_6]^{2-}$ (Fig. 8)[32]

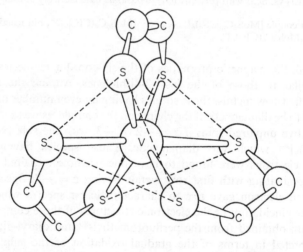

Fig. 8. The structure of $[VS_6C_6(CN)_6]^{2-}$ (CN omitted for simplicity).

cannot be described as a trigonal prism or as a trigonally distorted octahedron. Gray and his colleagues have suggested[32] that the corporate sulphur ligand system is seeking to impose a trigonal prismatic geometry and that the metal atoms prefers octahedral symmetry, the final result in this particular complex being a compromise. Bernal[33] has reported that the structure of $[FeS_6C_6(CN)_6]^{2-}$ is 'normal', in that it has D_3 symmetry.

Molecular orbital calculations performed on these oxidized complexes[30] were believed to show that there was a distinct possibility of delocalization involving the π- and σ-electron systems of the ligand, thereby accounting for the 'aromaticity' of some of the prismatic complexes ($[MoS_6C_6H_6]^0$ can be Friedel-Crafts-alkylated).[30] Delocalization in the planar bis-dithiolenes is believed to involve only the π-sulphur ligand systems. Attempts were also made to estimate the oxidation state of the metal ion in the highly oxidized species. Valence-bond resonance forms of $[MS_6C_6R_6]^z$ could be drawn which imply[30] an oxidation number of IV for M = Cr, Mo or W, and of III for M = V (Fig. 9 and 10). The dependence of $E_{1/2}$-values on ligand substituents

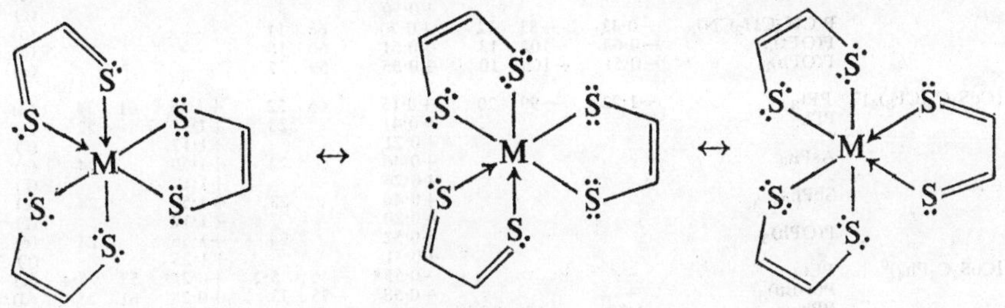

Fig. 9. Kékulé resonance forms for $[MS_6C_6R_6]^0$ (R omitted for simplicity; M = Cr, Mo or W).

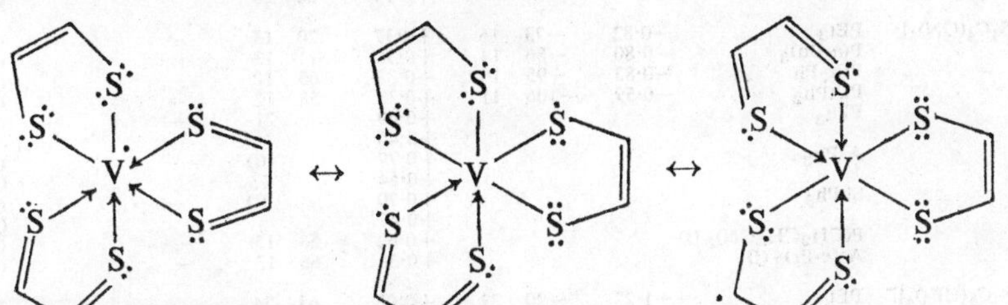

Fig. 10. Kékulé resonance forms for $[VS_6C_6R_6]^0$ (R omitted for simplicity).

was rationalized[17] in terms of the molecular orbital scheme devised for $[MS_6C_6Ph_6]^0$.[30]

Of further structural significance in the trigonal prismatic complexes is the near constancy of the interligand S—S distance (ca 3·05 Å). It was suggested[34] that this could be taken as evidence for some form of relatively strong interligand bonding which clearly should give added stability to the trigonal prismatic structures. However, there has not, as yet, been a convincing

Table 9. *Voltammetric Data Obtained from Stable Lewis Base Adducts of Cobalt and Iron Bis-1,2-Dithiolenes*
[*Data from References 35 and 36*]

		Electrode Processes									
		$-2 \rightleftharpoons -1$			$-1 \rightleftharpoons 0$			$0 \rightleftharpoons +1$			Ref.
Dithiolene	Adduct Ligand	$E_{1/2}$ (a)	R (b)	D (c)	$E_{1/2}$ (a)	R (b)	D (c)	$E_{1/2}$ (a)	R (b)	D (c)	
$[CoS_4C_4(CN)_4]^-$	PEt_3	-0.84	-75	17	$+0.46$	57	18	$+1.53$	69	27	(d)
	$P(i\text{-}Pr)_3$	-0.65	-120	15	$+0.43$	69	13	$+1.50$	75	24	(d)
	$As(n\text{-}Pr)_3$	-0.63	-107	15	$+0.42$	63	16	$+1.54$	55	30	(d)
	PEt_2Ph	-0.78	-80	10	$+0.44$	58	11	$+1.46$	64	20	(d)
	$PEtPh_2$	-0.59	-110	10	$+0.45$	65	11	$+1.51$	60	25	(d)
	PPh_3				$+0.77$		20				(e)
					$+0.54$						(f)
	$AsPh_3$				$+0.75$		20				(e)
					$+0.51$						(f)
	$SbPh_3$				$+0.73$		20				(e)
					$+0.50$						(f)
	$P(CH_2CH_2CN)_3$	-0.43	-83	12	$+0.70$	68	11	—			(d)
	$P(OEt)_3$	-0.68	-102	13	$+0.51$	60	18	—			(d)
	$P(OPh)_3$	-0.51	-105	10	$+0.55$	59	7	—			(d)
$[CoS_4C_4(CF_3)_4]^-$	PEt_3	-1.22	-99	20	$+0.15$	66	32	$+1.45$	61	34	(d)
	PPh_3				$+0.41$		23	$+1.50$		22	(e)
					$+0.21$			$+1.17$			(f)
	$AsPh_3$				$+0.46$		22	$+1.50$		24	(e)
					$+0.26$			$+1.17$			(f)
	$SbPh_3$				$+0.40$		23	$+1.50$		24	(e)
					$+0.20$			$+1.17$			(f)
	$P(OPh)_3$				$+0.52$		24	$+1.56$		24	(e)
					$+0.31$			$+1.23$			(f)
$[CoS_4C_4Ph_4]^0$	PEt_3	—			-0.55^*	-46	5·2	$+0.25^*$	57	5·4	(g)
	$P(n\text{-}Bu)_3$	—			-0.58	-75	33	$+0.28$	61	25	(d)
	PPh_3	-1.05	-68	26	-0.42	-68	20	$+0.30$	67	19	(d)
	$P(CH_2CH_2CN)_3$	-1.24^*	-106	6·7	-0.29^*	-59	5·2	$+0.44^*$	63	4·8	(g)
	$P(OPh)_3$ (h)	-1.15^*	-106	7·6	-0.44	-55	4·2	$+0.33$	58	4·2	(g)
$[Co(S_2C_6Cl_4)_2]^-$	PEt_3	—			$+0.14$	60	28				(d)
$[FeS_4C_4(CN)_4]^-$	PEt_3	-0.82	-73	16	$+0.37$	70	13	—			(d)
	$P(n\text{-}Bu)_3$	-0.80	-86	14	$+0.38$	71	13	—			(d)
	PEt_2Ph	-0.83	-95	16	$+0.38$	65	12	—			(d)
	$PEtPh_2$	-0.59	-106	11	$+0.38$	58	12	—			(d)
	PPh_3				$+0.69$		21	—			(e)
					$+0.46$						(f)
	$AsPh_3$				$+0.77$		(i)	—			(e)
					$+0.54$						(f)
	$SbPh_3$				$+0.70$		(i)	—			(e)
					$+0.47$						(f)
	$P(CH_2CH_2CN)_3$ (j)				$+0.62$	58	15	—			(d)
	$As(n\text{-}Pr)_3$ (j)				$+0.38$	66	13	—			(d)
$[FeS_4C_4(CF_3)_4]^-$	PEt_3	-1.22	-90	22	$+0.02$	63	24	—			(d)
	PPh_3				$+0.30$		23	—			(e)
					$+0.12$						(f)
	$AsPh_3$				$+0.47$		23	—			(e)
					$+0.26$						(f)
	$SbPh_3$				$+0.41$		23	—			(e)
					$+0.21$						(f)
	$P(OPh)_3$				$+0.37$		23	—			(e)
					$+0.18$						(f)
$[FeS_4C_4Ph_4]^0$	PEt_3				-0.71^*	-76	5·2	$+0.48^*$	66	5·1	(g)
	$P(n\text{-}Bu)_3$				-0.72	-60	19	$+0.5^*$	67	19	(d)
	PPh_3				-0.49^*	-62	4·7	$+0.50^*$	65	4·7	(g)
	$P(OPh)_3$				-0.39^*	-59	5·4	$+0.78$	77	11·4	(g)
	$P(CH_2CH_2CN)_3$				-0.43^*	-56	5·3	$+0.63$	32	5·0	(g)
Free Ligands	PEt_3	-0.56	-80	(k)							(d)
	PEt_2Ph	-0.70	-100	(k)							(d)
	$PEtPh_2$	-0.73	-160	(k)							(d)

explanation as to why these complexes adopt this geometry; cooperative interactions between ligand, and particularly between S, atoms may be a manifestation of attempts by the S atoms to relieve their 'electron deficiency' in these highly oxidized systems.

2. Lewis Base Adducts of Bis-1,2-Dithiolenes

The dianionic and neutral dimeric dithiolene complexes of cobalt and iron described earlier undergo dissociation in the presence of Lewis bases such as pyridines, phosphines, phosphites, arsines, NO, isonitriles, and CN^-, and usually form five-coordinate adducts with these bases.[35] Of particular interest, because of the variety of different compounds which could be prepared, were the phosphine and related Group V donor adducts, and a number of trialkyl, triaryl and mixed alkyl-aryl phosphine, arsine, phosphite and stibine complexes have been characterized.[27,35,36] These compounds undergo an extensive series of electron transfer reactions.

One of the major problems associated with these five-coordinate adducts is the apparent facility with which some of them dissociate in polar and non-polar solvents. Thus, they can either exist in an equilibrium

$$[M(L)\text{---}(S\text{---}S)_2]^z \rightleftharpoons [M\text{---}(S\text{---}S)_2]^z + L$$

or dissociate completely into the free Lewis base and what is presumed to be the monomeric precursor, $[M\text{---}(S\text{---}S)_2]^z$, $z = 0$ or -1. Most of the electrochemical studies have been carried out using species which were not obviously labile in solution, and the data was obtained by voltammetry in dichloromethane, the use of that solvent being particularly convenient because of its low polarity and therefore minimal tendency to promote dissociation of the

---

* Measured using DME in CH_2Cl_2, vs. SCE and $[Et_4N]ClO_4$ (0·05 M) as base electrolyte.

(a) Half-wave potential in volts, using RPE vs. SCE and $[Et_4N]ClO_4$ (0·05 M) as supporting electrolyte.

(b) Reversibility factor: $E_{3/4} - E_{1/4}$ in mV. ($R = 56$ mV for one electron reversible wave.)

(c) Diffusion current criterion: i_d/c in μamp/mmole: note variations due to different electrodes used in notes (d), (e), and (g).

(d) Reference 35.

(e) Reference 36; results vs. Ag/AgI, with $[n\text{-}Bu_4N][PF_6]$ (0·50 M) as supporting electrolyte.

(f) Calculated from the empirical formula reported in Reference 15, $E_{1/2}(CH_3CN)$ $\cong E_{1/2}(CH_2Cl_2) - 0·15$ (reference cell in CH_2Cl_2 Ag/AgI).

(g) Reference 37.

(h) Phosphite in excess.

(i) Presence of excess ligand obscured region above $+1·2$ V (note e).

(j) Waves not observed because of electrode coating.

(k) Multiple-electron step.

428 J. A. MCCLEVERTY

Table 10. *Polarographic and Voltammetric Data Obtained from Phosphine and Phosphite Adducts of Iron and Cobalt Bis-Dithiolene Complexes,* $[M(L)S_4C_4Ar_4]^z$

[*Data from Reference* 37]

			Electrode Processes								
			$+1 \leftrightarrow 0$			$0 \leftrightarrow -1$			$-1 \leftrightarrow -2$		
M	L	Ar	$E_{1/2}$ (a)	R (b)	D (c)	$E_{1/2}$ (a)	R (b)	D (c)	$E_{1/2}$ (a)	R (b)	D (c)
Fe	PEt₃	Ph	+0·48	+66	5·1	−0·71	−76	−5·2			
		4-MeC₆H₄	+0·42	+73	5·0	−0·77	−72	−5·4			
		4-MeOC₆H₄	+0·38	+62	5·5	−0·78	−51	−5·9			
		3-MeOC₆H₄	+0·51	+64	5·0	−0·69	−59	−5·6			
		2-MeOC₆H₄	+0·36	+61	5·1	−0·82	−66	−4·8	−1·17	−87	−4·2
		3,4-CH₂O₂C₆H₃	+0·46	+57	5·5	−0·70	−58	−5·3			
		2,5-(MeO)₂C₆H₃	+0·39	+64	4·7	−0·79	−79	−5·6	−1·26	−134	−5·1
	P(CH₂CH₂CN)₃	Ph	+0·63	+32	5·0	−0·43	−56	−5·3			
	PPh₃	Ph	+0·53	+70	4·3	−0·47	−77	−4·5			
		4-MeC₆H₄	+0·48	+64	3·5 (d)	−0·53	−78	−5·5			
		4-MeOC₆H₄	+0·43	+57	5·9	−0·48	−74	−5·5			
		3-MeOC₆H₄	+0·57	+61	4·6	−0·48	−77	−5·6			
	P(OPh)₃	Ph	+0·78	+77	11·4 (d)	−0·39	−59	−5·4			
		4-MeC₆H₄	+0·63	+50	3·1 (d)	−0·46	−48	−5·1			
		4-MeOC₆H₄	+0·64	+56	5·3	−0·47	−61	−5·7			
	P(NMe₂)₃	Ph	+0·42	+77	4·6	−0·60	−80	−5·0			
		4-MeC₆H₄	+0·37	+95	4·5	−0·66	−87	−5·3			
		4-MeOC₆H₄	+0·34	+121	4·0	−0·65	−78	−4·3			
		3-MeOC₆H₄	+0·65	+106	4·4	−0·59	−74	−4·5			
		2-MeOC₆H₄	+0·36	+86	4·1	−0·8		−10 (e)			
						(e)					
		3,4-CH₂O₂C₆H₃	+0·62	+73	4·6	−0·57	−73	−5·0			
Co	PEt₃	Ph	+0·25	+57	5·4	−0·55	−46	−5·2			
Co	PEt₃	4-MeC₆H₄	+0·20	+55	4·8	−0·56	−59	−4·9			
		4-MeOC₆H₄	+0·10	+55	4·7	−0·66	−44	−4·6			
		3-MeOC₆H₄	+0·27	+49	5·1	−0·55	−59	−5·5			
		2-MeOC₆H₄	+0·10	+62	4·4	−0·68	−60	−4·6			
		3,4-CH₂O₂C₆H₃	+0·20	+57	4·9	−0·59	−59	−5·0	−0·97	−77	−12·0 (e)
	P(CH₂CH₂CN)₃	2,5-(MeO)₂C₆H₃	+0·13	+64	4·8	−0·64	−65	−5·2			
	PPh₃	Ph	+0·44	+63	4·8	−0·29	−59	−5·2	−1·24	−106	−6·7
		Ph	+0·30	+67	19 (d)	−0·42	−68	−20 (d)	−1·05	−68	−26 (d)
		4-MeC₆H₄	+0·21	+57	4·7	−0·44	−71	−5·0	−1·25	−89	−16·4 (d)
		4-MeOC₆H₄	+0·13	+58	4·7	−0·49	−71	−5·0	−1·36	−66	−16·4 (d)
	P(OPh)₃ (f)	Ph	+0·33	+58	4·2	−0·44	−55	−4·2	−1·15	−106	−7·6 (d)
		4-MeC₆H₄	+0·22	+52	4·9	−0·38 (g)	−129	−4·9	−1·18	−56	−11·4 (d)
		4-MeOC₆H₄	+0·22	+61	5·4	−0·45	−72	−5·8			
	P(NMe₂)₃	Ph	+0·16	+86	5·3	−0·57	−74	−5·0			

(a) In volts, using a dropping mercury electrode in CH_2Cl_2 vs. SCE, with $[Et_4N][ClO_4]$ (0·05 N) as supporting electrolyte.

(b) Reversibility criterion; for reversible one-electron process, $R = E_{3/4} - E_{1/4} = 56$ mV.

(c) $D = i_d/c$ in μamp/mmole; for couple $[NiS_4C_4(CN)_4]^- + e^- \rightleftharpoons [NiS_4C_4(CN)_4]^{2-}$, $D = 5\cdot0$ μamp/mmole using Hg, and 12·0 μamp/mmole using Pt.

(d) At rotating Pt electrode.

(e) Ill-defined wave.

(f) Excess phosphite present in solution except at (g).

adducts (in contrast to DMSO, DMF or acetone). The number of electrons involved in each electron-transfer process was not directly measured, but, as in the case of other dithiolene systems, the relationships between the various redox reactions and the synthetic chemistry associated with these was such that in the majority of electrode processes observed, only one electron was involved.

It can be seen from the data assembled in Tables 9 and 10 that the phosphine and related adducts can undergo at least three electron transfer reactions, two of which certainly correspond to the generations of monoanions, neutral, and monocationic species, e.g.

$$([M(L)(S—S)_2]^{2-} \rightleftharpoons)[M(L)—(S—S)_2]^- \rightleftharpoons [M(L)—(S—S)_2]^0$$
$$\rightleftharpoons [M(L)—(S—S)_2]^+$$

Some of the complexes did not exhibit waves appropriate to the formation of all of these species, but, again, this was probably due to limitation in the voltage scan.

The half-wave potential data can be seen to depend upon three factors:

(a) the nature of the metal atom;

(b) the sulphur ligand substituents; and

(c) the nature of the substituents attached to the Lewis base donor atom.

Curiously, there is no obvious relationship between $E_{1/2}$ values and the Lewis base donor atoms themselves.

From the data in Table 9, it is apparent that the half-wave potentials of the reduction of the monoanions are independent of the metal ion, whereas those for the process

$$[M(L)—(S—S)_2]^- \rightleftharpoons [M(L)—(S—S)_2]^0 + e^-$$

are more positive for the cobalt complexes than for their iron analogues. The reverse is true for the oxidation process

$$[M(L)—(S—S)_2]^0 \rightleftharpoons [M(L)—(S—S)_2]^+ + e^-$$

The half-wave potentials for the reduction, and for the oxidation, of the neutral phosphine dithiolenes are dependent on the nature of the sulphur ligand substituent, as expected. Thus, the order of $E_{1/2}$-values obtained from these species is the same as that found in the bis- and tris-dithiolenes. A study of the effect of altering the position and nature of substituents on the phenyl rings of the bis-diphenyl-dithiolene adducts has recently been made.[37] Here, the $E_{1/2}$-values for the couple

$$[M(PR_3)S_4C_4Ar_4]^z + e^- \rightleftharpoons [M(PR_3)S_4C_4Ar_4]^{z-1}$$

(processes $z = 0 \rightarrow +1$ and $z = 0 \rightarrow -1$) became increasingly negative, with respect to a given R and to the substituent Ar, in the order

$$Ar = 3\text{-MeOC}_6H_4 < Ph < 4\text{-MeC}_6H_4 < 4\text{-MeOC}_6H_4 < 2\text{-MeOC}_6H_4$$

(Fig. 11), which correlated quite well with the known electronic directing effects of methyl and methoxy groups in the *ortho*, *meta*, and *para* positions

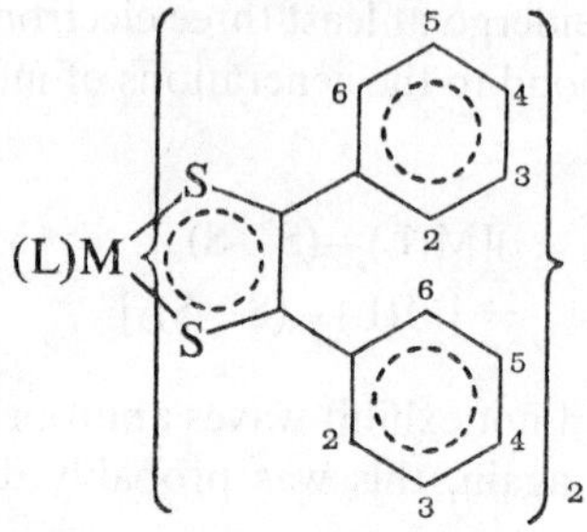

Fig. 11. The numbering system in ring-substituted diaryl-dithiolenes.

on a benzene ring. While the difference between $E_{1/2}$-values for each electrode process in the series $[M(PEt_3)S_4C_4Ar_4]^z$, for example, were small, they seemed real. Of the complexes containing disubstituted benzene rings, the $3,4\text{-CH}_2O_2C_6H_3$ derivatives had $E_{1/2}$-values reasonably close to the mean of the $E_{1/2}$-values of the 3- and 4-MeOC_6H_4-substituted species. However, the adducts with $2,5\text{-(MeO)}_2C_6H_3$ substituents had potentials closer to those of the 2-MeOC_6H_4-substituted complexes than to those of their 3-methoxy-substituted analogues.

The effect of the phosphine ligand substituents on the half-wave potentials for any one series of complexes having the same sulphur ligand substituents is considerable. In the dicyano-dithiolene complexes, this dependence was most marked on the potentials of the reduction of the monoanion. Here, $E_{1/2}$ became more negative, to a first approximation, as the basicity, or electron-releasing ability, of the phosphine increased, the order of increasingly negative potentials being

$$P(CH_2CH_2CN)_3 < P(OPh)_3 < PEtPh_2 < P(i\text{-Pr})_3$$
$$\simeq As(n\text{-Pr})_3 < P(OEt)_3 < PEt_2Ph < PEt_3$$

There was some evidence that this order was also followed in the half-wave potentials for the couple $[PR_3]^0 + ne^- \rightleftharpoons [PR_3]^{n-}$, as might be expected from the proposals of Vlček.[19] While the dependence of the value of the half-wave potentials on the donor atom substituents in the process $z = -1 \rightarrow 0$ (in the dicyano-dithiolenes) was not so marked, a classification could be made such that the $E_{1/2}$-values in the cobalt dicyano-dithiolenes averaged $+0.40$ V in the trialkyl and mixed alkyl-aryl phosphines, $+0.53$ V in the phosphites and $+0.70$ V in the $P(CH_2CH_2CN)_3$ adduct. Correspondingly, in the iron dicyano-dithiolene adducts, the potentials averaged $+0.38$ V in all of the phosphine adducts except that of the tricyanoethyl phosphine which occurred at $+0.62$ V. In these dicyano-dithiolenes, there was no apparent phosphine ligand effect on $E_{1/2}$ for the oxidation process $z = 0 \rightarrow +1$, but since these waves occurred very close to the oxidative breakdown of the medium, accurate determination of the electrochemical parameters was not possible. However, in the diaryl-dithiolene adducts there was a noticeable effect on $E_{1/2}$ for the processes $z = 0 \rightarrow -1$ and $z = 0 \rightarrow +1$ which was almost exactly parallel to that observed in the reduction processes of the cobalt dicyano-dithiolene species: the order of increasingly negative potentials was

$$P(CH_2CH_2CN)_3 < P(OPh)_3 < PPh_3 < PEt_3 < P(NMe_2)_3$$

The precise way in which the half-wave potentials for electron transfer reactions in these Lewis base adducts was influenced by the ligand substituents was of paramount importance in the context of their synthetic chemistry. Thus, the data in Tables 9 and 10 show quite clearly that combination of a strongly electron-releasing substituents on the sulphur ligand with a strongly electron-releasing phosphine should push the potentials for the couple

$$[M(L)S_4C_4R_4]^0 \rightleftharpoons [M(L)S_4C_4R_4]^+ + e^-$$

close to 0·0 V, thereby making oxidation of the neutral species relatively easy. Conversely, neutral phosphine adducts containing poorly electron-releasing substituents on the sulphur and phosphorus ligands should be relatively easy to reduce to the corresponding monoanions.

The synthetic applications of the electrochemical data obtained from Tables 9 and 10 are illustrated in Table 11. Thus, mono-cationic and anionic,

Table 11. *Synthetic Applications of $E_{1/2}$-values to Phosphine Dithiolene Complexes*

Oxidized or Reduced Species	$E_{1/2}$ (a)	Starting Material	Redox Rgt.	Solvent
$[Co(PEt_3)S_4C_4(CF_3)_4]^0$	$+0.15$	$[Co(PEt_3)S_4C_4(CF_3)_4]^-$	I_2	CH_2Cl_2
$[Co\{P(NMe_2)_3\}S_4C_4(4\text{-}MeC_6H_4)_4]^+$	$+0.10$	$[Co\{P(NMe_2)_3\}S_4C_4(4\text{-}MeC_6H_4)_4]^0$	I_2	CH_2Cl_2
$[Co\{P(NMe_2)_3\}S_4C_4(4\text{-}MeOC_6H_4)_4]^+$	$+0.01$	$[Co\{P(NMe_2)_3\}S_4C_4(4\text{-}MeOC_6H_4)_4]^0$	I_2	CH_2Cl_2
$[Co\{P(CH_2CH_2CN)_3\}S_4C_4Ph_4]^-$	-0.29	$[Co\{P(CH_2CH_2CN)_3\}S_4C_4Ph_4]^0$	N_2H_4	ethanol
$[Fe\{P(OPh)_3\}S_4C_4Ph_4]^-$	-0.39	$[Fe\{P(OPh)_3\}S_4C_4Ph_4]^0$	BH_4^-	ethanol

(a) In volts, vs. SCE, in CH_2Cl_2 with $[Et_4N][ClO_4]$ (0·05 M) as supporting electrolyte.

Table 12. *Potential Data Obtained from Nitric Oxide Adducts of Iron and Cobalt Bis-1,2-Dithiolenes*

[*Data from References* 39 *and* 40]

Complex	Couple	$E_{1/2}$ (a)	$E_{3/4} - E_{1/4}$ (b)	Slope (c)	i_d/c (d, e)
$[Fe(NO)S_4C_4(CN)_4]^{2-}$	$-1 \rightarrow -2$	$+0\cdot03$		62	-33
	$-2 \rightarrow -3$	$-1\cdot34$		72	33
$[Fe(NO)S_4C_4(CF_3)_4]^-$	$-2 \rightarrow -2$	$-0\cdot07$		58	18
	$-2 \rightarrow -3$	$-0\cdot36$		57	16
	$-1 \rightarrow \ 0$	$+0\cdot84$		53	117
$[Fe(NO)(S_2C_6Cl_4)_2]^-$	$-1 \rightarrow -2$	$-0\cdot24$		64	21
	$-1 \rightarrow \ 0$	$+0\cdot74$		58	-18
$[Fe(NO)(S_2C_6H_3Me)_2]^-$	$-1 \rightarrow -2$	$-0\cdot64$		59	37
	$-1 \rightarrow \ 0$	$+0\cdot27$		56	-32
$[Fe(NO)S_4C_4Ph_4]^0$	$0 \rightarrow -1$	$-0\cdot02$		58	20
	$-1 \rightarrow -2$	$-0\cdot83$		66	-20
	$0 \rightarrow +1$	$+0\cdot71$		56	-18
$[Fe(NO)S_4C_4(4\text{-}MeC_6H_4)_4]^0$	$0 \rightarrow -1$	$-0\cdot15$		-62	3
	$-1 \rightarrow -2$	$-0\cdot85$		-60	4
	$0 \rightarrow +1$	$+0\cdot54$		$+58$	5
$[Fe(NO)S_4C_4(4\text{-}MeOC_6H_4)_4]^0$	$0 \rightarrow -1$	$-0\cdot17$		-53	4
	$-1 \rightarrow -2$	$-0\cdot90$		-62	4
	$0 \rightarrow +1$	$+0\cdot46$		$+54$	5
$[Fe(NO)S_4C_4(3\text{-}MeOC_6H_4)_4]^0$	$0 \rightarrow -1$	$-0\cdot10$		-71	35
	(f)				
	$0 \rightarrow +1$	$+0\cdot62$		$+57$	39
$[Fe(NO)S_4C_4(2\text{-}MeOC_6H_4)_4]^0$	$0 \rightarrow -1$	$-0\cdot42$		-78	(h)
	(g)				
	$-1 \rightarrow -2$	$-0\cdot42$		-82	18
	(g)				
	$0 \rightarrow +1$	$+0\cdot45$		$+60$	17
	$+1 \rightarrow +2$	$+0\cdot95$		$+65$	11 (h)
$[Fe(NO)S_4C_4(3,4\text{-}CH_2O_2C_6H_3)_4]^0$	$0 \rightarrow -1$	$-0\cdot23$	-75		18
	(f)				
	$0 \rightarrow +1$	$+0\cdot88$	$+120$		33
$[Fe(NO)S_4C_4(2,5\text{-}(MeO)_2C_6H_3)_4]^0$	$0 \rightarrow -1$	$-0\cdot16$	-83		9
	$-1 \rightarrow -2$	$-0\cdot48$	-128		9
	$0 \rightarrow +1$	$+1\cdot31$	$+160$		12
$[Co(NO)S_4C_4(CN)_4]^{2-}$	$-1 \rightarrow -2$	$+0\cdot16$		86	-28
	$-2 \rightarrow -3$	$-1\cdot32$		69	31

and neutral species could be prepared if the choice of sulphur and phosphorus ligand was made carefully. As was mentioned earlier, many of the neutral and monoanionic dithiolene adducts exhibited voltammetric and polarographic reduction waves which could correspond to the formation of dianionic species. However, most of these processes were irreversible, and attempts to prepare the dianions chemically have so far failed.[35] Until more detailed electrochemical information is available, the existence of dianionic phosphine dithiolenes must be regarded with caution.

The monoanionic and monocationic iron phosphine adducts are paramagnetic and have one unpaired electron,[37] whereas the neutral species are diamagnetic. On the other hand, the cobalt mono-anions and cations are diamagnetic whereas the neutral species have one unpaired spin. Thus, at least three of the charge types detected in the voltammograms and polarograms of the cobalt and iron dithiolene adducts offered esr spectral data which has been used[35,38] in the elucidation of the electronic structure of these complexes. The iron monoanions displayed marked ^{31}P hyperfine splittings, whereas the monocations only showed sharp singlet esr signals; the cobalt esr spectra were characterized by eight-line multiplets, some of which showed a residual ^{31}P coupling. From a consideration of these data, it was concluded that these five-coordinate adducts could be described in molecular orbital terms in very much the same way as the planar bis-dithiolenes—namely, that they too had highly delocalized ground states. In the species $[Co(L)S_4C_4R_4]^0$, it was estimated that the orbital of the unpaired electron had only 20–25 per cent metal character. It was found that the sulphur ligand dependence of $E_{1/2}$ for the oxidation-reduction reactions could be easily rationalized using the proposed m.o. scheme, but that the dependence of these potentials on the Lewis base donor atom substituents could not be so explained.

Nitric oxide, like the phosphines previously described, also caused dissociation of the dimeric bis-dithiolene complexes of cobalt and iron. Thus, when NO was bubbled through solutions containing $[Fe—(S—S)_2]_2^{0,-2}$, the

---

(a) In volts; obtained using rotating Pt electrode unless otherwise stated; vs. SCE in CH_2Cl_2 with $[Et_4N][ClO_4]$ (0·1 M) as supporting electrolyte; estimated error ±10 mV.

(b) Reversibility criterion: for reversible one-electron wave the value should be ±56 mV.

(c) Reversibility criterion: slope obtained by plotting E_{obsd} vs. $\log i(I_d - i)^{-1}$ for anodic waves and $-E_{obsd}$ vs. $\log i(i_d - i)^{-1}$ for cathodic waves. For a reversible wave in either case, slope $= 0·059/n$ V, where $n =$ number of electrons involved (usually one).

(d) Diffusion current criterion, in μamp/mmole.

(e) Diffusion current proportional to surface area of Pt electrode; for the couple $[NiS_4C_4(CN)_4]^- + e^- \rightleftharpoons [NiS_4C_4(CN)_4]^{2-}$, $i_d/c = 5$, 18, or 30 μamp/mmole, depending on the length of Pt wire used in these experiments.

(f) Wave corresponding to couple $-1 \rightarrow -2$ not observed because of electrode coating.

(g) Recorded using dropping mercury electrode.

(h) Wave height reduced owing to electrode coating.

five-coordinate mononitrosyls, $[Fe(NO)—(S—S)_2]^{-,0}$, were formed.[39,40] Similarly, when $[CoS_4C_4(CN)_4]^{2-}$, or the corresponding dimeric dianions, were treated with NO, $[Co(NO)S_4C_4(CN)_4]^{2-}$ or $[Co(NO)—(S—S)_2]^-$ were produced; the latter were very unstable, readily dissociating into NO and $[Co—(S—S)_2]^-$.

The electrochemical data obtained from these mononitrosyls is summarized in Table 12. These results were obtained largely in dichloromethane, using a rotating Pt electrode, since it was discovered that polar solvents, particularly DMF, and mercury, reacted with the nitrosyls. The majority of the waves observed were reversible or nearly so, and from the related chemistry of the various nitrosyl species, it was apparent that each electrode process involved only one electron.

The iron complexes appear to exist as part of a five-membered electron transfer series

$$[Fe(NO)—(S—S)_2]^{3-} \rightleftharpoons [Fe(NO)—(S—S)_2]^{2-} \rightleftharpoons [Fe(NO)—(S—S)_2]^-$$
$$\rightleftharpoons [Fe(NO)—(S—S)_2]^0 \rightleftharpoons [Fe(NO)—(S—S)_2]^+$$

and there is some evidence[40] for the existence of a dicationic species, $[Fe(NO)S_4C_4(2\text{-MeOC}_6H_4)_4]^{2+}$ (the numbering in these complexes is the same as that shown in Fig. 11). The order of stability of the reduced species in the couples

$$[Fe(NO)—(S—S)_2]^z + e^- \rightleftharpoons [Fe(NO)—(S—S)_2]^{z-1}$$

was the same as or similar to that found in the bis-, tris-, and phosphine-dithiolene complexes discussed previously.

The use of the electrochemical data in the syntheses of various nitrosyls is summarized in Table 13, and it should be mentioned that some reducing

Table 13. *Synthetic Applications of $E_{1/2}$-values to Nitrosyl Bis-Dithiolene Complexes*

Oxidized or Reduced Species	$E_{1/2}$ (a)	Starting Material	Redox Reagent	Solvent
$[Fe(NO)S_4C_4(CN)_4]^{2-}$	+0·03	$[Fe(NO)S_4C_4(CN)_4]^-$	SO_3^{2-}	acetone/H_2O
$[Fe(NO)S_4C_4(CF_3)_4]^{2-}$	−0·07	$[Fe(NO)S_4C_4(CF_3)_4]^-$	SO_3^{2-}	acetone/H_2O
$[Fe(NO)(S_2C_6Cl_4)_2]^{2-}$	−0·24	$[Fe(NO)(S_4C_6Cl_4)_2]^-$	BH_4^-	THF/H_2O
$[Fe(NO)S_4C_4Ph_4]^{2-}$	−0·83	$[Fe(NO)S_4C_4Ph_4]^{2-}$	Na/Hg	THF
$[Fe(NO)S_4C_4Ph_4]^-$	−0·02	$[Fe(NO)S_4C_4Ph_4]^0$	N_2H_4	ethanol
$[Fe(NO)S_4C_4Ph_4]^0$	−0·02	$[Fe(NO)S_4C_4Ph_4]^-$	I_2	acetone/DMF
$[Co(NO)S_4C_4(CN)_4]^-$	+0·16	$[Co(NO)S_4C_4(CN)_4]^{2-}$	I_2	acetone

(a) In volts; obtained using rotating Pt electrode, vs. SCE. Solvent is CH_2Cl_2, and $[Et_4N][ClO_4]$ (0·1 M) supporting electrolyte.

agents, particularly amines and hydrazine, could not be used since they tended to displace NO from the complexes.

Because of the combination of spectral techniques which could be used in studying these compounds, a considerable amount of physical chemical data has been accumulated from them. Thus esr spectral parameters, NO stretching frequencies and electronic spectral information have been correlated. Molecular orbital schemes have been devised[35,38,39] to account for all of these data. The essential features which have to be explained are:

(a) the observation of ^{14}N hyperfine splittings in $[Fe(NO)-(S-S)_2]^{2-}$ but their absence in $[Fe(NO)-(S-S)_2]^0$ and $[Co(NO)-(S-S)_2]^-$;

(b) the shift in the g-factor from ca 2·027 in $[Fe(NO)-(S-S)_2]^{2-}$ to 2·009 in $[Fe(NO)-(S-S)_2]^0$, i.e. the trend towards the free-electron value on oxidation;

(c) the three-fold anisotropy of the g-tensor in the iron dianions, and the assignment of the tensors;

(d) the marked shift in ν_{NO} on moving from $[Fe(NO)-(S-S)_2]^{2-}$ to $[Fe(NO)-(S-S)_2]^-$ compared with the smaller shifts on moving from $[Fe(NO)-(S-S)_2]^-$ to $[Fe(NO)-(S-S)_2]^0$;

(e) the dependence of $E_{1/2}$ and ν_{NO} on the nature of the sulphur ligand substituents; and

(f) the occurrence of intense low-energy transitions in the electronic spectra of $[Fe(NO)-(S-S)_2]^0$ and $[Fe(NO)-(S-S)_2]^-$ and their virtual absence in all other nitrosyl dithiolene complexes of this type.

The first mo scheme[39] appeared to adequately explain virtually all of these points, but was inconsistent with the assignments of the g-tensors obtained from the low-temperature esr spectra. Subsequently, this difficulty was overcome,[35,38] and the revised mo scheme (Fig. 12) appears to satisfactorily account for all of the points mentioned above. There is clearly a relationship between these five-coordinate nitrosyls complexes and the phosphine adducts mentioned previously, and, indeed, the molecular orbital descriptions of the two groups of compounds are very similar.[35] It may be noted that in the nitrosyl complexes, the molecular orbitals involved in the electron transfer reactions are heavily sulphur ligand in character, thereby accounting nicely for the dependence of $E_{1/2}$ on the ligand substituents.

Dissociation of the dimeric iron and cobalt dicyano-, perfluoromethyl- and tetrachlorobenzene-dithiolenes occurred[35,41] in the presence of bidentate four-electron donor Lewis bases such as α,α'-dipyridyl, o-phenanthroline and bis(diphenylphosphino)ethane, and six-coordinate, monoanionic species, $[M(N-N)-(S-S)_2]^-$ or $[M(P-P)-(S-S)_2]^-$ (N—N and P—P represent bidentate nitrogen or phosphine ligands, respectively), were formed. The

15

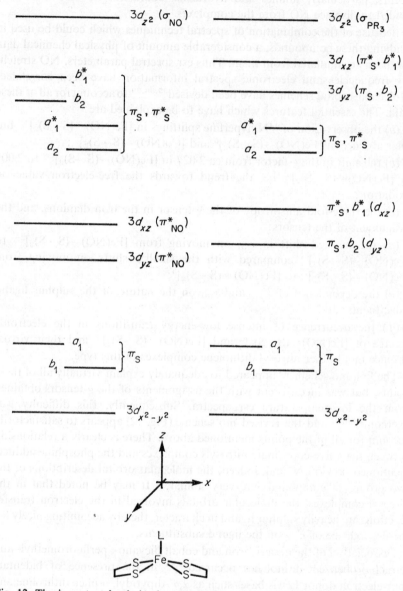

Fig. 12. The important levels in the proposed mo schemes for [Fe(NO)S₄C₄R₄]²⁻ and [Fe(PR₃)S₄C₄R₄]⁻.

cobalt species are diamagnetic whereas the iron complexes have one unpaired electron and exhibited a single line esr signal.

A voltammetric examination of the nitrogen adducts[35] established that they could undergo electron transfer reactions (Table 14) and it seems that the cobalt dicyano- and perfluoromethyl-dithiolenes are part of a four-membered electron transfer series:

$$[M(N\!-\!N)\!-\!(S\!-\!S)_2]^{2-} \rightleftharpoons [M(N\!-\!N)\!-\!(S\!-\!S)_2]^{-} \rightleftharpoons [M(N\!-\!N)\!-\!(S\!-\!S)_2]^{0}$$
$$\rightleftharpoons [M(N\!-\!N)\!-\!(S\!-\!S)_2]^{+}$$

although the same reservations which were expressed about the existence of some of the proposed dianionic and monocationic phosphine adducts of cobalt and iron bis-dithiolenes, must be voiced here. It is notable that i_d/c values (Table 14) are somewhat irregular, suggesting that some of the

Table 14. *Potential Data Obtained from Six-Coordinate Bis-Dithiolene Complexes Containing Bidentate Nitrogen Donor Ligands* [*Data from Reference* 35]

		Electrode Processes								
		$-1 \leftrightarrow -2$			$-1 \leftrightarrow 0$			$0 \leftrightarrow +1$		
Dithiolene	N-N (a)	$E_{1/2}$ (b)	R (c)	D (d)	$E_{1/2}$ (b)	R (c)	D (d)	$E_{1/2}$ (b)	R (c)	D (d)
$[CoS_4C_4(CN)_4]^-$	dipyr	-0.94	-94	11	$+0.59$	63	16	$+1.21$	85	14
	o-phen	-0.91	-82	14	$+0.60$	54	14	$+1.26$	85	23
$[CoS_4C_4(CF_3)_4]^-$	o-phen	-1.39	-95	11	$+0.25$	67	24	$+1.21$	66	31
$[Co(S_2C_6Cl_4)_2]^-$	o-phen	—			$+0.31$	56	23	—		
$[FeS_4C_4(CN)_4]^-$	dipyr	-0.64	-75	18	$+0.41$	55	25	—		
	o-phen	-0.68	-61	21	$+0.40$	63	22	—		
$[FeS_4C_4(CF_3)_4]^-$	o-phen	-1.06	-77	30	$+0.03$	56	22	—		
Free ligand	o-phen							$+1.21$		(e)

(a) Bidentate ligands: dipyr $= \alpha,\alpha'$-dipyridyl; o-phen $= o$-phenanthroline.
(b) In volts; recorded using rotating Pt electrode in CH_2Cl_2 vs. SCE, using $[Et_4N][ClO_4]$ (0·05 M) as supporting electrolyte; estimated error ±10 mV.
(c) Reversibility criterion, $E_{3/4} - E_{1/4}$ in mV; for one-electron process, $R = 56$ mV.
(d) $D = i_d/c$ in μamp/mmole.
(e) Multiple-electron oxidation step.

electrode processes may not be as simple as has been assumed. However, the generation of neutral species, $[M(N\!-\!N)\!-\!(S\!-\!S)_2]^{0}$, does not seem to be in doubt since treatment of $[Fe(o\text{-phen})S_4C_4(CF_3)_4]^-$ with I_2 resulted in the disappearance of the characteristic esr signal of the monoanion, which subsequently reappeared when the oxidized species was treated with sulphite ion. An X-ray structural examination of $[Co(o\text{-phen})S_4C_4(CN)_4]^-$,[42] has established that the compound has an octahedral, rather than a trigonal prismatic, geometry.

Treatment of the cobalt and iron dimeric bis-dicyano-dithiolenes with 1,1-dithiolato ligands[43] resulted in the formation of species such as $[M(S_2CNR_2)S_4C_4(CN)_4]^{2-}$ (M = Co or Fe and R = Me or Et) and $[M(S_2C{=}X)S_4C_4(CN)_4]^{3-}$ (M = Co or Fe and X = $C(CN)_2$, $CH(NO_2)$, $C(CN)(CO_2Et)$, $C(CN)(CONH_2)$, N(CN), etc.). The cobalt complexes are diamagnetic whereas the iron complexes have magnetic moments slightly greater than that expected for one unpaired electron (ca 2·2 BM). The dithiocarbamato complexes (containing S_2CNR_2) underwent two oxidation waves, the first of which apparently corresponded to the formation of a monoanion, $[M(S_2CNR_2)S_4C_4(CN)_4]^-$; the nature of the second wave is not yet understood but may involve two electrons. The trianionic species underwent at least one oxidation process, giving the corresponding dianion, and some species could be oxidized further, possibly giving a monoanion. A number of these oxidations could be effected using iodine, and the oxidized species isolated. The half-wave potentials showed a small but significant dependence on the nature of R and X, but this was altogether of a lower magnitude than the effect of the 1,2-dithiolato ligand substituents.

3. π-Cyclopentadienyl Dithiolene Complexes

The complexes which undergo interesting electron transfer reactions can be divided into two classes—those species containing π-C_5H_5 and the sulphur ligands alone, and those also containing NO.

The simple π-cyclopentadienyl dithiolenes are of several structural types—dimeric, e.g. $[\pi\text{-}C_5H_5MS_2C_2(CF_3)_2]_2$ (Fig. 13), monomeric, e.g. π-$C_5H_5MS_2C_2R_2$, and anionic, e.g. $[\pi\text{-}C_5H_5MS_4C_4R_4]^-$. All of these complexes have been studied voltammetrically[44] and polarographically,[45,46] but there have been few accompanying synthetic studies.

Dessy and his coworkers[45,46] have carried out a thorough electrochemical examination of a series of dimeric and monomeric bis-perfluoromethyl-dithiolene compounds. Of some interest was the reduction of $[\pi\text{-}C_5H_5VS_2C_2(CF_3)_2]_2$, which proceeded into two one-electron steps. The radical monoanion exhibited a complex esr spectrum which consisted of eight major lines, 75 gauss apart, each of which appeared to be further split into eight lines. These data suggested that a structural reorganization of the original dimer occurred on reduction, which removed the equivalency of the two vanadium atoms (^{51}V, $I = \frac{7}{2}$). The dianionic species was presumed to be diamagnetic since no resonances were detected from solutions containing it. The structurally similar molybdenum complex (Fig. 13), $[\pi\text{-}C_5H_5MoS_2C_2(CF_3)_2]_2$ was reduced irreversibly in a discrete two-electron process. The monomeric species $\pi\text{-}C_5H_5MS_2C_2R_2$ (R = CF_3, M = Co, Rh, Ir, or Ni; R = CN and M = Co) underwent at least one one-electron reduction, thereby generating

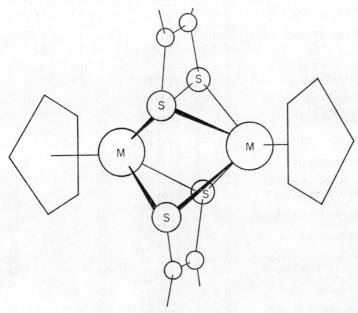

Fig. 13. Idealized structure of $[\pi\text{-}C_5H_5MS_2C_2(CF_3)_2]_2$ (CF_3 groups omitted for simplicity; M = V or Mo).

a radical monoanion, $[\pi\text{-}C_5H_5MS_2C_2R_2]^-$ (M = Co). Although none of these species have been isolated, there is considerable esr spectral evidence for their existence. Thus, the cobalt and rhodium monoanions exhibited characteristic esr signals. The corresponding nickel monoanion is diamagnetic, and the esr signal obtained from $\pi\text{-}C_5H_5NiS_2C_2(CF_3)_2$ was observed to disappear gradually during electrolytic reduction of the complex. Attempts to prepare these monoanions by treatment of the neutral species with ethanolic hydrazine resulted[66] in the formation of $[MS_4C_4(CF_3)_4]^{2-}$, but when $\pi\text{-}C_5H_5CoS_2C_2(CN)_2$, which is purple, was treated[44] with BH_4^-, brown solutions were formed which gave rise to esr signals identical to those of the electrochemically-generated $[\pi\text{-}C_5H_5CoS_2C_2R_2]^-$.

Systematic studies of nitrosyl π-cyclopentadienyl dithiolenes have been made with $[\pi\text{-}C_5H_4RMn(NO)\text{---}(S\text{---}S)]^z$ (R = H or Me),[44] and with

440 J. A. MCCLEVERTY

Table 15. *Potential Data Obtained from π-Cyclopentadienyl Dithiolene Complexes*
[Data from References 44, 45, 46, and 47]

Complex	Couple	$E_{1/2}$ (a)	R (b)	$E_{pc/2}$ (c)	$E_{pa/2}$ (d)	D (e)	C.R. (f)
$[\pi\text{-}C_5H_5VS_2C_2(CF_3)_2]_2$	$0 \to -1$	$-1 \cdot 3$ (g)		$1 \cdot 35$	$1 \cdot 25$		Yes
	$-1 \to -2$	$-2 \cdot 3$					
$[\pi\text{-}C_5H_5MoS_2C_2(CF_3)_2]_2$	$0 \to -2$	$-1 \cdot 7$		$1 \cdot 9$	$1 \cdot 7$		No
$\pi\text{-}C_5H_5CoS_2C_2(CF_3)_2$	$0 \to -1$	$-1 \cdot 1$ (h)		$1 \cdot 2$	$0 \cdot 9$		Yes
	$-1 \to -2$	$-2 \cdot 9$					
$\pi\text{-}C_5H_5RhS_2C_2(CF_3)_2$	$0 \to -1$	$-1 \cdot 4$		$1 \cdot 5$	$1 \cdot 2$		Yes
$\pi\text{-}C_5H_5NiS_2C_2(CF_3)_2$	$0 \to -1$	$-0 \cdot 96$ (i)		$0 \cdot 8$			
	$-1 \to -2$	$-2 \cdot 4$					
$\pi\text{-}C_5H_5CoS_2C_2(CN)_2$	$0 \to -1$	$-0 \cdot 45$ (jk)	-59				
$(\pi\text{-}C_5H_5)_2TiS_2C_2(CN)_2$	$0 \to -1$	$-0 \cdot 73$ (k)	-80				
$[\pi\text{-}C_5H_5TiS_4C_4(CN)_4]^-$	$-1 \to -2$	$-0 \cdot 73$ (k)	-68				
	$-1 \to \ 0$	$+0 \cdot 98$ (k)	$+125$				
$[\pi\text{-}C_5H_5MoS_4C_4(CN)_4]^-$	$-1 \to -2$	$-1 \cdot 42$ (k)	-70				
	$-1 \to \ 0$	$+0 \cdot 78$ (k)	$+70$				
$[\pi\text{-}C_5H_5WS_4C_4(CN)_4]^-$	$-1 \to -2$	$-1 \cdot 70$ (k)	-60				
	$-1 \to \ 0$	$+0 \cdot 78$ (k)	$+70$				
$[\pi\text{-}C_5H_4MeMn(NO)S_2C_2(CN)_2]^-$	$-1 \to \ 0$	$+0 \cdot 09$ (kl)	$+54$			$4 \cdot 7$	
$[\pi\text{-}C_5H_4MeMn(NO)S_2C_6Cl_4]^0$	$0 \to +1$	$+0 \cdot 91$ (km)	$+54$			$7 \cdot 8$	
	$0 \to -1$	$-0 \cdot 32$ (km)	-61			$8 \cdot 6$	
$[\pi\text{-}C_5H_4MeMn(NO)S_2C_6H_3Me]^0$	$0 \to +1$	$+0 \cdot 54$ (k)	$+60$			$10 \cdot 0$	
	$0 \to -1$	$-0 \cdot 52$ (kn)	-66			$8 \cdot 0$	
$[\pi\text{-}C_5H_4MeMn(NO)S_2C=$ $C(CN)_2]^-$	$-1 \to \ 0$	$+0 \cdot 10$ (kp)	$+58$			$5 \cdot 8$	
$[\pi\text{-}C_5H_4MeMn(NO)S_2CNMe_2]^0$	$0 \to +1$	$+0 \cdot 38$ (k)	$+55$			$6 \cdot 0$	
$[\pi\text{-}C_5H_4MeMn(NO)S_2CNEt_2]^0$	$0 \to +1$	$+0 \cdot 32$ (k)	$+60$			$7 \cdot 8$	
$[\pi\text{-}C_5H_5Mn(NO)S_2C_2(CN)_2]^-$	$-1 \to \ 0$	$0 \cdot 00$ (kq)					
$[\pi\text{-}C_5H_5Mn(NO)S_2C_6Cl_4]^0$	$0 \to +1$	$+0 \cdot 90$ (k)	$+58$				
	$0 \to -1$	$-0 \cdot 34$ (kr)	-57				
$[\pi\text{-}C_5H_5Mn(NO)S_2C_6H_3Me]^0$	$0 \to +1$	$+0 \cdot 56$ (k)	$+52$				Yes
	$0 \to -1$	$-0 \cdot 51$ (k)	-60				Yes
$[\pi\text{-}C_5H_5Mo(NO)S_2C_2(CN)_2(I)]^-$	$-1 \to \ 0$	$+0 \cdot 60$	$+59$			22	
$[\pi\text{-}C_5H_5Mo(NO)S_2C_6Cl_4(I)]^-$	$-1 \to \ 0$	$+0 \cdot 48$	$+58$			19	
$[\pi\text{-}C_5H_5Mo(NO)S_2C_2(CF_3)_2(I)]^-$	$-1 \to \ 0$	$+0 \cdot 40$	$+54$			21	

(a) In volts, obtained with dropping mercury electrode in dimethoxyethane vs. Ag/AgClO$_4$ reference cell using $[n\text{-}Bu_4N][ClO_4]$ (0·1 M) as supporting electrolyte. Unless otherwise stated (see (k)) the complex concentration is 2×10^{-3} M.

(b) Reversibility criterion; $E_{3/4} - E_{1/4}$ in mV, obtained using system in (k).

(c) Potentials at half-peak height observed by triangular voltammetry at a mercury microelectrode, sweep rates of 1·0 V/sec; cathodic wave.

(d) As (c); anodic wave.

(e) Diffusion current criterion; i_d/c in μamp/mmole; D depends on surface area of electrode, being either 5·0, 8·6, or 20 μamp/mmole depending on electrode used.

(f) Chemical reversibility.

(g) Monoanion had 52-line multiplet at $\langle g \rangle = 1 \cdot 990$ (esr spectrum in solution).

(h) Monoanion had eight-line multiplet (Co^{57}, I = 7/2) at $\langle g \rangle = 2 \cdot 454$.

(i) On reduction, esr signal of neutral species ($\langle g \rangle = 2 \cdot 0479$) disappears.

(j) Monoanion had eight-line multiplet at $\langle g \rangle = 2 \cdot 5$.

(k) Measured using rotating Pt electrode in CH_2Cl_2 vs. SCE with $[Et_4N][ClO_4]$ (0·05 M) as supporting electrolyte; estimated error ± 10 mV.

$[\pi\text{-}C_5H_5Mo(NO)\text{---}(S\text{---}S)I]^z.$[47] The voltammetric data was obtained in chloroform (Table 15), and the majority of the waves appeared to be one-electron processes. Both π-cyclopentadienyl- and π-methylcyclopentadienyl manganese complexes were investigated, and there appeared to be no significant difference between $E_{1/2}$-values obtained from the two groups containing identical sulphur ligands. In the manganese system, it appeared that the 1,2-dithiolene complexes existed as part of a three-membered electron transfer series:

$$[\pi\text{-}C_5H_4RMn(NO)\text{---}(S\text{---}S)]^- \rightleftharpoons [\pi\text{-}C_5H_4RMn(NO)\text{---}(S\text{---}S)_2]^0$$
$$\rightleftharpoons [\pi\text{-}C_5H_4RMn(NO)\text{---}(S\text{---}S)]^+$$

The voltammograms of the 1,1-dithiolenes, $[\pi\text{-}C_5H_4RMn(NO)S_2C\text{=}C(CN)_2]^-$, displayed only one wave, similar to that obtained from the isomeric $[\pi\text{-}C_5H_4RMn(NO)S_2C_2(CN)_2]^-$, corresponding to the formation of a neutral species, but this limited electrochemical behaviour was probably due to the restricted voltage scan.* The dithiocarbamates, $[\pi\text{-}C_5H_4RMn(NO)S_2CNR_2]^0$, which were isoelectronic with the monoanionic dithiolenes, exhibited one oxidation wave, corresponding to the formation of a monocation. The synthetic chemistry associated with the electrochemical data obtained from these compounds is summarized in Table 16. The neutral 1,2-dithiolenes and 1,1-dithiolene exhibited characteristic esr spectra (six-line multiplets, ^{53}Mn, $I = \frac{5}{2}$) and ir spectra (ν_{NO}). The $E_{1/2}$-values and ν_{NO} were sensitive, in the expected way, to the substituents on the 1,2-dithiolato ligands, and ν_{NO} also depended on the overall charge on the complexes.

The molybdenum complexes, $[\pi\text{-}C_5H_5Mo(NO)\text{---}(S\text{---}S)I]^-$,[47] exhibited an apparent one-electron oxidation waves, corresponding to the formation of the neutral species, but attempts to prepare these were unsuccessful. As

---

* The complexes $[\pi\text{-}C_5H_5Mn(NO)S_2C\text{=}X]^-$, where $X = C(CN)CO_2Et$, $C(CN)CONH_2$, and $N(CN)$, undergo reversible one-electron oxidation, and the half-wave potentials were dependent on the nature of the sulphur ligand substituents, X. The existence of $[\pi\text{-}C_5H_5Mn(NO)S_2C\text{=}X]^0$ was confirmed by e.s.r. spectral studies, and the g-tensors and $\langle a_{Mn}\rangle$ were significantly different from those of $[\pi\text{-}C_5H_5Mn(NO)S_2C_2R_2]^0$ ($R = CN$ or CF_3) and $[\pi\text{-}C_5H_5Mn(NO)S_2C_6Cl_4]^0$. The perthiocarbonate, $[\pi\text{-}C_5H_5Mn(NO)S_3C\text{=}S]^0$, obtained by oxidation of the corresponding monoanion, had a g-tensor and $\langle a_{Mn}\rangle$ value significantly different from those of $[\pi\text{-}C_5H_5Mn(NO)S_2C\text{=}X]^0$. (J. A. McCleverty and D. G. Orchard, *J. Chem. Soc.*, A, 3315 (1971)).

(l) Neutral species had six-line multiplet (Mn^{53}, $I = 5/2$) at $\langle g\rangle = 2\cdot01$.
(m) Neutral species had six-line multiplet (Mn^{53}, $I = 5/2$) at $\langle g\rangle = 2\cdot016$.
(n) Neutral species had six-line multiplet (Mn^{53}, $I = 5/2$) at $\langle g\rangle = 2\cdot013$.
(p) Neutral species had six-line multiplet (Mn^{53}, $I = 5/2$) at $\langle g\rangle = 2\cdot019$.
(q) Neutral species had six-line multiplet (Mn^{53}, $I = 5/2$) at $\langle g\rangle = 2\cdot015$.
(r) Neutral species had six-line multiplet (Mn^{53}, $I = 5/2$) at $\langle g\rangle = 2\cdot017$.
(s) Neutral species had six-line multiplet (Mn^{53}, $I = 5/2$) at $\langle g\rangle = 2\cdot016$.

Table 16. *Application of $E_{1/2}$-values to Synthesis of π-Cyclopentadienyl Metal Dithiolene Complexes*

Oxidized or Reduced Species	$E_{1/2}$ (a)	Starting Material	Redox Reagent	Solvent
$[\pi\text{-}C_5H_4MeMn(NO)S_2C_2(CN)_2]^0$	+0·09	$[\pi\text{-}C_5H_4MeMn(NO)S_2C_2(CN)_2]^-$	I_2	acetone/ ethanol
$[\pi\text{-}C_5H_4MeMn(NO)S_2C_6Cl_4]^-$	−0·32	$[\pi\text{-}C_5H_5MeMn(NO)S_2C_6Cl_4]^0$	Na/Hg	ethanol
$[\pi\text{-}C_5H_4MeMn(NO)S_2C_6H_3Me]^-$ (b)	−0·52	$[\pi\text{-}C_5H_4MeMn(NO)S_2C_6H_3Me]^0$	Na/Hg	THF
$[\pi\text{-}C_5H_5CoS_2C_2(CN)_2]^-$ (b)	−0·45	$[\pi\text{-}C_5H_5CoS_2C_2(CN)_2]^0$	BH_4^-	ethanol

(a) In Ch_2Cl_2 vs. SCE, using rotating Pt electrode, with $[Et_4N][ClO_4]$ (0·05 M) as supporting electrolyte.
(b) Complexes not isolated, but characterized spectrally.

expected, the half-wave potential values were dependent on the nature of the sulphur ligand.

II. ANALOGUES OF THE BIS-1,2-DITHIOLENES

In 1965, interest in the general scope of the electron transfer reactions was extended from complexes having the coordination unit $[M\text{—}S_4]^z$ to those having the units $[M\text{—}N_4]^z$, $[M\text{—}N_2S_2]^z$, $[M\text{—}N_2O_2]^z$, $[M\text{—}O_2S_2]^z$ and $[M\text{—}O_4]^z$. It was suggested by Balch, Holm, and Röhrscheid[48] that if species derived from 1,2-disubstituted benzenes, having these coordination units, could be prepared, five-membered series related by one-electron transfer reactions should, in principle, exist (Fig. 14). The terminal members of the

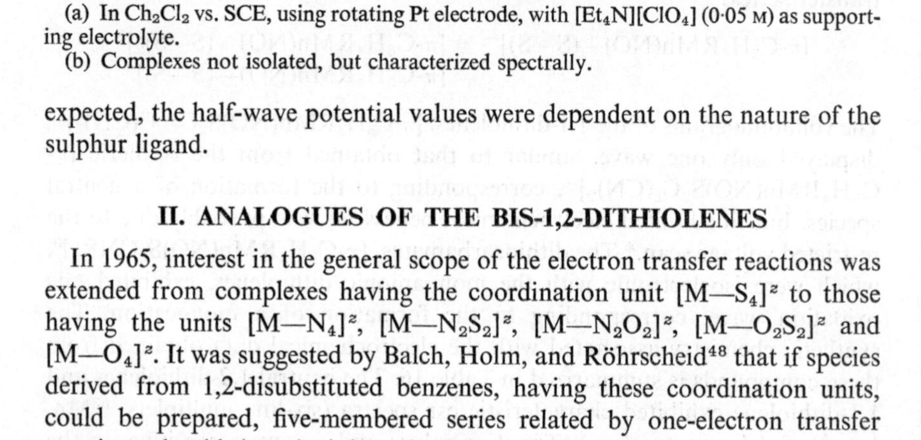

Fig. 14. The idealized electron transfer series for $[M(XYC_6H_4)_2]^z$.

series were expected to be defined as metal complexes of the potentially stable dianions, $\{C_6H_4XY\}^{2-}$ and the 'quinonoidal' species $\{C_6H_4XY\}^0$.

Feigl and Fürth, in 1927,[49] had shown that Ni^{2+} ions reacted with o-phenylenediamine in concentrated ammonia in the presence of air giving a purple complex $NiC_{12}H_{12}N_4$. This empirical formula was confirmed[50,51] by mass spectrometry, which also indicated that the compound had the structure shown in Fig. 15, corresponding to the middle member of the potential electron transfer series, $[Ni\text{—}(N\text{—}N)_2]^0$. Similar compounds of palladium,

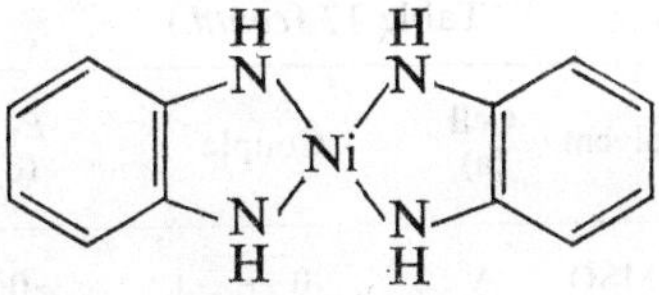

Fig. 15

platinum, and cobalt could also be prepared.[50] For each of the o-phenylenediimine complexes of the nickel group, voltammetric oxidation and reduction waves were observed which linked the $[M—(N—N)_2]^{2-}$ and $[M—(N—N)_2]^{2+}$ species by one-electron steps, and no other waves were observed in the voltammograms or polarograms of these species. Oxidation of the neutral nickel group complexes to the monocations was achieved[50] using iodine, as the potentials suggested would be feasible (Table 17). Thus, $[M—(N—N)_2]I$

Table 17. *Polarography and Voltammetry of Metal Dithiolene Analogues; Couple* $[ML_2]^z + e^- \rightleftharpoons [ML_2]^{z-1}$

[*Data from References* 13, 48, 50, 57, 59 *and* 62]

M	Ligand, L	Solvent	Cell (a)	Couple	$E_{1/2}$ (b)	I (c)	i_d/c (d)	Slope (e)
Zn	cis-N_2S_2 (f)	DMSO	A	$0 \rightarrow -1$	$-1\cdot08$		6·4	
				$-1 \rightarrow -2$	$-1\cdot44$		5·3	
		CH_3CN	A	$0 \rightarrow -1$	$-1\cdot10$		25	
		CH_2Cl_2	B	$0 \rightarrow -1$	$-1\cdot31$		23	
	gma (g)	DMSO	A	$0 \rightarrow -1$	$-0\cdot75$		6·5	
				$-1 \rightarrow -2$	$-1\cdot15$		5·6	
	$O_2C_6H_4$	CH_2Cl_2	B	$-2 \rightarrow -1$	$+0\cdot1$ (h)			
	$O_2C_6Cl_4$	CH_2Cl_2	B	$-2 \rightarrow -1$	$+0\cdot5$ (h)			
Cd	cis-N_2S_2 (f)	DMSO	A	$0 \rightarrow -1$	$-1\cdot13$		6·1	
				$-1 \rightarrow -2$	$-1\cdot55$		5·3	
	gma (g)	DMSO	A	$0 \rightarrow -1$	$-0\cdot79$		6·5	
				$-1 \rightarrow -2$	$-1\cdot33$		5·2	
	$O_2C_6Cl_4$	CH_2Cl_2	B	$-2 \rightarrow -1$	$+0\cdot3$ (h)			
Cu	cis-N_2S_2 (f)	DMSO	A	$0 \rightarrow -1$	$-0\cdot15$		6·3	
		CH_3CN	A	$0 \rightarrow -1$	$-0\cdot23$		29	
		CH_2Cl_2	B	$0 \rightarrow -1$	$-0\cdot09$		26	
				$0 \rightarrow +1$	$+1\cdot15$		25	
	cis-N_2O_2 (i)	CH_2Cl_2	B	$0 \rightarrow -1$	$-0\cdot49$		19	
	SOC_6H_4	DMSO	A	$-2 \rightarrow -1$	$-0\cdot19$		4·5	
	pt (j)	DMSO	A	$0 \rightarrow +2$	$+1\cdot03$		12	
	$O_2C_6H_4$	CH_2Cl_2	B	$-2 \rightarrow -1$	$+0\cdot07$		21 (k)	
	$O_2C_6Cl_4$	CH_2Cl_2	B	$-2 \rightarrow -1$	$+0\cdot60$		20	
Ni	cis-N_2S_2 (f)	DMSO	A	$0 \rightarrow -1$	$-0\cdot53$		7·0	
				$-1 \rightarrow -2$	$-1\cdot24$		6·9	
		CH_3CN	A	$0 \rightarrow -1$	$-0\cdot55$		33	
				$-1 \rightarrow -2$	$-1\cdot32$		23	
		CH_2Cl_2	B	$0 \rightarrow -1$	$-0\cdot48$		28	
				$-1 \rightarrow -2$	$-1\cdot26$		23	

Table 17 (contd.)

M	Ligand, L	Solvent	Cell (a)	Couple	$E_{1/2}$ (b)	I (c)	i_d/c (d)	Slope (e)
	$S(NH)C_6H_4$	DMSO	A	$0 \rightarrow -1$	-0.19		6·4	
				$-1 \rightarrow -2$	-1.04		6·2	
		CH_2Cl_2	B	$0 \rightarrow +1$	$+1.05$		27	
				$0 \rightarrow -1$	-0.03		27	
				$-1 \rightarrow -2$	-0.93		22	
	$SC(Ph)NNH^1$	DMSO	A	$0 \rightarrow -1$	-0.14		6·6	
				$-1 \rightarrow -2$	-1.13		6·4	
	gma (g)	DMSO	A	$0 \rightarrow -1$	-0.30		6·5	
				$-1 \rightarrow -2$	-1.05		6·9	
	cis-N_2O_2 (i)	DMSO	A	$0 \rightarrow -1$	-0.92		6·1	
				$-1 \rightarrow -2$	-1.67		5·8	
		CH_3CN	A	$0 \rightarrow -1$	-1.00		32	
		CH_2Cl_2	B	$0 \rightarrow -1$	-0.86		30	
	SOC_6H_4	DMSO	A	$-2 \rightarrow -1$	-0.43		6·9	
				$-1 \rightarrow 0$	$+0.38$		7·5	
		CH_2Cl_2	B	$-2 \rightarrow -1$	-0.32		24	
				$-1 \rightarrow 0$	$+0.53$ (n)			
	$SOC_{10}H_6^0$	DMSO	A	$-2 \rightarrow -1$	-0.49		6·5	
				$-1 \rightarrow 0$	$+0.22$		6·8	
		CH_2Cl_2	B	$-2 \rightarrow -1$	-0.37		20	
				$-1 \rightarrow 0$	$+0.37$		20	
	pt (j)	DMSO	A	$0 \rightarrow +1$	$+0.78$		7·5	
	$(NH)_2C_6H_4$	DMSO	A	?	$+0.23$		-15.4	82
				$0 \rightarrow -1$	-0.88	1·1	7·7	63
				$-1 \rightarrow -2$	-1.59	1·0	7·5	61
		acetone	A	$+2 \rightarrow +1$	$+0.73$	(m)	(m)	(n)
				$+1 \rightarrow 0$	$+0.14$	(m)	(m)	53
				$0 \rightarrow -1$	-0.89	(m)	(m)	66
				$-1 \rightarrow -2$	-1.43	(m)	(m)	67
	$(NH)_2C_6H_3(4\text{-}i\text{-Pr})$ (p)	acetone	A	$+2 \rightarrow +1$	$+0.81$	$+0.81$	-24	87
				$+1 \rightarrow 0$	$+0.09$	-2.7	-31	53
				$0 \rightarrow -1$	-0.98	3·0	32	66
				$-1 \rightarrow -2$	-1.51	2·6	30	67
		CH_2Cl_2	B	$+2 \rightarrow +1$	$+1.32$		-20	
				$+1 \rightarrow 0$	$+0.44$		-25	
				$0 \rightarrow -1$	-0.77		28	
				$-1 \rightarrow -2$	-1.4		26	
	phenanthren (q)	CH_3CN	A	$-2 \rightarrow 0$	-0.65		45	
		CH_2Cl_2	B	$-2 \rightarrow -1$	-0.38		19	
	$O_2C_6H_4$	CH_3CN	A	$-2 \rightarrow -1$	-0.38		24	
		CH_2Cl_2	B	$-2 \rightarrow -1$	-0.06		20	
	$O_2C_6Cl_4$	CH_3CN	A	$-2 \rightarrow 0$	$+0.15$		41	
		CH_2Cl_2	B	$-2 \rightarrow -1$	$+0.48$		26	
	phenanthroq (r)	CH_3CN	A	$0 \rightarrow -1$	-0.65			
				$-1 \rightarrow -2$	-1.21			
		CH_2Cl_2	B	$0 \rightarrow -1$	-0.51			
				$-1 \rightarrow -2$	-1.3			
	benzoquin (t)	CH_3CN	A	$0 \rightarrow -1$	-0.31			
				$-1 \rightarrow -2$	-0.90			
	tetrachlor (u)	CH_3CN	A	$0 \rightarrow -1$	$+0.14$			
				$-1 \rightarrow -2$	-0.61			
		CH_2Cl_2	B	$0 \rightarrow -1$	$+0.30$		25	
				$-1 \rightarrow -2$	-0.7			

Table 17 (contd.)

M	Ligand, L	Solvent	Cell (a)	Couple	$E_{1/2}$ (b)	I (c)	i_d/c (d)	Slope (e)
Pd	SOC$_6$H$_4$	DMSO	A	$-2 \rightarrow -1$	-0.13		6.2	
	(NH)$_2$C$_6$H$_4$	DMSO	A	$+2 \rightarrow +1$	$+0.78$		-5.6	56
				$+1 \rightarrow 0$	$+0.10$		-7.2	65
				$0 \rightarrow -1$	-0.89	1.1	7.5	67
				$-1 \rightarrow -2$	-1.44	1.0	6.7	67
Pt	(NH)$_2$C$_6$H$_4$	DMSO	A	$+2 \rightarrow +1$	$+0.78$		-5.3	53
				$+1 \rightarrow 0$	$+0.23$		-6.7	59
				$0 \rightarrow -1$	-1.01	1.0	6.6	65
				$-1 \rightarrow -2$	-1.70	0.92	6.2	67
Co	(NH)$_2$C$_6$H$_4$	DMSO	A	?	$+0.61$		-9.9	134
				$+1 \rightarrow 0$	-0.17	0.84	-7.3	58
				$0 \rightarrow -1$	-0.80	1.0	8.1	71
				$-1 \rightarrow -2$	-1.83	0.92	6.4	61
	SOC$_6$H$_4$	DMSO	A	$-2 \rightarrow -1$	-0.62		6.5	
				$-1 \rightarrow 0$	$+0.29$		6.8	
		CH$_2$Cl$_2$	B	$-2 \rightarrow -1$	-0.60		19	
				-2 (v) $\rightarrow -1$ (v)	$+0.46$		10	
				-1 (v) $\rightarrow 0$ (v)	$+0.62$		11	

(a) Reference cell: A. SCE, generally with [n-Pr$_4$N][ClO$_4$] (0·1 M) as supporting electrolyte; B. Ag/AgI electrode with [n-Bu$_4$N][PF$_6$] (0·1 M) as supporting electrolyte.

(b) In volts.

(c) $i_d/m^{2/3}t^{1/6}c$, where i_d is in μamp, m in mh/sec, t in sec, and c in mmoles/l.; if there is an entry in this column, the corresponding $E_{1/2}$ and slope values were determined at a dropping mercury electrode; all waves less positive than ca $+0.1$ V were measured at both a dropping Hg and rotating Pt electrode and good agreement (± 10 mV) was obtained; waves obtained using Hg electrode were close to reversibility, and the $E_{1/2}$-values and slopes using this electrode are given in columns six and nine.

(d) In μamp/mmole, obtained using rotating Pt electrode.

(e) Slope of plot of log $i(i_d - i)^{-1}$ vs. E (59 mV for reversible one-electron transfer).

(f) Tetradentate macrocyclic ligand biacetylbis (thiobenzoylhydrazone).

(g) Tetradentate macrocyclic ligand glyoxal mercaptoanil.

(h) Poorly defined waves with no plateaux.

(i) Tetradentate tricyclic ligand biacetyl bis(benzoylhydrazone).

(j) Anion derived from 1-hydroxy-2-pyridinethione.

(k) $E_{3/4} - E_{1/4} = 100$ mV.

(l) Derived from thiobenzoylhydrazide.

(m) Saturated solutions; ratio of diffusion currents essentially 1:1:1:1.

(n) Poorly defined waves.

(o) Derived from 1-mercapto-2-naphthol.

(p) Derived from 4-isopropyl-o-phenylenediamine.

(q) 9,10-phenanthrenediolate.

(r) 9,10-phenanthroquinone.

(s) Irreversible wave.

(t) 1,2-benzoquinone.

(u) o-chloranil = tetrachloro-1,2-benzoquinone.

(v) Dimeric species with the charge per dimeric unit indicated.

was isolated, and the voltammograms of these species were very similar to those of their neutral precursors. However, the half-wave potentials were very dependent on the solvent, and this seemed to be related to the instability of the $[M—(N—N)_2]^+$ salts in solution. The oxidation waves of these monocations were very close to, or coincident with, the oxidation waves for the couple $2I^- \rightleftharpoons I_2 + 2e^-$. Further oxidation of these neutral species could be accomplished using bromine, although the dications so formed were only partially characterized. Voltammetric reduction waves corresponding to the process $[M—(N—N)_2]^{2+} + e^- \rightleftharpoons [M—(N—N)_2]^+$ were not detected, but may have been obscured by waves arising from the reduction of bromine, in Br_3^- or Br_5^-. Dicationic species containing Pd and Pt could not be prepared chemically, but their existence was definitely established electrochemically. Because of the very negative half-wave potentials for the couples

$$[M—(N—N)_2]^0 + e^- \rightleftharpoons [M—(N—N)_2]^-$$

and

$$[M—(N—N)_2]^- + e^- \rightleftharpoons [M—(N—N)_2]^{2-},$$

isolation of the mono- and di-anions was not attempted. However, the authenticity of the monoanions was assured by the observation of characteristic esr signals in solutions containing electrolytically generated $[Ni—(N—N)_2]^-$. These esr signals were similar to those detected in solutions containing $[Ni(S_2C_6H_3Me)_2]^-$ and related complexes. The monocationic species also gave rise to esr signals, although the values of the g-tensor were very much closer to the free-electron value than those obtained from the monoanions. This would seem to indicate that the unique electron in the positively charged species is in an orbital of considerably more ligand character than that in the uninegatively charged species.

The voltammetric examination[50] of $[Co\{(NH)_2C_6H_4\}_2]^0$, and the corresponding mono-iodide obtained from it by reaction with I_2, provided clear evidence for an incomplete electron-transfer series with $z = 0$, ± 1, and -2. Unambiguous evidence for $[Co\{(NH)_2C_6H_4\}_2]^{2+}$ could not be obtained since the number of electrons involved in the oxidation wave at $+0.61$ V (Table 17) in DMSO was not directly ascertained.

The bis-acetylanil nickel complexes, $[Ni\{(MeCNPh)_2\}_2]X_2$ (X = I or NO_3) and $[Ni\{(MeCNPh)_2\}_2]^0$ had electrochemical properties[50] similar to their o-phenylenediimine analogues. Thus the dicationic species underwent a two-electron reduction giving the neutral species, which could then be further reduced in two one-electron steps giving the mono- and di-anion.

An interesting series of complexes containing the $[MN_4]^z$ coordination unit have been obtained from the deprotonated Schiff base, butane-2,3-bis(2'-pyridylhydrazine).[52] It was postulated that a potential five-membered

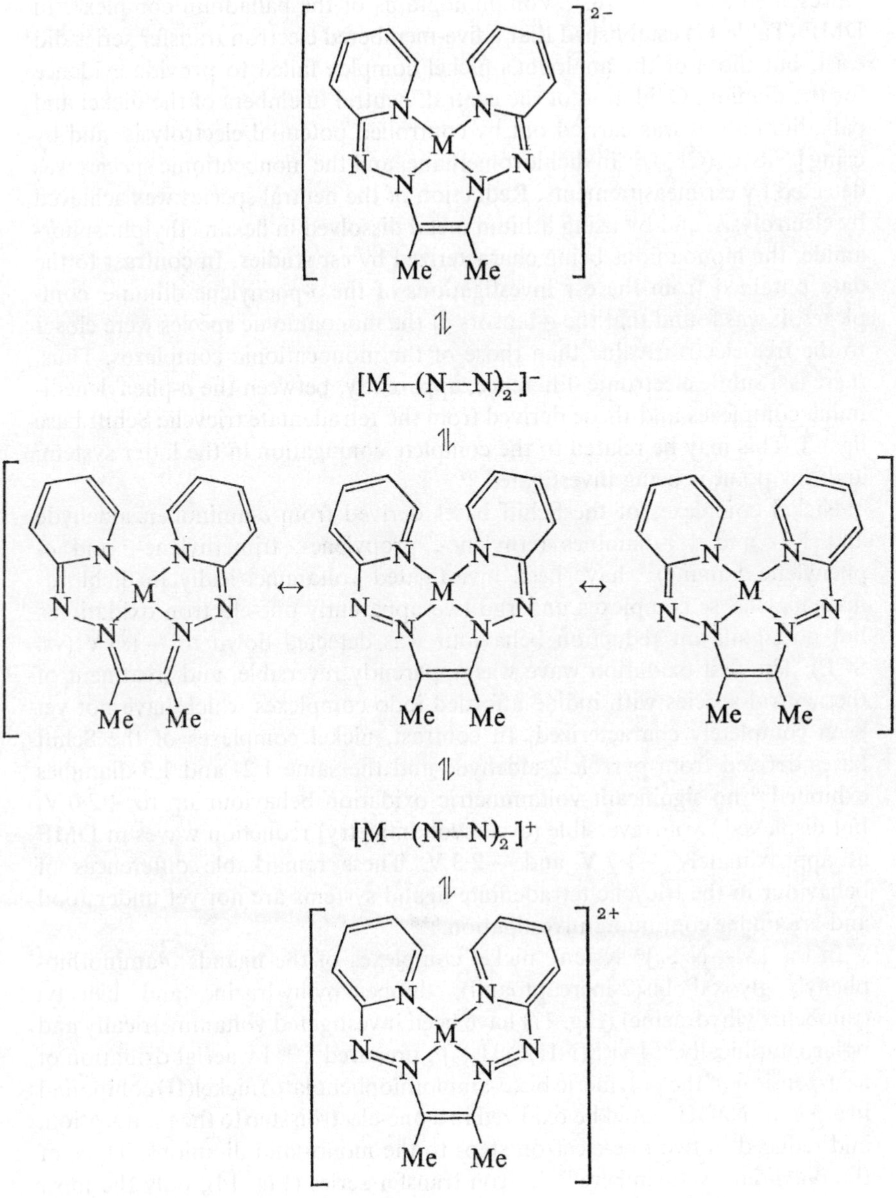

$$[M—(N—N)_2]^-$$

$$[M—(N—N)_2]^+$$

Fig. 16

electron-transfer series involving metal complexes of this ligand could be represented as in Fig. 16.[53] Voltammograms of the palladium complex[53] in DMF (Table 17) established that a five-membered electron transfer series did exist, but those of the analogous nickel complex failed to provide evidence for the dianion. Oxidation of the central, neutral, members of the nickel and palladium series was carried out by controlled potential electrolysis, and by using $[NiS_4C_4(CF_3)_4]^0$ in dichloromethane, and the monocationic species was detected by esr measurements. Reduction of the neutral species was achieved by electrolysis, and by using lithium metal dissolved in hexamethylphosphoramide, the monoanions being characterized by esr studies. In contrast to the data obtained from the esr investigations of the o-phenylene diimine complexes, it was found that the g-tensors of the monoanionic species were closer to the free-electron value than those of the monocationic complexes. Thus, there is a subtle electronic difference, apparently, between the o-phenylenediimine complexes and those derived from the tetradentate tricyclic Schiff base ligand. This may be related to the complete conjugation in the latter system, and this point is being investigated.[54]

Nickel complexes of the Schiff bases derived from o-aminobenzaldehyde and 1,2- and 1,3-diamines (ethylene-, propylene-, trimethylene-, and o-phenylene-diamine)[55] have been investigated voltammetrically in dichloromethane. These complexes undergo two apparently one-electron oxidations, but no significant reduction behaviour was detected down to -1.5 V (vs. SCE). The first oxidation wave was apparently reversible, and treatment of the neutral species with iodine afforded iodo-complexes which have not yet been completely characterized. In contrast, nickel complexes of the Schiff bases derived from pyrrole-2-aldehyde and the same 1,2- and 1,3-diamines exhibited[56] no significant voltammetric oxidation behaviour up to $+2.0$ V, but displayed two irreversible (cyclic voltammetry) reduction waves in DMF at approximately -1.7 V and -2.3 V. These remarkable differences of behaviour in the tricyclic tetradentate ligand systems are not yet understood and are under continuing investigation.[54]*

In the $[M—N_2S_2]^z$ system, nickel complexes of the ligands o-aminothiophenyl, glyoxal bis(2-mercaptoanil), thiobenzoylhydrazine and biacetyl (thiobenzoylhydrazine) (Fig. 17) have been investigated voltammetrically and polarographically.[57] $[Ni\{S(NH)C_6H_4\}_2]^0$, prepared[57,58] by aerial oxidation of a suspension of the polymeric bis(o-aminothiophenolato) nickel(II) compound in aqueous NaOH, could be oxidized in a one-electron step to the monocation, and reduced in two one-electron steps to the mono- and di-anions. Thus, of the potential five-membered electron transfer series (Fig. 14), only the most

* The species derived from 1,3-diaminopropane and DL-2,4-diaminopentane exhibited reversible one-electron reduction behaviour.

oxidized member ($z = +2$) was not detected. This may have been due to the instability of the contributing o-thionequinoneimine structure (Fig. 14; X = NH, Y = S). Chemical reduction of $[Ni\{S(NH)C_6H_4\}_2]^0$ to the green monoanion was achieved using zinc in pyridine, but the salt was not isolated. The nickel bis-thiobenzoylhydrazide, $[Ni\{SC(Ph)NNH\}_2]^0$, could also be reduced in DMSO to give a mono- and a di-anion, at potentials similar to those of the iminothiophenolate. The g-tensors obtained from the esr spectra of the $[Ni-(N-S)_2]^-$ species were similar to those observed in the spectra of the $[Ni-(S-S)_2]^-$ analogues.

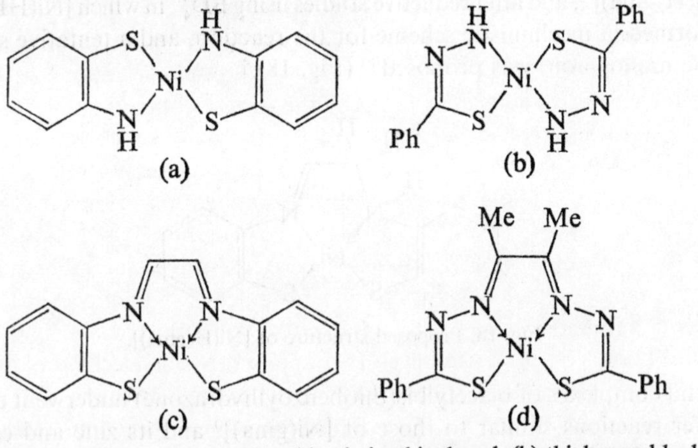

Fig. 17. Nickel complexes derived from (a) o-iminothiophenol, (b) thiobenzoyl hydrazine, (d) glyoxal mercaptoanil and (e) biacetyl bis(thiobenzoylhydrazone).

The nickel, zinc, and cadmium complexes of glyoxal bis(2-mercaptoanil), $[M(gma)]^0$, could be voltammetrically[59] and polarographically[59] reduced in two one-electron steps (Table 17) corresponding to the formation of $[M(gma)]^-$ and $[M(gma)]^{2-}$. However, the potentials of the reduction waves obtained from the zinc and cadmium complexes were significantly more negative than those of their nickel analogue, and no attempts were made to isolate the monoanions. However, these radical monoanions were generated[59] electrolytically, and their esr spectra consisted of a single, isotropic, line (DMF/CHCl$_3$ glasses), the g-values being extremely close to that of the free electron value. It was suggested that these monoanionic species could only be reasonably formulated as containing Zn(II) and Cd(II) with closed-shell d^{10} configurations and a coordinated radical anion ligand. In other words, on reduction, the incoming electron was added to a delocalized molecular orbital almost exclusively of ligand character. Reduction of $[Ni(gma)]^0$ to the monoanion was achieved by controlled potential electrolysis,[59,60] and also by using sodium amalgam in THF under scrupulously anaerobic conditions; $[Ni(gma)]^-$

exhibited an esr signal with $\langle g \rangle = 2\cdot0041$ (which had three-fold anisotropy, like $[\text{Ni—(S—S)}_2]^-$, in rigid glasses at 100°K).[67] If the reduction of $[\text{Ni(gma)}]^0$ was carried out using $[n\text{-Bu}_4\text{N}][\text{BH}_4]$ under nitrogen,[60] a different compound was formed, whose nature was only discovered by a combination of esr spectral, electrochemical and chemical studies. Thus, while $[\text{Ni(gma)}]^-$ was dark red-brown and had electrochemical properties not significantly different to $[\text{Ni(gma)}]^0$, the new complex was dark green, displayed an esr signal with $\langle g \rangle = 2\cdot046$, and had electrochemical properties markedly different to $[\text{Ni(gma)}]^-$. Eventually, the constitution of the green species was established as $[\text{Ni(H}_2\text{gma)}]^-$, and after reductive studies using BD_4^-, in which $[\text{Ni(HDgma)}]^-$ was formed, a mechanistic scheme for the reaction, and a tentative structure for the monoanion, was proposed[60] (Fig. 18).†

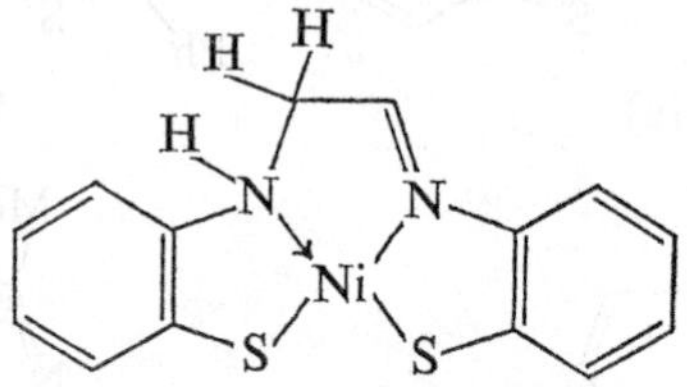

Fig. 18. Proposed structure of $[\text{Ni(H}_2\text{gma)}]$.

Metal complexes of biacetylbis(thiobenzoylhydrazone) underwent electron-transfer reactions similar to those of $[\text{Ni(gma)}]^0$ and its zinc and cadmium analogues.[57] However, the copper complex (Table 17), underwent a one-electron oxidation and reduction, and no evidence for a dianionic species could be obtained. The electrochemical properties of a series of nickel complexes of the thiosemicarbazones derived from thiosemicarbazide or N-methylthiosemicarbazide have been investigated.[61] These complexes which have been studied are shown in Fig. 19, and the electrochemical data is given in Table 18. In DMF, the complexes exhibited an ill-defined voltammetric oxidation wave (multi-electron process) beginning at ca $+1\cdot00$ V (SCE), and two reversible polarographic reduction waves. The reversibility of these processes was confirmed by cyclic voltammetry. The data show that the $E_{1/2}$-values depend on:

(a) the substituents on the azomethine bridge (a linear correlation of $E_{1/2}$ with Taft's σ^* inductive constant can be made);

† The crystal structure determination of the reduced form of the diacetylmercaptoanil nickel complexes, $[\text{Ni(H}_2\text{dma)}]^-$, has confirmed that hydrogenation of the azomethine bridge occurred, although the structure is different from that proposed in Fig. 18. Thus, the bridging group is N–CHMe–CHMe–N. (Z. Dori, R. Eisenberg, E. I. Stiefel, and H. B. Gray, *J. Amer. Chem. Soc.*, **92**, 1506 (1970).)

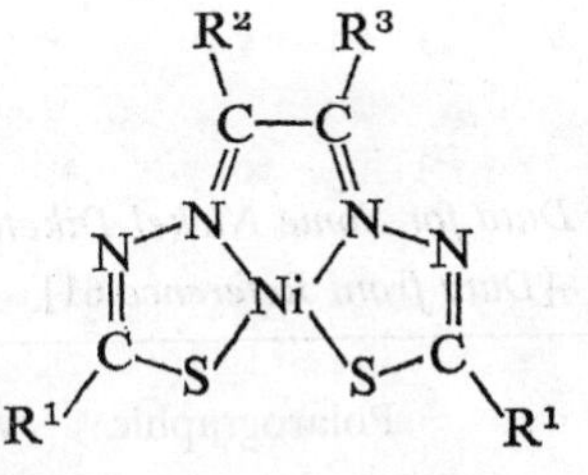

(I)

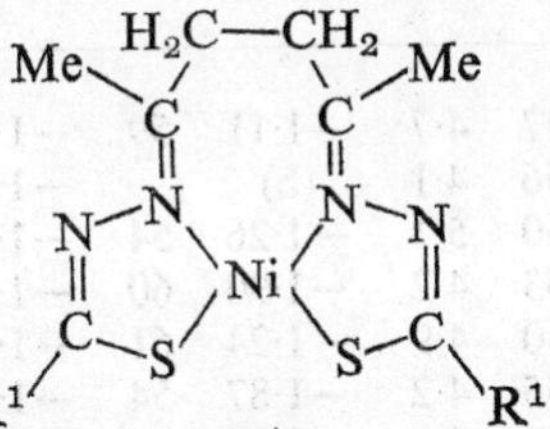

(II)

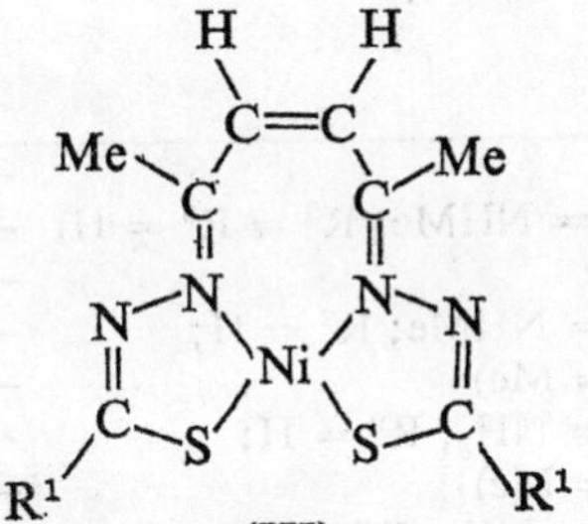

(III)

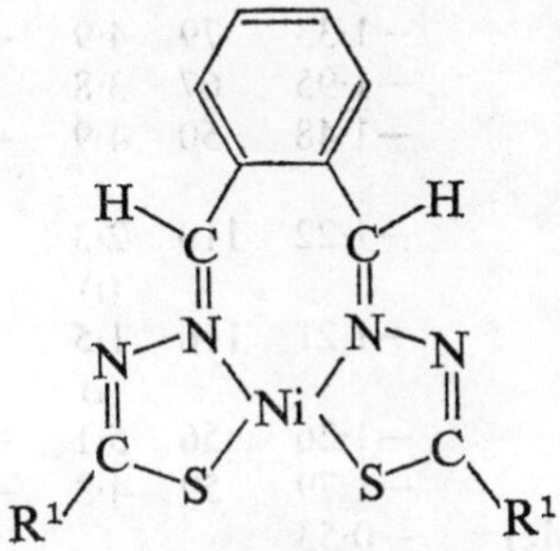

(IV)

Fig. 19

(b) the substituents on the NNCS ring fragment (compare $E_{1/2}$-values of [I, R = NHMe, R' = R'' = Me] and [I, R = Ph, R' = R'' = Me]); and

(c) the extent of conjugation around the ligand framework (compared II with I, III or IV, or $[Ni\{SC(Ph)NNH\}_2]^0$ with its biacetylbis(thiobenzoyl-hydrazone) analogue, i.e. [I, R = Ph, R' = R'' = Me]).

A comparison of the data obtained from II with those from III and IV, which have 5:7:5-membered chelate ring systems, and with those obtained from I (5:5:5), shows that $E_{1/2}$ for the first wave in II is ca 100 mV more negative

Table 18. *Electrochemical Data for Some Nickel Diketone Bisthiosemicarbazones*
[*Data from Reference* 61]

Complex (a)	Polarographic			Voltammetric		CV (g)	
	$E_{1/2}$ (b)	R (c)	D (d)	$E_{1/2}$ (e)	R (c)	$E_{1/2}$ (e)	P (f)
(I; R^1 = NHMe; R^2 = R^3 = H)	−1·10	57	4·7	−1·11	59	−1·12	100
	−1·80	56	4·1	(h)		−1·80	105
(I; R = NHMe; R^2 = H;	−1·22	60	5·1	−1·26	54	−1·23	100
R^3 = Me)	−1·86	53	4·2	−1·90	60	−1·88	110
(I; R = NH_2; R^2 = H;	−1·21	50	4·9	−1·24	61	−1·22	100
R^3 = Me)	−1·82	55	4·2	−1·87	54	−1·84	100
(I; R^1 = NH_2; R^2 = R^3 = Me)	−1·31	58	4·8	−1·33	59	−1·34	105
	−1·89	55	4·2	−1·93	53	−1·91	110
(I; R^1 = NHMe; R^2 = R^3 = Me)	−1·37	58	5·2	−1·35	56	−1·35	105
	−1·94	58	4·2	(h)		−1·95	120
(I; R^1 = NHMe; R^2, R^3 = $[CH_2]_4$)	−1·33	79	4·9	−1·38	60	−1·32	120
	−1·95	67	3·8	(h)		−1·88	115
(II; R = NHMe)	−1·48	50	4·9	−1·48	55	−1·49	115
						(−2·09	145)
(III; R = NH_2)	−1·22	110	2·3			−1·19	210
			(i)				
(III; R = NHMe)	−1·21	116	1·5	(h)			
			(i)				
(IV; R = NHMe)	−1·26	56	5·1	−1·27	50	−1·27	100
	−1·79	59	4·3	−1·85	50	−1·81	110
(I; R^1 = Ph; R^2 = R^3 = Me)	−0·53						
	−1·24						

(a) For abbreviations see Fig. 19.
(b) In DMF, with dropping mercury electrode, vs. SCE using [Et₄N][ClO₄] (0·05 M) as base electrolyte; estimated error ±10 mV.
(c) $E_{3/4} - E_{1/4}$ in mV; for reversible one-electron wave, R = 56 mV.
(d) i_d/c in µamp/mmole.
(e) In DMF using rotating Pt electrode; vs. SCE; estimated error ±10 mV.
(f) Peak height separation; for reversible one-electron wave, P = 90 mV.
(g) CV = cyclic voltammetry.
(h) Electrode coated.
(i) Saturated solutions.

than that in [I, R = NHMe, R' = R" = Me] whereas $E_{1/2}$ for the corresponding waves in III and IV is very similar to, or only 40 mV more negative than that in [I, R = NHMe, R' = R" = Me]. This would seem to imply that IV, the best models of which indicated that the benzene ring was steeply tilted out of the NiN_2S_2 coordination plane, thereby destroying full conjugation, behaved as though it was planar and conjugated, like I but unlike II. These suggestions were confirmed by the X-ray crystal structural determination[61] of IV, which showed that the molecule was essentially planar.

A very interesting point arises from a comparison of the potential data obtained from the various $[Ni—(N—S)_2]^z$ species. Considering the couple $[Ni—(N—S)_2]^0 + e^- \rightleftharpoons [Ni—(N—S)_2]^-$, the half-wave potentials in the $[Ni(gma)]^z$ system are some 100 mV more negative than in the supposedly non-conjugated $[Ni(H_2gma)]^z$ system. However, in the thiosemicarbazone series, the conjugated species have potentials at least 100 mV more positive than their non-conjugated analogue, II (Fig. 19). This apparent contradiction in potential trend is not easily rationalized, and obviously requires more detailed investigation. Furthermore, the half-wave potentials for the first reduction process in $[Ni\{S(NH)C_6H_4\}_2]^0$ and $[Ni(H_2gma)]^0$ are apparently very similar, which is puzzling since we might reasonably have expected a substituent effect in the $[Ni(H_2gma)]^z$ system (it can, after all, be regarded as a N-methyl substituted o-aminothiophenolate species) to push the $E_{1/2}$-values to a more negative point relative to $[Ni\{S(NH)C_6H_4\}_2]^z$. This assumes, of course, that the electronic descriptions of the two molecules are basically similar. One difference between the complexes is obviously that the donor atom pairs are mutually cis in the tricyclic species and $trans$ in the bis-bidentate complex; whether the relative positions of donor atoms make any significant difference to redox properties in these complexes is not known, but seems, intuitively, unlikely.

Several complexes of biacetyl bis(benzoylhydrazone), $[M—(N—O)_2]^z$, have been investigated electrochemically[57] and may be compared with their $[M—(N—S)_2]^z$ analogues. The nickel complex underwent two one-electron reductions in DMSO where the complex appeared to exist as a paramagnetic, solvated species (with two coordinated DMSO molecules). In dichloromethane and acetonitrile, however, where the molecule is probably planar, only the first reduction wave was detected, and this occurred at a potential significantly more negative than that of the $[Ni—(N—S)_2]^z$ analogue in the same solvent. The copper and zinc complexes did not undergo well-defined one-electron reduction processes. The nickel complex could be reduced electrolytically, or by using sodium amalgam in 2-methyltetrahydrofuran; for $[Ni—(N—O)_2]^-$, $\langle g \rangle = 2 \cdot 0009$, noticeably different to that obtained from $[Ni—(N—S)_2]^-$ ($\langle g \rangle = 1 \cdot 9979$).

A polarographic investigation,[62] in DMSO and dichloromethane (Table 17) of complexes derived from o-mercaptophenol, $[M(OSC_6H_4)_2]^z$, where $M = Cu$, Ni or Pd, revealed the existence of charged species with $z = -2$ or -1. In addition, $[Ni(OSC_6H_4)_2]^-$, and its analogue derived from 1-mercapto-2-naphthol, could be oxidized further to give neutral species. In DMSO, the mercapto-naphtholate underwent additional oxidation apparently corresponding to the process $z = 0 \rightarrow +2$. The half-wave potentials for equivalent processes in the nickel mercaptophenolate and -naphtholate differed by 60–120 mV, being more cathodic in the latter system. Voltammetric oxidation of $[Co(OSC_6H_4)_2]^{2-}$ in DMSO revealed two waves of equal diffusion current corresponding to the formation of monoanionic and neutral species. In dichloromethane, however, voltammetric oxidation of $[Co(OSC_6H_4)_2]^-$ revealed two waves of diffusion current about one-half that of the diffusion current of the cathodic wave appropriate to the formation of $[Co(OSC_6H_4)_2]^{2-}$. The latter data was interpreted as evidence for the existence of dimeric species, $[Co(OSC_6H_4)_2]_2^-$ and $[Co(OSC_6H_4)_2]_2^0$. From the relevant electrochemical data, it is evident that several di- and mono-anionic species could be synthesized, and this was achieved[62] with nickel, and cobalt as the central metal atoms. Some limited voltammetric data has been obtained from copper and nickel complexes of 1-hydroxy-2-pyridinethione, which indicated that, in this system, an electron transfer series analogous to that in the mercapto-phenolate species could exist.

The voltammetric and polarographic oxidation waves obtained[13] from $[M—(O—O)_2]^z$ species in acetonitrile (Table 17) were generally indicative of irreversible electrode processes, but a slight improvement in wave profile could be made by using dichloromethane as the solvent. Even so, the oxidation waves were not completely reversible. It was not established unequivocally that only one electron was involved in these reactions, but by comparison with the electrochemistry of $[NiS_4C_4(CN)_4]^{2-}$ and $[Ni(S_2C_6H_3Me)_2]^-$, the diffusion current of which for the known one-electron oxidations were very similar to those in the $[M—(O—O)_2]^{2-}$ series, this seemed very likely. Thus, the dianionic nickel complexes derived from phenanthrene-9, 10-diol, catechol and tetrachlorocatechol (1,2-dihydroxy-tetrachlorobenzene), and their copper analogues, underwent probable one-electron oxidations to the corresponding monoanions. Attempts to prepare monoanionic nickel species by chemical or electrolytic methods were unsuccessful. Some esr evidence for the existence of oxidized zinc and cadmium species, e.g. $[M(O_2C_6Cl_4)_2]^-$, was obtained when the dianionic metal complexes were treated, in solution, with o-chloranil, tetracyanoethylene or $[NiS_4C_4(CF_3)_4]^0$. The central members of the electron transfer series, $[M—(O—O)_2]^0$, could be obtained by direct chemical methods, usually not involving oxidation of the dianions, $[M—(O—O)_2]^{2-}$

(an exception was the oxidation of $[Zn(O_2C_6H_4)_2]^{2-}$ to $[Zn(O_2C_6H_4)_2]^0$ using aqueous persulphate ion). The only information pertaining to monocationic species was obtained from the voltammograms of $[Ni(O_2C_6Cl_4)_2]^{2-}$ in DMSO, which exhibited waves at $+0.17$ and $+0.78$ V. These waves appeared to correspond to a two- and a one-electron transfer process, respectively, appropriate to the generation of $[Ni(O_2C_6Cl_4)_2]^0$ and $[Ni(O_2C_6Cl_4)_2]^+$; the second oxidation wave was not observed in acetonitrile or dichloromethane. There was no electrochemical evidence for $[Ni-(O-O)_2]^{2+}$, and compounds such as $[Ni(o\text{-phenanthroquinone})_2Br_2]$ contained the octahedral coordination unit $[NiO_4Br_2]$.

III. FACTORS INFLUENCING THE REDOX BEHAVIOUR OF DITHIOLENE COMPLEXES AND THEIR ANALOGUES

Four major factors seem to influence the redox behaviour of dithiolene complexes and their analogues. These are the nature of the solvent in which the electron transfer reactions are taking place, the central metal atom, the substituents or donor atom sets in the principal ligands, the extent of conjugation in the latter, and the substituents on the donor atoms of additional (Lewis base) ligands.

The effect of solvents on half-wave potential behaviour is common enough in electrochemistry and need not be elaborated here. However, the failure to observe some oxidation waves in such solvents as DMF or DMSO must be related to comments (ii) and (iii) made on p. 407. The effect of the central metal atom is such that in the general couple

$$[M(L)_m-(X-Y)_n]^z + e^- \rightleftharpoons [M(L)_m-(X-Y)_n]^{z-1}$$

the reduced species tends to be oxidatively stabilized when the metal atom comes from the first row of the transition series in the Periodic Table, whereas the oxidized species is usually more stable with second or third row metals.

The influence of principal ligand substituents has been discussed in detail already, as has the effect of Lewis base donor atom substituents. In summary, substituents exerting an electron-releasing effect tend to push $E_{1/2}$-values relatively to negative values. The effects of conjugation are not yet fully documented, but most of the evidence seems to suggest that in tricyclic tetradentate ligand complexes, conjugation has the effect of moving $E_{1/2}$-values in the positive direction relative to those of the non-conjugated species.

It is clear that the nature of the donor atoms in the systems $[M-(X-Y)_2]^z$ has a profound effect on the half-wave potentials of a series of complexes where M remains constant and X and Y can be O, N, or S. Because of the close geometrical and electronic similarities of complexes of the type $[M(XYC_6H_4)_2]^z$, meaningful comparison of the potentials of the couple $[M—(X—Y)_2]^z + e^- \rightleftharpoons [M—(X—Y)_2]^{z-1}$ could be made, and the order of oxidative stability was, for M = Ni,

$$[Ni—(N—N)_2] > [Ni—(N—S)_2] > [Ni—(S—S)_2]$$
$$> [Ni—(S—O)_2] > ([Ni—(O—O)_2])$$

These results can be rationalized in the light of Vlček's proposals[19] (p. 409). Holm and his colleagues have suggested[50] that when the free ligands are most stable as dianions, then dianionic and monoanionic (and possibly neutral) bis-substituted species of the five-membered electron transfer series would be stabilized, whereas when the ligand was most stable in the neutral (or oxidized) form, the metal complexes tended to be stabilized as neutral, mono-, or di-cationic species.

IV. TRANSITION METAL DITHIOTROPOLONATES

An interesting development in the chemistry of transition metal dithiolene and related complexes has been the formation and characterization[63] of metal complexes of dithiotropolone (Fig. 20). These complexes, whose

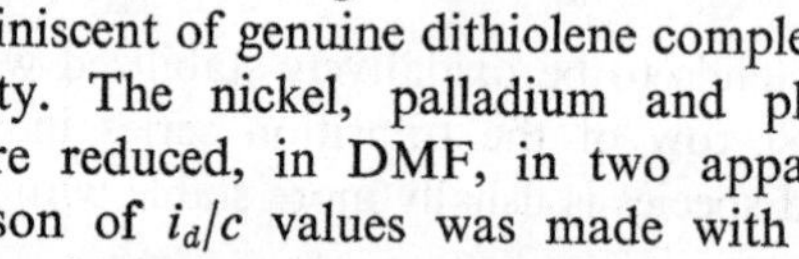

Fig. 20

electronic spectra in the 'visible' region are characterized by intense absorptions, a feature reminiscent of genuine dithiolene complexes, exhibit electrochemical reducibility. The nickel, palladium and platinum complexes, $[M\{S_2C_7H_5\}_2]^0$, were reduced, in DMF, in two apparently one-electron processes (comparison of i_d/c values was made with $[NiS_4C_4Me_2]^0$ and $[Ni(S_2C_6H_3Me)_2]^-)$, which were not reversible by cyclic voltammetry, and at potentials apparently considerably more negative than the corresponding dithiolene complexes $[MS_4C_4R_4]^0$ (R = H, alkyl or aryl; M = Ni, Pd or Pt). These data suggested that the dithiotropolonates do not have the high electron affinities of $[MS_4C_4R_4]^0$.

A comparison of the electrochemical properties of the dithiotropolonates with those of their analogues containing N and O donor atoms, e.g. $[M\{SOC_7H_5\}_2]^0$, $[M\{S(NMe)C_7H_5\}_2]^0$, and $[M\{(NMe)_2C_7H_5\}_2]^0$, was also made. The half-wave potentials for the general couple

$$[M\{XYC_7H_5\}_2]^0 + e^- \rightleftharpoons [M\{XYC_7H_5\}_2]^{-1}$$

follow the order Pt (most positive) $>$ Pd $>$ Ni, Cd $>$ Zn, and $S_2C_7H_5$ $>$ SOC_7H_5 $>$ $S(NMe)C_7H_5$ $>$ $(NMe)_2C_7H_5$. A number of complexes also exhibit oxidation waves corresponding to the formation of $[M\{XYC_7H_5\}_2]^+$.

No details of calculations of the electronic structures of dithiotropolonates are yet available, but perhaps one significant theoretical difference between this class of molecules and the dithiolenes is that a set of contributing valence bond structures, which furnish a satisfactory simple ground state description of $[MS_4C_4R_4]^0$ etc.,[3] cannot be written for the metal dithiotropolonates.

Table 19. *Voltammetric Data Obtained from Dithiotropolonato Metal Complexes* [*Data from Reference* 63]

Complex	Electrode Processes (a)					
	$z = -2 \rightleftharpoons -1$		$z = -1 \rightleftharpoons 0$		$z = 0 \rightleftharpoons +1$	
	$E_{1/2}$	i_d/c	$E_{1/2}$	i_d/c	$E_{1/2}$	i_d/c
$[Ni\{S_2C_7H_5\}_2]^z$	$-1\cdot34$	12	$-0\cdot91$	15		
$[Ni\{SOC_7H_5\}_2]^z$	$-1\cdot58$ (b)	17	$-1\cdot14$ (b)	18	$+0\cdot72$ (b)	-22
$[Ni\{S(NMe)C_7H_5\}_2]^z$ (c)	$-1\cdot85$	16	$-1\cdot42$ (b)	19	$+0\cdot62$	-32
$[Ni\{(NMe)_2C_7H_5\}_2]^z$			$-1\cdot65$ (b)	22	$+0\cdot24$ (b)	-18
$[Pd\{S_2C_7H_5\}_2]^z$	$-1\cdot25$	14	$-0\cdot89$	17		
$[Pt\{S_2C_7H_5\}_2]^z$	$-1\cdot26$	12	$-0\cdot85$	16		
$[Cu\{S_2C_7H_5\}_2]^z$	$-1\cdot54$	15	$-0\cdot39$ (b)	17	$+0\cdot50$ (b)	-15
$[Zn\{S_2C_7H_5\}_2]^z$	$-1\cdot21$	11	$-0\cdot89$	16		
$[Zn\{SOC_7H_5\}_2]^z$	$-1\cdot47$ (b)	14	$-1\cdot28$	14		
$[Zn\{S(NMe)C_7H_5\}_2]^z$	$-1\cdot75$ (b)	18	$-1\cdot53$ (b)	19	$+1\cdot06$	-20
$[Zn\{(NMe)_2C_7H_5\}_2]^z$			$-2\cdot16$	19		
$[Cd\{S_2C_7H_5\}_2]^z$	$-1\cdot16$	8	$-0\cdot90$	15		
$[NiS_4C_4Me_4]^z$	$-1\cdot08$ (b)	19	$+0\cdot09$ (b)	20		
$[Ni(S_2C_6H_3Me)_2]^z$	$-0\cdot61$ (b)	18	$+0\cdot43$	-19		

(a) In DMF solution, using rotating Pt electrode; solutions 10^{-3} M in complex and $0\cdot05$ M in $[nPr_4N][ClO_4]$ as supporting electrolyte; vs. SCE, in volts.
(b) Reversible by cyclic voltammetry.
(c) Further reduction wave at $-2\cdot06$ V, $i_d/c = 16$, reversible by cyclic voltammetry.

V. LIGAND EXCHANGE REACTIONS

Earlier, we described how, in dithiolene chemistry, reduced forms in couples more positive than ca $+0\cdot50$ V could only be oxidized by such special reagents such as $[NiS_4C_4(CF_3)_4]^0$ and it analogues. Such an oxidation could be visualized as:

$$[MS_4C_4R_4]^{2-} + [NiS_4C_4(CF_3)_4]^0 \rightarrow [MS_4C_4R_4]^- + [NiS_4C_4(CF_3)_4]^-$$

and, generally, the products of these and similar reactions could be separated by fractional crystallization. The use of this type of reaction presupposed that very little, or no, ligand exchange took place, and, indeed, early synthetic work showed that this assumption was essentially correct. However, later work[64] showed that such reactions between neutral and dianionic species, or between pairs of monoanionic dithiolenes, under certain conditions, did result in the formation of mixed-ligand species:

$$\left.\begin{array}{l} [NiS_4C_4R_4]^{2-} + [NiS_4C_4R_4']^0 \\ \\ [NiS_4C_4R_4]^- + [NiS_4C_4R_4']^- \end{array}\right\} \rightarrow 2[NiS_4C_4R_2R_2']^-$$

Voltammetry proved to be a particularly convenient technique in detecting the occurrence and extent of the ligand reorganization. Initially, the characteristic voltammograms of the reactants in the chosen solvent (dichloromethane or acetonitrile) were observed but, gradually, intermediate waves due to the formation of the mixed-ligand species appeared which grew at the expense of the initial waves (Table 20). The mixed-ligand complexes were formed to a much greater extent than that expected for random reorganization of the

Table 20. *Mixed Ligand Dithiolene Complexes of Nickel:*
Couples [MLL']^z + e^-[MLL']$^{z-1}$

[*Data from Reference* 64]

L	Ligands L'	$z = -2 \rightarrow -1$ CH$_3$CN (a)	CH$_2$Cl$_2$ (b)	$z = -1 \rightarrow 0$ CH$_3$CN (a)	CH$_2$Cl$_2$ (b)
S$_2$C$_2$(C$_6$H$_5$)$_2$	S$_2$C$_2$(C$_6$H$_5$)$_2$	-0.82	-0.83	$+0.12$	-0.01
S$_2$C$_2$(C$_6$H$_5$)$_2$	S$_2$C$_2$(CF$_3$)$_2$	-0.55	-0.55	$+0.43$	$+0.35$
S$_2$C$_2$(C$_6$H$_5$)$_2$	S$_2$C$_2$(CN)$_2$	-0.35	-0.38	$+0.57$	$+0.43$
S$_2$C$_2$(CF$_3$)$_2$	S$_2$C$_2$(CF$_3$)$_2$	-0.12	-0.22	$+0.92$	$+0.80$
S$_2$C$_2$(CF$_3$)$_2$	S$_2$C$_2$(CN)$_2$	$+0.06$ (c)	-0.01	—	—
S$_2$C$_2$(CN)$_2$	S$_2$C$_2$(CN)$_2$	$+0.23$	$+0.12$	—	—

(a) In volts, with rotating Pt electrode vs. SCE and [n-Pr$_4$N][ClO$_4$] (0·042 M) as supporting electrolyte.
(b) In volts, with dropping mercury electrode vs. SCE and [n-Bu$_4$N][PF$_6$] (0·1 M) as supporting electrolyte.
(c) Measured using dropping mercury electrode with [n-Pr$_4$N][ClO$_4$] (0·1 M) as supporting electrolyte.

Table 21. *Ligand Exchange Reactions in Solution*
[*Data from Reference* 64]

Mixed Species	Solvent	Equilibrium Time (a)	% Mixed Species (b)
$[NiS_4C_4(C_6H_5)_2(CF_3)_2]^-$	CH_3CN	13 days at 25°	92
	CH_2Cl_2	6 days at 40°	77
$[NiS_4C_4(C_6H_5)_2(CN)_2]^-$	CH_3CN	4 hr. at 82°	88
	CH_2Cl_2	6 days at 40°	81
$[NiS_4C_4(CF_3)_2(CN)_2]^-$	CH_2Cl_2	37 days at 40°	74

(a) Time required for voltammograms of a solution containing equimolar amounts (ca 10^{-3} M) of parent compounds to show no further change.
(b) Equilibrium values determined from voltammograms recorded at 25°; estimated error ±3 per cent.

Table 22. *Voltammetric Data for Ligand-Exchange Reactions Between Iron Nitrosyl Bis-Dithiolenes*
[*Data from Reference* 65]

Time (a)	$E_{1/2}$ (b)	i_d/c (c)	Assignment of Process (d)	
0	−1·58	9·5	$[Fe(NO)S_4C_4(CN)_4]^{2-}$	→ $[Fe(NO)S_4C_4(CN)_4]^{3-}$
	−1·02	9·0	$[Fe(NO)S_4C_4Ph_4]^-$	→ $[Fe(NO)S_4C_4Ph_4]^{2-}$
	−0·14	16·5	$[Fe(NO)S_4C_4(CN)_4]^-$	→ $[Fe(NO)S_4C_4(CN)_4]^{2-}$
			$[Fe(NO)S_4C_4Ph_4]^-$	→ $[Fe(NO)S_4C_4Ph_4]^0$
	+0·70	9·0	$[Fe(NO)S_4C_4Ph_4]^0$	→ $[Fe(NO)S_4C_4Ph_4]^+$
20	−1·60	12·0	$[Fe(NO)S_4C_4(CN)_4]^{2-}$	→ $[Fe(NO)S_4C_4(CN)_4]^{3-}$
	−1·08	10·0	$[Fe(NO)S_4C_4Ph_4]^-$	→ $[Fe(NO)S_4C_4Ph_4]^{2-}$
	−0·70	9·0	$[Fe(NO)S_4C_4(CN)_2Ph_2]^-$	→ $[Fe(NO)S_4C_4(CN)_2Ph_2]^{2-}$
	(e)			
	−0·14	15·5	$[Fe(NO)S_4C_4(CN)_4]^-$	→ $[Fe(NO)S_4C_4(CN)_4]^{2-}$
			$[Fe(NO)S_4C_4Ph_4]^-$	→ $[Fe(NO)S_4C_4Ph_4]^0$

(a) In hours.
(b) In volts, in CH_2Cl_2 vs. SCE with $[Et_4N][ClO_4]$ (0·05 M) as base electrolyte, using rotating Pt electrode; recorded with fast scan and not corrected for iR drop.
(c) In μamp/mmole.
(d) All complex ions equimolar and ca 10^{-3} M.
(e) Corrected $E_{1/2} = -0.55$, $E_{3/4} - E_{1/4} = 56$ mV.

ligands, and it was established that the rate of reorganization was both temperature and solvent dependent (Table 21). It was possible, therefore, to establish the conditions under which ligand exchange proceeded to near completion so that isolation of the complexes could be effected.

Mechanistic information about ligand exchanges between iron nitrosyl bis-dithiolene species has been obtained[65] by a combination of voltammetry, esr and ir spectroscopy. Equimolar amounts of $[Fe(NO)S_4C_4(CN)_4]^-$ and $[Fe(NO)S_4C_4Ph_4]^-$ were allowed to react together in dichloromethane for 20 h. After this time, it was found that the voltammograms (Table 22) of the mixture contained waves originating from the starting materials and also a new wave which was due to the reaction

$$[Fe(NO)S_4C_4(CN)_2Ph_2]^- + e^- \rightleftharpoons [Fe(NO)S_4C_4(CN)_2Ph_2]^{2-}$$

However, esr and ir spectral studies established that the initial reactants underwent a very fast electron transfer reaction:

$$[Fe(NO)S_4C_4(CN)_4]^- + [Fe(NO)S_4C_4Ph_4]^-$$
$$\rightarrow [Fe(NO)S_4C_4(CN)_4]^{2-} + [Fe(NO)S_4C_4Ph_4]^0$$

The last two species then underwent a relatively slow ligand exchange reaction accompanied by a further electron transfer reaction, giving $[Fe(NO)S_4C_4(CN)_2Ph_2]^-$. Unfortunately, attempts to isolate the mixed ligand species were unsuccessful.

VI. THE ELECTROCHEMISTRY OF ORGANOMETALLIC COMPOUNDS

Until recently, the electrochemical study of transition metal organo-metallic compounds has been a relatively neglected field, but interest in the subject has been stimulated by a remarkable series of papers published since 1966 by R. E. Dessy and his coworkers. Dessy's approach has been to survey the electrochemical properties of a wide range of complexes and to follow this up by a study in depth of a selection of organometallic systems. These systems have included metal-metal bonded species, acetylene and olefin metal carbonyl derivatives, and bridged bimetallic compounds containing, as the bridging atoms, S, P, or As. This section of the review has been so arranged as to permit discussion of a number of these systems and to outline their most important and interesting features.

Dessy and his colleagues have systematically investigated the organo-metallic compounds firstly by polarography in dimethoxyethane (DME), normally using $[n\text{-}Bu_4N][ClO_4]$ as the base electrolyte, and secondly by multiple triangular sweep studies, also in DME, to establish chemical and electrochemical reversibility in the system. This was followed by exhaustive controlled potential electrolysis at the appropriate potential and determination of the number of electrons involved in the polarographic step. Having reduced the organometallic species in solution, polarography of this reduced species was examined primarily to establish, by another method, reversibility of the system and to discover whether any reorganization of the ligands in the reduced species had taken place (if some structural reorganization had taken place, a change in half-wave potential was usually evident, *vide infra*). A similar technique was employed if the electrochemical process was oxidative. ESR, uv, and occasionally ir, spectral examinations were made of the reduced or oxidized solutions where appropriate, and then attempts were made to electrolyse the species in the oxidized or reduced solutions back to the starting materials. Finally, these re-electrolysed solutions were examined again by polarography, uv, ir, and esr spectroscopy for comparison with the initial species in solution.

Experimentally, the polarographic and controlled potential electrolysis work was carried out using an H-type cell.[68] The reference electrode was $Ag/AgClO_4$ and the base electrolyte, as already mentioned, $[n\text{-}Bu_4N][ClO_4]$. Cyclic voltammetry was effected using a hanging or sheared mercury drop assembly.

1. The General Electrochemical Behaviour of Organometallic Complexes

In this section, only those complexes are discussed which have not been investigated in detail. The electrochemistry of these compounds, however, do provide some interesting features.

In general, the radical anion, denoted RmQ^{-}, formed by the addition of an electron to the organometallic species RmQ, can react in nine different ways, as summarized in Fig. 21. In the majority of cases investigated by Dessy and his coworkers, the lower paths were followed.

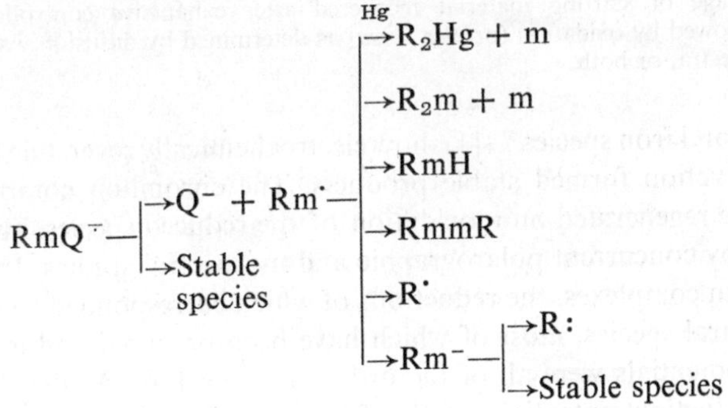

$$RmQ^{-} \begin{cases} \xrightarrow{} Q^{-} + Rm^{\cdot} \begin{cases} \xrightarrow{Hg} R_2Hg + m \\ \xrightarrow{} R_2m + m \\ \xrightarrow{} RmH \\ \xrightarrow{} RmmR \\ \xrightarrow{} R^{\cdot} \\ \xrightarrow{} Rm^{-} \begin{cases} \xrightarrow{} R: \\ \xrightarrow{} \text{Stable species} \end{cases} \end{cases} \\ \xrightarrow{} \text{Stable species} \end{cases}$$

Fig. 21

The electrochemical data obtained from a series of chromium, manganese, iron, titanium and vanadium compounds[69,70] containing π-arene or π-C_5H_5 ligands are summarized in Table 23. The chromium complexes 1–4, and the

Table 23. *Electrochemical Data Obtained from Some π-Arene and π-Cyclopentadienyl Complexes*

[*Data from References 69 and 70*]

Complex	$E^1_{1/2}$ (a)	$E^2_{1/2}$ (a)	n (b)	E_{pc} (c)	E_{pa} (c)	% OR recov. (d)
1. $[(\pi\text{-}C_6H_6)_2Cr]^+$	$-1\cdot3$		1	$-1\cdot4$	$-1\cdot1$	100
2. $[(\pi\text{-}C_5H_5)(\pi\text{-}C_7H_7)Cr]^+$	$-1\cdot2$		1	$-1\cdot4$	$-1\cdot1$	100
3. $[(\pi\text{-}C_6H_5Ph)(\pi\text{-}C_6H_6)Cr]^+$	$-1\cdot3$	$-3\cdot3$	1			100
4. $[(\pi\text{-anthracene})_2Cr]^+$	$-1\cdot4$		1	$-1\cdot6$	$-1\cdot3$	100
5. anthracene	$-2\cdot6$	$-3\cdot1$				
6. biphenyl	$-3\cdot3$					
7. $[(\pi\text{-mesitylene})Mn(CO)_3]^+$	$-0\cdot9$		1	$-1\cdot0$	$-0\cdot8$	50
8. $[(\pi\text{-mesitylene})(\pi\text{-}C_5H_5)Fe]^+$	$-2\cdot0$			$-2\cdot1$	$-1\cdot9$	
9. $[(\pi\text{-}C_6H_6)(\pi\text{-}C_5H_5)Fe]^+$	$-1\cdot9$	$-2\cdot9$	1	$-2\cdot0$	$-1\cdot8$	40
10. $[(\pi\text{-}C_6H_7)(\pi\text{-}C_5H_5)Fe]^0$	$-0\cdot6$		-1	none	$-0\cdot3$	
11. $(\pi\text{-}C_5H_5)_2TiCl_2$	$-1\cdot4$	$-2\cdot7$	1	$-1\cdot5$	$-1\cdot3$	
12. $(\pi\text{-}C_5H_5)_2VCl_2$	$-1\cdot0$	$-2\cdot3$	1			

(a) In volts, using three-electrode potentiostatic geometry, in dimethoxyethane vs. $Ag/AgClO_4$ with [n-Bu_4N][ClO_4] (0·1 M) as supporting electrolyte, at 25°C. Substrate concentration 2×10^{-3} M.

(b) Number of electrons established by controlled potential electrolysis and coulometry; positive values represent reductions and negative values oxidations.

(c) Peak currents, cathodic and anodic, respectively, at 1 V/sec triangular sweeps on hanging Hg drop using the starting material.

(d) Percentage of starting material recovered after exhaustive controlled potential reduction followed by oxidation (or *vice versa*) as determined by diffusion wave height or ultraviolet spectra, or both.

manganese and iron species 7–11, show electrochemically reversible behaviour, and on reduction formed stable products. The chromium complexes were shown to be regenerated on reoxidation of the reduced species, and this was confirmed by concurrent polarographic and uv spectral studies. In the series of chromium complexes, the reductions of which corresponded to the formation of neutral species, most of which have been prepared and isolated, the reduction potentials were all of the order of $-1\cdot3 \pm 0\cdot1$ V, despite the fact that the reduction potentials of the free arene ligands alone varied from

-2.6 V to beyond the wave caused by the breakdown of the base electrolyte. Although anthracene itself exhibited two reduction waves, $[(\pi\text{-anthracene})_2\text{Cr}]^+$ displayed only one reduction wave at a potential well removed from those of the hydrocarbon. Diphenyl was reduced in a one-electron step (at -3.3 V) whereas $[(\pi\text{-}C_6H_6)(\pi\text{-}C_6H_5Ph)\text{Cr}]^+$ could be reduced in two one-electron steps (at -1.3 and -3.3 V). There was a degree of similarity between the systems $[(\pi\text{-mesitylene})\text{Mn}(CO)_3]^+$ and $[(\pi\text{-arene})(\pi\text{-}C_5H_5)\text{Fe}]^+$: both species obeyed the rare gas rule (that is, the total number of electrons in the metal ion plus the number of electrons donated by the ligands to the metal equals the number of electrons in the succeeding rare gas, in this case, krypton (36 electrons)) and are therefore, in a sense, 'isoelectronic'. Both complexes underwent one-electron reductions affording $[(\pi\text{-mesitylene})\text{Mn}(CO)_3]^0$ and $[(\pi\text{-arene})(\pi\text{-}C_5H_5)\text{Fe}]^0$. ESR spectral examination of the latter revealed broad signals which were not calibrated. It is interesting that reaction of both of these cationic species with BH_4^-,[71,72] normally a mild reducing agent, resulted in the formation not of the neutral species but of π-cyclohexadienyl species, e.g. $[\pi\text{-}C_6Me_3H_4\text{Mn}(CO)_3]$ and $[(\pi\text{-}C_6H_7)(\pi\text{-}C_5H_5)\text{Fe}]$. The electrochemistry of the latter (10) revealed that it could be oxidized in a one-electron step, at a potential significantly different from (9), but the nature of this oxidized species was not determined.

The bis-π-cyclopentadienyl dichlorides of titanium and vanadium both underwent two one-electron reductions and it appeared that the sequence of events was

$$(\pi\text{-}C_5H_5)_2\text{MCl}_2 \rightarrow (\pi\text{-}C_5H_5)_2\text{MCl} \rightarrow (\pi\text{-}C_5H_5)_2\text{M}$$

Thus, $(\pi\text{-}C_5H_5)_2\text{TiCl}_2$, which is orange red in solution, was reduced to $(\pi\text{-}C_5H_5)_2\text{TiCl}$, which was pale green, and the esr signals obtained from the electrochemically generated species were similar to those obtained from the chemically prepared Ti^{III} complex. Further reduction afforded a blue solution which displayed weak esr signals probably arising from small amounts of $(\pi\text{-}C_5H_5)_2\text{TiCl}$: $(\pi\text{-}C_5H_5)_2\text{Ti}$ is blue and is believed to be dimeric and diamagnetic.[73] The reduction of the vanadium complexes proceeded similarly, although characterization of the various reduced products by esr techniques was not unequivocal.

Metal carbonyls and some of their derivatives containing Lewis bases underwent one- or two-electron reductions[69,70] (Table 24) at potentials averaging -2.8 ± 0.2 V; the exceptions to this were $[(\text{dipyridyl})\text{Mo}(CO)_4]$, whose first reduction wave occurred at -2.2 V, and $\text{Fe}(CO)_5$, where the potential was -2.14 V. The chromium complexes appeared to be reduced in a two-electron step, and it should be noted that $\text{Cr}(CO)_6$ can be reduced to $[\text{Cr}(CO)_5]^{2-}$ and $[\text{Cr}_2(CO)_{10}]^{2-}$ using sodium in liquid ammonia or sodium

Table 24. *Electrochemical Data Obtained from Some Transition Metal Carbonyl Complexes*

[*Data from References* 69 *and* 70]

Complex	$E^1_{1/2}$ (a)	$E^2_{1/2}$ (a)	n (b)	E_{pc} (c)	E_{pa} (c)	% OR recov. (d)
$Cr(CO)_6$	$-2 \cdot 7$		?	$-3 \cdot 3$	$-2 \cdot 0$	0
$\pi\text{-}C_6H_6Cr(CO)_3$	$-3 \cdot 0$		2	(e)		100
$Cr(CO)_5P(NMe_2)_3$	$-2 \cdot 9$		2			0
$Mo(CO)_6$	$-2 \cdot 7$		1	$-3 \cdot 2$	$-2 \cdot 0$	50
$Mo(CO)_5PPh_3$	$-2 \cdot 7$	$-3 \cdot 7$	2	$-3 \cdot 2$	$-2 \cdot 1$	0
$Mo(CO)_5P(NMe_2)_3$	$-2 \cdot 7$		1	$-3 \cdot 1$	$-0 \cdot 7$	0
$Mo(CO)_4(Me_2NCH_2CH_2NMe_2)$	$-3 \cdot 0$		1	$-3 \cdot 2$	none	0
$Mo(CO)_4$dipyridyl	$-2 \cdot 2$	$-2 \cdot 8$	1	$-2 \cdot 3$	$-1 \cdot 9$	80
$W(CO)_6$	$-2 \cdot 6$		1	$-3 \cdot 0$	$-2 \cdot 0$	20
$Fe(CO)_5$	$-2 \cdot 4$		1	$-2 \cdot 8$	$-1 \cdot 8$	5
$Ni(CO)_4$	$-2 \cdot 9$		1	$-3 \cdot 2$	$-1 \cdot 5$	0

(a) In volts, using three-electrode potentiostatic geometry, in dimethyoxyethane vs. $Ag/AgClO_4$ with $[n\text{-}Bu_4N][ClO_4]$ (0·1 M) as supporting electrolyte, at 25°C.

(b) Number of electrons established by coulometry and controlled potential electrolysis; positive values represent reductions.

(c) Peak currents, cathodic and anodic, respectively, at 1 V/sec triangular sweeps on hanging Hg drop using the starting material.

(d) Percentage of starting material recovered after exhaustive controlled potential reduction followed by oxidation (or *vice versa*) as determined by diffusion wave height or ultraviolet spectra, or both.

(e) Ill-defined wave.

amalgam in THF.[74] DipyridylMo(CO)$_4$ was reduced in two discrete one-electron steps, and the addition of the first electron resulted in the formation of a stable radical anion; the esr and ir spectra of such species are discussed later. Reduction of $(\pi\text{-}C_6H_6)Cr(CO)_3$ led to a stable yellow solution from which the starting material was recovered quantitatively on reoxidation. However, reoxidation of reduced solutions originally containing Mo(CO)$_6$, W(CO)$_6$, dipyridylMo(CO)$_4$, and Fe(CO)$_5$ (most of which can be smoothly reduced to carbonylate anions by chemical means) resulted in only partial recovery of the starting materials. This is not unreasonable since formation of carbonylate anions was usually accompanied by loss of CO, and reconstitution of the starting material (i.e. CO gain) could only be achieved at the expense of some of the reduced species.

The electrochemical data obtained from some π-cyclopentadienyl metal

Table 25. *Electrochemical Data Obtained from Some π-Cyclopentadienyl Metal Carbonyl Species*
[*Data from References* 69 *and* 70]

Complex	$E^1_{1/2}$ (a)	$E^2_{1/2}$ (a)	n (b)	E_{pc} (c)	E_{pa} (c)	% OR recov. (d)
π-C$_5$H$_5$Mn(CO)$_3$	$-3{\cdot}0$			$-3{\cdot}5$	(e)	0
π-C$_5$H$_5$Re(CO)$_3$	$-3{\cdot}3$			$-3{\cdot}5$	$-1{\cdot}5$	0
[π-C$_5$H$_5$Fe(CO)$_3$]$^+$	$-1{\cdot}4$	$-2{\cdot}2$	1	$-1{\cdot}6$	$-0{\cdot}2$	
[π-C$_5$H$_5$Fe(CO)$_2$NCMe]$^+$	$-0{\cdot}8$		1	$-1{\cdot}2$	$-0{\cdot}2$	
[π-C$_5$H$_5$Fe(CO)$_2$PPh$_3$]$^+$	$-1{\cdot}4$		1	$-1{\cdot}6$	$-0{\cdot}2$	
[π-C$_5$H$_5$Fe(CO)(PPh$_3$)$_2$]$^+$	$-1{\cdot}4$		1	$-1{\cdot}7$	none	
π-C$_5$H$_5$Mo(CO)$_3$CH$_2$Ph	$-2{\cdot}1$		1	$-2{\cdot}5$	$-1{\cdot}0$	0 (f)
π-C$_5$H$_5$Mo(CO)$_2$CH$_2$CO$_2$Me	$-2{\cdot}0$		1	$-2{\cdot}2$	$-1{\cdot}0$	0 (f)
π-C$_5$H$_5$Mo(CO)$_3$CF$_3$	$-2{\cdot}1$		?	$-2{\cdot}5$		0 (f)
π-C$_5$H$_5$Fe(CO)$_2$COMe	$-2{\cdot}5$		1	$\sim -2{\cdot}0$	(g)	0
π-C$_5$H$_5$Fe(CO)$_2$CH$_2$Ph	$-2{\cdot}5$		2			0
π-C$_5$H$_5$Fe(CO)$_2$C$_6$F$_5$	$-2{\cdot}3$	$-3{\cdot}0$	?	(g)	(g)	0
π-C$_5$H$_5$Fe(CO)$_2$COCH=CHPh	$-2{\cdot}4-$ $-2{\cdot}7$	(e)	2	$-2{\cdot}4$	$-1{\cdot}9$	

(a) In volts, using three-electrode potentiostatic geometry, in dimethoxyethane vs. Ag/AgClO$_4$ with [n-Bu$_4$N][ClO$_4$] (0·1 M) as supporting electrolyte, at 25°C.

(b) Number of electrons established by coulometry; positive values represent reduction.

(c) Peak currents, cathodic and anodic, respectively at 1 V/sec triangular sweeps on hanging Hg drop using the starting material.

(d) Percentage of starting material recovered after exhaustive controlled potential reduction followed by oxidation (or *vice versa*) as determined by diffusion wave height or ultraviolet spectra, or both.

(e) Ill-defined wave.

(f) New wave at $-1{\cdot}3$ V appeared after oxidation-reduction cycle.

(g) Complex wave.

carbonyl species[69,70] is recorded in Table 25. It may be seen that the $E_{1/2}$-values for the reduction of π-C$_5$H$_5$M(CO)$_3$ (M = Mn or Re) decreased as the metal becomes heavier and, while [π-C$_5$H$_5$Fe(CO)$_3$]$^+$ is formally iso-electronic with π-C$_5$H$_5$Mn(CO)$_3$, the reduction potential of the former was 1·6 V more anodic than that of the latter, possibly because the iron species was positively charged. The most noticeable feature about the series [π-C$_5$H$_5$Fe(CO)$_n$L$_{2-n}$]$^+$ was the similarity between reduction potentials of the various members, and also that the product on reoxidation of the reduced species was invariably [π-C$_5$H$_5$Fe(CO)$_2$]$_2$. The molybdenum complexes, π-C$_5$H$_5$Mo(CO)$_3$R, which formally contains Mo^{II} and σ-bonded R (alkyl) groups, had $E_{1/2} = -2\cdot0 \pm 0\cdot1$ V. These reductions occurred in one-electron steps and the product in each case was $[\pi\text{-}C_5H_5Mo(CO)_3:]^-$ and R$\cdot$. The analogous $\pi\text{-}C_5H_5Fe(CO)_2R$ behaved similarly, $E_{1/2}$ averaging $-2\cdot4$ $\pm0\cdot1$ V, but the number of electrons involved in the reduction steps varied, being either 1 or 2. However, the product was always $[\pi\text{-}C_5H_5Fe(CO)_2:]^-$.

Reduction of the Group VI metal tetracarbonyl olefin complexes[69,70] (Table 26) appeared to occur near $-2\cdot8$ V (although $C_7H_8Mo(CO)_4$ had

Table 26. *Electrochemical Data Obtained from Some Cyclic Olefins and Their Transition Metal Carbonyl Complexes*
[*Data from References 69 and 70*]

Complex	$E^1_{1/2}$ (a)	$E^2_{1/2}$ (a)	n (b)	E_{pc} (c)	E_{pa} (c)	% OR recov. (d)
$C_7H_8Cr(CO)_4$ (e)	$-2\cdot8$		2	$-3\cdot2$	$-0\cdot8$	0
$C_7H_8Mo(CO)_4$ (e)	$-2\cdot3$			(f)	(f)	
$C_7H_8W(CO)_4$ (e)	$-2\cdot8$		1	$-3\cdot1$	(g)	0
$C_7H_8Mo(CO)_3$ (h)	$-2\cdot3$		1	$-2\cdot5$	$-1\cdot0$	0
$C_8H_{12}W(CO)_4$ (i)	$-2\cdot9$		1	$-3\cdot2$	(g)	0
$C_8H_8Fe(CO)_3$ (j)	$-2\cdot0$	$-2\cdot5$	1	$-2\cdot2$	$-1\cdot7$	30
$C_6H_8Fe(CO)_3$ (k)	$-2\cdot7$		1	$-3\cdot2$	$-1\cdot2$	70
$C_4H_6Fe(CO)_3$ (l)	$-2\cdot6$		1	$-2\cdot7$	(g)	0
norbornadiene, C_7H_8	$-3\cdot1$					
cycloheptatriene, C_7H_8	$-3\cdot2$					
cycloöcta-1,5-diene, C_8H_{12}	(m)					
cycloöctatetraene, C_8H_8	$-2\cdot5$					

(a) In volts, using three-electrode potentiostatic geometry, in dimethyoxyethane vs. $Ag/AgClO_4$ with $[n\text{-}Bu_4N][ClO_4]$ (0·1 M) as base electrolyte at 25°C.
(b) Number of electrons established by coulometry; positive values represent reduction.
(c) Peak currents, cathodic and anodic, respectively at 1 V/sec triangular sweeps on hanging Hg drop using the starting material.
(d) Percentage of starting material recovered after exhaustive controlled potential reduction followed by oxidation (or *vice-versa*) as determined by diffusion wave height or ultraviolet spectra, or both.
(e) Norbornadiene.
(f) Decomposed in the solvent.
(g) Ill-defined waves.
(h) Cycloheptatriene.
(i) Cycloöctadiene (1:5).
(j) Cycloöctatetraene.
(k) Cyclohexa-1,3-diene.
(l) Buta-1,3-diene.
(m) No wave observed.

Table 27. *Electrochemical Data Obtained from Bimetallic and Polymetallic Species*
[*Data from Reference* 76]

Complex (abbreviation)	$E_{1/2}$ (a)	n (b)	Products
$[\pi\text{-}C_5H_5Mo(CO)_3]_2$ (mo-mo)	$-1\cdot4$	2	2 mo:$^-$
$\pi\text{-}C_5H_5Mo(CO)_3Fe(CO)_2\pi\text{-}C_5H_5$ (mo-fe)	$-1\cdot4$	1	mo:$^-$ + ?
$\pi\text{-}C_5H_5Mo(CO)_3SnMe_3$ (mo-sn)	$-1\cdot9$	1	mo:$^-$ + sn$^\cdot$
$\pi\text{-}C_5H_5Mo(CO)_3SnPh_3$ (mo-sn)	$-2\cdot4$	1	mo:$^-$ + sn$^\cdot$
$\pi\text{-}C_5H_5Mo(CO)_3PbPh_3$ (mo-pb)	$-2\cdot2$	2	mo:$^-$ + pb:$^-$
$Mn_2(CO)_{10}$ (mn-mn)	$-1\cdot7$	2	2 mn:$^-$
$Re_2(CO)_{10}$ (re-re)	$-2\cdot3$	2	2 re:$^-$
$\pi\text{-}C_5H_5Fe(CO)_2Mn(CO)_5$ (fe-mn)	$-1\cdot6$	1	mn:$^-$ + fe$^\cdot$
$Mn(CO)_5SnMe_3$ (mn-sn)	$-1\cdot9$	1	mn:$^-$ + sn$^\cdot$
$Mn(CO)_5SnPh_3$ (mn-sn)	$-2\cdot5$	1	mn:$^-$ + sn$^\cdot$
$Re(CO)_5SnPh_3$ (re-sn)	$-2\cdot5$	1	unknown
$Mn(CO)_5PbEt_3$ (mn-pb)	$-1\cdot8$	1	mn:$^-$ + pb:$^-$
$Mn(CO)_5PbPh_3$ (mn-pb)	$-2\cdot1$	2	mn:$^-$ + pb:$^-$
$Re(CO)_5PbPh_3$ (re-pb)	$-2\cdot4$	1	unknown
$[\pi\text{-}C_5H_5Fe(CO)_2]_2$ (fe-fe)	$-2\cdot2$	2	2 Fe:$^-$
$[\pi\text{-}C_5H_5Ru(CO)_2]_2$ (ru-ru)	$-2\cdot6$	2	2 ru:$^-$
$\pi\text{-}C_5H_5Fe(CO)_2SnPh_3$ (fe-sn)	$-2\cdot6$	1	fe:$^-$ + sn$^\cdot$
$\pi\text{-}C_5H_5Fe(CO)_2PbPh_3$ (fe-pb)	$-2\cdot1$	2	fe:$^-$ + pb:$^-$
$Co_2(CO)_8$ (co-co)	$-0\cdot9$	2	2 co:$^-$
$Co(CO)_4SnMe_3$ (co-sn)	$-1\cdot6$	1	co:$^-$ + sn$^\cdot$
$Co(CO)_4SnPh_3$ (co-sn)	$-1\cdot6$	1	co:$^-$ + sn$^\cdot$
$[\pi\text{-}C_5H_5Ni(CO)]_2$ (ni-ni)	$-2\cdot4$	2	2 ni:$^-$
$Ph_3SnSnPh_3$ (sn-sn)	$-2\cdot9$	2	2 sn:$^-$
$Ph_3PbPbPh_3$ (pb-pb)	$-2\cdot0$	2	2 pb:$^-$
$[\pi\text{-}C_5H_5Mo(CO)_3]_2SnMe_2$ (mo-sn-mo)	$-1\cdot8$	2	2 mo:$^-$ + Me$_2$Sn
$[\pi\text{-}C_5H_5Fe(CO)_2]_2SnMe_2$ (fe-sn-fe)	$-2\cdot7$	2	2 Fe:$^-$ + Me$_2$Sn
$[\pi\text{-}C_5H_5Mo(CO)_3]_2Sn[\pi\text{-}C_5H_5Fe(CO)_2]_2$	$-1\cdot8,$ $-2\cdot0,$ $-2\cdot5$		
$[\pi\text{-}C_5H_5Cr(CO)_3]_2Hg$ (cr-hg-cr)	$-1\cdot3$	2	2 cr:$^-$ + Hg
$[\pi\text{-}C_5H_5Mo(CO)_3]_2Hg$	$-1\cdot3$	2	2 mo:$^-$ + Hg
$[Mn(CO)_5]_2Hg$ (mn-hg-mn)	$-1\cdot1$	2	2 mn:$^-$ + Hg
$[\pi\text{-}C_5H_5Fe(CO)_2]_2Hg$	$-2\cdot0$	2	2 fe:$^-$ + Hg
$Fe_2(CO)_9$ (fe-(CO)$_3$-fe)	$-2\cdot4$	2	stable dianion
$[\pi\text{-}C_5H_5Co(CO)]_3$	$-1\cdot6$	1	stable anion
$(\pi\text{-}C_5H_5)_3Ni_3(CO)_2$	$-1\cdot5$	1	stable anion

(a) In volts, in dimethoxyethane vs. Ag/AgClO$_4$ with [n-Bu$_4$N][ClO$_4$] (0·1 M) as supporting electrolyte at 25°C.

(b) Number of electrons established by coulometry; positive values represent reduction.

16

$E_{1/2} = -2 \cdot 3$ V) and was an irreversible process; [cycloheptatrieneMo(CO)$_3$] and [cycloöcta-1,5-dieneW(CO)$_4$] behaved similarly. These potentials were somewhat anodic when compared with the reduction potentials of the free olefins. Cycloöctatetraene iron tricarbonyl, $C_8H_8Fe(CO)_3$ and its 1,3-cyclohexadiene (C_6H_8) analogue both provided stable radical anions on reduction, and these were characterized by their esr spectra. It is noteworthy that $C_8H_8Fe(CO)_3$ (in which one butadiene fragment is not involved in bonding to the Fe(CO)$_3$ group) was reduced in two one-electron steps and that the second reduction potential was identical to that of the free hydrocarbon.

2. Metal-Metal Bonded Systems

There has been considerable interest in the formation of metal–metal bonded species,[75] both of the homobimetallic and heterobimetallic type. There are three major problems associated with a synthetic approach to the synthesis of metal–metal bonds. These are that in a reaction of m–x with m′ : ‾ or m′–x with m: ‾, which is designed to produce m–m′, there must be a favourable ΔF^0, or a strong metal–metal bond. An approach to evaluation of bond strength could be made by investigating the electrochemical fission of such bonds in the hope that there may be a correlation between $E_{1/2}$ (reduction) and bond strength. Dessy and his colleagues studied[76] the electrochemistry of a number of metal–metal bonded organometallics with this end in view.

In systems containing metal–metal bonds (Table 27), there were two common and exclusive routes for the electrochemical reduction of homo- and hetero-bimetallic species (Fig. 22). These involved either addition of two

$$m-m' \begin{cases} \xrightarrow{2e^-} m:^- + m': \\ \xrightarrow{1e^-} m:^- + m'' \end{cases}$$

Fig. 22

electrons to the bimetallic system followed by scission to give two monometallic anions, or addition of only one electron leading to the formation of one anion and a monometallic radical species. Several clear patterns of electrochemical behaviour in these bimetallic systems could be detected. Thus, as the periodic table was descended, the reduction of homobimetallic systems required increasingly negative potentials, and, in heterobimetallic species, the half-wave potentials for the reduction of the species fell into three distinct groups—those which reduced at potentials more anodic than either of their parent homobimetallic species, those which reduced at potentials more cathodic, or those which reduced at values between those of their homobimetallic parents.

Systems reducing at potentials more anodic than either of their homobimetallic parents included

$$\pi\text{-}C_5H_5Mo(CO)_3Fe(CO)_2\pi\text{-}C_5H_5 \quad \text{and} \quad \pi\text{-}C_5H_5Fe(CO)_2Mn(CO)_5$$

The reduction processes involved only one electron and the residual radical in the latter system was $[\pi\text{-}C_5H_5Fe(CO)_2]^-$, but the fate of the radical in the former system, where $[\pi\text{-}C_5H_5Mo(CO)_3:]^-$ was formed, was unknown. $\pi\text{-}C_5H_5Mo(CO)_3PbPh_3$ and $Mn(CO)_5PbPh_3$ were reduced at potentials more cathodic than either of their homobimetallic parents, and the processes involved two electrons. The large majority of complexes, however, underwent reduction at potentials intermediate to those of their parent homobimetallics. In all of these electrode reactions only one electron was involved, except in the case of $\pi\text{-}C_5H_5Fe(CO)_2PbPh_3$, where $n = 2$, and it was interesting that the reduction potentials of $[\pi\text{-}C_5H_5Fe(CO)_2]_2$ and Ph_6Pb_2 differed by only 0·2 V. In the $\pi\text{-}C_5H_5Mo(CO)_3L$ and $Mn(CO)_5L$ systems, where L = $SnMe_3$ or $SnPh_3$, it was found that the reduction potentials of the trimethyl derivatives were positive relative to those of the triphenyl tin species. By extending the ideas of Lewis and Nyholm[77] on the stability of metal–metal bonds, Dessy and his colleagues suggested that in heterobimetallic compounds, where non-bonding d-electron interaction could lead to metal–metal bond instability, electron-withdrawing groups should reduce such interactions and thereby lead to increased stability, whereas electron-releasing groups would tend to increase such interactions and destabilize the m–m' bond.

Reductions of the linear trimetallic systems, m–Hg–m led to the formation of $2m:^-$ and to the release of mercury at potentials some 0·2–0·4 V more anodic than the $E_{1/2}$-values of the parent m–m systems. In the trimetallic tin complexes, $[\pi\text{-}C_5H_5Mo(CO)_3]_2SnMe_2$ and $[\pi\text{-}C_5H_5Fe(CO)_2]_2SnMe_2$, two-electron reduction took place, releasing $[\pi\text{-}C_5H_5M(CO)_n:]^-$ and $SnMe_2$, and the reduction potentials occurred at values very close to those in $\pi\text{-}C_5H_5Mo(CO)_3SnMe_3$ and its iron analogue. Reduction of the pentametallic species $[\pi\text{-}C_5H_5Mo(CO)_3]_2Sn[Fe(CO)_2\pi\text{-}C_5H_5]_2$ occurred in discrete steps, at $-1\cdot8$, $-2\cdot0$, and $-2\cdot5$ V, and it was suggested that the processes corresponded to 'electrochemical amputation' of the transition metal organometallic sidearms of the tin. Certainly, the reduction potentials occurred at values similar to those in molybdenum and iron tin derivatives.

Three of the four cluster compounds listed in Table 27 could be reduced reversibly in one-electron steps, and the radical anions so formed were apparently stable.

From the accumulated data, it seems that there may be a relationship between bond strength and reduction potential. While actual bond strengths of transition metal species are unknown, it seems likely, particularly from mass spectral data, that bond stability increases as the metals within any one group become heavier, and there is a definite parallel between this and the half-wave (reduction) potentials.

Another very important factor involved in a successful preparation of a bimetallic species, m–m', is that there should be a reasonable rate of reaction between $m:^-$ and m'–x or $m':^-$ and m–x (i.e. $\Delta F^{\pm}_{m-m'}$ low). This means, among other things, that the organometallic anion, $m:^-$, must have a relatively high nucleophilicity. In order to assess the nucleophilic character of the transition metal species discussed in this section, the rates of reaction of $m:^-$ with simple aliphatic halides, RX, were measured electrochemically.[78] By means of the group reactions $m:^- +$ RX and R'X, and $m':^- +$ R'X and R"X, the relative nucleophilicities of the series of organometallic anions were estimated semiquantitatively.

The experimental results were obtained by first dissolving the parent homo-bimetallic complex in DME containing the supporting electrolyte at milli-molar concentrations. This solution was added to the cathode compartment of the electrolytic cell (of H-type) and, after carrying out a polarographic measurement to establish the reduction potential, a voltage was chosen, which was well up on the diffusion plateau, so that controlled potential electrolysis was effected (a large mercury pool was used in the cathode compartment). After electrolysis was complete, a rotating platinum electrode was inserted into the cathode compartment and well-defined limiting oxidative currents for each anion could be observed, and $E_{1/2}$-values easily determined. It was shown that, with the organometallic anions $[\pi\text{-}C_5H_5Fe(CO)_2:]^-$, $[\pi\text{-}C_5H_5Mo(CO)_3:]^-$, and $[Mn(CO)_5:]^-$ over the range 5×10^{-3} to 0.5×10^{-3} M concentrations, the current varied linearly with concentration (these data were obtained using an XY recorder). For kinetic experiments, a magnetic stirrer was inserted into the cathode compartment and the limiting current was set at a full-scale deflection on the y-axis of the XY recorder. The desired substrate (RX) was then rapidly injected into the cathode compartment and simultaneously the time base on the x-axis of the recorder was started. In this way, a plot of anion concentration vs. time was obtained. In almost all cases, the recorded currents fell smoothly to zero, and the calculations, using the appropriate rate equations, showed that most of the reactions obeyed second order kinetics. Under pseudo-first order conditions (excess of RX), plots within a run showed good first-order behaviour and a constant k_2 was recorded between runs.

Table 28 shows the observed k_2 values for the anions studied and Table 29 shows a series of second order rate constants, k_2', normalized relative to $[Co(CO)_4:]^-$, using paired groups as previously described. The rate ratios for calculation of k_2' values from the observed k_2 values were Br:I = 1:50 and

Table 28. *Rate Constant Data Obtained for the Reactions between Organometallic Anions and Alkyl Halides* (e)

[*Data from Reference* 78]

Organometallic Anion	Alkyl Halide	Concentration (a)	k_2 (b)
$[\pi\text{-}C_5H_5Fe(CO)_2:]^-$	EtBr	$2{\cdot}0 \times 10^{-3}$ M (A & O)	139
	$i\text{-}PrBr$	$2{\cdot}0 \times 10^{-3}$ M (O, A)	0·17
$[\pi\text{-}C_5H_5Ru(CO)_2:]^-$	EtBr	$2{\cdot}0 \times 10^{-3}$ M (A & O)	15
$[\pi\text{-}C_5H_5Ni(CO):]^-$	EtBr	$2{\cdot}0 \times 10^{-3}$ M (A & O)	10·9
$[Re(CO)_5:]^-$	MeI	$2{\cdot}0 \times 10^{-3}$ M (O, A)	254
$[\pi\text{-}C_5H_5W(CO)_3:]^-$	MeI	$2{\cdot}0 \times 10^{-3}$ M (O, A)	(5) (c)
$[Mn(CO)_5:]^-$	MeI	$2{\cdot}0 \times 10^{-3}$ M (O, A)	0·77
$[\pi\text{-}C_5H_5Mo(CO)_3:]^-$	MeI	$2{\cdot}0 \times 10^{-3}$ M (O, A)	0·67
$[\pi\text{-}C_5H_5Cr(CO)_3:]^-$	MeI	$4{\cdot}0 \times 10^{-3}$ M (O)	0·04
		$4{\cdot}0 \times 10^{-1}$ M (A)	
$[Co(CO)_4:]^-$	MeI	$5{\cdot}0 \times 10^{-3}$ M (O)	0·01
		$5{\cdot}0 \times 10^{-1}$ M (A)	
$[Cr(CO)_5CN:]^-$	MeI	$4{\cdot}0 \times 10^{-3}$ M (O)	(d)
		$4{\cdot}0 \times 10^{-1}$ M (A)	
$[Ph_3Sn:]^-$	EtBr	$2{\cdot}0 \times 10^{-3}$ M (A & O)	22
$[Ph_3Pb:]^-$	MeI	$2{\cdot}0 \times 10^{-3}$ M (A & O)	490
	EtI	$2{\cdot}0 \times 10^{-3}$ M (O, A)	2·64
	EtBr	$2{\cdot}0 \times 10^{-3}$ M (O, A)	0·06

(a) Abbreviations: A & O = alkyl halide and organometallic anion concentration, O, A = organometallic anion concentration, alkyl halide concentration, A = alkyl halide concentration, O = organometallic concentration.

(b) Rate constant, whose values represent averages for three or more runs usually differing by 10 per cent or less; in $M^{-1}\,sec^{-1}$.

(c) Because of overlap of two oxidative polarographic waves of this anion, an accurate value was not determined.

(d) Too slow to measure, as also for Mo and W analogues.

(e) For reaction $m:^- + RX \to m - R + X^-$ in dimethoxyethane at $25°$ with $0{\cdot}1$ M $[n\text{-}Bu_4N][ClO_4]$.

Me:Et:i-Pr = 100:1:0·001. The span of rates was so large that no single substrate could cover the series.

The reactions were summarized as m:$^-$ + RX → mR + X:$^-$, and although the products of these reactions were not isolated, gas chromatographic analyses of the solutions after a considerable time following reaction commencement revealed no trace of RX when equal concentrations of m:$^-$ and RX were used. Furthermore, there was no evidence for products from R˙ due to abstraction, disproportionation and/or coupling. Polarographic investigation disclosed no evidence of m–m or m–H and it was therefore concluded

Table 29. *Relative Rate Constant Data Obtained for the Reactions Between Organometallic Anions and Alkyl Halides* (a)

[*Data from Reference* 78]

Organometallic Anion	k_2' (b)	$E_{1/2}$ (c)	$E_{1/2}$ (d)
$[\pi\text{-}C_5H_5Fe(CO)_2:]^-$	$7{\cdot}0 \times 10^7$	$-2{\cdot}2$	$-1{\cdot}6$
$[\pi\text{-}C_5H_5Ru(CO)_2:]^-$	$7{\cdot}5 \times 10^6$	$-2{\cdot}6$	$-1{\cdot}5$
$[\pi\text{-}C_5H_5Ni(CO):]^-$	$5{\cdot}5 \times 10^6$	$-2{\cdot}4$	$-1{\cdot}4$
$[Re(CO)_5:]^-$	$2{\cdot}5 \times 10^4$	$-2{\cdot}3$	$-0{\cdot}8$ to $-1{\cdot}4$ (e)
$[\pi\text{-}C_5H_5W(CO)_3:]^-$	~500		$-1{\cdot}0$
$[Mn(CO)_5:]^-$	77	$-1{\cdot}7$	$-0{\cdot}55$
$[\pi\text{-}C_5H_5Mo(CO)_3:]^-$	67	$-1{\cdot}4$	$-0{\cdot}55$
$[\pi\text{-}C_5H_5Cr(CO)_3:]^-$	4	$-1{\cdot}3$	$-0{\cdot}80$
$[Co(CO)_4:]^-$	1	$-1{\cdot}0$	$-0{\cdot}20$
$[Cr(CO)_5(CN):]^-$	$<0{\cdot}01$ (f)		ca.$-0{\cdot}2$
$[Ph_3Sn:]^-$	$1{\cdot}1 \times 10^8$	$-2{\cdot}9$	$-0{\cdot}95$
$[Ph_3Pb:]^-$	$2{\cdot}6 \times 10^4$	$-2{\cdot}0$	$-0{\cdot}90$

(a) For reaction $m:^- + RX \rightarrow m - R + X^-$ at 25° in dimethoxyethane, with 0·1 M $[n\text{-}Bu_4N][ClO_4]$; normalized to $[Co(CO)_4:]^-$ being unit rate.
(b) Relative rate constant.
(c) Reduction potential (dropping mercury electrode) for $m - m$, in volts, in dimethoxyethane vs. $Ag/AgClO_4$, with $[n\text{-}Bu_4N][ClO_4]$ (0·1 M) as base electrolyte.
(d) Oxidation potential (rotating Pt electrode), in volts, as (c).
(e) Several pauses in step.
(f) Too slow to measure accurately, as for Mo and W analogues.

that the rate data represented a reasonable measurement of the nucleophilicity of the anions m:⁻.

The range of relative rates exceeded 10^9, with $[Co(CO)_4:]^-$ near one extreme and $[\pi\text{-}C_5H_5Fe(CO)_2:]^-$ near the other. From the data it was clear that in the Group VI and VII transition metal series, the nucleophilicity of the organometallic anion increased as the metal, within each group, became heavier. This was entirely consistent with the $E_{1/2}$-values for the reduction of m–m, which, it may be recalled, also became more cathodic as the metal became heavier. However, the reverse is true of the kinetic data for $[\pi\text{-}C_5H_5Fe(CO)_2:]^-$ and its ruthenium analogue. These data may be correlated with several chemical experiments, notably those of Parshall[79] who attempted to establish a qualitative series of nucleophilicities by using the equilibrium constant of the reaction

$$[M(CO)_n:]^- + BH_3 \overset{K}{\rightleftharpoons} [M(\rightarrow BH_3)(CO)_n]^-$$

Although no quantitative measurement of K could be made, a relative order of nucleophilicity was found to be $[Re(CO)_5]^- > [Mn(CO)_5]^- \gg [Co(CO)_4]^-$. Stone and his colleagues have suggested[80] that in the reactions of $[\pi\text{-}C_5H_5Fe(CO)_2:]^-$ and $[Re(CO)_5:]^-$ with fluoroolefins and fluoroaromatics, in which $F:^-$ is displaced, the nucleophilicity of the former is greater than the latter. The correspondence of these chemical facts with the electrochemically measured nucleophilicities is encouraging, and suggests that the latter could be used practically.

The final important factor involved in metal–metal bond formation is that the free energy of activation for the process $m:^- + m'x \rightarrow m\text{--}m' + x:^-$ or $m':^- + mx$ must be lower than the activation free energy for any other process that $m:^-$ and $m'x$ (or $m':^-$ and mx) may utilize in the interaction which leads to any product other than $m\text{--}m'$. In an attempt to elucidate this factor, the interaction between $m:^-$, generated by electrochemical scission of $m\text{--}m$ at low concentration (using $[n\text{-}Bu_4N]^+$) as counter-ion and $m'x$, also at low concentration, was studied[81] in order to establish what product, if any, was formed. Some of the possibilities of reaction are given in Fig. 23. The

$$m:^- + m'x \rightarrow \begin{cases} m\text{--}m' + x^- & S_N \text{ substitution (x displacement)} \\ m^{\cdot} + m'' + x^- & \text{Electron transfer} \\ mx + m':^- & \text{Charge–x interchange} \\ m\text{--}m'\text{--}x & S_N \text{ substitution (ligand displacement)} \end{cases}$$

Fig. 23

identity of the heterobimetallic products from direct displacement of x could be established at low concentrations (10^{-3} M) from the extensive polarographic data already accrued from the study of the species, and they could also be characterized by their uv spectra. The reaction may, in general, proceed in two ways, depending on the lives of the radicals formed in the electron transfer process (Fig. 24). If the radicals were short-lived, and the

$$m:^- + m'x \rightarrow m^{\cdot} + m'' + x^- \xrightarrow{\text{caged process}} m\text{--}m'$$

$$\downarrow \text{ non-caged process}$$

$$m\text{--}m', m\text{--}m, m'\text{--}m' \text{ and/or m–H type products}$$

Fig. 24

process was 'caged', the heterobimetallic species were readily formed. If, however, the radicals were long-lived, then a variety of products ensued. A further complication might arise because of a redistribution reaction

$$m:^- + m'x \rightarrow m\text{--}m' \xrightarrow{m:^-} m\text{--}m + m':^-$$

Chart 1

	$[\pi\text{-}C_5H_5Fe(CO)_2\text{:}]^-$	$[Ph_3Sn\text{:}]^-$	$[Ph_3Pb\text{:}]^-$	$[Mn(CO)_5\text{:}]^-$	$[\pi\text{-}C_5H_5Mo(CO)_3\text{:}]^-$
$\pi\text{-}C_5H_5Fe(CO)_2I$ (feI) Ph_3SnCl (snCl)	fe-fe fe-sn	sn-fe sn-sn	pb-fe ?	no rn. (a) mn-mn sn-sn mn-sn	no rn. (a) mo-sn
$Ph_3PbOCOMe$ (pbOAc)	fe-pb	sn-sn pb· Ph_4Pb	pb· Ph_4Pb	mn-pb	not mo-pb
$Mn(CO)_5Cl$ (mn-Cl)	(b)	sn-sn "mn"	not mn-pb	no rn. (a)	no rn. (a)
$\pi\text{-}C_5H_5Mo(CO)_3Cl$ (moCl)	no rn. (a)	sn-sn mo:$^-$ moCl	pb-pb mo:$^-$ moCl	mn-mn mo:$^-$ moCl	no rn. (a)

(a) No reaction.
(b) Product not characterized.

Table 30. *Characterization of Some Bimetallic Species*

Complex	$E_{1/2}$ (a)	n (b)	λ_{max} (c)
$[\pi\text{-}C_5H_5Fe(CO)_2]_2$	$-2\cdot2$	2	344
Ph_6Sn_2	$-2\cdot9$	2	247
Ph_6Pb_2	$-2\cdot0$	2	
$[Mn(CO)_5]_2$	$-1\cdot7$	2	342, 276
$[\pi\text{-}C_5H_5Mo(CO)_3]_2$	$-1\cdot4$	2	386
$\pi\text{-}C_5H_5Fe(CO)_2I$	$-1\cdot2, -2\cdot2$	1	350
Ph_3SnCl	$-1\cdot6$	1	
$Ph_3PbOCOCH_3$	$-1\cdot4, -2\cdot2$	1	
$Mn(CO)_5Cl$	$-2\cdot1 (-2\cdot4)$	2	280
$\pi\text{-}C_5H_5Mo(CO)_3Cl$	$-1\cdot3$	2	234, 324
$\pi\text{-}C_5H_5Fe(CO)_2SnPh_3$	$-2\cdot6$	2	298
$\pi\text{-}C_5H_5Fe(CO)_2PbPh_3$	$-2\cdot1$	2	327
$\pi\text{-}C_5H_5Fe(CO)_2Mn(CO)_5$	$-1\cdot5, -2\cdot1$	1	384
$\pi\text{-}C_5H_5Fe(CO)_2Mo(CO)_3\pi\text{-}C_5H_5$	$-1\cdot4$	1	382, 286, 252
$\pi\text{-}C_5H_5Mo(CO)_3SnPh_3$	$-2\cdot4$	1	282
$\pi\text{-}C_5H_5Mo(CO)_3PbPh_3$	$-2\cdot2$	2	
$Mn(CO)_5SnPh_3$	$-2\cdot5$	1	
$Mn(CO)_5PbPh_3$	$-2\cdot1$	2	295
Ph_3SnH	$-3\cdot1$	1	
$Mn(CO)_5H$	No wave		
Ph_4Pb	$-3\cdot2$		

(a) In volts, in dimethoxyethane, vs. $Ag/AgClO_4$ with $[n\text{-}Bu_4N][ClO_4]$ (0·1 M) as base electrolyte at 25°C.
(b) Number of electrons at $E_{1/2}$ established by coulometry.
(c) In mμ, ultraviolet spectrum recorded in dimethoxyethane.

This topic was investigated[82] separately from the main study of the m–m′ bond formation. Some light could be thrown on the nature of the reactions by investigating the commutation reactions m:⁻ + m′x and m′: + mx, and from this, some data leading to the ideal method of synthesis of m–m′ might be expected.

The experimental approach to this study[81] was similar to that used in the kinetic investigations. In the cathode compartment, the homobimetallic species m–m was reduced polarographically (to establish $E_{1/2}$) and was then reduced electrolytically to produce m:⁻. A polarogram of the reduced solution confirmed that only m:⁻ was present and then the other partner of the series, m′x, was rapidly injected. The polarogram of the solution mixture was then recorded, and if the reaction was not complex, a new wave corresponding to the quantitative formation of m–m′, appeared. Occasionally, small waves, often due to mx and m′–m′, were evident.

In Chart 1, the conclusions of this research are presented, and the supporting data is given in Tables 30 and 31. In the symmetrical reactions m:⁻ + mx, it was evident that the drop in nucleophilic character in the sequence

$$[\pi\text{-}C_5H_5Fe(CO)_2:]^-$$
$$\cong ([Ph_3Sn:]^-) > ([Ph_3Pb:]^-) > [Mn(CO)_5:]^- > [\pi\text{-}C_5H_5Mo(CO)_3:]^-$$

was largely responsible for the failure of the reactions

$$[\pi\text{-}C_5H_5Mo(CO)_3:]^- + \pi\text{-}C_5H_5Mo(CO)_3Cl$$

and

$$[Mn(CO)_5:]^- + Mn(CO)_5Cl$$

In the reactions involving $[\pi\text{-}C_5H_5Fe(CO)_2:]^-$ and $[Ph_3Sn:]^-$, product formation was very rapid, as expected from nucleophilicity studies ($k_2 > 10$). The reactions between $[\pi\text{-}C_5H_5Fe(CO)_2:]^-$ and Ph_3SnCl and Ph_3PbOAc (Ac = acyl), and between $[Ph_3Sn:]^-$ and $[Ph_3Pb:]^-$ and $\pi\text{-}C_5H_5Fe(CO)_2I$ commuted, the heterobimetallic complexes being formed either by an S_N displacement or a caged reaction involving electron transfer pathways; neither alternative could be distinguished. It is interesting that $\pi\text{-}C_5H_5Fe(CO)_2PbPh_3$ and $\pi\text{-}C_5H_5Fe(CO)_2SnPh_3$ were prepared in large scale electrolytic reactions. The hexaphenyldilead was reduced in DME containing $[n\text{-}Bu_4N][ClO_4]$ in a large cell containing a mercury pool and a stirrer. The metal compound was exhaustively electrolysed in the anode compartment (at $-2\cdot1$ V), thereby generating $2[Ph_3Pb:]^-$, and then $\pi\text{-}C_5H_5Fe(CO)_2I$, dissolved in degassed DME, was injected using a syringe. After 30 min of stirring, the mercury pool was discarded, ether added to the DME to effect precipitation of $[n\text{-}Bu_4N]$ $[ClO_4]$, and the organometallic species recovered from the DME/ether

Table 31. *Formation of Bimetallic Species in Commutation Experiments and the Identification of the Products*

Reaction (a)	$E_{1/2}$ (b)	λ_{max} (c)	Other Observations	Conclusion
fe:$^-$ + feI	$-1\cdot2$			feI (unreacted)
	$-2\cdot2$	344		fe-fe
	(d)			
sn:$^-$ + snCl	$-2\cdot9$	247		sn-sn
pb:$^-$ + pbOAc	$-2\cdot2$		initial wave at $-2\cdot2$ slowly	
	$-3\cdot2$		disappears as wave at $-3\cdot2$ appears	PbPh$_4$
mn:$^-$ + mnCl	$-2\cdot1$		mn:$^-$ wave undisturbed	no reaction
mo:$^-$ + moCl	$-1\cdot3$		mo:$^-$ wave undisturbed	no reaction
fe:$^-$ + snCl	$-2\cdot6$	298		fe-sn
sn:$^-$ + feI	$-2\cdot6$	298	fe-sn isolated in large scale run	fe-sn
fe:$^-$ + pbOAc	$-2\cdot1$	327		fe-pb
pb:$^-$ + feI	$-2\cdot1$	327	fe-pb isolated in large scale run	fe-pb
fe:$^-$ + mnCl	$-2\cdot1$	290		?
		310		?
		340		fe-fe
mn:$^-$ + feI	$-1\cdot2$	350	mn:$^-$ wave undisturbed	no reaction
	$-2\cdot2$			
fe:$^-$ + moCl	$-1\cdot3$		fe:$^-$ wave undisturbed	no reaction
mo:$^-$ + feI	$-1\cdot2$	350	mo:$^-$ wave undisturbed	no reaction
sn:$^-$ + mnCl	$-2\cdot4$			?
	$-2\cdot9$	244	authentic mn-sn shows only small wave in this region	sn-sn
mn:$^-$ + snCl	$-1\cdot7$	340		mn-mn
	$-2\cdot5$			mn-sn
	$-2\cdot9$			sn-sn
	(e)			
sn:$^-$ + moCl	$-1\cdot3$		only 1/2 halide (mx) required to produce all mo-mo	moCl
	$-2\cdot9$	244		sn-sn
mo:$^-$ + snCl	$-2\cdot4$			mo-sn
pb:$^-$ + mnCl	$-2\cdot2$	268		not pb-mn
mn:$^-$ + pbOAc	$-2\cdot1$	296		mn-pb
pb:$^-$ + moCl	$-1\cdot3$	308	only 1/2 halide (mx) required to produce all pb-pb	
	$-2\cdot0$			pb-pb
mo:$^-$ + pbOAc	$-1\cdot6$	312		not mo-pb
	$-2\cdot0$			
	(d)			
mn:$^-$ + moCl	$-1\cdot3$		only 1/2 halide (mx) required to produce all mn-mn	moCl
	$-1\cdot7$	276		mn-mn
mo:$^-$ + mnCl	$-2\cdot0$		mo+$^-$ wave undisturbed	no reaction
	$-2\cdot4$			

mixture by evaporation *in vacuo*. A 95 per cent yield of the iron-lead heterobimetallic complex was obtained. π-$C_5H_5Fe(CO)_2SnPh_3$ was produced in a similar fashion, except that $[\pi$-$C_5H_5Fe(CO)_2]_2$ was reduced (at $-2\cdot4$ V) and the anion treated with Ph_3SnCl.

Lack of commutation in a number of experiments (Chart 1) may have been due to a low specific reaction rate constant, under the conditions of these experiments, for one of the members of the reaction pairs. This seemed to be the case particularly in the reactions between $[\pi$-$C_5H_5Fe(CO)_2:]^-$ and $Mn(CO)_5Cl$ and between $[Mn(CO)_5:]^-$ and π-$C_5H_5Fe(CO)_2I$ (the latter had $k_2 < 0\cdot01$). In the reaction between $[\pi$-$C_5H_5Fe(CO)_2:]^-$ and $Mn(CO)_5Cl$, the expected heterobimetallic, $[\pi$-$C_5H_5Fe(CO)_2Mn(CO)_5]$ was not detected, although it has been chemically synthesized.[83] The electrochemical data preclude the formation, via one-electron transfer pathways, of significant amounts of $Mn_2(CO)_{10}$ and $[\pi$-$C_5H_5Fe(CO)_2]_2$, but it was suggested that nucleophilic attack by $[\pi$-$C_5H_5Fe(CO)_2:]^-$ on $Mn(CO)_5Cl$ had occurred, and that CO had been expelled leading to a new, and as yet uncharacterized, type of anion, $[\pi$-$C_5H_5Fe(CO)_2Mn(CO)_4Cl]^-$. It is worth noting that heterobimetallic carbonylate anions are known, e.g. $[FeMn(CO)_9]^-$,[84] but related species containing halogen have not yet been reported. The reaction of $[Ph_3Pb:]^-$ with $Mn(CO)_5Cl$ is somewhat similar, namely in that the expected heterobimetallic species was not detected, and it seemed possible that $[Mn(CO)_5PbPh_3Cl]^-$ may have been formed: the commutation experiment, however, afforded the desired $Ph_3PbMn(CO)_5$. Non-commutation occurred in the system $[Mn(CO)_5:]^-$ with π-$C_5H_5Mo(CO)_3Cl$ and $[\pi$-$C_5H_5Mo(CO)_3:]^-$ with $Mn(CO)_5Cl$. The latter had $k_2 < 0\cdot01$ and it is pertinent that π-$C_5H_5Mo(CO)_3Mn(CO)_5$ has not been isolated as a result of chemical experiments. When two moles of $[Mn(CO)_5:]^-$ were allowed to react with π-$C_5H_5Mo(CO)_3Cl$, only $Mn_2(CO)_{10}$ was detected electrochemically. Further addition of the molybdenum chloride made no significant change, other than to provide polarographic waves derived from the reduction of the halide. This reaction could be interpreted as

$$[Mn(CO)_5:]^- + \pi\text{-}C_5H_5Mo(CO)_3Cl \rightarrow (CO)_5MnMo(CO)_3\pi\text{-}C_5H_5$$

$$\downarrow {}^{[Mn(CO)_5:]^-}$$

$$_2(CO)_{10} + [\pi\text{-}C_5H_5Mo(CO)_3:]^-$$

(a) Abbreviations: fe $= \pi$-$C_5H_5Fe(CO)_2$; mn $= Mn(CO)_5$; mo $= \pi$-$C_5H_5Mo(CO)_3$; sn $= Ph_3Sn$; pb $= Ph_3Pb$; OAc $=$ acetate.

(b) In volts, in dimethoxyethane, vs. $Ag/AgClO_4$ with $[n$-$Bu_4N][ClO_4]$ (0·1 M) as base electrolyte at 25°C.

(c) Ultraviolet spectrum in mμ, recorded in dimethoxyethane.

(d) Small wave.

(e) Strong wave.

that is, as a redistribution reaction, rather than as the sequence

$$[Mn(CO)_5:]^- + \pi\text{-}C_5H_5Mo(CO)_3Cl \rightarrow Mn(CO)_5Cl + [\pi\text{-}C_5H_5Mo(CO)_3:]^-$$
$$[Mn(CO)_5:]^- + Mn(CO)_5Cl \rightarrow Mn_2(CO)_{10} + Cl^-$$

since that latter reaction was too slow. Although it was not possible to verify these suggestions, some relevant experiments involving the molybdenum species with $[Ph_3Sn:]^-$ were carried out. Thus, in the reactions between two moles of $[Ph_3Sn:]^-$ and $[\pi\text{-}C_5H_5Mo(CO)_3]_2$, hexaphenylditin and two moles of $[\pi\text{-}C_5H_5Mo(CO)_3:]^-$ were formed. When $[Ph_3Sn:]^-$ was allowed to react with $\pi\text{-}C_5H_5Mo(CO)_3SnPh_3$, however, the products were again Ph_6Sn_2 and the molybdenum anion, which strongly suggested that the second reaction was involved in the first. This combination of experiments implied that the mechanistic suggestions relating to the reactions of $[Mn(CO)_5:]^-$ with $\pi\text{-}C_5H_5Mo(CO)_3Cl$ were essentially correct. The reactions between $[Ph_3Sn:]^-$ and $\pi\text{-}C_5H_5Mo(CO)_3Cl$ behaved similarly to the reaction involving the manganese carbonylate anion and the molybdenum chloride, but the commutation experiments afforded the expected $\pi\text{-}C_5H_5Mo(CO)_3SnPh_3$. The system $[Ph_3Pb:]^-/\pi\text{-}C_5H_5Mo(CO)_3Cl$ also behaved in the same way as that involving $[Ph_3Sn:]^-$, but the commutation experiment did not afford the heterobimetallic.

The systems $[Ph_3Sn:]^-/Mn(CO)_5Cl$ and $[Mn(CO)_5:]^-/Ph_3SnCl$ did not commute. The former did not afford the heterobimetallic and the products of the reaction were Ph_6Sn_2, and an uncharacterized manganese complex (not $HMn(CO)_5$) which had electrochemical properties similar to the product obtained by electrochemically reducing $Mn(CO)_5Br$. The latter reaction afforded a mixture of $Mn_2(CO)_{10}$, Ph_6Sn_2 and $Ph_3SnMn(CO)_5$, probably as a result of a non-caged process occurring subsequent to a one-electron transfer reaction, i.e.

$$[Mn(CO)_5:]^- + Ph_3SnCl \rightarrow Cl^- + Mn(CO)_5^{\cdot} + Ph_3Sn^{\cdot}$$

random coupling

It is interesting to discover that the above reaction occurred in the presence of $[n\text{-}Bu_4N]^+$ as the supporting cation, and that yields of the heterobimetallic species were 30–40 per cent[81] but that when Na^+ was used as counter ion,[85] the yields of $Ph_3SnMn(CO)_5$ were increased to ca 80 per cent.

The previous discussion has shown that redistribution reactions among the organometallic species are by no means uncommon. That this phenomenon is significant has been shown by a study[82] of reactions of the type

$$m:^- + m\text{-}m' \rightarrow m\text{-}m + m':^-$$

In Chart 2 the results of the interaction between $[\pi\text{-}C_5H_5Fe(CO)_2:]^-$, $[Ph_3Sn:]^-$, $[Ph_3Pb:]^-$, $[Mn(CO)_5:]^-$, and $[\pi\text{-}C_5H_5Mo(CO)_3:]^-$ with the homo- and hetero-bimetallic species derived from these ions, are presented. In this chart, the anions are listed according to their oxidation potential at a

Chart 2

	$[\pi\text{-}C_5H_5Fe(CO)_2:]^-$ fe:$^-$; $-1\cdot60$ V (a)	$[Ph_3Sn:]^-$ sn:$^-$; $-0\cdot95$ V	$[Ph_3Pb:]^-$ pb:$^-$; $-0\cdot90$ V	$[Mn(CO)_5:]^-$ mn:$^-$; $-0\cdot55$ V	$[\pi\text{-}C_5H_5Mo(CO)_3:]^-$ mo:$^-$; $-0\cdot55$ V
fe-fe $-2\cdot2$ V (b)		sn-fe	no rn.	no rn.	no rn.
sn-sn $-2\cdot9$ V	fe-sn		no rn.	no rn.	no rn.
pb-pb $-2\cdot0$ V	fe-pb	sn-sn		no rn.	no rn.
mn-mn $-1\cdot7$ V	fe-fe	sn-sn	?		no rn
mo-mo $-1\cdot4$ V	fe-fe	sn-sn	pb-pb	no rn.	
fe-sn $-2\cdot6$ V	no rn.	no rn.	no rn.	no rn.	
fe-pb $-2\cdot1$ V	no rn.	fe-sn	no rn.	no rn.	
fe-mo $-1\cdot4$ V	fe-fe	sn-sn + fe-sn	fe-pb	no rn.	
mo-sn $-2\cdot4$ V	fe-sn	sn-sn	?	no rn.	
mo-pb $-2\cdot2$ V	fe-pb	sn-sn	pb-pb	no rn.	

(a) In volts, half-wave of oxidation potential, in DME vs. Ag/AgClO$_4$ with $[n\text{-}Bu_4N]$ [ClO$_4$] ($0\cdot1$ M) as supporting electrolyte at 25°C.
(b) In volts, reduction potential, as (a).

platinum electrode, a parameter which is associated with the nucleophilicity of these anions.

In the reactions between $[\pi\text{-}C_5H_5Fe(CO)_2:]^-$ and Ph_6Sn_2, and between $[Ph_3Sn:]^-$ and $[\pi\text{-}C_5H_5Fe(CO)_2]_2$, the products were $\pi\text{-}C_5H_5Fe(CO)_2SnPh_3$, and $[Ph_3Sn:]^-$ and $[\pi\text{-}C_5H_5Fe(CO)_2:]^-$. Under the conditions of the experiment, it seemed that the rate constant for the redistribution must have been greater that 10 M^{-1} sec^{-1}, and in both cases, there was no evidence for either homobimetallic parent species. Algebraic summation of the two commutations (which gave $Ph_6Sn_2 + [\pi\text{-}C_5H_5Fe(CO)_2]_2 \rightarrow 2\pi\text{-}C_5H_5Fe(CO)_2SnPh_3$), suggested that, since solvation effects should not have been large and entropy changes should have been small, the strength of the tin-iron bond, in these systems, was greater than the average strength of the iron–iron and tin–tin bonds. It was not possible to predict from the oxidation potentials of these anions that such redistribution reactions would occur, but it was shown that the linear correlation between oxidation potential of these anions and their nucleophilicities, which occurred for most of the anions, was not valid for

$[Ph_3Sn:]^-$. The reactions of $[\pi\text{-}C_5H_5Fe(CO)_2:]^-$ with $Mn_2(CO)_{10}$ and $[\pi\text{-}C_5H_5Mo(CO)_3]_2$ both led to the dimer $[\pi\text{-}C_5H_5Fe(CO)_2]_2$ and the corresponding manganese and molybdenum anions; in the commutation experiments, no reaction occurred, as was expected from the nucleophilicities of the respective anions.

Reaction of $[\pi\text{-}C_5H_5Fe(CO)_2:]^-$ with $\pi\text{-}C_5H_5Mo(CO)_3MPh_3$ (M = Sn or Pb) gave only $\pi\text{-}C_5H_5Fe(CO)_2MPh_3$ and the molybdenum carbonylate anion, results which are entirely in accordance with the nucleophilicities of the respective anions. Similarly, treatment of $\pi\text{-}C_5H_5Fe(CO)_2PbPh_3$ with $[Ph_3Sn:]^-$ afforded the iron–tin heterobimetallic complexes, whereas no redistribution apparently occurred when the last was treated with $[Ph_3Sn:]^-$; there was no observable reaction between the iron–tin complex and $[\pi\text{-}C_5H_5Fe(CO)_2:]^-$. It was apparent from these data that

$$\pi\text{-}C_5H_5Fe(CO)_2Mo(CO)_3\pi\text{-}C_5H_5 \quad \text{and} \quad [\pi\text{-}C_5H_5Mo(CO)_3]_2$$

would both react with $[\pi\text{-}C_5H_5Fe(CO)_3:]^-$ giving the di-iron compound.

It seems from the data presented in Chart 2 that, in the redistribution reactions, the anion expelled is nearly always the least neucleophilic of the two possible anions which can be derived from a heterobimetallic. There are, of course, certain exceptions to this, exemplified by the reaction of $\pi\text{-}C_5H_5Fe(CO)_2Mo(CO)_3\pi\text{-}C_5H_5$ with $[Ph_3Sn:]^-$, in which, as predicted, the iron–tin heterobimetallic and the molybdenum carbonylate anion were formed, but also Ph_6Sn_2 was detected. Furthermore, some reactions involving $[Mn(CO)_5:]^-$ seem to involve displacement of non-metallic groups, e.g. CO, when treated with a nucleophile.

In practical terms, the foregoing discussion would seem to suggest that preparation of a metal–metal bonded compound would be best achieved by choosing the weaker nucleophile (m:⁻) of the pair for the reaction m:⁻ + m′x, and by mixing them by inverse addition, that is m:⁻ to m′x rather than m′x to m:⁻, where the halide is always deficient. A rationalization, based on these electrochemical data, may therefore be presented for the fact that many reactions involving a metal halide and a (different) organometal anion result only in the formation of homobimetallic species.

3. Bimetallic Species Containing Bridging Atoms

Associated with the oxidation and reduction of homo- and hetero-bimetallic species is the electrochemical behaviour of bimetallic systems containing μ-bridging groups. These bridges may be mercapto (SR), phosphido (PR$_2$) or arsenido (AsR$_2$) groups, and the complexes containing them may be of two

Table 32. *Electrochemical Data Obtained from Bridged Bimetallic Complexes [Data from Reference 86]*

Complex	$E^1_{1/2}$ (a)	$E^2_{1/2}$ (a)	n (b)	$E_{pc/2}$ (c)	$E_{pa/2}$ (c)	Chem. Rev. (d)	g (e)
[Cr(CO)₄PMe₂]₂ (j)	−1·85		2	−1·9	−1·7	yes	1·993 (g)
[W(CO)₄PMe₂]₂ (j)	−1·9		2	−1·9	−1·6	yes	1·994 (g)
[W(CO)₄AsMe₂]₂ (j)	−1·8		2	−1·9	−1·4	yes	no signal
[π-C₅H₅Fe(CO)PPh₂]₂	−0·2		−1	−0·1	−0·2	yes	
(k)	(f)	−0·5	−1	−0·7	−0·5	yes	1·977 (h)
[Fe(CO)₃PMe₂]₂ (l)	−2·1		2	−2·2	−1·9	yes	1·999 (g)
[Fe(CO)₃AsMe₂]₂ (l)	−0·2		−1	−0·2	−0·1		weak signal (h)
		−1·9	1	−2·0	−1·8	yes	2·064 (g)
[Fe(NO)₂PPh₂]₂ (j)	−1·7		1	−1·75	−1·65	yes	1·937 (g)
		−2·1	1	−2·15	−2·05	yes	
[π-C₅H₅Fe(CO)SMe]₂	−0·6		−1	−0·7	−0·5	yes	1·998 (h)
(k)							
[π-C₅H₅CoPPh₂]₂ (l)	−0·3			(i)	(i)	no	2·003 (g)
		−2·6	1	−2·65	−2·55	yes	
[π-C₅H₅NiPPh₂]₂ (k)	−2·3		1	−2·35	−2·25	yes	2·060 (g)
[π-C₅H₅CoSMe]₂ (l)	−0·5		−1	−0·5	−0·4	yes	2·120 (h)
		−2·4	1	−2·45	−2·35	yes	2·105 (g)
[π-C₅H₅NiSMe]₂ (k)	−0·5		−1	−0·6	−0·4	yes	2·007 (h)
		−1·8	1	−2·0	−1·7	slight	

(a) In volts, in dimethoxyethane, vs. Ag/AgClO₄ with [n-Bu₄N][ClO₄] (0·1 M) as base electrolyte at 25°C.

(b) Number of electrons determined by coulometry, positive values represent reductions and negative values oxidations.

(c) Potentials at half-peak height observed by triangular voltammetry at a mercury microelectrode sweep rate of 1 V/sec.

(d) Established by controlled-potential electrolysis at Ag pool.

(e) g-factor in solution, obtained by controlled potential electrolysis.

(f) Measured on rotating Pt electrode.

(g) For monoanionic species.

(h) For monocationic species.

(i) Irreversible process.

(j) Linear metal-metal bond.

(k) No metal-metal bond.

(l) Bent metal-metal bond assumed.

types—those having, in addition to the bridges, a metal–metal bond, and those having no metal–metal interaction.

A series of Group VI metal complexes, $[M(CO)_4LMe_2]_2$, where $L = P$ or As, were examined electrochemically.[86] All species exhibited (Table 32) a reversible two-electron reduction wave corresponding to the formation, presumably, of $[M(CO)_4LR_2]_2^{2-}$. In the chromium system, the neutral species was yellow whereas the dianion, which was diamagnetic, was red. During the electrolysis of these compounds, however, transient formation of a dark green solution was observed at the electrolysis mid-point (this corresponded to the addition of only one electron). This dark green species was apparently $[Cr(CO)_4LMe_2]_2^{-}$, which was generated quantitatively by mixing equimolar amounts of the neutral and dianion species in solution. The reactions were summarized as

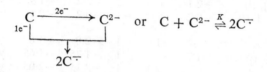

The monoanionic species were paramagnetic and afforded characteristic esr signals in solution. The infrared spectra of the three species (neutral, mono-, and di-anion) indicated that the binuclear molecular structure remained intact upon reduction. In the analogous tungsten system, similar behaviour was observed, but from a partial analysis of the esr spectra, it seemed that K, the equilibrium constant for the reaction giving the monoanion, was less than unity and that the dianion was the more favoured species, in contrast the chromium system, where $K = 10$. On reduction of $[W(CO)_4AsMe_2]_2$, a dianion was formed, but admixture of this dianion and the neutral species did not lead to the appearance of esr signals appropriate to the monoanion. It was suggested that in bridged bimetallic species of this type, increased polarizability of the metal atom or of the bridging atom led to changes in the equilibrium constant thereby favouring the formation of the dianion.

The iron phosphide, $[\pi\text{-}C_5H_5Fe(CO)PPh_2]_2$, which is not believed to contain an iron–iron bond, could be oxidized in two, apparently reversible, one-electron steps. Admixture of the neutral and dicationic species led to the formation of the blue monocation, which exhibited a broad unresolved esr signal. The related carbonyl complex, $[Fe(CO)_3PMe_2]_2$, which does contain an Fe—Fe bond, could be reduced in one two-electron step, although admixture of the neutral and dianionic species afforded the expected mono-anion, $[Fe(CO)_3PMe_2]_2^{-}$. While this phosphido complex did not oxidize polarographically its arsenido analogue could be oxidized reversibly in a one-electron step, but reduced only in a one-electron process. The monocation exhibited a weak esr signal in solution, whereas the signal from the mono-anion was much stronger, although it was not resolved. The formally iso-electronic nitrosyl, $[Fe(NO)_2PPh_2]_2$, could be reduced in two one-electron steps affording a mono- and a di-anion. The monoanion was paramagnetic and exhibited a triplet esr signal (^{31}P hyperfine splitting; ^{14}N ($I = 1$) hyperfine interactions were not detected). From the infrared, nmr and esr data which could be obtained from these various iron complexes, it was apparent that no major structural reorganization occurred when the neutral bridged binuclear species were oxidized or reduced.

$[\pi\text{-}C_5H_5Fe(CO)SMe]_2$ could be oxidized electrochemically to a stable, paramagnetic monoanion, and chemical oxidation, using $AgSbF_6$,[66] afforded $[\pi\text{-}C_5H_5Fe(CO)SMe]_2^+[SbF_6^-]$. The infrared data obtained from this compound indicated that, again, no structural reorganization had occurred on oxidation.

The electrochemical behaviour of $[\pi\text{-}C_5H_5MPPh_2]_2$ and $[\pi\text{-}C_5H_5MSMe]_2$, $M = Co$ or Ni, was strikingly similar. The phosphine complexes could both be reduced to monoanions, which were paramagnetic. However, only the cobalt complexes exhibited resolvable ^{31}P hyperfine interactions, in their esr spectra, and this may have been a reflection of the structural differences (see Table 32). The mercaptide species, however, were oxidized and reduced in one-electron steps, thereby affording mono-anions and cations. ESR signals were obtained from the cobalt species, and from $[\pi\text{-}C_5H_5NiSMe]_2^+$, but not from $[\pi\text{-}C_5H_5NiSMe]_2^-$.

From the data presently available, there appears to be little correlation between geometry of the compounds and their electrochemical properties. It seems that the presence of metal–metal bonds certainly appears to guarantee the formation of a stable radical monoanion, and possibly a dianion, but the lack of a metal–metal interaction correlated only poorly with the ability of the system to yield a radical cation.

4. Metal Olefin Complexes and Related Species

Many olefin metal carbonyl complexes, often obtained by cyclization of acetylenes in the presence of metal carbonyls, exhibited electrochemical behaviour[87] (Table 33). Many of these, however, were relatively uninteresting since either no evidence of transiently stable radical ions, formed on oxidation or reduction, could be found, or the lifetimes of these radical ions were too short to permit their observation by normal controlled-potential reductive or oxidative generation, esr and other spectral techniques. However, some

Table 33. *Electrochemical Data Obtained from Transition Metal Olefin, Acetylene and Related Complexes*
[*Data from Reference* 87]

Complex	$E^1_{1/2}$	$E^2_{1/2}$	n	$E_{pc/2}$	$E_{pa/2}$	Chem rev.	Comment
	(a)	(a)	(b)	(c)	(c)	(d)	
1. R, R ring with Fe—Fe(CO)₃(CO)₃, R = p-C₆H₄Cl	-1.24	-1.87	1 1	-1.25 -1.4	-1.2 -1.7	yes no	$E_{1/2}^{ox} = -0.5$ V.; radical anion yellow
2. Ph, Ph ring, Me, Me, Fe—Fe(CO)₃(CO)₃	-1.50	-2.12	1 1	-1.5 -2.2	-1.4 -2.1	yes no	$E_{1/2}^{ox} = -0.84$, -1.32 V; radical anion dark red brown
3. (PhCCPh)₂Fe₃(CO)₈	-1.3	-2.0	1	-1.3	-1.2	yes	radical anion dark green
4. (2,4-diphenyltropone) Fe(CO)₃	-1.61	-1.92	1 1	-1.6 -2.0	-1.5 -1.8	yes yes	radical anion brown
5. (2,4,6-triphenyltropone) Fe(CO)₃	-1.62	-1.98	1 1	-1.6 -1.9	-1.5 -1.8	yes yes	radical anion brown-red
6. thiophene ring, Fe(CO)₃, S—Fe(CO)₃	-1.4	-2.0	1 1	-1.5 -2.0	-1.4 -1.9	yes yes	$E_{1/2}^{ox} = -0.61$, -1.4 V.
7. benzo ring, Fe(CO)₃, S—Fe(CO)₃	-1.64	-2.50	1 1	-1.5 -2.0	-1.4 -1.9	yes 	$E_{1/2}^{ox} = -1.40$, -1.56 V.
8. (π-C₅H₅Ni)₂PhC≡CPh	-2.2		1	-2.2	-2.1	yes	
9. (π-C₅H₅Ni)₂HC≡CH	-2.2		2	-2.5	-1.9		
10. (π-C₈H₈) (π-duroquinone)Ni (e)	-2.0		1	(f)	(f)	no	

complexes did provide relatively stable radical ions, and these have been investigated by esr spectroscopy, and by ir techniques.

Reduction of (1), formed by reaction of $p\text{-ClC}_6\text{H}_4\text{C}\equiv\text{CC}_6\text{H}_4p\text{-Cl}$ with $\text{Fe}_3(\text{CO})_{12}$, occurred in two reversible one-electron processes. Controlled potential electrolysis of the solution gave a radical anion which had an oxidation potential of -0.5 V, compared to the reduction potential of -1.2 V of the parent species (1). Reoxidation of the solution at -0.4 V afforded a neutral material which was evidently identical to (1). The apparent inconsistency of these data was interpreted in terms of a reorganizational process which occurred relatively slowly after reduction, i.e.

$$(1)_1 \underset{E_{1/2}{}^1 + E_{1/2}{}^2}{\overset{\pm e^-}{\rightleftharpoons}} (1)_{\bar{1}} \overset{\text{slow}}{\rightleftharpoons} (1)_{\bar{2}}$$

with the overall process $-e^-\,(E_{1/2}{}^1 + E_{1/2}{}^2)$ from $(1)_1$ to $(1)_{\bar{2}}$.

The complex (2) underwent an electrochemically reversible one-electron reduction at -1.5 V, and the resulting reduced species exhibited oxidation waves at -0.8 and -1.3 V. Controlled potential oxidation of this species at -0.7 V afforded the starting material, (2), but controlled potential electrolysis at -1.2 V caused both oxidative waves to disappear, and also led to the regeneration of (2). An explanation of this electrochemical behaviour is shown below.

$$(2)_1 \underset{E_{1/2}{}^1}{\overset{\pm e^-}{\rightleftharpoons}} (2)_{\bar{1}} \overset{\text{slow}}{\longrightarrow} (2)_{\bar{2}} \underset{}{\overset{K \simeq 1}{\rightleftharpoons}} \text{A}_{\bar{3}}$$

with $-e^-,\,E_{1/2}{}^2$ and $-e^-,\,E_{1/2}{}^3$ oxidation processes.

The complexes (3), (4), and (5), could be reduced in two reversible one-electron steps. The radical anions were paramagnetic whereas the dianions were diamagnetic: admixture of the neutral species with equimolar amounts of the dianions afforded the monoanionic species.

The sulphur compounds (6) and (7) underwent two reversible one-electron

(a) In volts, in dimethoxyethane, vs. Ag/AgClO$_4$ with [n-Bu$_4$N][ClO$_4$] (0·1 M) as supporting electrolyte at 25°.

(b) Number of electrons established by coulometry; positive values represent reductions.

(c) Potential at half-height, cathodic and anodic, respectively, at 1 V/sec sweeps on hanging Hg drop employing the starting material.

(d) Chemical reversibility based on recovery of starting material after redox-cycle.

(e) C$_8$H$_8$ = cycloöctatetraene.

(f) Irreversible.

processes. The polarograms of the reduced solutions exhibited a new oxidation wave which occurred at a potential different to the reduction potential of the starting material, but reoxidation of the reduced species afforded the original (6) and (7). It was suggested that reduction led to an equilibrium condition between two anionic radicals as shown below.

$$M_1 \underset{E_{1/2}^1}{\overset{\pm e^-}{\rightleftharpoons}} M_1^- \overset{-e^-,\, E_{1/2}^1}{\underset{}{\rightleftharpoons}} M_2^-$$

with overall $-e^-, E_{1/2}^2$

$$[M = (6) \text{ or } (7)]$$

Of the two nickel complexes examined (Table 33) [π-cycloöctatetraene-π-duroquinoneNi] was irreversibly reduced in a one-electron step giving the radical anion of duroquinone, whereas $(\pi\text{-}C_5H_5Ni)_2PhC\equiv CPh$, (8), underwent a reversible, one-electron, reduction at -2.2 V. The analogous $(\pi\text{-}C_5H_5Ni)_2 HC\equiv CH$, displayed a two-electron reduction process at -2.2 V, but when controlled potential reductive electrolysis was limited to the passage of current equivalent to only one electron per molecule, the resulting solution exhibited a single sharp esr signal, presumably due to $[(\pi\text{-}C_5H_5Ni)_2HC\equiv CH]^-$.

VII. SPECTRAL STUDIES OF ORGANOMETALLIC RADICAL ANIONS

As indicated throughout the discussion of the electrochemistry of organo-metallic species, many of the radical anions which have been produced have also been studied by esr and ir spectral methods. Although these studies have not led, as yet, to detailed discussion of the bonding in these organometallic species, some general points have been made relating to the structures of the reduced species and the way in which negative charge is distributed about the complexes.

Most of the spectral work of relevance to this discussion has been carried out with bridged homobimetallic species and it was assumed[88] that electro-chemical reduction of such compounds did not lead to structural change, relative to the starting material, but did give species in which the added elec-tron, or electrons, were in an orbital which had considerable π^*-ligand character. ESR, nmr, and ir spectral data provided evidence to justify these assumptions.

The nmr spectrum of $[Cr(CO)_4PMe_2]_2^0$ showed that the methyl groups on each P atom, and the P nuclei, were equivalent, and when this species was reduced to the radical anion, the esr spectrum, which consisted of ^{31}P and ^{1}H, but no metal, hyperfine interactions, indicated that the methyl group and P nuclei were still equivalent. The infrared spectrum of the radical anion showed that ν_{CO} had decreased relative to the neutral starting material but, in all other ways, was identical to $[Cr(CO)_4PMe_2]_2^0$, i.e. no structural change had occurred on reduction. Similar data was obtained from $[Fe(CO)_3PMe_2]_2$, whose nmr spectrum revealed that there were two sets of equivalent methyl groups, consistent with a 'bent' FeP_2Fe arrangement (Fig. 25). The esr spectrum

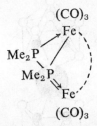

Fig. 25. The idealized structure of $[Fe(CO)_3PMe_2]_2$.

obtained from the reduced species, $[Fe(CO)_3PMe_2]_2^-$, also indicated that the P nuclei remained equivalent and that the methyl groups existed in two equivalent sets. The analysis of the esr spectrum obtained from the monomeric $[\text{bipyridylMo(CO)}_4]^-$ showed that the multiplets (45 lines) arose entirely from ^{1}H and ^{14}N hyperfine interactions; no ^{95,97}Mo couplings were resolved. The hyperfine coupling constants of the free bipyridyl radical anion were very similar to those obtained from the organometallic anion.

It was concluded that, on reduction of an organometallic species, the alteration of ν_{CO} occurred primarily as a result of transmission of charge via the σ-bonding system rather than the π-bonding system as usually proposed. Evidence for this was obtained from the force constant calculations, which indicated that the transmission was partially isotropic since both force constants in species such as $[\text{bipyridylMo(CO)}_4]^{0,-}$, $[Cr(CO)_4PMe_2]_2^{0,-}$, etc., decreased on reduction, and also that the transmission had directional properties. The latter caused the force constant of the carbonyl groups *trans* to the other ligand (i.e. bipyridyl, PMe$_2$, etc.) to decrease more than the force constant of the *cis* CO group.

VIII. π-DICARBOLLYL DERIVATIVES OF TRANSITION METALS

An extremely interesting development in the chemistry of polyhedral organo-boranes has been the discovery of a wide variety of π-dicarbollyl derivatives of transition metals.[89] Typical of these compounds are the iron, cobalt and nickel derivatives of the (3)-1,2-$B_9C_2H_{11}$ and (3)-1,7-$B_9C_2H_{11}$ dianions (the 'dicarbollide' nomenclature is not strictly rigorous).[89,90] An electrochemical study of these compounds has established[89] that they undergo discrete one-electron transfer reactions which are reminiscent of the simple metallocenes, $(\pi\text{-}C_5H_5)_2M$, M = Fe, Co or Ni.

The electrochemical data obtained by polarographic and cyclic voltammetric studies of the bis-π-dicarbollides (Fig. 26) and their π-cyclopentadienyl

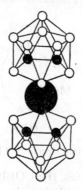

Fig. 26. The idealized structure of $[(\pi\text{-}(3)\text{-}1,2\text{-}B_9C_2H_{11})_2M]^z$, a bis-$\pi$-dicarbollide metal complex.

π-dicarbollide analogues are summarized in Table 34. The half-waves obtained from the bis-substituted species depend on three factors, the metal atom ($E_{1/2}$ becoming increasingly negative in the order Fe < Ni < Co), dicarbollide substituent, and relative position in the polyhedral carborane framework of the C atoms. In the iron complexes, the dicarbollide substituent effect was such that $E_{1/2}$ became more negative in the order

$$B_9C_2H_{11} < B_9C_2H_{10}Ph < B_9C_2H_9Me_2$$

but this was reversed in the cobalt series; the effect of bromine atoms on the carborane framework in the latter series was to raise the half-wave potentials for the various couples to more positive values. There was a significant difference between $E_{1/2}$-values obtained from the 1,2- and 1,7-dicarbollide species, but insufficient data was available for meaningful comparisons.

The iron complexes were most often isolated as monoanions, which could be reduced, either in acetonitrile or acetone using sodium amalgam, to the corresponding dianions; reoxidation was accomplished using air. Aerial oxidation of $[\pi\text{-}C_5H_5Fe(\pi\text{-}(3)\text{-}1,2\text{-}B_9C_2H_{11})]^-$ afforded the neutral species, and reduction of the latter was effected using sodium amalgam. The nickel

Table 34. *Voltammetric and Polarographic Data Obtained from Metal Carborane Complexes*

[Data from Reference 89]

Complex	Method (a)	Half-wave potential (b) for process			
		$+1 \leftarrow 0$	$0 \leftarrow -1$	$-1 \rightarrow -2$	$-2 \rightarrow -3$
$[(\pi\text{-}1,2\text{-}B_9C_2H_{11})_2Fe]^-$	pol			-0.424	
$[(\pi\text{-}1,2\text{-}B_9C_2H_9Me_2)_2Fe]^-$	pol			-0.538	
$[(\pi\text{-}1,2\text{-}B_9C_2H_{10}Ph)_2Fe]^-$	pol			-0.464	
$\pi\text{-}C_5H_5Fe(\pi\text{-}1,2\text{-}B_9C_2H_{11})$	cv		-0.08		
$[(\pi\text{-}1,2\text{-}B_9C_2H_{11})_2Co]^-$	pol			-1.42	
	cv		$+1.57$	-1.46	
$[(\pi\text{-}1,2\text{-}B_9C_2H_8Br_3)_2Co]^-$	cv		irrev	-0.48	-1.58
$[(\pi\text{-}1,2\text{-}B_9C_2H_9Me_2)_2Co]^-$	pol			-1.16	
	cv		irrev	-1.13	
$[(\pi\text{-}1,2\text{-}B_9C_2H_{10}Ph)_2Co]^-$	pol			-1.28	
$\pi\text{-}C_5H_5Co(\pi\text{-}1,2\text{-}B_9C_2H_{11})$	cv	irrev	-1.25		
$[(\pi\text{-}1,7\text{-}B_9C_2H_{11})_2Co]^-$	cv		irrev	-1.17	
$[(\pi\text{-}1,2\text{-}B_9C_2H_{11})_2Ni]^-$	pol			-0.63	
	cv		$+0.25$	-0.59	
$[(\pi\text{-}1,7\text{-}B_9C_2H_{11})_2Ni]^-$	cv		$+0.55$	-0.91	

(a) Pol: using dropping mercury electrode in 50 per cent aq. acetone, vs. SCE with 0·1 M LiClO₄ as base electrolyte. Cv: cyclic voltammetry with rotating Pt electrode in CH₃CN with 0·1 M [Et₄N][ClO₄] as base electrolyte.

(b) In volts.

bis-π-dicarbollides were obtained as monoanions which could be oxidized, in aqueous solution, to the corresponding neutral species using one equivalent of ferric ion. Reduction of the neutral species to the monoanion was achieved with 'mossy cadmium' in acetone, and of the monoanion to the dianion using hydrazine and potassium hydroxide.

IX. CONCLUSION

The electrochemical study of transition metal dithiolene and organo-metallic complexes has obviously been very fruitful and has awakened in the inorganic chemist an awareness of the uses of electrochemical techniques.

The particular importance of the dithiolene system is the accessibility of the various members of an electron transfer series, and in the ease with which many of these members could be isolated and characterized. It seems likely that further studies of these and related compounds will lead to a better understanding of the origin of the 'facile electron transfer reactions', but it seems less probable that there will be many structural surprises, either molecular or electronic. Unless a new ligand system is discovered, the further extension of electrochemical studies into classical coordination chemistry is not likely to develop at the pace of recent years.

The electrochemical investigation of organometallic compounds is only beginning. The number and variety of systems which can be studied is very large. It seems likely that future developments will be associated with organo-electrochemical problems, and some interesting and novel results may be expected.

REFERENCES

1. McCLEVERTY, J. A.: *Progr. in Inorg. Chem.*, **10**, 49 (1968).
2. GRAY, H. B. 'Electronic Structures of Square Planar Complexes', in *Transition Metal Chemistry*, Vol. 1, p. 240, Edited by R. L. Carlin (Edward Arnold, London, 1965).
3. SCHRAUZER, G. N.: 'Coordination Compounds of Unsaturated 1,2-Dithiols and 1,2-Dithioketones', in *Transition Metal Chemistry*, Vol. 4, p. 299 (Edward Arnold, London, 1968).
4. COATES, G. C., GREEN, M. L. H., and WADE, K.: *Organometallic Compounds*, Vol. 2, *Transition Metals* (Methuen & Co. Ltd., London, 1968).
5. MAKI, A. H., EDELSTEIN, N., DAVISON, A., and HOLM, R. H.: *J. Amer. Chem. Soc.*, **86**, 4580 (1964).
6. SCHMITT, R. D., and MAKI, A. H.: *J. Amer. Chem. Soc.*, **90**, 2288 (1968).
7. SCHRAUZER, G. N., and MAYWEG, V. P.: *J. Amer. Chem. Soc.*, **87**, 3585 (1965).
8. SHUPACK, S. I., BILLIG, E., CLARK, R. J. H., WILLIAMS, R., and GRAY, H. B.: *J. Amer. Chem. Soc.*, **86**, 4594 (1964).
9. SCHRAUZER, G. N., and MAYWEG, V. P.: *J. Amer. Chem. Soc.*, **84**, 3221 (1962).
10. GRAY, H. B., WILLIAMS, R., BERNAL, I., and BILLIG, E.: *J. Amer. Chem. Soc.*, **84**, 3596 (1962).
11. DAVISON, A., EDELSTEIN, N., HOLM, R. H., and MAKI, A. H.: *J. Amer. Chem. Soc.*, **85**, 2029 (1963).
12. GRAY, H. B., and BILLIG, E.: *J. Amer. Chem. Soc.*, **85**, 2019 (1963).
13. ROHRSCHEID, F., BALCH, A. L., and HOLM, R. H.: *Inorg. Chem.*, **5**, 1542 (1966).
14. DAVISON, A., and HOLM, R. H.: *Inorg. Synth.*, **10**, 8 (1967).
15. BALCH, A. L., DANCE, I. G., and HOLM, R. H.: *J. Amer. Chem. Soc.*, **90**, 1139 (1968).
16. McCLEVERTY, J. A., and PALMER, J.: Unpublished work.
17. OLSON, D. C., MAYWEG, V. P., and SCHRAUZER, G. N.: *J. Amer. Chem. Soc.*, **88**, 4876 (1966).
18. TAFT, R. W., JR.: *Steric Effects in Organic Chemistry*, Chapter 13, Edited by M. S. Newman (Wiley, New York, 1965).
19. VLČEK, A. A.: *Progr. in Inorg. Chem.*, **5**, 211 (1963); *Z. Anorg. u. Allgem. Chem.*, **304**, 109 (1960).
20. BAKER-HAWKES, M. T., BILLIG, E., and GRAY, H. B.: *J. Amer. Chem. Soc.*, **88**, 4870 (1966).
21. WHARTON, E. J., and McCLEVERTY, J. A.: *J. Chem. Soc.*, A, 2258 (1969).
22. WEIHER, J. F., MELBY, L. R., and BENSON, R. E.: *J. Amer. Chem. Soc.*, **86**, 4329 (1964).
23. FRITCHIE, C. J.: *Acta Cryst.*, **20**, 107 (1966).
24. DAVISON, A., EDELSTEIN, N., HOLM, R. H., and MAKI, A. H.: *Inorg. Chem.*, **3**, 814 (1964).

25. ENEMARK, J. H., and LIPSCOMB, W. N.: *Inorg. Chem.*, **4**, 1729 (1965).
26. DAVISON, A., HOWE, D. V., and SHAWL, E. T.: *Inorg. Chem.*, **6**, 458 (1967).
27. SCHRAUZER, G. N., MAYWEG, V. P., FINCK, H. W., and HEINRICH, W.: *J. Amer. Chem. Soc.*, **88**, 4604 (1966).
28. KING, R. B.: *Inorg. Chem.*, **2**, 641 (1963).
29. DAVISON, A., EDELSTEIN, N., HOLM, R. H., and MAKI, A. H.: *J. Amer. Chem. Soc.*, **86**, 2799 (1964).
30. SCHRAUZER, G. N., and MAYWEG, V. P.: *J. Amer. Chem. Soc.*, **88**, 3235 (1966).
31. EISENBERG, R., STIEFEL, E. I., ROSENBERG, R. C., and GRAY, H. B.: *J. Amer. Chem. Soc.*, **88**, 2874 (1966).
32. STIEFEL, E. I., DORI, Z., and GRAY, H. B.: *J. Amer. Chem. Soc.*, **89**, 3353 (1967).
33. BERNAL, I.: Personal communication; Proceedings of 155th A.C.S. Meeting, San Francisco, 1968.
34. STIEFEL, E. I., EISENBERG, R., ROSENBERG, R. C., and GRAY, H. B.: *J. Amer. Chem. Soc.*, **88**, 2956 (1966).
35. MCCLEVERTY, J. A., ATHERTON, N. M., CONNELLY, N. G., and WINSCOM, C. J.: *J. Chem. Soc.*, A, 2242 (1969).
36. BALCH, A. L.: *Inorg. Chem.*, **6**, 2158 (1967).
37. MCCLEVERTY, J. A., and RATCLIFF, B.: *J. Chem. Soc.*, A, 1631 (1970).
38. WINSCOM, C. J.: Ph. D. Thesis, University of Sheffield, 1968; N. M. Atherton and C. J. Winscom, to be published.
39. MCCLEVERTY, J. A., ATHERTON, N. M., LOCKE, J., WHARTON, E. J., and WINSCOM, C. J.: *J. Amer. Chem. Soc.*, **89**, 6082 (1967).
40. MCCLEVERTY, J. A., and RATCLIFF, B.: *J. Chem. Soc.*, A, 1627 (1970).
41. MCCLEVERTY, J. A., and ORCHARD, D. G.: *J. Chem. Soc.*, A (1971), in press.
42. KHARE, G. P., PIERPOINT, C. G., and EISENBERG, R.: *Chem. Commun.*, 1692 (1968).
43. MCCLEVERTY, J. A., ORCHARD, D. G., and SMITH, K.: *J. Chem. Soc.*, A (1971), in press.
44. MCCLEVERTY, J. A., JAMES, T. A., and WHARTON, E. J.: *Inorg. Chem.*, **8**, 1340 (1969).
45. DESSY, R. E., STARY, F. E., KING, R. B., and WALDROP, M.: *J. Amer. Chem. Soc.*, **88**, 471 (1966).
46. DESSY, R. E., KING, R. B., and WALDROP, M.: *J. Amer. Chem. Soc.*, **88**, 5112 (1966).
47. MCCLEVERTY, J. A., and JAMES, T. A.: *J. Chem. Soc.*, A, 3308 (1970).
48. BALCH, A. L., RÖHRSCHEID, F., and HOLM, R. H.: *J. Amer. Chem. Soc.*, **87**, 2301 (1965).
49. FEIGL, F., and FÜRTH, M.: *Monatsh*, **48**, 445 (1927).
50. BALCH, A. L., and HOLM, R. H.: *J. Amer. Chem. Soc.*, **88**, 5201 (1966).
51. STIEFEL, E. I., WATERS, J. H., BILLIG, E., and GRAY, H. B.: *J. Amer. Chem. Soc.*, **87**, 3016 (1965).
52. CHISWELL, B., and LIONS, F.: *Inorg. Chem.*, **3**, 490 (1964).
53. GANSOW, O. A., OLCOTT, R. J., and HOLM, R. H.: *J. Amer. Chem. Soc.*, **89**, 5470 (1967).
54. MCCLEVERTY, J. A., MCKENZIE, E. D., JAMES, T. A., and BRAY, J.: To be published.
55. HIGSON, B., and MCKENZIE, E. D.: *Inorg. Nucl. Chem. Lett.*, **6**, 209 (1970).
56. MCCLEVERTY, J. A., and JONES, C. J.: *J. Chem. Soc.*, A (1971), in press.
57. HOLM, R. H., BALCH, A. L., DAVISON, A., MAKI, A. H., and BERRY, T. E.: *J. Amer. Chem. Soc.*, **89**, 2866 (1967).
58. HIEBER, W., and BRÜCK, R.: *Z. Anorg. u. Allgem. Chem.*, **269**, 13 (1952).
59. MAKI, A. H., BERRY, T. E., DAVISON, A., HOLM, R. H., and BALCH, A. L.: *J. Amer. Chem. Soc.*, **88**, 1080 (1966).
60. LALOR, F., HAWTHORNE, M. F., MAKI, A. H., DAVISON, A., GRAY, H. B., DORI, Z., and STIEFEL, E. I.: *J. Amer. Chem. Soc.*
61. BAILEY, N. A., HULL, S. E., JONES, C. J., and MCCLEVERTY, J. A.: *Chem. Commun.*, 124 (1970).
62. BALCH, A. L.: *J. Amer. Chem. Soc.*, **91**, 1948 (1969).
63. FORBES, C. E., and HOLM, R. H.: *J. Amer. Chem. Soc.*, **92**, 2297 (1970).
64. DAVISON, A., MCCLEVERTY, J. A., SHAWL, E. T., and WHARTON, E. J.: *J. Amer. Chem. Soc.*, **89**, 830 (1967).

65. WHARTON, E. J., WINSCOM, C. J., and MCCLEVERTY, J. A.: *Inorg. Chem.*, **8**, 393 (1969).
66. KING, R. B., and BISNETTE, M. B.: *Inorg. Chem.*, **6**, 469 (1967).
67. MAKI, A. H., EDELSTEIN, N., DAVISON, A., and HOLM, R. H.: *J. Amer. Chem. Soc.*, **86**, 4580 (1964).
68. DESSY, R. E., KITCHING, W., and CHIVERS, T.: *J. Amer. Chem. Soc.*, **88**, 453 (1966).
69. DESSY, R. E., STARY, F. E., KING, R. B., and WALDROP, M.: *J. Amer. Chem. Soc.*, **88**, 471 (1966).
70. DESSY, R. E., KING, R. B., and WALDROP, M.: *J. Amer. Chem. Soc.*, **88**, 5112 (1966).
71. WINKHAUS, G., PRATT, L., and WILKINSON, G.: *J. Chem. Soc.*, 3807 (1961).
72. JONES, D., PRATT, L., and WILKINSON, G.: *J. Chem. Soc.*, 4458 (1962).
73. WATTS, G. W., BAYE, L. J., and DRUMMOND, F. O.: *J. Amer. Chem. Soc.*, **88**, 1138 (1966); F. Calderazzo, J. J. Salzmann, and P. Mosimann, *Inorg. Chim. Acta*, **1**, 65 (1967).
74. BEHRENS, W., and HAAG, W.: *Chem. Ber.*, **94**, 312 (1961).
75. BAIRD, M. C.: *Progr. in Inorg. Chem.*, **9**, 1 (1968).
76. DESSY, R. E., WEISSMAN, R. M., and POHL, R. L.: *J. Amer. Chem. Soc.*, **88**, 5117 (1968).
77. LEWIS, J., and NYHOLM, R. S.: *Chemistry*, 557 (1963).
78. DESSY, R. E., POHL, R. L., and KING, R. B.: *J. Amer. Chem. Soc.*, **88**, 5121 (1966).
79. PARSHALL, G.: *J. Amer. Chem. Soc.*, **86**, 361 (1964).
80. BRUCE, M. I., and STRONE, F. G. A.: *Angew. Chem. Intern. Edit. Engl.*, **7**, 747 (1968).
81. DESSY, R. E., and WEISSMAN, P. M.: *J. Amer. Chem. Soc.*, **88**, 5124 (1966).
82. DESSY, R. E., and WEISSMAN, P. M.: *J. Amer. Chem. Soc.*, **88**, 5129 (1966).
83. KING, R. B., TREICHEL, P. M., and STONE, F. G. A.: *Chem. and Ind.*, 747 (1961).
84. RUFF, J. K.: *Inorg. Chem.*, **7**, 1878 (1968).
85. GORSICH, R. D.: *J. Amer. Chem. Soc.*, **84**, 2486 (1962).
86. DESSY, R. E., KORNMANN, R., SMITH, C., and HAYTOR, R.: *J. Amer. Chem. Soc.*, **90**, 2001 (1968).
87. DESSY, R. E., and POHL, R. L.: *J. Amer. Chem. Soc.*, **90**, 1995 (1968).
88. DESSY, R. E., and WIECZOREK, L.: *J. Amer. Chem. Soc.*, **91**, 4963 (1969).
89. HAWTHORNE, M. F., YOUNG, D. C., ANDREWS, T. D., HOWE, D. V., PILLING, R. L., PITTS, A. D., REINTJES, M., WARREN, L. F., and WEGNER, P. A.: *J. Amer. Chem. Soc.*, **90**, 879 (1968).
90. ADAMS, R. M.: *Inorg. Chem.*, **2**, 1087 (1963).
91. DAVISON, A., EDELSTEIN, N., HOLM, R. H., and MAKI, A. H.: *Inorg. Chem.*, **2**, 1227 (1963).
92. MCCLEVERTY, J. A., LOCKE, J., WHARTON, E. J., and GERLOCH, M.: *J. Chem. Soc.*, A, 816 (1968).

INDEX